Studies in Computational Intelligence

Volume 1022

The series "Studies in Computational Intelligence" (SCI) publishes new developments and advances in the various areas of computational intelligence—quickly and with a high quality. The intent is to cover the theory, applications, and design methods of computational intelligence, as embedded in the fields of engineering, computer science, physics and life sciences, as well as the methodologies behind them. The series contains monographs, lecture notes and edited volumes in computational intelligence spanning the areas of neural networks, connectionist systems, genetic algorithms, evolutionary computation, artificial intelligence, cellular automata, self-organizing systems, soft computing, fuzzy systems, and hybrid intelligent systems. Of particular value to both the contributors and the readership are the short publication timeframe and the world-wide distribution, which enable both wide and rapid dissemination of research output.

Indexed by SCOPUS, DBLP, WTI Frankfurt eG, zbMATH, SCImago.

All books published in the series are submitted for consideration in Web of Science.

More information about this series at https://link.springer.com/bookseries/7092

Michael Zgurovsky · Nataliya Pankratova
Editors

System Analysis & Intelligent Computing

Theory and Applications

Springer

Editors
Michael Zgurovsky
Igor Sikorsky Kyiv Polytechnic Institute
Kyiv, Ukraine

Nataliya Pankratova
Igor Sikorsky Kyiv Polytechnic Institute
Kyiv, Ukraine

ISSN 1860-949X ISSN 1860-9503 (electronic)
Studies in Computational Intelligence
ISBN 978-3-030-94909-9 ISBN 978-3-030-94910-5 (eBook)
https://doi.org/10.1007/978-3-030-94910-5

This Springer imprint is published by the registered company Springer Nature Switzerland AG
The registered company address is: Gewerbestrasse 11, 6330 Cham, Switzerland

Preface

The International Scientific Conference *System Analysis & Intelligent Computing* (SAIC) is a series of conferences conducted on the base of Igor Sikorsky Kyiv Polytechnic Institute in Kyiv, Ukraine. The topics of the conference cover the modern directions in the field of system analysis of complex systems; computational intelligence; intelligent computing technologies; and data science and risk management in financial world. The aim of the conference is to exchange the most recent developments between researchers from different countries and organizations in various fields related to theoretical and applied system analysis and intelligent computing technologies.

The latest SAIC 2020 Conference was held from October 5–9, 2020, as the second conference in the series started in 2018. SAIC attracted hundreds of researchers and professionals working in the field of system analysis, decision making, computational intelligence, and related areas. Overall more than 70 papers were presented, demonstrating a wide range of the newest methods suitable for dealing with a multitude of various urgent problems in all fields of human activity.

This book contains extended versions of the 20 selected papers that reflect the latest theoretical and practical scientific approaches to the modern scope of global issues, including COVID-19 pandemic, urban underground construction, social systems and networks, and many others. The chapters of the book are grouped into two large parts:

Part I: System Analysis of Complex Systems

- Methods, models, and technologies of system analysis of complex systems of different nature in conditions of uncertainty and risks
- System methodology of foresight in the tasks of planning and making strategic decisions
- Problem-oriented methods of analysis, diagnostics, safety, and security of complex systems in conditions of uncertainty and risks
- Cyber-physical systems and control
- System methodology of sustainable development
- Cognitive modelling of complex systems

- Nonlinear problems of system analysis
- Decision making and decision support systems

Part II: Computational Intelligence and Intelligent Computing Technologies

- Fuzzy logic systems, fuzzy neural networks and applications
- Neural networks, deep learning neural networks
- Machine learning and self-learning
- Intellectual decision-making systems
- Pattern recognition, image processing, automatic speech recognition
- Bioinformatics and bio-inspired computing
- Fractal analysis

The purpose of the book is to present and discuss the systematized description of the new theoretical results and practical applications of the system analysis methods and intelligent computing technologies. The selected papers provide both an overview of some modern problems, and the recent approaches and techniques designed to deal with them. We hope that the broad scope of topics covered in this book will convey to the reader the modern state of the system analysis methods.

We would like to express our sincere appreciation to all authors for their contributions as well as to reviewers for their timely and interesting comments and suggestions. We certainly look forward to working with all contributors again in nearby future.

Kyiv, Ukraine
November 2021

Michael Zgurovsky
Nataliya Pankratova

Contents

Part I
System Analysis of Complex Systems

Chapter 1
Cyclic Regularities of the Covid-19 Spread and Vaccination Effect on Its Further Reduction

Michael Z. Zgurovsky, Pavlo O. Kasyanov, Olha P. Kupenko, Kostiantyn V. Yefremov, Nataliia V. Gorban, and Mariya M. Perestyuk

Abstract The patterns of periodic occurrence of infectious diseases pandemics over the past twenty years and their impact on the economy and society are studied in this paper. The analysis of the spread of the COVID-19 coronavirus pandemic during the last 1.5 years in the global context is carried out. The impact of vaccination on the reduction of the coronavirus pandemic in different countries and regions of the world has been analyzed. It is noted that regardless of the region of the world, the risks of disease recurrence do not disappear even after achieving collective immunity. The impact of vaccination on the further spread of the COVID-19 pandemic has been studied and its likely mitigation in different countries and regions of the world has been predicted. This aspect of the study aims to identify vaccination trends and, on their basis, predict the approximate time horizons of the COVID-19 pandemic in Ukraine and other countries.

Keywords COVID-19 pandemic · Cyclic regularities · Vaccination trends · Systems analysis · Cluster and regression-correlation analysis · Big data mining ·

M. Z. Zgurovsky · K. V. Yefremov (✉) · N. V. Gorban · M. M. Perestyuk
National Technical University of Ukraine "Igor Sikorsky Kyiv Polytechnic Institute", Kyiv 03056, Ukraine
e-mail: k.yefremov@wdc.org.ua

M. Z. Zgurovsky
e-mail: mzz@kpi.ua

N. V. Gorban
e-mail: nata_gorban@i.ua

M. M. Perestyuk
e-mail: maria@perestyuk.com

P. O. Kasyanov
Institute for Applied System Analysis, National Technical University of Ukraine "Igor Sikorsky Kyiv Polytechnic Institute", Kyiv 03056, Ukraine
e-mail: kasyanov@i.ua

O. P. Kupenko
Dnipro University of Technology, Dnipro 49005, Ukraine

M. Zgurovsky and N. Pankratova (eds.), *System Analysis & Intelligent Computing*, Studies in Computational Intelligence 1022, https://doi.org/10.1007/978-3-030-94910-5_1

The method of similarity in mathematical modeling · Methods of technical analysis of time series

1 Introduction

The World Data Center "Geoinformatics and Sustainable Development" that works at the Igor Sikorsky Kyiv Polytechnic Institute (Ukraine), serves as an analytical platform for the presented study. The Ukrainian Center is a part of the network of 53 global data centers located in 13 countries around the world and coordinated by the International Scientific Council (ISC) headquartered in Paris. The Center gives the access to various big data stored in this network.

Using a powerful supercomputer to store and process big data, since 2006 the team of highly qualified systems analysts of the Ukrainian Data Center conducts annual research in the field of sustainable development and security for countries and regions of the world. Therefore, from the first weeks of the COVID-19 pandemic in Europe, the World Data Center "Geoinformatics and Sustainable Development" (hereinafter referred to as the Data Center) began to study the spread of the disease and its impact on the economy and society of European countries. Over the past year and a half, the Center has conducted more than 20 such studies. They concerned the advanced modeling and foresight of the disease spread in the long-term (several years), medium-term (several months) and short-term (5–7 days) time horizons.

The methods of systems analysis, cluster and regression-correlation analysis, big data mining, the method of similarity in mathematical modeling, and other approaches were used in the study.

All outcomes of the studies are presented on the Data Center portal (http://wdc.org.ua/). They are widely reflected in national and foreign media coverage (http://wdc.org.ua/uk/covid19-media-publications).

2 Cyclic Regularity of the Spread of Pandemic Infectious Diseases at the Beginning of the 21st Century

Over the past two decades, large-scale infectious diseases (pandemics) were more frequent; they had a significant impact on human health, social development, and the economies of countries and regions of the world.

2.1 The Cyclic Nature of Pandemic Infectious Diseases Over the Past Two Decades

During this time period, the following four pandemics can be identified:

1. The outbreak of severe acute respiratory syndrome (SARS) occurred in 35 countries from November 2002 to May 2004. More than 8000 SARS cases were reported. The mortality rate was about 11%;
2. From January 2009 to August 2010, the majority of countries in the world suffered from the Swine flu pandemic. 700 million to 1.4 billion people were affected;
3. In 2014–2015, West Africa, the United States of America and Europe suffered from the Ebola pandemic. The mortality rate reached 50%;
4. Unfortunately, the beginning of 2020 was marked by the fastest and most widespread outbreak of the COVID-19 pandemic almost all over the world. As of the end of September 2021, the disease affected more than 230 million people worldwide, 4.7 million died. The average global mortality rate in this period was about 2.1%.

It can be seen that the emergence of pandemics during this time is of cyclic nature with an approximate recurrence period of about five to six years. In order to analyze the impact of these pandemics on the world economy, we compare them along the time axis with the following fundamental periodic processes (Fig. 1):

1. Nikolai Kondtatiev's 40–50-year economic cycles based on changes of the technological structure of the society (blue line);

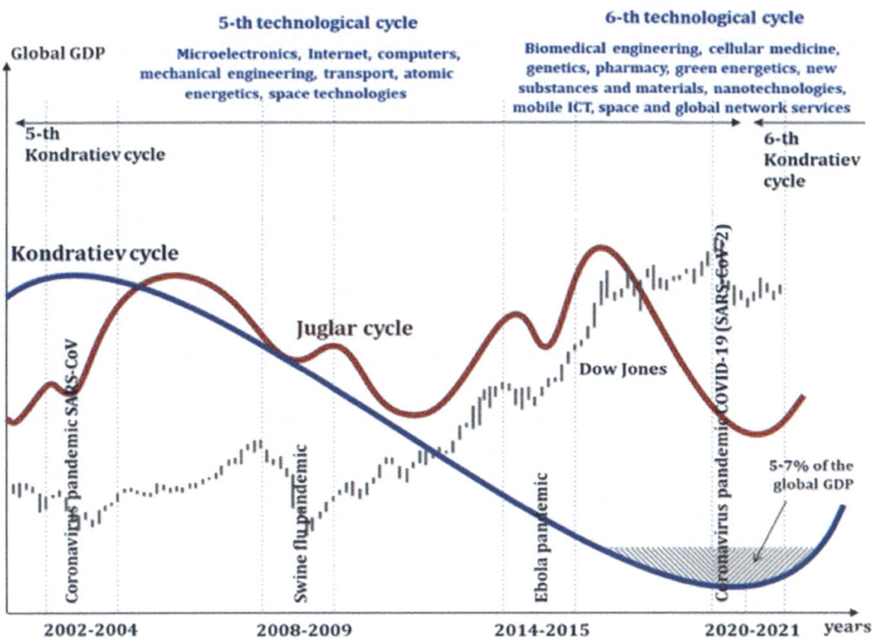

Fig. 1 Impact of contagious disease pandemics on the economy and society development

2. Clement Juglar's 7–11-year cycles associated with fluctuations in the volumes of investments in fixed capital (orange line);
3. Dow Jones industrial average that reflects the general capitalization of 30 biggest American companies whose activities in total set the trend for the global economy (dotted lines).

Figure 1 shows that Kondratiev's cycle passes through the bottom of the cycle in 2020–2021 and begins to grow. Therefore, in fact, there are objective preconditions for further long-term growth of the world economy.

The intensity of such growth during the current period is significantly reduced due to the COVID-19 pandemic. It entails the bottom of the Juglar cycle and decrease in the Dow Jones index by 30–40%.

According to Juglar, this recession will last for about a year and a half, during this period investments will be redirected to the 6th technological structure. Economic revival must begin in accordance with both Kondratiev and Juglar theories. Many experts predict that from the middle of 2021 there will be a significant recovery of the world economy due to the 6th technological structure. It will cause a reversal in the world economy development which will start to grow.

Figure 2 shows the drop in the global GDP due to the multiple impact of four pandemics, namely SARS, Swine flu, Ebola and COVID-19.

Now let's analyze the nature and features of the COVID-19 pandemic spread in the global extent and in Ukraine.

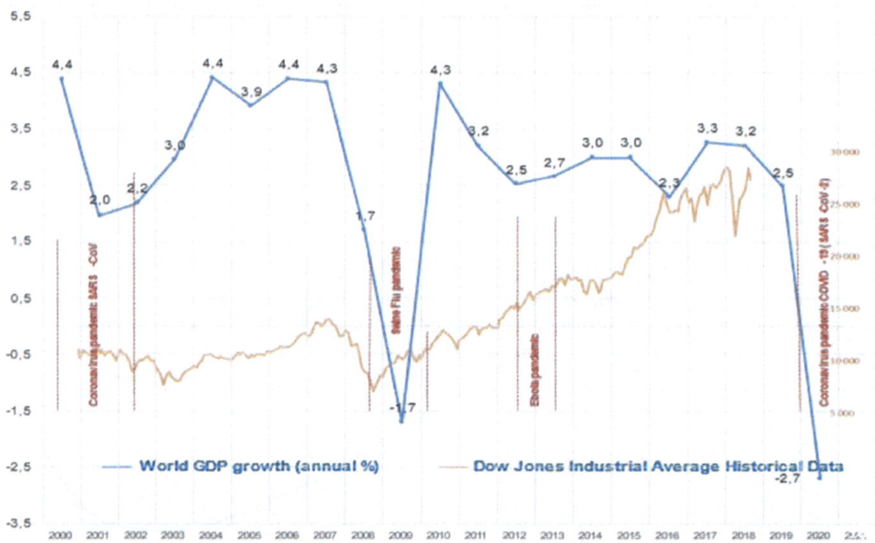

Fig. 2 Impact of four pandemics (SARS-CoV, Swine flu, Ebola, and COVID-19) on the global GDP dynamics

2.2 *Analysis of the Cyclic Nature of the COVID-19 Pandemic Spread in Global and Regional Contexts*

It is known that the disease incidence in Europe started in northern Italy, in the province of Lombardy. From there, the virus began to spread rapidly to neighboring Western European countries—France, Spain, Germany, which had the largest number of businesses, trade and transportation contacts both with China and with each other. Then, with a lag of 2–3 weeks, the pandemic covered the countries of Central and Eastern Europe—Poland, Romania, Ukraine and the Balkans—Croatia, Serbia, Slovenia, Montenegro and others. Now, for more than 1.5 years the COVID-19 global pandemic pressures people to live in quarantine restrictions and has a global impact on all spheres of life.

The goal of this section of the study is to identify the nature of the COVID-19 pandemic spread for the timespan (January 2020–September 2021) in order to predict the possible development of this process in the near future and assess the approximate horizon of the pandemic remission.

As the main source of coronavirus pandemic spread data, we will use the global database [1]. The nonstationary behavior and heteroskedasticity properties of these data does not allow us to effectively use regression models to work with them. Therefore, based on the significant heterogeneity and non-stationarity of coronavirus distribution, their stochastic nature and high volatility, existence of so-called "heavy tails" of COVID-19 spread, methods of technical analysis of time series based on basic indicators were used in order to identify stable trends; these methods were also used to track major trends and identify "trading signals" in stock markets [2–4]. The following basic indicators were used in the study:

- "***Supertrend***" *indicator* is used to detect a trend for highly volatile data. When there is a growing trend on the output curve, the values of the "supertrend" indicator are below it, and during a dropping trend the values of this indicator are above the original data curve. The intersection of the indicator and the data curves may demonstrate a break in the previous trend. Frequent intersections of these curves indicate that there is no express trend for the data;
- "***Zigzag***" *indicator* combines the most significant local extreme points on the data diagram and is insensitive to small fluctuations. It is convenient to use this indicator for the analysis of previous data fluctuations;
- *Technical indicator based on the* ***Hurst exponent*** is one of the key fractal characteristics [5–7]. Sequences for which this value is greater than 0.5 are considered persistent, they mostly maintain the actual trend, i.e., growth in the past is more likely to lead to growth in the future, and vice versa. If the value of the Hurst exponent is equal to 0.5, it means that there is no express tendency (trend). Values which are less than 0.5 indicate the anti-persistence of the series, when the actual trend is more likely to change to the opposite in the future;
- *Adaptive version of the "****supertrend****" indicator* allows to detect the presence of weekly seasonal effects in the number of reported new cases of the disease that directly affect the level of volatility of these data.

2.3 *The Cyclic Nature of COVID-19 Spread in the Global Extent*

Researchers from around the world and well-known international organizations come to the conclusion that the fourth wave of COVID-19 continues currently in the global extent. In the study, this fact is confirmed by the behavior of the "zigzag" indicator, built for a smoothed curve of new reported cases (Fig. 3):

The *first* express wave took place from March to May 2020 and was caused by the epidemic spread in Italy, Spain, Germany and other Western European countries.

The *second* express wave, which was on the rise from late September 2020 to the end of December 2020, reflected a sharp increase in morbidity in Canada, the United States of America, the United Kingdom and Europe.

The *third* wave, whose escalation lasted from March to May 2021, was caused by a rapid increase in the number of new COVID-19 cases in South America, Great Britain, where a Beta strain was detected—a new coronavirus mutation called the British strain, and in Europe, which was also affected by this strain.

Researchers attribute the *fourth* wave of the disease, which lasts from early July 2021 up to date, to the appearance of the next virus mutation, so-called Delta strain.

Four global waves of the COVID-19 morbidity are expressly shown in the diagram of global mortality (Fig. 4). The "zigzag" indicator does not fix the peak of the last wave, as at the moment there is only a slight decline in global mortality, which may be temporary.

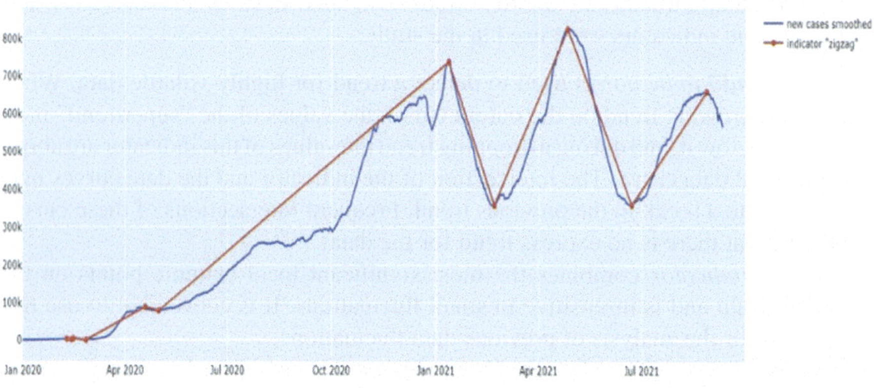

Fig. 3 Analysis of the dynamics of new identified patients with COVID-19 in the world based on the use of the "zigzag" indicator

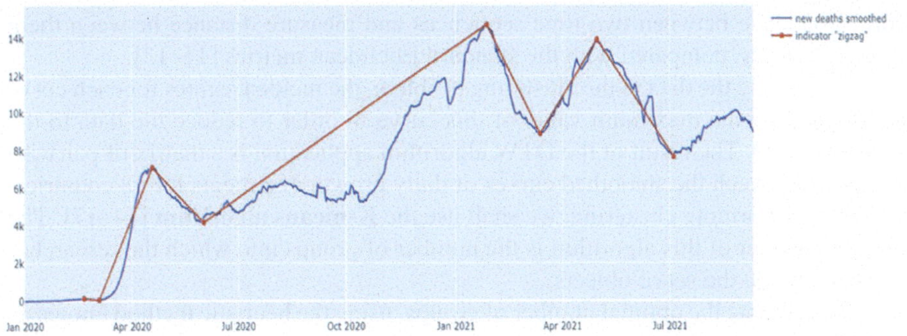

Fig. 4 Analysis of the dynamics of mortality caused by COVID-19 in the world, based on the use of the "zigzag" indicator

3 Analysis of Trends in Pandemic Processes in the World Based on the Use of the Clusterization Method

The basic study precondition is the affirmation that a reduction in the number of new COVID-19 cases is possible provided the formation of collective immunity when more than 50% of the population have been exposed to the virus in the past, and reduction in severe and fatal cases if at least 30% of the population is vaccinated with at least one dose [8]. According to the latest data from the World Health Organization, with the rise of the Delta strain of the COVID-19 virus, which affects even fully vaccinated people, expectations from the vaccination campaign have rather changed. Now the main positive effect expected from vaccination is reduction in the incidence of severe disease and complications, hospitalizations and deaths.

The world leaders in the vaccinated population, Israel, the United States of America, the United Kingdom, the United Arab Emirates, Chile, Canada, Denmark, Norway, and other Western European countries, can be used to study the impact of vaccination on the pandemic, in particular, on the number of daily cases and daily mortality [9].

3.1 Countries Clustering by COVID-19 Incidence Rate

Hereinafter we will limit ourselves to the countries with the population not less than 5 million persons where the epidemic proceeds not less than a year. This selected subset consists of 48 countries. Data for analysis were obtained from the database [10] as of mid-September 2021.

The use of cluster analysis in order to solve this problem requires measuring the pairwise distances between objects, which are the smoothed curves of daily growth of patients in this case. Let's involve the **Dynamic Time Warping** technique (DTW, algorithm of the time scale dynamic transformation) which allows to find optimum correspondence between two time sequences and measure distance between them more relevantly, compared with the standard Euclidean metrics [11–13].

After scaling the data in the clustering problem, the incidence rates for each curve are divided by the maximum value of this curve in order to reduce the data to the segment [0, 1]. The result of the DTW algorithm application is a matrix of pairwise distances between the smoothed curves of daily growth of patients for 48 countries.

When performing clustering, we shall use the **K-means algorithm** [14–17]. The input parameter of this algorithm is the number of groups into which the researcher wants to divide the set of objects.

Let's estimate the optimal number of groups, using the heuristic method known as the "**elbow method**" [18]. It consists in repeated performance of clustering using the chosen algorithm and estimation of the clustering quality. The result of the algorithm application is the function of the clustering quality versus the clusters number. This method allows to choose the optimal number of clusters (groups), where the curve of the obtained function demonstrates the most significant inflection.

In Fig. 5, the most noticeable breaks of the blue curve occur for two and four country groups. However, the division into 4 clusters is more evened and informative.

The diagram given in Fig. 6 shows the centers of the groups obtained by using the **K-means** algorithm. We see that the results look relevant.

The **first group** includes 13 countries: Austria, Belgium, Bulgaria, the Czech Republic, Hungary, India, Italy, Poland, Portugal, Romania, Slovakia, Sweden and Ukraine, where the morbidity rates decreased significantly since the last wave and then slightly increased.

In Austria, Bulgaria, Poland, Hungary, Romania, Slovakia, Ukraine and Sweden, the wave of morbidity caused by the Delta strain is just beginning (Fig. 7). The morbidity in the Czech Republic is low throughout the summer. In India, after a rapid and deadly wave that lasted from April to June this year and was caused by the

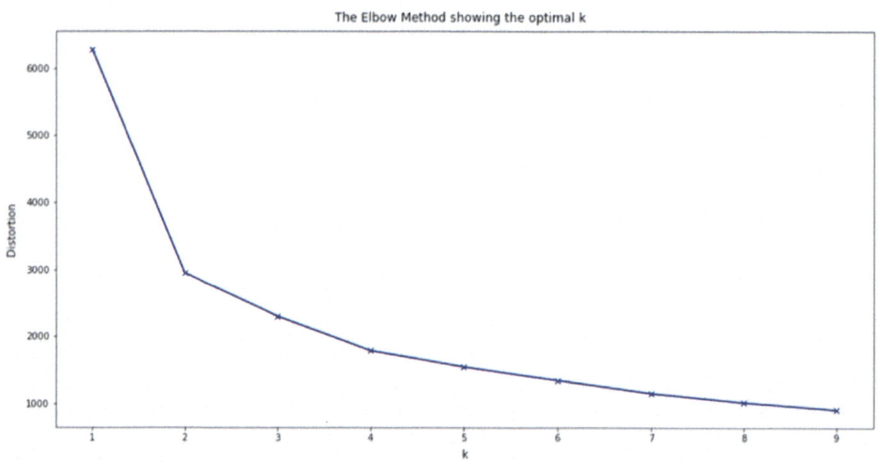

Fig. 5 Result of the "elbow method" algorithm application

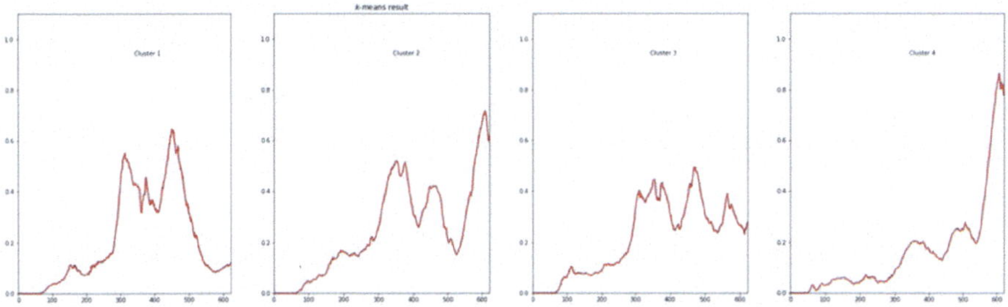

Fig. 6 Result of the K-means algorithm application: centers of obtained 4 groups of countries

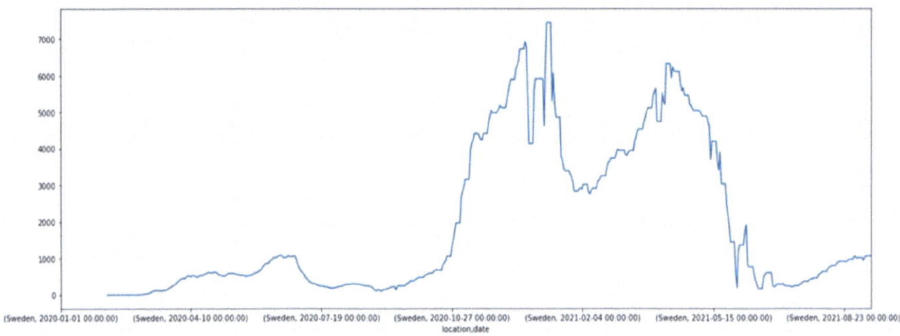

Fig. 7 The diagram of the smoothed dynamics of COVID-19 new cases in Sweden, which is a typical representative of the first group of countries

Delta strain, the morbidity is relatively low now. In Italy and Portugal, after a slight rise in morbidity in August and early autumn, the situation began to stabilize.

The **second group** includes 15 countries: Azerbaijan, Finland, Greece, Iraq, Israel, Morocco, Norway, Pakistan, Palestine, Russia, Syria, the United Kingdom, the United States of America, Belarus, where the last wave was intensive and continues to develop at high levels of daily morbidity (Fig. 8, the example of Israel).

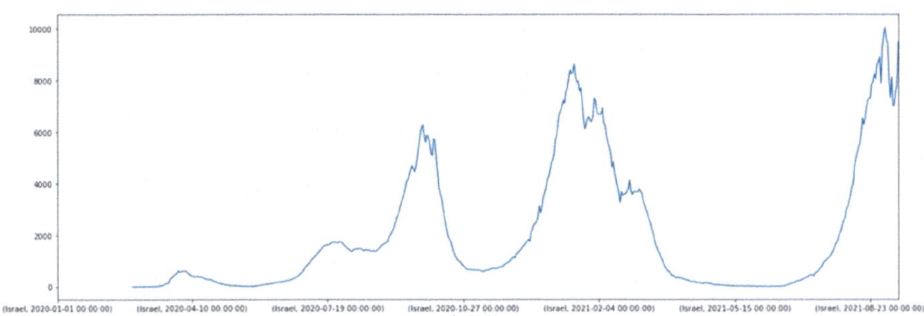

Fig. 8 The diagram of the smoothed dynamics of COVID-19 new cases in Israel

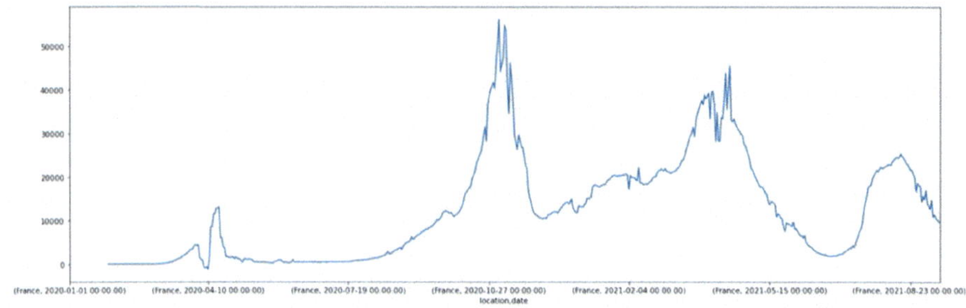

Fig. 9 The diagram of the smoothed dynamics of COVID-19 new cases in France

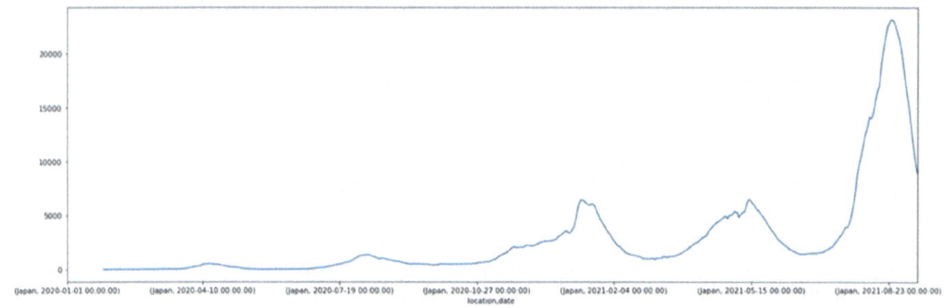

Fig. 10 The diagram of the smoothed dynamics of COVID-19 new cases in Japan

The **third group** includes 11 countries: Canada, Denmark, France, Germany, the Netherlands, Serbia, Spain, Switzerland, Tunisia, Turkey, the United Arab Emirates, where the last wave of morbidity is clearly visualized, however, it does not exceed the intensity of previous waves (Fig. 9).

The **fourth group** includes 9 countries: Australia, Iran, Japan, Malaysia, the Philippines, South Korea, Sri Lanka, Thailand, Vietnam, where the last wave of morbidity significantly exceeds the intensity of the previous ones (Fig. 10).

3.2 Clustering by the Mortality Dynamics Level

Let's analyze the dynamics of mortality for the selected group of countries. We apply the DTW method to the curves of mortality dynamics in order to obtain a matrix of pairwise distances between them and determine the optimal number of subgroups (clusters) using the elbow method (Fig. 11).

We see that the most significant break points of the curve occur for the number of clusters 2, 3 and 4. We choose the case of 4 clusters, whose centers are shown in Fig. 12.

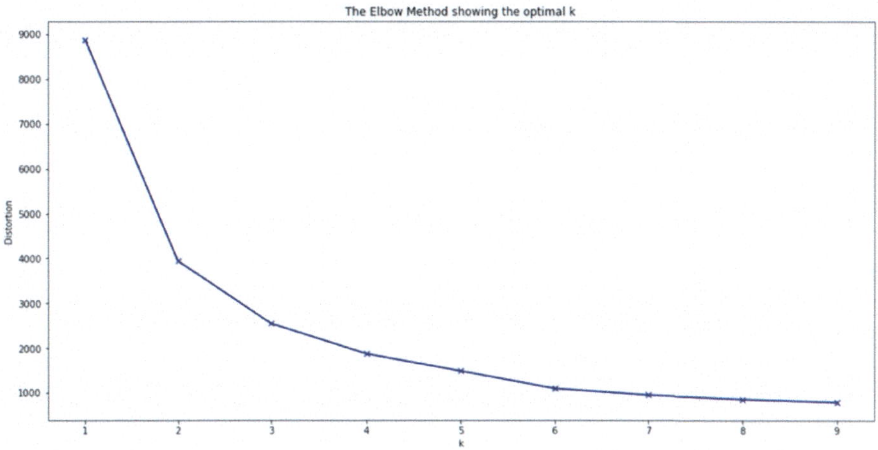

Fig. 11 Application of the "elbow method" for clustering the mortality curves for the selected group of countries

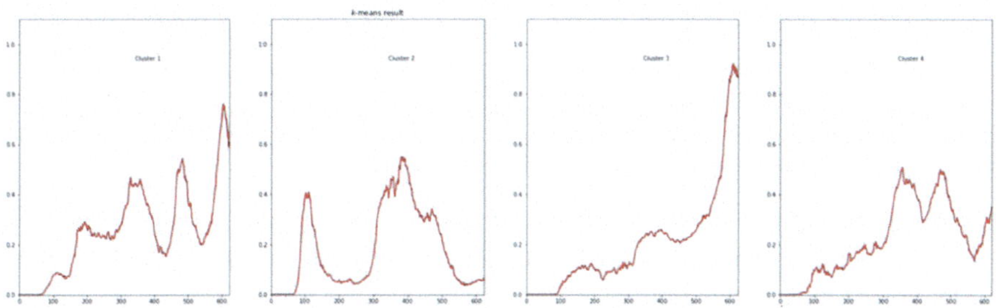

Fig. 12 Results of the K-means algorithm application

The **first cluster** includes countries where the last wave in terms of mortality is similar and close to previous waves. The countries included in this cluster (the level of fully vaccinated population is indicated in parentheses) are: Azerbaijan (30%), Iran (14%), Iraq (5%), Morocco (43%), Pakistan (10%), Philippines (15%), Turkey (48%). It is easy to note that the vaccination rate is less than 60% in these countries. For Turkey, the diagram of the smoothed dynamics of new deaths is shown in Fig. 13.

The **second cluster** includes a group of countries where the daily mortality has significantly decreased since the last wave and continues to be at a relatively low level. These are the following countries (the level of fully vaccinated population is indicated in parentheses): Austria (58.61%), Belgium (71%), Canada (68.5%), Czech Republic (54.6%), Denmark (74%), Finland (56.8%), France (65.3%), Germany (61.7%), Hungary (57.7%), India (13%), Italy (64.5%), the Netherlands (63.4%), Norway (63.4%), Chile (72%), Poland (50.6%), Portugal (81.6%), Slovakia (40.5%), Spain (75.7%), Sweden (61%), Switzerland (52%), the United Arab Emirates (78.4%), the United Kingdom (64.6%, Fig. 14).

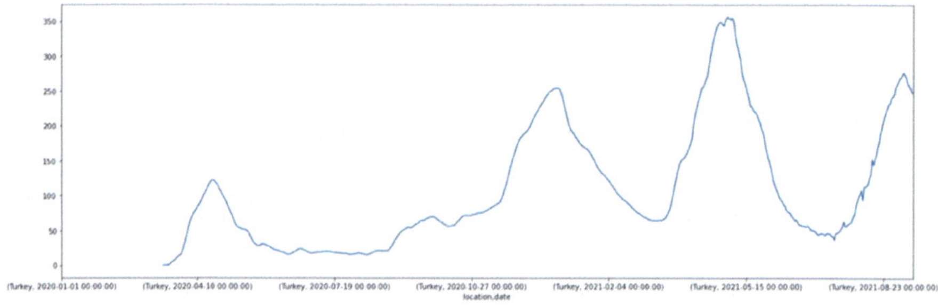

Fig. 13 The diagram of smoothed dynamics of new deaths caused by COVID-19 in Turkey

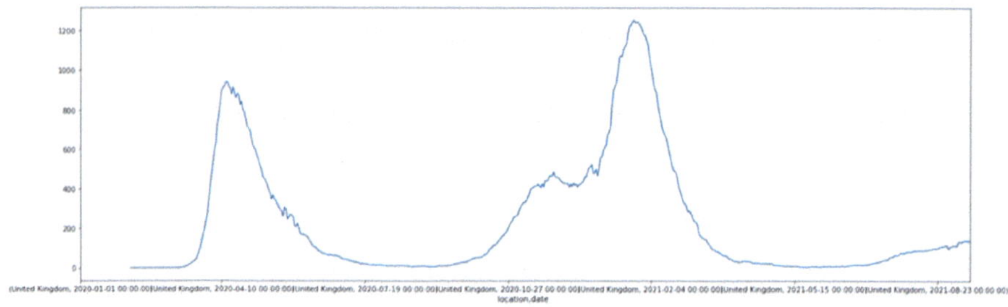

Fig. 14 The diagram of smoothed dynamics of new deaths caused by COVID-19 in the United Kingdom, a typical representative of the second group of countries

It is easy to note that except for India, other countries have a really high level of vaccination of their population.

In this context, the majority of countries of this cluster, namely Canada, Denmark, France, Germany, the Netherlands, Spain, Switzerland, the United Arab Emirates, Finland, Norway and the United Kingdom in terms of morbidity were classified in the previous section as countries in which the Delta strain has significantly raised the level of daily new cases of COVID-19. **Hence, we can conclude that mass-scale vaccination does have a positive effect on reducing mortality caused by the virus**.

The **third cluster** includes countries where the dynamics of mortality during the last wave significantly exceeds the levels that were observed during the previous waves (Fig. 15). These are Belarus (14.7%), Russia (27.3%), Malaysia (53.7%, Fig. 15), Sri Lanka (49.6%), Thailand (17.6%), Vietnam (5.7%).

Although Malaysia and Sri Lanka have about 50% of the vaccinated population, now these countries demonstrate relatively high mortality rates.

The **fourth cluster** includes countries where the dynamics of mortality has recently begun to grow: Australia (34%), Bulgaria (18%), Greece (56.7%), Israel (59.6%), Japan (52.4%), Palestine (9%), Romania (27%), Serbia (32.7%), South Korea (40%), Syria (1%), Tunisia (25.5%), Ukraine (12%), the USA (53.7%). Diagrams of the smoothed dynamics of new deaths caused by COVID-19 in Romania

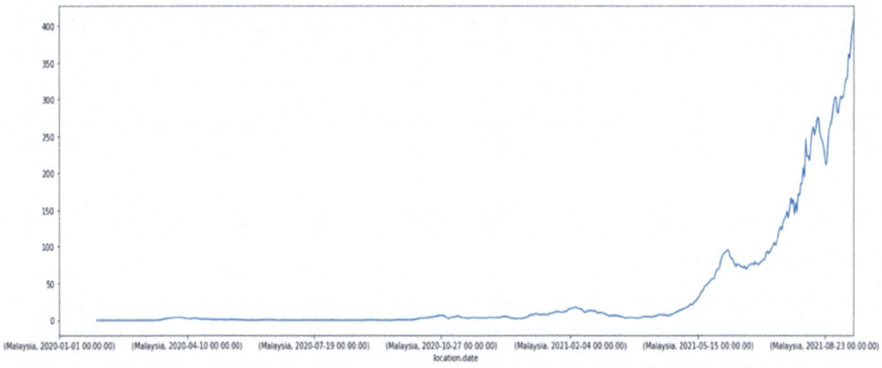

Fig. 15 The diagram of smoothed dynamics of new deaths caused by COVID-19 in Malaysia, a typical representative of the third group of countries

and Ukraine, which are typical representatives of the fourth group, are shown in Fig. 16.

In terms of morbidity, Ukraine is included in the group of the countries where the new wave caused by the Delta strain has not reached its maximum yet. Mortality in September 2021 is growing faster than in the same period in 2020. Ukraine has a low level of fully vaccinated population at present and may face severe consequences

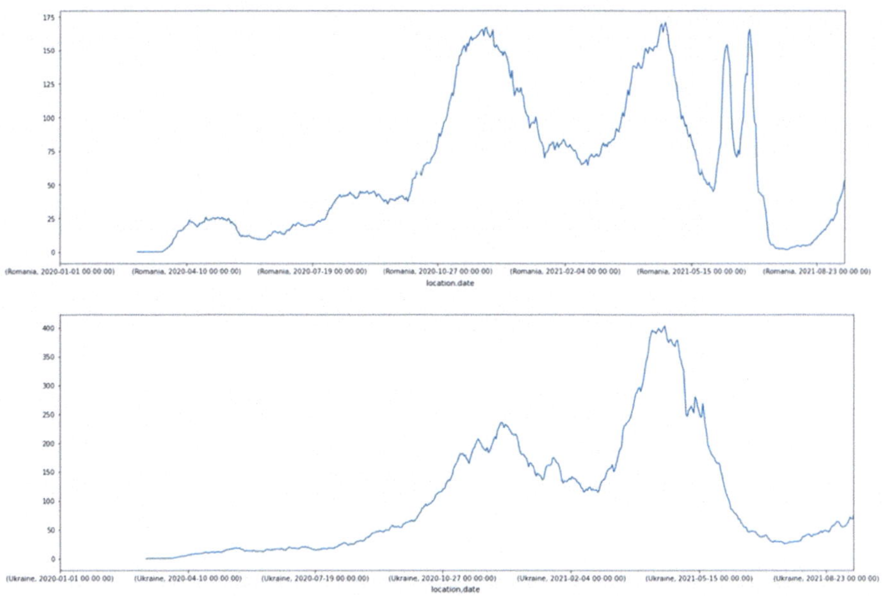

Fig. 16 The diagrams of smoothed dynamics of new deaths caused by COVID-19 in Romania and Ukraine, typical representatives of the fourth group of countries

of the pandemic in the near future. It is now extremely important to accelerate the rate of vaccination of the population. However, despite the fact that vaccination is free-of-charge and vaccines are supplied in sufficient quantities, the level of public distrust in COVID-19 vaccinations and non-compliance with safety rules in public places leave the country without adequate protection against a new wave of disease.

4 Conclusions

1. The impact of pandemics, namely, severe acute respiratory syndrome (SARS), Swine flu, Ebola and COVID-19 that have occurred over the past 20 years, on the development of the global economy and society is analyzed in the study. It is shown that these pandemics have cyclic nature with a recurrence period of approximately five to six years. They significantly affect the global economy and social relations, cause the rupture of economic chains and slowing down the economy and society development.
2. Using the methods of technical analysis, the cyclic nature of the COVID-19 pandemic process over a period of time (from January 2020 to September 2021) is shown, when four waves of the disease spread are revealed. The obtained results show a significantly lower level of new morbidity and mortality cases during the current fourth wave of the pandemic process compared with the second and third waves, in the countries with a high percentage of vaccinated population [19–25].
3. Based on the use of big data mining methods and in order to further predict and draw up possible scenarios for the pandemic process development, the countries are clustered according to the main signs of the disease spread, vaccination rates and the rate of pandemic remission.

   (a) In terms of morbidity, clusters are formed as shown below:

      (i) The first group includes countries where morbidity levels decreased significantly after the last third wave, and at the beginning of the fourth wave these levels began to slightly increase. These are 13 countries: Austria, Belgium, Bulgaria, Czech Republic, Hungary, India, Italy, Poland, Portugal, Romania, Slovakia, Sweden and Ukraine.

      (ii) The second group includes countries where the last third wave was intensive and now continues to grow demonstrating high levels of daily morbidity. These are Azerbaijan, Finland, Greece, Iraq, Israel, Morocco, Norway, Pakistan, Palestine, Russia, Syria, the United Kingdom, the United States of America, Belarus.

      (iii) The third group includes countries where the last wave of morbidity is expressly manifested but does not exceed the parameters of the previous waves. This group includes 11 countries:

Canada, Denmark, France, Germany, the Netherlands, Serbia, Spain, Switzerland, Tunisia, Turkey, the United Arab Emirates.

(iv) The fourth group includes countries where the last wave of morbidity expressly exceeds the intensity of the previous ones. This group includes Australia, Iran, Japan, Malaysia, the Philippines, South Korea, Sri Lanka, Thailand, Vietnam.

(b) In terms of mortality dynamics which was caused by COVID-19, the following clusters are obtained:

(i) The first cluster includes countries where the last wave in terms of mortality is close to previous waves. These are Azerbaijan (30%), Iran (14%), Iraq (5%), Morocco (43%), Pakistan (10%), the Philippines (15%), Turkey (48%). In these countries, the vaccination rate is less than 60%.

(ii) The second cluster includes the group of countries where the daily mortality has significantly decreased since the previous wave and remains at relatively low level. These are the following countries (the level of fully vaccinated population is indicated in parentheses): Austria (58.61%), Belgium (71%), Canada (68.5%), Czech Republic (54.6%), Denmark (74%), Finland (56.8%), France (65.3%), Germany (61.7%), Hungary (57.7%), India (13%), Italy (64.5%), the Netherlands (63.4%), Norway (63.4%), Chile (72%), Poland (50.6%), Portugal (81.6%), Slovakia (40.5%), Spain (75.7%), Sweden (61%), Switzerland (52%), the United Arab Emirates (78.4%), the United Kingdom (64.6%). Except for India, the countries of this cluster have a high level of vaccination of their population. Therefore, we can conclude that mass-scale vaccination does have a positive effect on reducing mortality caused by the infection.

(iii) The third cluster includes countries where the dynamics of mortality during the last wave significantly exceeds the mortality levels observed during the previous waves. These countries are Belarus (14.7%), Russia (27.3%), Malaysia (53.7%), Sri Lanka (49.6%), Thailand (17.6%), Vietnam (5.7%).

(iv) The fourth cluster includes countries where the dynamics of mortality is currently beginning to increase: Australia (34%), Bulgaria (18%), Greece (56.7%), Israel (59.6%), Japan (52.4%), Palestine (9%), Romania (27%), Serbia (32.7), South Korea (40%), Syria (1%), Tunisia (25.5%), Ukraine (12%), the United States of America (53.7%). In this group, Israel should be singled out, where the mortality rate is one of the lowest in the world, and the wave of the Delta strain caused mortality twice lower than the previous wave.

4. Ukraine is in the group of countries where the new wave has not reached its maximum yet. The country has a low level of fully vaccinated population as of

mid-September 2021 (about 12%) and therefore may face severe consequences of the pandemic in the near future. As one of the indicators that characterize the complexity of the pandemic course in the country is the number of hospitalizations and mortality level, it is important to protect with vaccination the population, which is the most vulnerable to the adverse course of COVID-19, namely, the age group of 60 + and persons with the comorbidities, and to do it as soon as possible.

References

1. Our World in Data. https://covid.ourworldindata.org; https://covid.ourworldindata.org/data/owid-covid-data.csv. Last accessed 17 Sept 2021
2. Navarro, P.: When the Market Moves, Will You Be Ready? McGraw-Hill Education, NY (2003)
3. Meucci, A.: Risk and Asset Allocation. Springer, NY (2009)
4. Lopez de Prado, M.M.: Advances in Financial Machine Learning. Wiley, Hoboken (2018)
5. Yu, S., Piao, X., Hong, J., Park, N.: Bloch-like waves in random-walk potentials based on supersymmetry. Nat. Commun. **6**, 8269 (2015)
6. Hurst exponent evaluation and R/S-analysis. https://github.com/Mottl/hurst. Last accessed 17 Sept 2021
7. Multifractal Detrended Fluctuation Analysis. https://github.com/LRydin/MFDFA. Last accessed 17 Sept 2021
8. When will the COVID-19 pandemic end? https://www.mckinsey.com/industries/healthcare-systems-and-services/our-insights/when-will-the-covid-19-pandemic-end. Last accessed 26 Sept 2021
9. Coronavirus in the U.S.: Latest map and case count. https://www.nytimes.com/interactive/2021/us/covid-cases.html. Last accessed 26 Sept 2021
10. The project of the Global Change Data Lab Our World in Data. https://ourworldindata.org/. Last accessed 26 Sept 2021
11. Al-Naymat, Gh., Sanjay, Ch., Javid, T.: Sparsedtw: A novel approach to speed up dynamic time warping. In: AusDM'09—Conferences in Research and Practice in Information Technology, pp. 117–127. AusDM 2009; Melbourne (2012)
12. Keogh, E.J., Pazzani, M.J.: Derivative dynamic time warping. In: Proceedings of the 2001 SIAM International Conference on Data Mining, pp. 1–11. Society for Industrial and Applied Mathematics, Philadelphia (2001)
13. Salvador, S., Chan, P.: Toward accurate dynamic time warping in linear time and space. Intell. Data Anal. **11**(5), 561–580 (2007)
14. Lee, S., Kim, J., Jeong, Y.: Various validity indices for fuzzy K-means clustering. Korean Manage. Rev. **46**(4), 1201–1226 (2017)
15. Pigott, T.D.: A review of methods for missing data. Educ. Res. Eval. **7**(4), 353–383 (2001)
16. Steinhaus, H.: Sur la division des corps matériels en parties. Bull. Acad. Polon. Sci. **1**(804), 801 (1956)
17. Lloyd, S.P.: Least squares quantization in PCM. IEEE Trans. Inform. Theor. **28**(2), 129–137 (1982)
18. Bholowalia, P., Kumar, A.: EBK-means: a clustering technique based on elbow method and k-means in WSN. Int. J. Comput. Appl. **105**(9), 17–24 (2014)
19. Auer, P., Burgsteiner, H., Maass, W.: A learning rule for very simple universal approximators consisting of a single layer of perceptrons. Neural Netw. **21**(5), 786–795 (2008)
20. Saadat, S., Rikhtegaran, T.Z., Logue, J., et al.: Binding and neutralization antibody titers after a single vaccine dose in health care workers previously infected With SARS-CoV-2. JAMA **325**(14), 1467–1469 (2021)

21. Robert Koch-Institut: COVID-19 und Impfen: Antworten auf häufig gestellte Fragen (FAQ). Impfstofftypen. https://www.rki.de/SharedDocs/FAQ/COVID-Impfen/gesamt.html. Last accessed 24 Sept 2021
22. [Ministero della Salute] Circolare: Vaccinazione dei soggetti che hanno avuto un'infezione da SARS-CoV.2. https://www.salute.gov.it/portale/vaccinazioni/archivioNormativaVaccina zioni.jsp. Last accessed 29 July 2021
23. Standard country or area codes for statistical use, UN Statistics Divisionhttps. https://unstats.un.org/unsd/methodology/m49/. Last accessed 17 Sept 2021
24. COVID-19 Data Repository by the Center for Systems Science and Engineering (CSSE) at Johns Hopkins University. https://github.com/CSSEGISandData/COVID-19. Last accessed 26 Sept 2021
25. Anders Tegnell and the Swedish Covid Experiment, FT. https://www.ft.com/content/5cc92d45-fbdb-43b7-9c66-26501693a371. Last accessed 26 Sept 2021

Chapter 2
Cyber-Physical Systems Operation with Guaranteed Survivability and Safety Under Conditions of Uncertainty and Multifactor Risks

Nataliya Pankratova, Alexey Malishevsky, and Vladimir Pankratov

Abstract The strategy of operation of cyber-physical systems under conditions of uncertainty and multifactor risks on the basis of the general problem of multifactor risks, acceptable margin of permissible risk, forecasting of destabilizing dynamics of risk factors, principles, hypotheses, axioms is proposed. It allows to timely analyze emergency situations, accidents and disasters with guaranteed survivability and safety. As an example of the cyber-physical system functioning, the problem of simulating the operation of an electric refrigerator truck which delivers perishable goods is considered. The principle of guaranteed functioning has been implemented, which makes it possible to control the parameters of the system in real time, identify abnormal situations and determine their causes, and, if possible, ensure the survivability of functioning. The developed visualization of the process of predicting the functioning allows us to visually observe the future values of the system parameters, and, therefore, promptly make changes to the control if necessary, which allows, in particular, to respond to the occurrence of emergency situations and make a decision on their timely prevention.

Keywords Strategy · Uncertainty · Multifactor risk · Margin of permissible risk · General problem · Survivability · Safety

1 Introduction

Advances in information technology, computing, storage, telecommunications, automation, and other areas have opened pathways for the new technological revolution—Cyber-Physical Systems (CPS). Cyber-Physical Systems (CPS) has emerged as a unifying name for systems where the cyber parts, i.e., the computing and communication parts, and the physical parts are tightly integrated, both at the design time and during operation. Such systems use computations and communication deeply embedded in and interacting with physical processes to add new capabilities to physical systems. These cyber-physical systems range from miniscule (pace makers)

N. Pankratova · A. Malishevsky (✉) · V. Pankratov
Institute for Applied System Analysis, Igor Sikorsky Kyiv Polytechnic Institute, Kyiv, Ukraine

M. Zgurovsky and N. Pankratova (eds.), *System Analysis & Intelligent Computing*, Studies in Computational Intelligence 1022, https://doi.org/10.1007/978-3-030-94910-5_2

to large-scale (a national power-grid). There is an emerging consensus that new methodologies and tools need to be developed to support cyber-physical systems.

There are many definitions for CPS in the literature.

In [1], cyber-physical systems are defined as "physical, biological and engineered systems whose operations are monitored, coordinated, controlled and integrated by computing and communication core". In [2], authors define cyber-physical systems as "tightly coupled cyber and physical systems" that exhibit the high level of "integrated intelligence" and where computational processes interact with physical components.

A cyber-physical system can represent a single object with many tightly coupled components like an electric vehicle where a bunch of physical components work together, each being controlled by the vehicle's software. On the other hand, a cyber-physical system can represent a set of electric vehicles cooperating together to achieve a certain goal, or a set of smart buildings in the electric power grid.

The next section will list some of the examples of cyber-physical systems and their applications.

2 Related Work

2.1 Applications of CPS

There are many applications for cyber-physical systems including transportation, medical systems, health care, smart homes, social networks, gaming, power management, power grids, networks, etc.

Transportation enjoys a lot of attention from CPS research. There are two main areas where CPS are employed: on a single vehicle level and on a system of vehicles. Architectures have been developed such as the Automotive Open System Architecture (AUTOSAR) with its timing models, fault tolerance, instrumentation support, feedback, and verification support [3]; frameworks have been specifically developed or adapted for transportation CPS such as SysWeaver framework extensions (a model-based design, integration, and analysis framework) [3]; formal models have been developed such as a model for a distributed car control system where each car is controlled by adaptive cruise control [4]; other applications even include Cyber-Physical bicycle systems where video processing and computational capabilities are added to the bicycle to monitor the environment behind the biker in order to detect rear-approaching vehicles and alert the biker [5].

Medical systems have gained a lot of attention from CPS research including frameworks for interacting medical devices such as Medical Device Plug-and-Play (MD PnP) Interoperability initiative [6]; design of an interactive cyber-physical system for people with disability and frail elderly people for providing health-care services them that observe the motion and activities of the users and provide the services to them at the desired location [7]; models and tools for designing and analyzing of body area networks (BANs) such as an abstract cyber-physical model of BANs, called

BAN-CPS, capturing the undesirable side-effects of the medical devices (cyber) on the human body (physical), and BAND-AiDe tool [8].

Another application area for CPS is smart communities. Cyber-physical systems can be applied to individual homes or to the whole communities. For example, in Neighborhood Watch application, individual homes monitor the surroundings and detect suspicious events; Pervasive Healthcare application monitors body sensors and sends information to healthcare providers if such a need arises [9]. Another example is considering buildings as a cyber-physical energy system where the joint optimization of energy use by its occupants and information processing equipment is performed [10]. Another example is CPS consisting of smart buildings autonomously interacting with the grid and participating in real-time electricity markets and day-ahead markets [11].

Social networking is one of the application areas of CPS such as CenceMe application which represents the system that combines the inference of the presence of individuals using mobile phones with sharing of this information through social networking applications such as Facebook and MySpace [12].

Another application of CPS is gaming where cyber-physical gaming environment is used to improve user experience such as in the distributed gaming in 3D tele-immersive environments of a virtual lightsaber duel game [13] or in a video game using body-area inertial sensor networks positioned on players as inputs creating a cyber-physical system to play Quid ditch sports from "Harry Potter" where the player rides a flying broom in a practice field [14].

CPS can be applied to integrated circuits, for example, to a three-dimensional integration of multi-core processor-cache system to solve the problem of die thermal run-away by building a 3D thermal model and performing the thermal management in a cyber-physical fashion [15].

CPS can be used for data centers where each data center is modeled as two interacting dynamic networks: a computational (cyber) network that represents the distribution and flow of computational tasks and a thermal (physical) network that characterizes the distribution and flow of thermal energy, and controlled with the goal of optimizing the trade-off between the quality of computational services and energy costs [16].

Cyber-physical systems for energy systems and electric power grids have gained popularity due to an increased interest in the renewable energy. For example, cyber-physical energy systems (CPES) can include renewable energy sources and gridable vehicles (gridable vehicles can be used as loads, sources and energy storages in CPES) and use intelligent scheduling and control of CPES elements to evolve a sustainable integrated electricity and transportation infrastructure [17]. Another example is a cyber-physical system for the electric power grid where infrastructure security is of a paramount importance [18]. Also, future energy systems can be modeled as the intertwined physical-cyber network interconnections of many non-uniform components where all physical components are represented as modules interconnected by means of an electric network [19].

Networking systems can be represented by CPS where, for example, AnySense, a network architecture that supports video communication between 3G phones and

Internet hosts in Cyber-Physical Systems, is used and can support a class of ubiquitous Cyber-Physical Systems that require video-based information collection and sharing [20].

Alarm systems can be represented by CPS, for example, a high-confidence cyber-physical alarm system (CPAS) which establishes the connection of the Internet through GPRS/CDMA/3G and enables the transformation in alarm mode from traditional one-way alarm to two-way alarm while achieving mutual communication control among terminal equipments, human machine interfaces and users by using the existing mobile communication network [21].

2.2 *Modeling, Simulation, and Verification of CPS*

"A CPS is a 'system of systems', where complex and heterogeneous systems interact in a continuous manner and proper regulation of it necessitates careful co-design of the overall architecture of CPSs" [22]. Thus, several modeling techniques have proposed, along with semantics and programming tools for design of CPSs. Quality models are of paramount importance for building CPSs; however, there are many challenges in building them due to CPSs' heterogeneous nature, concurrency of physical processes, and their time sensitiveness.

Many modeling approaches and tools have been suggested frequently offering notations and concepts and specific to certain application domains. Modeling tools include the meta-modeling techniques and metaprogrammable tools, formal semantics approaches, multi-agent semantic models, event based semantic models, actor oriented design approach [23]; an Adaptive Discrete Event (ADE) model which uses Discrete Event Calculus while defining abnormal event rules to handle unanticipated events [24]; a layered spatiotemporal event model where events are represented as functions of attribute-based, temporal, and spatial event conditions [25]; modeled as hybrid automata where the physical dynamics within modes are combined with discrete switching behavior between them [26]; the formal logic that expresses safety properties by combining probabilities with epistemic operators and, together with a reference model, form a formal framework for multi-agent cyber physical systems [27]; formal models concerning policy-based coordination and Interactive Agents using event-based semantics for cyber physical systems [28]; models that use the situation calculus to handle control problems in cyber-physical environments [29]; formal methods of security specification and verification to describe confidentiality in CPSs and a general approach to specify and verify information flow properties in cyber-physical systems using bisimulation techniques [30].

There are methods and software tools to model, build, and verify cyber-physical systems such as architecture tools to model physical systems, their interconnections, and the interactions between physical and cyber components [31], techniques that help to synthesize algorithmically cyber-physical architectural models with real-time constraints [32], approaches that automate the mapping from analytical models to simulation codes to bridge the gap between analytical modeling and building running simulators for cyber-physical systems [33], model-checking techniques that verify the correctness of the cyber-physical composition by constructing the model and checking it with the help of RT-PROMELA and RT-SPIN, respectively [34].

3 Cyber-Physical Systems with Guaranteed Survivability and Safety

3.1 Introduction

In this work, we consider a type of cyber-physical systems where computational elements interact with sensors, while monitoring the indicators and supporting the functioning of complex technical objects (CTO) [35, 36].

The analytical platform was proposed for supporting the functioning of such systems includes the model, principles, statements, axioms, methods, a system of sensors, software for the data analysis with the goal of controlling physical elements [37, 38]. The software for guaranteed survivability and safety of the functioning of CPS is implemented as an informational platform for technical diagnostics. This tool allows timely and reliably determine, evaluate and forecast risk factors and based on that taking into account the feedback timely identify causes of emergency situations before failures can occur and other undesirable consequences. The functioning of such cyber-physical systems is based on the general problem of multifactor risks, acceptable margin of permissible risk, forecasting of destabilizing dynamics of risk factors, principles, hypotheses, axioms, which are directly related to the analysis of emergency situations, accidents and disasters [39, 40].

The main idea of the strategy is to ensure, in real conditions of the functioning of a complex system, the timely and reliable detection, assessment of risk factors, forecasting of their development over a given time of operation, and on this basis, the timely elimination of the causes of emergency situations before failures and other undesirable consequences [41].

3.2 Modified Information Platform for Technical Diagnostics of Functioning CPS

Let us briefly consider a diagnostic unit, constituting the basis of the algorithm for managing the security of cyber-physical systems in abnormal situations, which is implemented as an information platform, containing the following modules [37]:

1. Receiving and processing of initial information in the process of functioning CPS.

2. Restoring functional dependencies (FD) and identifying patterns based on empirical discretely specified samples.
3. Quantizing the original variables.
4. Forecasting of non-stationary processes.
5. Reliability of information transmitted from sensors.
6. Construction of the process of technical diagnostics.

Let us consider in more detail the just mentioned 5 module of the information platform for technical diagnostics (IPTD).

Evaluation of the Reliability of the Information Transmitted from the Sensors. To obtain information, the CPS uses sensors that continuously provide computing systems with data received from the environment. Both wired and wireless data transmission methods can be used to exchange data from the CPS sensors. On the basis of the obtained data, computing systems can control the physical elements of the system, or supporting their process of functioning. One of the important issues in this interconnection is the reliability of the transmitted information from various sensors and transmitting devices. One of the types of unreliability of the transmitted information during the functioning of the CPS is the failure or malfunction of the sensors. The fundamental complexity of this problem lies in the fact that a priori it is difficult to identify this failure directly without cross-checking sensors, the installation of which is usually unprofitable. Moreover, each investigated CPS has its own peculiarities of recording critical parameters and their tracking.

Evaluation of the reliability of the information transmitted from the sensors can be implemented using various techniques, in particular, the Chauvenet's criterion is used. Based on this criterion on the test sample, as well as at each later step when receiving a new element from the set of incoming indicators, for each coordinate, estimates of the expected value and variance are calculated separately from the set of all previous values of these indicators. Then these estimates are substituted into the inverse distribution function of the Gaussian random variable and the probability is determined for each coordinate. Then multiplication is performed by the size of the sample considered up to the current moment, and if the obtained value is greater than 0.5 for at least one coordinate of the indicators by which the sensor is monitored, then it is considered that the transmitted information from the corresponding sensor is incorrect.

Detection of random sensor failures can be based on the construction of Bollinger bands [42] and step functions of the 1st and 2nd level [37]. The use of both step functions and Bollinger bands, which characterize a technical analysis tool and a technical indicator reflecting the current deviations of the observed value, provides a decrease in the level of dependence on the error of the measured indicators. Bollinger Bands are plotted as upper (moving average plus 2 standard deviations) and lower (moving average minus 2 standard deviations) boundaries around the moving average, but the band width is proportional to the standard deviation from the moving average over the analyzed time period. The moving average period can be selected arbitrarily, but it should be noted that: the longer the period of the moving average, the less sensitive it will be to changes in the observed value; a moving average with a very small period will generate a large number of false signals; a moving average with a very long period will be constantly late.

Taking into account these factors, by empirical considerations, the period of the moving average can be equal to 10. Thus, for 10 dimensions of a given sample, a moving exponentially smoothed average and average deviation are calculated for this interval. The upper and lower Bollinger bands are formed: an exponentially smoothed mean plus two standard deviations and an exponentially smoothed mean minus two standard deviations.

A genetic algorithm can be used to parameterize the model. The procedure for detecting possible sensor malfunctions is based on the following considerations. If the sensor functions are normal, each of its readings does not go beyond the threshold level. Any indication can be confirmed by the previous and subsequent values. First, this is due to the nature of the monitored processes: most changes in the status of a process do not occur instantly. Therefore, an abrupt change in sensor readings can be taken as evidence of failure of the measuring instruments. This approach is implemented as follows. At each step, the arithmetic average of the previous and subsequent measurements is calculated. Then this value is compared with the current value. If the deviation exceeds the threshold level, the operator displays a message about a possible sensor malfunction. Sensor failure can also be tracked by comparing predicted and actual measurements. Since the prediction follows the general behavior of the system based on the latest measurements, a deviation in the actual value could indicate a sensor malfunction. Therefore, the system performs a regular comparison of forecasts and their respective recovered values. As in the previous case, a deviation exceeding the threshold level gives a message about a possible sensor malfunction.

There are general problems with the operation of sensors, that is, deviations of the recorded values from the true ones. An exponential smoothing method can be used to smooth out these deviations. This method is well suited for working with dynamically changing quantities, since the most recent measurements have the greatest weight. Thus, if at some stage there is a significant deviation of the value of a certain parameter from the previous one, and at the next step the value of this parameter returns to its previous level, then, most likely, the sensor has failed and the exponential smoothing method will ensure the elimination of this obstacle. If at some stage there really was a leap, then this will be reflected in all subsequent measurements and the model will quickly pass into a new dynamic state.

Let us consider the construction of the technical diagnostics process using the functioning of the cyber-physical system of an electric refrigerator truck as an example.

3.3 Case Study

The Subject of the Study. As an example of the cyber-physical system used on our case study was an electric refrigerator truck used to deliver cargo.

The vehicle must distribute perishable goods around the city. The load is distributed in equal parts at four points with different distances between them. Electric vehicle capacity is 800 kg. Each consumer receives the same amount of cargo, weighing 200 kg.

In normal mode, taking into account the terrain, the vehicle can perform this work when powered only from a fully charged battery. The movement of the vehicle is monitored by the operator of the dispatch center, which has an emergency forecasting system, in order to make a timely decision on changing the route to the next customer.

Under the terms of the contract between the carrier and the consumer of the goods, the carrier pays a penalty for delay in delivery, which is proportional to the time of delay.

Thus, the first critical parameter that needs to be controlled is the profit from the transportation. The profit can be determined using the formula:

$$Q = Q_{in} - W_{AB}C_{kWh} - V_B C_B - k_{n_1}\tilde{t}_1 - k_{n_2}\tilde{t}_2 - k_{n_3}\tilde{t}_3 - k_{n_4}\tilde{t}_4 - Q_a,$$

where Q is profit, Q_{in} is amount received from the customer, W_{AB} is the amount of electricity received from the grid to charge the battery (BAT), C_{kWh} is the electricity price (per kWh), V_B is the amount of consumed gasoline in liters, C_B is the gasoline price per liter, k_{nj} are coefficients for computing the penalty for each destination, $j = 1, \ldots, 4$, $\tilde{t}_1, \ldots, \tilde{t}_4$—delivery delays of goods to the corresponding destinations, Q_a is fixed expenses for depreciation of equipment, salaries, etc.

With normal delivery, there are no delays, i.e. $\tilde{t}_1 = \tilde{t}_2 = \tilde{t}_3 = \tilde{t}_4 = 0$; gasoline consumption for charging the battery does not exceed a certain level. In this case, the main expenses include the initial charge of the battery, a small amount of consumed gasoline, the inevitable depreciation costs, and other fixed expenses.

The situation becomes abnormal when due to traffic conditions or other factors, the delivery of goods is not done on time or the gasoline consumption is too high. This leads to a sharp decrease in profitability due to the payment of penalties to the customer, payment for gasoline.

The situation also becomes abnormal if the estimated remaining range is significantly lower than the remaining travel distance, or the energy reserve in the battery is lower than the permissible one either for a long time or at the beginning of the trip.

In the event of a delay due to traffic congestion, or other factors, the battery can be recharged from a gasoline generator or a stationary charging station.

Battery charging is also possible due to the recuperative braking or descending from the hill. Since the charging current of the battery is limited, the peak bursts of energy are accumulated in a special storage device (the high-capacity capacitor bank), and then distributed among potential consumers.

An emergency situation can happen when losses occur as a result of a late delivery of goods. Abnormal and emergency situations are relative to the profit from the transportation *980 price units* and *850 price units*, the range of *13,000 m* and *1000 m*, the energy stored in the battery *20 MJ* and *5 MJ*.

Evaluation of the Guaranteed Functioning. The approach of system guaranteed functioning assessment and rational coordination of resources of acceptable risk is used to study the process of functioning of the CPS using an electric refrigerator truck as an example. The state of the vehicle at any time from the given interval is characterized by values of indicators of the profit from deliveries y_1, estimated power reserve y_2, the amount of energy y_3 in the rechargeable battery. True functional dependencies y_1, y_2, y_3 are given as discretely specified samples Y_1, Y_2, Y_3, which are functions of discretely specified parameters x_{ij}, shown in Table 1. In accordance with the IPTG procedures, the input data were pre-processed and transformed into the form required to restore the FD. Functional dependencies $y_i = f_i(x_1, x_2x_3),\ i = \overline{1, m}$ were restored in terms of functions Y_1, Y_2 and Y_3, discretely specified. An assumption was made about the existence of a relationship between the variable parameters on the basis of the fact that many variables are technologically related, for example,

Table 1 Description of dependencies Y_1, Y_2, Y_3 and their variable parameters x_{ij}

Function	Description	Arguments	Description
Y_1	Profit from deliveries	x_{11}	Average speed
		x_{12}	Current speed
		x_{13}	Distance to point 4
		x_{14}	Range
		x_{15}	The amount of gasoline consumed
		x_{16}	expenses $(-W_{AB}C_{kWh} - Q_a)$
		x_{17}	The amount received from the customer for the transportation of the goods
Y_2	Range Estimated power reserve	x_{21}	Average travel speed
		x_{22}	Average battery discharge power
		x_{23}	The amount of stored energy in the battery
		x_{24}	The total weight of the vehicle
		x_{25}	Current speed
		x_{26}	Current battery discharge/charge power
Y_3	The amount of energy in the battery	x_{31}	Mechanical shaft power on traction motors
		x_{32}	Charge power provided by the generator
		x_{33}	The power consumed by the refrigerator
		x_{34}	The power consumed by other equipment
		x_{35}	Traction motors current

the current speed, average speed and distance to travel, or mechanical shaft power and traction motors current. In connection with this assumption, the restoration of functional dependencies in a multiplicative form was chosen [37], since in this case the possible relationship between the variables is taken into account.

A linear regression model was chosen as a forecasting model, the parameters for which were calculated using the least squares method. This model was chosen due to the fact that, with the relative simplicity of implementation and low computation costs, it gives fairly accurate results comparable to the results of more costly methods, for example, methods of time series analysis.

Prediction of the values of functions is performed at each step of observation, while $N_0 = 50$ last measurements of variables are used for forecasting, with the help of which prediction is made for $N_1 = 10$ steps ahead. Thus, the dynamic sample for prediction is $N_D = N_0 + N_1 = 60$ values. Prediction dynamically changes at each iteration, so that in the developed program you can monitor the indicators as they change dynamically and their predicted values in 10 steps.

The procedure for detecting an abnormal n_j (or emergency a_j) situation is based on the use of threshold values $y_i^{n_j}$ for the current *j-th* sensor (and not only the current value is used, but also the predicted ones). An abnormal situation occurs if at least one of the sensors satisfies the inequality $\exists j\left(y_i^j < (>)\, y_i^{n_j}\right)$, depending on whether the minimum or maximum threshold is used. The detection of an emergency situation is similarly accomplished: threshold values $y_i^{a_j}$ are used for sensor j, an emergency situation occurs when the inequality $\exists j\left(y_i^j < (>)\, y_i^{a_j}\right)$ is satisfied, depending on whether the minimum or maximum threshold is used.

According to the statement of the problem, an emergency is a situation in which losses were incurred as a result of untimely delivery of the cargo. Thus, the only established threshold value for an emergency will be $Y1 < 0$, which has practical meaning, since from the point of view of the company organizing the delivery of the cargo, the complete failure of the task will only be the occurrence of losses as a result; while any other situation can still be turned into profit, for example, by using the economy mode of an electric vehicle, or even by using other cars. Although such a situation, of course, will not be normal, the company will still make a profit as a result.

In the developed software, when an emergency situation is detected, this is immediately reported to the operator by a corresponding message on the screen, and the type (emergency situation, emergency situation in several parameters, etc.) and a description of this emergency situation including which specific threshold has been exceeded/dropped below in the corresponding sensor. The value of the risk factor is also calculated in parallel. This is done according to the formula:

$$\rho_i^j = \frac{\left|y_j^i - y_i^{n_j}\right|}{\left|y_i^{n_j} - y_i^{a_j}\right|}.$$

where y_i^j is the current j sensor value, $y_i^{n_j}$ is the threshold value for an abnormal situation to occur, $y_i^{a_j}$ is the threshold value for an emergency situation to occur. In accordance with the obtained value of the function, the current level of danger is determined and its classification is performed.

Confidence intervals were used to identify the reliability of the information transmitted from the sensors due to sensor failure. The predicted values y_{i_f} of the critical variables y_i are used to construct the confidence interval I. If, during further operation, the values y_i do not fall into this interval, this indicates a sensor failure.

$$I = \left[y_{i_f} - \Delta, y_{i_f} + \Delta\right], \Delta = t_\alpha S_y,$$

where S_y is mean square error of the forecast model; t_α is value for t-statistic for probability α, the probability that a value will fall within the confidence interval; in this problem $\alpha = 0.8$. With this method, sensor failures can be detected with a high probability and the error can be prevented from affecting decision making.

The margin of permissible risk, as the duration of functioning of a complex system in a certain mode, during which the degree and level of risk as a result of the possible impact of risk factors do not exceed a priori specified permissible values, is calculated in this work according to the formulas [39]:

$$\Delta_i[s] = \max_k\left\{|y_i[t_{k-1}] - y_i[t_k]|, k < s + n_f\right\},$$

$$T[s] = \min_i\left\{\frac{y_i[t_s] - y_{iw}}{\Delta_i[s]}\right\} \times T_0,$$

where: $T[s]$ is the margin of permissible risk of an abnormal situation at time s, c; y_{iw} is the value for an abnormal situation level; n_f is number of forecasting steps.

If there is an abnormal situation on the forecast interval, then the resource of acceptable risk is 0. That is, first, the maximum change in each indicator y_i for one time interval in an abnormal mode is found for all the data that are available at the moment, as well as the predicted ones. Then the resource of admissible risk is the minimum of the partial difference of the current and abnormal parameters and its maximum change in the abnormal mode. Simply put, the resource of acceptable risk is the time during which an abnormal situation can be controlled in a manageable manner, ensuring the guaranteed survivability of functioning.

Some results of the functioning of an electric refrigerator truck are shown in Figs. 1 and 2 in the form of the distribution of critical variables during operation.

As a result of monitoring the functioning of the electric refrigerator truck for a normal situation, the value of the hazard was 0, the level of the hazard—0, the margin of permissible risk 3333 s (Fig. 1); for an abnormal situation, the value of the hazard was 0.2004, the level of the hazard was 1 (an abnormal situation based on one parameter), the *margin of permissible risk* was 0 s (Fig. 2). The situation corresponds to the battery energy being too low.

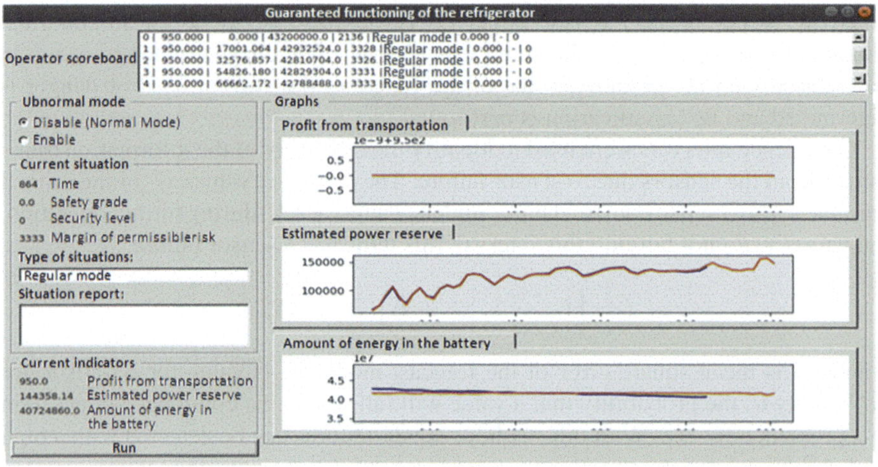

Fig. 1 Monitoring results in the normal mode of operation of the electric refrigerator truck

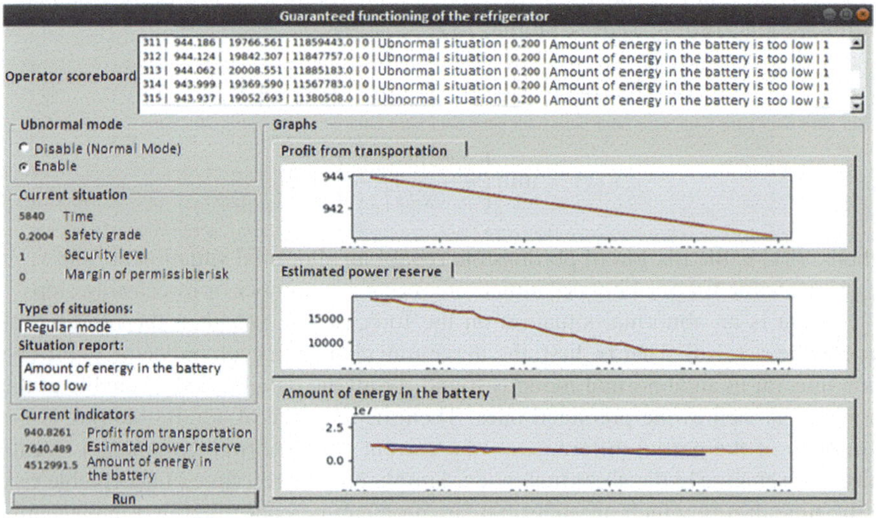

Fig. 2 Monitoring results in the abnormal mode of operation of the electric refrigerator truck

4 Conclusions

A survey of the modeling and functioning of cyber-physical systems is given, from which follows the need to develop new concepts, models, methods, approaches to support the functioning of CPS.

As an example of the functioning of the CPS, we considered the problem of simulating the operation of an electric refrigerator truck in order to deliver perishable goods. Situations are considered that can cause the system to exit from the normal mode. The principle of guaranteed functioning has been implemented, which makes it possible to control the parameters of the system in real time, identify abnormal situations and determine their causes, and, if possible, ensure the survivability of functioning. The developed visualization of the process of predicting the functioning allows you to visually observe the future values of the system parameters, and, therefore, promptly make changes to the control if necessary, which allows, in particular, to respond to the occurrence of emergency situations and make a decision on their timely prevention.

Combining a number of similar CPS models into a single network will allow online rational distribution of the required resources among various consumers. Modern technologies make it possible to provide distributed computing and crowdsourcing, information exchange between users and the shaping of collective knowledge.

Given that CPSs are the driving force behind innovative transformations, many complex problems still need to be addressed. The heterogeneity of data from different applications and devices should be considered. It is necessary to develop models and techniques for collecting, storing and processing big data coming from various network devices, analyze the obtained results and make decisions in a timely manner. The safety and reliability of the functioning of subsystems and CPS systems should be taken into account, since all actions are coordinated between devices in real time. Natural and situational uncertainty due to environmental variability should be taken into account. CPSs have enormous potential, helping to solve critical problems for our society and surpassing modern distributed systems in terms of safety, performance, efficiency, reliability and many other properties.

Acknowledgements The presented results were obtained in the National Research Fund of Ukraine project 2020.01/0247 «System methodology-based tool set for planning underground infrastructure of large cities providing minimization of ecological and technogenic risks of urban space».

References

1. Dumitrache, I.: The next generation of cyber-physical systems. J. Control Eng. Appl. Inf. **12**(2), 3–4 (2010)
2. Foundations for Innovation: Strategic R&D Opportunities for 21st Century Cyber-Physical Systems. https://www.nist.gov/el/upload/12-Cyber-Physical-Systems020113_final.pdf. Last accessed 1 Aug 2021
3. Bhatia, G., Lakshmanan, K., Rajkumar, R., Bhatia, G.: An end-to-end integration framework for automotive cyber-physical systems using Sysweaver. In: The First Analytic Virtual Integration of Cyber-Physical Systems Workshop, pp. 23–30, San Diego, CA (2010)
4. Loos, S., Platzer, A., Nistor, L.: Adaptive cruise control: hybrid, distributed, and now formally verified. In: Butler, M., Schulte, W. (eds.) FM 2011: Formal Methods. Lecture Notes in Computer Science, vol. 6664, pp. 42–56. Springer, Berlin, Heidelberg (2011)
5. Smaldone, S., Tonde, C., Ananthanarayanan, V., Elgammal, A., Iftode, L.: The cyber-physical bike: A step towards safer green transportation. In: Proceedings of the 12th Workshop on Mobile

Computing Systems and Applications, pp. 56–61. Association for Computing Machinery, Phoenix, AZ (2011)
6. Goldman, J.M., Schrenker, R.A., Jackson, J.L., Whitehead, S.F.: Plug-and-play in the operating room of the future. Biomed. Instrum. Technol. **39**(3), 194–199 (2005)
7. Lim, S., Chung, L., Han, O., Kim, J.: An interactive cyber-physical system (CPS) for people with disability and frail elderly people. In: Proceedings of the 5th International Conference on Ubiquitous Information Management and Communication, pp. 1–8. ACM, Seoul, Korea (2011)
8. Banerjee, A., Kandula, S., Mukherjee, T., Gupta, S.: BAND-AiDe: a tool for cyber-physical oriented analysis and design of body area networks and devices. ACM Trans. Embedded Comput. Syst. **11**(S2), 1–29 (2012)
9. Li, X., Lu, R., Liang, X., Shen, X., Chen, J., Lin, X.: Smart community: an internet of things application. Commun. Mag. **49**(11), 68–75 (2011)
10. Kleissl, J., Agarwal, Y.: Cyber-physical energy systems: focus on smart buildings. In: DAC, pp. 749–754. ACM, Anaheim, CA, USA (2010)
11. Savvides, A., Paschalidis, I., Caramanis, M.: Cyber-physical systems for next generation intelligent buildings. In: WiP session at the ACM/IEEE Second International Conference on Cyber-Physical System. ACM/IEEE, Chicago, IL, USA (2011)
12. Miluzzo, E., Lane, N.D., Fodor, K., Peterson, R., Lu, H., Musolesi, M., Eisenman, S.B., Zheng, X., Campbell, A.T.: Sensing meets mobile social networks: the design, implementation and evaluation of the CenceMe application. In: Proceedings of the 6th ACM Conference on Embedded Network Sensor Systems, pp. 337–350. ACM, Raleigh, NC, USA (2008)
13. Wu, W., Arefin, A., Huang, Z., Agarwal, P., Shi, S., Rivas, R., Nahrstedt, K.,: I'm the Jedi!—a case study of user experience in 3D tele-immersive gaming. In: 12th IEEE International Symposium on Multimedia, pp. 220–227. IEEE, Taichung, Taiwan (2010)
14. Wu, C.H., Chang, Y.T., Tseng, Y.C.: Multi-screen cyber-physical video game: an integration with body-area inertial sensor networks. In: 8th IEEE International Conference on Pervasive Computing and Communications Workshops, pp. 832–834. IEEE, Mannheim, Germany (2010)
15. Qian, H., Huang, X., Yu, H., Chang, C.H.: Cyber-physical thermal management of 3D multi-core cache-processor system with microfluidic cooling. J. Low Power Electron. **7**(1), 110–121 (2011)
16. Parolini, L., Tolia, N., Sinopoli, B., Krogh, B.H.: A cyber-physical systems approach to energy management in data centers. In: Proceedings of the 1st ACM/IEEE International Conference on Cyber-Physical Systems, pp. 168–177. ACM/IEEE, Stockholm, Sweden (2010)
17. Saber, A.Y., Venayagamoorthy, G.K.: Efficient utilization of renewable energy sources by gridable vehicles in cyber-physical energy systems. Syst. J. **4**(3), 285–294 (2010)
18. Sridhar, S., Govindarasu, M.: Cyber–physical system security for the electric power grid. Proc. IEEE **100**(1), 210–224 (2012)
19. Ilic, M.D., Xie, L., Khan, U.A., Moura, J.M.F.: Modeling future cyber-physical energy systems. In: IEEE Power & Energy Society 2008 General Meeting—Conversion and Delivery of Electrical Energy in the 21st Century, pp. 1–9. IEEE, Pittsburgh, PA, USA (2008)
20. Xing, G., Jia, W., Du, Y., Tso, P., Sha, M., Liu, X.: Toward ubiquitous video-based cyber-physical systems. In: IEEE International Conference on Systems, Man and Cybernetics, pp. 48–53. IEEE, Singapore (2008)
21. Ma, L., Yuan, T., Xia, F., Xu, M., Yao, J., Shao, M.: A high-confidence cyber-physical alarm system: Design and implementation. In: 2010 IEEE/ACM International Conference on Green Computing and Communications and International Conference on Cyber, Physical and Social Computing, pp. 516–520. IEEE, Hangzhou, China (2010)
22. Khaitan, S.K., Mccalley, J.: Design techniques and applications of cyberphysical systems: a survey. IEEE Syst. J. **9**(2), 1–16 (2014). https://doi.org/10.1109/JSYST.2014.2322503
23. Lee, W.A., Neuendorffer, S., Wirthlin, M.J.: Actor-oriented design of embedded hardware and software systems. J. Circ. Syst. Comput. **12**(03), 231–260 (2003)
24. Yue, K., Wang, L., Ren, S., Mao, X., Li, X.: An adaptive discrete event model for cyber-physical system. In: Analytic Virtual Integration of Cyber-Physical Systems Workshop, pp. 9–15. San Diego, CA, USA (2010)

25. Tan, Y., Vuran, M., Goddard, S.: Spatio-temporal event model for cyber-physical systems. In: 29th IEEE International Conference on Distributed Computing Systems Workshops, pp. 44–50. IEEE, Montreal, Quebec, Canada (2009)
26. Jha, S., Gulwani, S., Seshia, S., Tiwari, A.: Synthesizing switching logic for safety and dwell-time requirements. In: Proceedings of the 1st ACM/IEEE International Conference on Cyber-Physical Systems, pp. 22–31. ACM, Stockholm, Sweden (2010)
27. Bujorianu, M., Bujorianu, M., Barringer, H.: A formal framework for user-centric control of multi-agent cyber-physical systems. In: Fisher, M., Sadri, F., Thielscher, M. (eds.) Computational Logic in Multi-agent Systems, CLIMA 2008, Lecture Notes in Computer Science, vol. 5405, Springer, Berlin, Heidelberg (2009)
28. Talcott, C.: Cyber-physical systems and events. In: Wirsing, M., Banâtre, J.P., Hölzl, M., Rauschmayer, A. (eds.) Software-Intensive Systems and New Computing Paradigms. Lecture Notes in Computer Science, vol. 5380, Springer, Berlin, Heidelberg (2008)
29. Singh, V., Jain, R.: Situation based control for cyber-physical environments. In: IEEE Military Communications Conference, pp. 1–7. IEEE, Boston, MA, USA (2009)
30. McMillin, B., Akella, R.: Verification of information flow properties in cyber-physical systems. In: Workshop on Foundations of Dependable and Secure Cyber-Physical Systems, pp. 37–40. Chicago, IL, USA (2011)
31. Rajhans, A., Cheng, S.W., Schmerl, B., Garlan, D., Krogh, B.H., Agbi, C., Bhave, A.: An architectural approach to the design and analysis of cyber-physical systems. Electron. Commun. EASST **21**, 1–10 (2009)
32. Hang, C., Manolios, P., Papavasileiou, V.: Synthesizing cyber-physical architectural models with real-time constraints. In: Gopalakrishnan, G., Qadeer, S. (eds.) Computer Aided Verification 2011. Lecture Notes in Computer Science, vol. 6806, pp. 441–456. Springer, Berlin, Heidelberg (2011)
33. Zhu, Y., Westbrook, E., Inoue, J., Chapoutot, A., Salama, C., Peralta, M., Martin, T., Taha, W., O'Malley, M., Cartwright, R., Ames, A., Bhattacharya, R.: Mathematical equations as executable models of mechanical systems. In: Proceedings of the 1st ACM/IEEE International Conference on Cyber-Physical Systems, pp. 1–11. ACM, Stockholm, Sweden (2010)
34. Sun, Y., McMillin, B., Liu, X., Cape, D.: Verifying noninterference in a cyber-physical system the advanced electric power grid. In: Seventh International Conference on Quality Software, pp. 363–369. IEEE, Portland, OR, USA (2007)
35. Henning, S.: Diagnostic Analysis for Mechanical Systems. In: Proceedings of the 2000 ASME Design Theory and Methodology Conference, pp. 11–24. ASME, Baltimore, MD, USA (2000)
36. Huang, S.R., Huang, K.H., Chao, K.-H., Chiang, W.T.: Fault analysis and diagnosis system for induction motors. Comput. Electr. Eng. **54**(C), 195–209 (2016)
37. Pankratova, N.D.: System strategy for guaranteed safety of complex engineering systems. Cybern. Syst. Anal. **46**(2), 243–251 (2010)
38. Pankratova, N.D.: System analysis in dynamics of diagnosing of complex engineering systems. Syst. Res. Inf. Technol. **1**, 33–49 (2008)
39. Pankratova, N.D.: The integrated system of safety and survivability complex technical objects operation in conditions of uncertainty and multifactor risks. In: 2017 IEEE First Ukraine Conference on Electrical and Computer Engineering (UKRCON), pp. 1135–1140. IEEE, Kyiv, Ukraine (2017)
40. Pankratova, N.D., Pankratov, V.: Development of the technical object analytical platform for cyber-physical systems. Syst. Anal. Eng. Control **XXIII**(1), 321–332 (2019)
41. Pankratova, N.D.: Creation of the physical model for cyber-physical systems. In: Arseniev, D., Overmeyer, L., Kalviainen, H., Katalinic, B. (eds.) Cyber-Physical Systems and Control 2019, Lecture Notes in Networks and Systems, vol. 95, pp. 68–77. Springer, Cham (2019). https://doi.org/10.1007/978-3-030-34983-7
42. Colby, R.: The Encyclopedia of Technical Market Indicators, 2nd edn. McGraw-Hill Education (2002)

Chapter 3
System Approach to Control-Oriented Mathematical Modeling of Thermal Processes of Buildings

Alexander Kutsenko, Sergii Kovalenko, Vladimir Tovazhnyanskyy, and Svitlana Kovalenko

Abstract A heated building is presented as a complex system consisting of elements separated from each other and from the environment by heat-conducting surfaces. The thermal process of a building consists of heat and mass transfer processes in the walls of the building elements and associated changes in the parameters of the internal air. The given mathematical model is a system of ordinary differential equations for internal air and partial differential equations that model the processes of heat transfer through internal and external walls. It is proposed to consider internal and external walls in the form of a discrete system of interconnected heat-transferring flat elements. This allows us to represent a mathematical model of the thermal process of a building in the form of a finite-dimensional system of ordinary linear differential equations, focused on the application of methods of modern control theory. To assess the adequacy of a finite-dimensional model the analytical solution of the problem of two-sided symmetric heat supply is compared with the numerical solution of the corresponding system of ordinary differential equations at various degrees of discreteness. The task of reducing a multilayer wall with a uniform wall with a minimum number of layers has been set and solved for multilayer walls. Computational experiments have shown high accuracy of the two-dimensional reduced model under harmonic perturbations of the ambient temperature.

Keywords Thermal process of a building · Mathematical model · Systems approach · Electrical analogy · Finite-dimensional approximation · Model reduction

---

A. Kutsenko (✉) · S. Kovalenko · V. Tovazhnyanskyy · S. Kovalenko
National Technical University "Kharkiv Polytechnic Institute",
2, Kyrpychova Str., Kharkiv 61002, Ukraine
e-mail: kuzenko@kpi.kharkov.ua

S. Kovalenko
e-mail: adbc@ukr.net

M. Zgurovsky and N. Pankratova (eds.), *System Analysis & Intelligent Computing*,
Studies in Computational Intelligence 1022, https://doi.org/10.1007/978-3-030-94910-5_3

1 Introduction

Despite the fact that the process of heat supply of buildings for various purposes is one of the main processes that provide the vital functions of mankind, and its scientific and technical support in the form of such disciplines as technical thermodynamics and heat engineering has more than two hundred years of history, interest in the heat supplying process has increased in the last few decades. This is due both to the natural growing needs of mankind in a comfortable habitat, and to the economic and political pressure of the monopolist countries that control energy sources. The problem of energy conservation is particularly acute for countries with the pronounced continental climate, such as Ukraine. Reducing the hydrocarbon energy consumption for heating buildings can be achieved in various ways. Firstly, it is the use of new building materials and technologies that allows increasing the thermal resistance of building external walls. Secondly, it is the use of alternative energy sources in distributed energy systems. Thirdly, it is the widespread use of automatic heat supply control systems for buildings. The use of automatic heat supply control systems, along with maintaining a comfortable premises temperature, also makes it possible to implement program control of the thermal regime of a building depending on the time of day, in order to provide minimum energy consumption.

2 Analysis of Research and Publications

The problem of automating heat supply processes due to its relevance over the last few decades has been the subject of many publications. An in-depth analysis of the problem is contained in the review [1]. The basis for creating any effectively functioning control system is an adequate mathematical model of the controlled process. The process of heat supply of buildings is a complex dynamic distributed process of heat-mass transfer, occurring in conditions of uncertainties of a large number of physical and constructive parameters, as well as in conditions of external disturbing climatic and technological factors. These circumstances do not allow designing a detailed distributed mathematical model of thermal processes in a building suitable for the analysis and synthesis of heat supply control systems. Among the many studies on mathematical and computer modeling of the thermal state of buildings, the publications [2, 3] can be highlighted as differing by a systems approach to the problem and by an attempt to take into account the largest possible number of factors affecting the process of building heat supply. The quasi-static mathematical model of thermal processes which is applied in practice [4, 5], does not allow adequately predicting the temperature regimes of building elements in the conditions of daily and seasonal fluctuations of external climatic conditions, as well as in case of changes of parameters of the heat carrier.

The mathematical models of the thermal process in the building external walls, which allow to take into account the processes of heat accumulation, on the one hand,

and to avoid integration of partial differential equations by approximating them using a system of ordinary differential equations, on the other hand, are considered in [6, 7]. At the same time, the authors of the studies consider the obtained approximating system together with the differential equations of heat accumulation with internal air and the contents of the rooms. This leads to a rigid system of differential equations due to a wide variation of the time constants of heat accumulation by external enclosure of a building, the indoor air and the useful contents of the room.

In [8], a simplified electrical analogy of a heat supply system was proposed that takes into account the accumulation of heat by all elements of a building, as well as the ventilation system, glazing, and adjustable heating devices. In [9], the minimum allowable number of discretization layers of the flat barrier was assessed by the criterion of layer temperature deviations from the theoretical continuous temperature distribution obtained as a result of an analytical solution of the heat equation for symmetric boundary conditions. The principal constrain of problems statements in [8] and [9] is the homogeneity of the wall material by its thickness.

The most advanced in this direction of research should be considered the study [10] in which a detailed analysis of possible approaches to the synthesis of mathematical models of thermal processes of buildings aimed at solving problems of controlling thermal state was carried out.

The paper discusses discrete multi-element models of buildings walls, generally consisting of layers with different thermophysical properties. Such an approach leads to a mathematical model in the form of a system of discrete linear equations of high dimensionality, which makes it difficult to use it for the synthesis of real systems of automatic heat supply control.

The paper [11] experimentally confirmed the adequacy of the discrete representation of thermal processes, focused on solving problems of controlling the thermal state of buildings.

3 Goals and Objectives of the Study

The goal of this study is to develop a formal approach to the synthesis of a mathematical model of thermal processes of a building that is well adapted to the methods of modern control theory. To achieve this goal, it is proposed to consider, by following a systems approach, a heated building as a system of interconnected rooms (system elements) interacting with each other, as well as with the environment and heating devices—heat sources, and on this basis to formalize the principles of construction of a mathematical model of thermal processes of a building.

Since the building construction elements are objects with distributed parameters, whose mathematical models are partial differential equations, it is proposed to represent distributed building elements in the form of finite-dimensional systems of interconnected elements with lumped parameters and reduce the system to the minimum feasible dimension. To develop and justify the procedure for reducing the multidimensional model to the system of the minimum possible order according to the criterion of accuracy of the results of numerical modelling.

4 Systems Approach to Mathematical Modelling of Thermal Processes of a Building

At the heart of the systems approach to the study of any complex processes is the representation of the original system as a set of interconnected elements, each of which has a fairly simple and physically grounded mathematical description. In addition to the mathematical models of individual elements, for a complete mathematical description of the system, it is necessary to specify the rules of interaction between the elements, which are also derived from different physical laws.

The main elements of the system in question are the internal air of separate rooms of the building, formed by a set of partition walls and floor structures, which are separate elements of the system: external walls of the building, glazing, heating devices, and ventilation system.

The interaction between these elements and the environment is based on the laws of heat-mass transfer. Thus the heat exchange between the air and the surfaces of the external and partition walls is based on convection and radiation, and the heat exchange between the surfaces of the external walls is performed on the basis of heat transfer. Mass transfer processes between the rooms and the environment are performed by forced ventilation as well as diffusion and filtration processes.

Following [9], the separate rooms of a building will be represented by the vertices X of some graph. Its set of edges U corresponds to heat or material flows between the individual elements. Let assign to each vertex $x_k \in X$ $\left(k = \overline{0, N}\right)$ its physical characteristics of internal air such as T_k—the air temperature and V_k—the volume of the premises. As the vertex x_0 of the graph, we take ambient air with the temperature T_0. Each edge $u_{ij} \in U$ is associated with a vector P_{ij} of parameters that determines the process of heat transfer between i-th and j-th elements of the building.

$$P_{ij} = \left(H_{ij}, F_{ij}, c_{ij}, \rho_{ij}, \lambda_{ij}\right),$$

where H_{ij} is the thickness of the partition or external wall, F_{ij} is the contact surface; c_{ij}, ρ_{ij}, λ_{ij} are the specific heat capacity, density and thermal conductivity coefficient of the material of the partition or external wall, assuming that the structure of separating constructions is homogeneous in the direction orthogonal to their surfaces.

The mass transfer processes caused by ventilation and the possible transfer of the air between adjacent rooms have a significant impact on the thermal condition of the building rooms. The intensity of mass transfer will be set by a skew-symmetric matrix G_{ij} which is a mass or volume air flow rate between the rooms of a building.

It is not difficult to see that the matrix G_{ij} is a skew-symmetric one, i.e. $G_{ij} = -G_{ji}$, $G_{ii} = 0, \forall i, j \in 0, N$.

To find the heat flow between rooms i and j, use the basic laws of heat transfer according to which a non-stationary temperature field $T_{ij}(x, t)$, described by a differential heat equation, is formed inside partition or external walls.

$$c_{ij}\rho_{ij}\frac{\partial T_{ij}}{\partial t} = \lambda_{ij}\frac{\partial^2 T_{ij}}{\partial x^2}. \tag{1}$$

On the surfaces of the partition (external) wall $x = 0, x = H_{ij}$ there are boundary conditions of the 3rd kind.

$$\alpha_i\big(T_i(t) - T_{ij}(0, t)\big) = -\lambda_{ij}\left(\frac{\partial T_{ij}(0, t)}{\partial x}\right), \tag{2}$$

$$\alpha_j\big(T_{ij}\big(H_{ij}, t\big) - T_j(t)\big) = -\lambda_{ij}\left(\frac{\partial T_{ij}\big(H_{ij}, t\big)}{\partial x}\right), \tag{3}$$

where α_i and α_j are convective heat transfer coefficients.

The temperature of the indoor air of i-th room, under the assumption of its complete mixing, satisfies the heat balance equation in differential form, which corresponds to the first law of thermodynamics for open systems:

$$dU_i = \sum_{j=1}^{N}\Delta Q_{ij} + \sum_{j=1}^{N}c_p T_j \Delta m_j - c_p T_i \Delta m_i + \Delta Q_i, \tag{4}$$

where dU_i is increment of internal energy, ΔQ_{ij} is a heat flow of the j-th element, Δm_j is the air mass increment, caused by j-th element, c_p is the isobaric heat capacity of the air, Δm_i is the air mass flow of the i-th room, ΔQ_i is a heat flow of the internal heat source.

Then for the temperature T_i of the indoor air of i-th room the following differential equation, corresponding to Eq. (4), is true:

$$c_V\rho_B V_i\frac{dT_i}{dt} = \alpha_i\sum_{\substack{j=0\\ j\neq i}}^{N}F_{ij}\big(T_{ij}(0, t) - T_i\big) + c_p\sum_{\substack{j=0\\ j\neq i}}^{N}G_{ij}\big(T_j - T_i\big) + Q_i(t), \quad i = \overline{1, N}, \tag{5}$$

where c_V is the isochoric specific heat of the air, ρ_B is the air density, $Q_i(t)$ is the power of sources of heat.

Thus, the mathematical model of the thermal processes of a building is the system of N ordinary differential equations of the first law of thermodynamics (5), as well as a system of $\frac{N(N-1)}{2}$ differential equations of heat conduction of the form (1) with boundary conditions (2) and (3). To integrate the complete system (1)–(5), it is necessary to set the initial conditions for the temperature of rooms $T_i(0)$ and for the initial distribution of temperature fields $T_{ij}(x, 0)$, as well as to set the laws of heat supply $Q_i(t)$ and mass transfer $G_{ij}(t)$.

This mathematical model assumes the absence of any material contents of the rooms of the modeled building. This is also true for filling buildings with elements with the mass much smaller than the mass of partition walls and floor structures. If it is necessary to consider the accumulation of heat by the internal contents, the system of heat balance Eq. (5) must be supplemented with a system of differential equations of the form

$$C_{ai}\frac{dT_{ai}}{dt} = \alpha F_{ai}(T_i - T_{ai}), \quad i = \overline{1, N}, \tag{6}$$

where C_{ai} is the heat capacity of the contents of the i-th room, T_{ai} and F_{ai} is the average temperature and thermal contact surface of accumulating elements of i-th room, α is the heat transfer coefficient.

Since the mass of the indoor air is negligible compared to the mass of external and partition walls, the system of Eqs. (1), (5), (6) belongs to the Tikhonov class. This allows, instead of the system of differential Eq. (5), to consider a system of algebraic equations resulting from equating the right-hand sides of (5) to zero.

5 Discrete Mathematical Model of Flat Elements of Buildings

When considering building as a system of interconnected elements, a system of differential equations was obtained, composed of both ordinary and partial differential equations. Such a mathematical model, due to its complexity, aimed at solving problems of analyzing the thermal state of a building.

The distributed model is unsuitable for solving control problems, since the main mathematical apparatus of the theory of automatic control is focused on mathematical models in the form of systems of ordinary differential equations.

To move from a distributed system to a system with lumped parameters, represent the physical body under consideration (in our case, an external or partition wall) as a combination of homogeneous elements A_i with lumped thermophysical properties, separated by heat-conducting partition walls.

Then the internal energy U_t of each of the elements A_i will satisfy the differential equation of the first law of thermodynamics

$$\frac{dU_i}{dt} = \sum_{j=1}^{n} Q_{ij}, \tag{7}$$

where Q_{ij} is a skew-symmetric matrix of heat flows between elements of the system.

Considering the assumption made above with respect to homogeneity of the individual elements, formula (7) can be transformed to the form:

$$C_i \frac{dT_i}{dt} = \sum_{j=1}^{n} \alpha_{ij} F_{ij} (T_i - T_j), \tag{8}$$

where α_{ij} are the elements of a symmetric matrix of heat transfer coefficients, F_{ij} is a symmetric matrix of the surface areas of thermal contact between the elements A_i and A_j, C_i is the heat capacity of the elements A_i.

5.1 Heat Transferring Elements

In the analysis and synthesis of thermal processes in heated rooms, the heat transferring elements are of particular interest. The element A will be called heat transferring one if it is in thermal contact with the elements B and C, and their temperatures T_A, T_B and T_C satisfy the condition $T_B > T_A > T_C$.

Generally, instead of a single element B or C, there are the sets of elements $B_1, B_2, \ldots, B_n$ and $C_1, C_2, \ldots, C_m$. These sets are in thermal contact with an element A and satisfy the conditions $T_B > T_A > T_C$, $i = \overline{1, n}$, $j = \overline{1, m}$.

From the basic relations of the 1st principle of thermodynamics (7) and (8), the differential equation for the heat transferring element can be written as

$$\frac{dU_A}{dt} = Q_{BA} - Q_{AC},$$

where Q_{XY} is the value of the heat flow from the element X to the element Y.

The corresponding differential equation for the temperature of the heat transferring element will take the form

$$C_A \frac{dT_A}{dt} = \alpha_{AB} F_{AB} (T_B - T_A) - \alpha_{AC} F_{AC} (T_A - T_C).$$

5.2 Electrical Analogy

The method of analogy of physical processes of different nature is widely used both in theoretical studies and in the practical implementation of mathematical models. The electrical analogy is the most popular because of a visual representation of the characteristics of the elements and variables describing the process (current and voltage), as well as ease of implementation by analog elements. The use of electrical analogies for the analysis of thermal processes in a continuous medium has been the subject of numerous scientific studies.

The main elements of the electrical model of thermal processes are the resistance R and the capacitance C, the transverse and longitudinal variables are current and voltage. Thermal analogues of electrical processes are the thermal resistance and the heat capacity, and corresponding variables are the heat flow and the temperature.

The value of the thermal resistance for a lumped element is defined as

$$R = \frac{1}{\alpha F},$$

where α is the heat transfer coefficient, F is the thermal contact surface.

Then the ratio between the heat flow and the temperatures of the elements in thermal contact can be represented in the form that is similar to Ohm's law for an electrical circuit

$$Q = \frac{T_2 - T_1}{R}.$$

The heat capacity C characterizes the ability of a thermal element to accumulate internal energy in a manner similar to an electrical capacitance that accumulates electrical energy. The relationship between the heat flow and the temperature of the element is written similarly to the equation of a capacitor for an electrical circuit and looks like

$$C\frac{dT}{dt} = Q.$$

Thus, the electrical analogy of the heat transferring element can be represented by the following scheme.

In Fig. 1 T_1 and T_2 are the temperatures of elements involved in the heat transfer process, R_1 and R_2 are the thermal resistances, T is the temperature of the heat

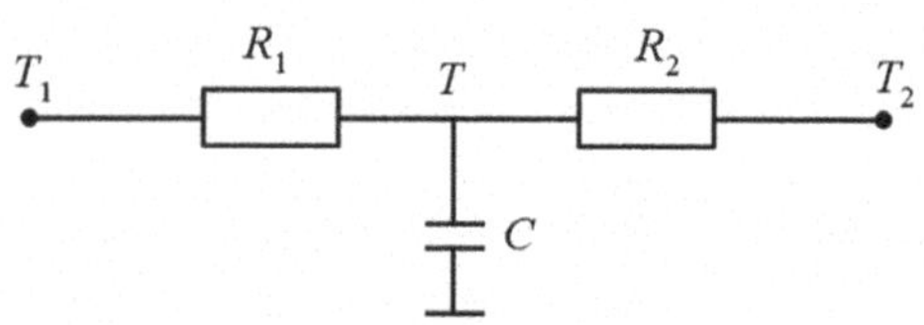

Fig. 1 Electrical analogy of the heat transferring element

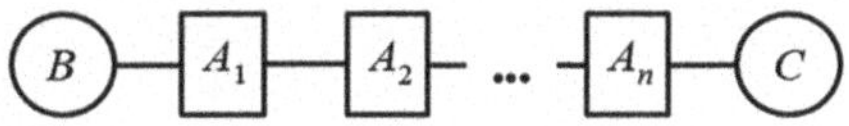

Fig. 2 One-dimensional chain of heat transferring elements

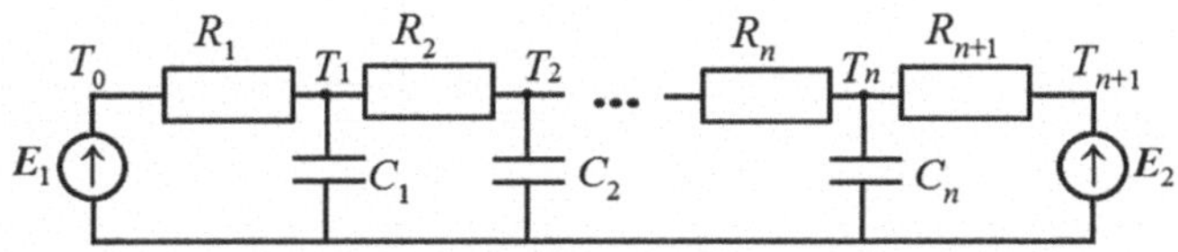

Fig. 3 Electrical analogue of one-dimensional circuit of heat transferring elements

transfer element. The corresponding differential equation of the heat transferring element on the basis of Kirchhoff's law for a node with the temperature T takes the form

$$C\frac{dT}{dt} = \frac{T_1 - T}{R_1} - \frac{T - T_2}{R_2}.$$

5.3 Chains of Heat Transferring Elements

In terms of mathematical modeling of heat transfer processes, the linear (one-dimensional) circuits of "flat" heat transferring elements are of particular interest.

In Fig. 2 A_1, A_2, ..., A_n are heat transferring elements that are in thermal contact with each other. The chain, shown in Fig. 2, will be the heat transferring one if the following conditions are met $T_B > T_{A_1} > T_{A_2} > \cdots > T_{A_n} > T_C$.

The electrical analogue of such a chain is shown in Fig. 3.

In Fig. 3 E_1 and E_2 are voltage sources that are similar to sources of a given temperature T_0 and T_{n+1}, respectively. For the circuit shown in Fig. 3 on the basis of the first Kirchhoff's law for nodes it is possible to write the corresponding system of differential equations for temperatures T_k

$$C_k\frac{dT_k}{dt} = \frac{T_{k-1} - T_k}{R_k} - \frac{T_k - T_{k+1}}{R_{k+1}}, \quad k = \overline{1, n}. \tag{9}$$

At the first stage consider a plate uniform in thickness and physical parameters and represent it as a chain of n identical heat transfer elements (Fig. 3), for which $R_1 = R_2 = \cdots = R_{n+1}$ and $C_1 = C_2 = ... = C_n$. Let us compare the heat process in such a circuit with the known solution of the symmetric one-dimensional problem of heat conduction (Fig. 4). This problem is formulated as follows: find a solution to the heat equation

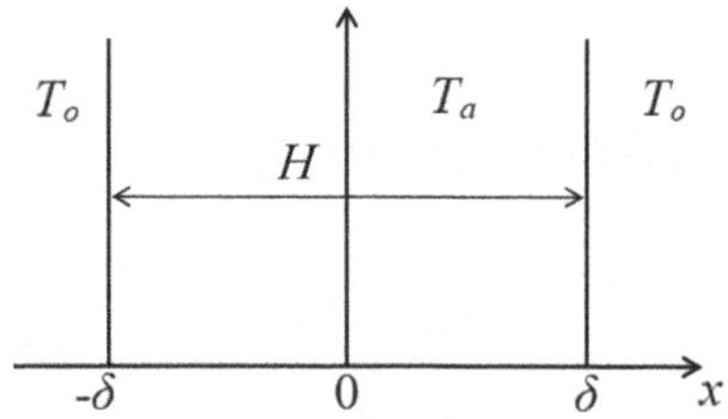

Fig. 4 Statement of the symmetric problem of heat conduction

$$c\rho\frac{\partial\theta}{\partial t} = \lambda\frac{\partial^2\theta}{\partial x^2},$$

where $\theta(x,t) = T(x,t) - T_0$ is a relative temperature, $x \in [-\delta, \delta]$, $\delta = \frac{H}{2}$, $T(x,0) = T_a\ \forall x \in [-\delta, \delta]$, $T_0 = \text{const}$ is an ambient temperature, $T_a \neq T_0$.

Boundary conditions are those of the 3rd kind

$$\lambda\left|\frac{\partial\theta}{\partial x}\right|_{x=\pm\delta} = -\alpha\theta|_{x=\pm\delta}.$$

The analytical (exact) solution of the formulated problem of bilateral symmetric heat supply to the plate takes the form [12]

$$\theta(x,t) = \theta(x,0)\sum_{i=1}^{\infty}\frac{2\sin\mu_i}{\mu_i + \sin\mu_i\cos\mu_i}\cos\left(\mu_i\frac{x}{\delta}\right)e^{-\mu_i^2 Fo}, \tag{10}$$

where μ_i are the roots of the characteristic Eq. (11), Fo is the Fourier number.

$$\mu \mathrm{tg} \mu = Bi. \tag{11}$$

Criteria Bi and Fo are determined based on the physical parameters of the plate and the conditions of convective heat transfer at the boundaries

$$Bi = \frac{\alpha\delta}{\lambda},\quad Fo = \frac{\lambda t}{c\rho\delta^2}.$$

Corresponding finite-dimensional system of ordinary differential Eqs. (9) can be written in the form

$$\begin{aligned}
\frac{d\theta_1}{d\tau_n} &= -(L_n + 1)\theta_1 + \theta_2,\\
\frac{d\theta_k}{d\tau_n} &= \theta_{k-1} - 2\theta_k + \theta_{k+1},\quad k = \overline{2, n-1},\\
\frac{d\theta_n}{d\tau_n} &= \theta_{n-1} - \theta_n,
\end{aligned} \tag{12}$$

where $\tau_n = \frac{\lambda n^2}{c\rho\delta^2}t$ is the relative time, $\theta_k = T_k - T_0$ is the relative temperature, $L_n = \frac{2nBi}{Bi+2n}$.

In the case of $n = 1$, the system (12) degenerates into one differential equation

$$\frac{d\theta_1}{d\tau_1} = -L_1\theta_1. \tag{13}$$

Numerical experiments to substantiate the minimal dimension n of the approximating system (12) were carried out for various values of the criterion Bi by integrating system (12) under the initial conditions $\theta_1(0) = \theta_2(0) = \cdots = \theta_n(0) = 1$ and comparing the results of integration with the analytical solution (10). In relation (10), the first 4 terms were taken into account, since the contribution of the subsequent ones to the parameters of the transient process was extremely insignificant.

In Figs. 5, 6 and 7 the results of the analytical solution of the heat conduction equation and the results obtained by numerical integration of system (12) at $Bi = 10$ and different values of n are shown.

The estimation of the accuracy of finite-dimensional discretization was carried out according to the criterion

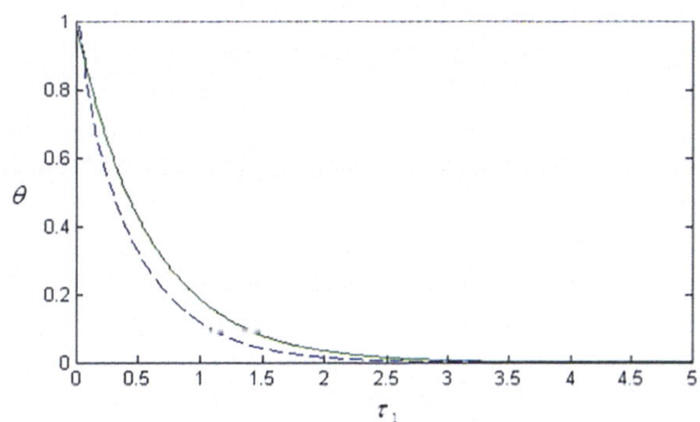

Fig. 5 Comparison of the analytical solution of the heat equation with the result of numerical integration ($n = 1$): - - - - - - - - -—analytical solution; ————————numerical solution

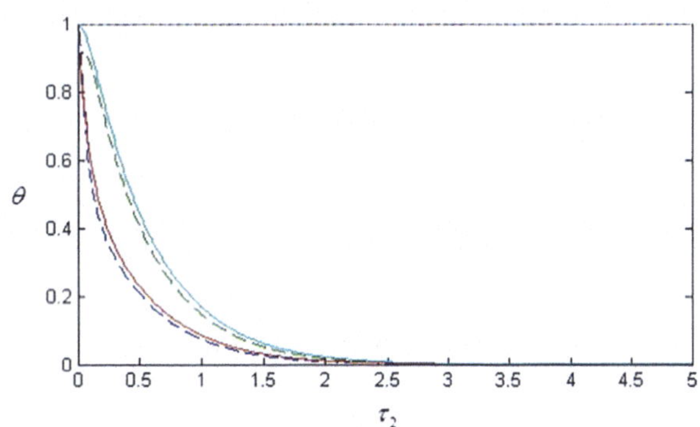

Fig. 6 Comparison of the analytical solution of the heat equation with the result of numerical integration ($n = 2$)

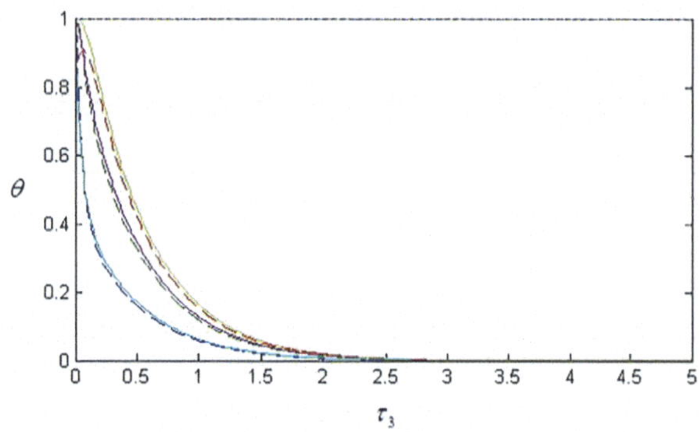

Fig. 7 Comparison of the analytical solution of the heat equation with the result of numerical integration ($n = 3$)

$$\varepsilon = \frac{1}{N}\sum_{i=1}^{N}\left|\frac{\theta_i^* - \theta_i}{\theta_i^*}\right|, \tag{14}$$

where θ_i^* is the theoretical value of temperature, θ_i is the value of temperature obtained from a finite-dimensional model, is the number of comparison points. The error values ε for different n and Bi are given in the table.

It follows from Table 1 that the accuracy acceptable for practice $(10 \div 12)\%$ for any values of the Biot criterion is achieved at $n = 2$. Increasing the order of approximation to 3 improves the accuracy to $(3 \div 7)\%$.

Thus, based on the results of a computational experiment, it is proved that the third order of the approximating system of differential equations is sufficient accuracy for practical purpose.

To improve the accuracy of the approximation at $n = 1$, we introduce a correction coefficient $k \in (0.9 \div 1.4)$ and instead of the value L_1 in the differential Eq. (13) for the wall temperature, we will consider kL_1. As a result of optimization according to the criterion of average deviation (14), optimal values k were obtained for various Bi. The optimization results are shown in Table 2.

The obtained result can be considered as a substantiation of the possibility of a one-dimensional approximation of the distributed heat transfer process through a flat wall.

Table 1 Relative error ε depending on n and Bi

n	Bi				
	1	4	10	20	50
1	0.1680	0.3388	0.4222	0.4504	0.4566
2	0.0584	0.1078	0.1231	0.1242	0.0512
3	0.0377	0.0645	0.0709	0.0401	0.0264
4	0.0304	0.0494	0.0637	0.0303	0.0141
5	0.0270	0.0425	0.0331	0.0255	0.0077

Table 2 Optimal values k for various Bi

Bi	1	4	10	20	50
k	1.11	1.23	1.28	1.3	1.3
ε	0.0041	0.0145	0.0138	0.0119	0.0107

6 Reduction of a Multizone Mathematical Model

The results of mathematical modeling of the thermal process by integrating of (9) with the increasing number of discretization layers will approach to the results of a distributed mathematical model of the form (1). At the same time, the volumes of calculations naturally grow and the reliability of the obtained results decreases. Thus, there is a natural problem: to construct a mathematical model of the thermal process of the form (9) of the minimum dimension $n^* \leq n$, where n is some specified initial dimension of the system, determined by the construction of the external walls; n^* is the dimension of the reduced system in which the deviation of thermal processes from thermal processes of the original one does not exceed a specified value.

Transform the system of Eq. (9) to a non-dimensional form. To do this, introduce the following notation

$$R = \sum_{i=1}^{n+1} R_i, \mathrm{C} = \sum_{j=1}^{n} C_j, \tau = RC.$$

Multiplying the left and right side of each of the equations of system (9) by τ and dividing k-th equation by C_k, we get

$$\tau \frac{dT_k}{dt} = \frac{\tau}{C_k R_k}(T_{k-1} \quad T_k) \quad \frac{\tau}{C_k R_{k+1}}(T_k \quad T_{k+1}), k - \overline{1, n}.$$

Introduce notations:

$$r_k = \frac{R_k}{R}, \; s_k = \frac{C_k}{C}, \; \tau_{ij} = s_i r_j, \; \vartheta = \frac{t}{\tau}, \; \xi_{ij} = \frac{1}{\tau_{ij}}, \quad i, j = \overline{1, n},$$

where r_k, s_k are relative resistances and capacitance of the, τ_{ij} is relative time constant, ϑ is non-dimensional time.

Taking into account the introduced notation, the system of differential equations of the thermal process in the chain of elements will take the form

$$\frac{dT_k}{d\vartheta} = \xi_{kk} T_{k-1} - (\xi_{kk} + \xi_{kk+1}) T_k + \xi_{kk+1} T_{k+1}, \quad k = \overline{1, n}. \tag{15}$$

This system is a n-dimensional linear stationary inhomogeneous system of differential equations for n state variables $T_1, T_2, \ldots, T_n$. To solve it, it is necessary to specify n initial conditions for $T_k(0)$ and functions $T_0(\vartheta)$ and $T_{n+1}(\vartheta)$.

As an integral indicator of thermal processes of a multizone thermal system, we take the heat flows at the chain outputs

$$q_1 = \frac{T_0 - T_1}{R_1}, \; q_{n+1} = \frac{T_{n+1} - T_n}{R_{n+1}}. \tag{16}$$

As the input actions we take $T_0 = \text{const}$—the indoor temperature and $T_{n+1}(\vartheta)$—the ambient temperature:

$$T_{n+1}(\vartheta) = T^* + \Delta T \sin \omega \vartheta,$$

where T^* is the average daily temperature, ΔT is the amplitude of daily temperature fluctuations, ω is the circular frequency of fluctuations in the scale of non-dimensional time.

As a criterion for comparing the heat flows (16), total relative deviations were taken

$$\varepsilon_1 = \frac{1}{N} \sum_{p=1}^{N} \frac{\left|q_1(t_p) - q_1^*(t_p)\right|}{\left|q_1(t_p)\right|},$$

$$\varepsilon_2 = \frac{1}{N} \sum_{p=1}^{N} \frac{\left|q_{n+1}(t_p) - q_{n+1}^*(t_p)\right|}{\left|q_{n+1}(t_p)\right|}, \tag{17}$$

where $q_1(t_p), q_{n+1}(t_p), q_1^*(t_p)\; q_{n+1}^*(t_p)$ are the values of the heat flows for compared models at time points $t_p\ \left(p = \overline{1, N}\right)$, obtained as a result of integrating the systems of differential Eq. (15) for dimensions n and n^*.

For each selected value $n^* < n$, it is necessary to solve the optimization problem for one of the criteria (17).

The relative resistances and capacitances r_k and s_k (10) are the optimizing parameters, the choice of which is determined by the conditions $\sum_k r_k = 1, \sum_k s_k = 1$. Thus, the number of variable parameters in the optimization problem is $2n^* - 1$.

Computer simulation of discrete thermal processes in the form of a numerical solution of systems of linear differential equations, as well as the solution of criteria optimization problems (17) was carried out in the MATLAB–Simulink environment.

Numerical experiments performed when reduction of the original system (15) when $n = 10$ with randomly selected parameters r_k and s_k show that the optimal two-dimensional reduction gives almost indistinguishable thermal processes ($\varepsilon < 0.1\%$) on both of the indicators (17). One-dimensional reduction allows us to achieve similar accuracy on one of indicators (17). On the second indicator, the accuracy is within 10%.

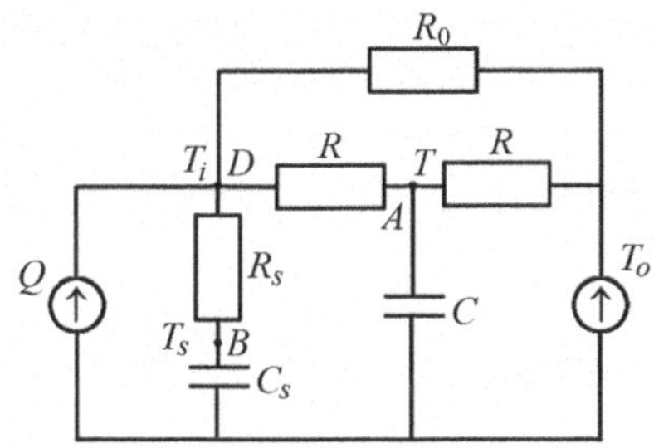

Fig. 8 Electrical analogy of the thermal process of a building

The obtained results allow us to propose a simplified electrical (Fig. 8) and the corresponding mathematical model of thermal processes in a building [13].

In the Fig. 8 R is a thermal resistance of half of the thickness of an outer wall; R_0, R_s are thermal resistances of glazing and internal walls; C, C_s are heat capacity of the outer and inner walls, respectively; T, T_s are average temperatures of the outer and inner walls; T_i, T_o are indoor and outdoor temperatures; Q is a heat flow of heating devices.

Applying Kirchhoff's 1st Law to the nodes A, B and D, after a series of simple transformations, we obtain a simplified mathematical model of the thermal process in the form

$$\begin{aligned} \frac{dT}{d\vartheta} &= -[1+2(\rho+\omega)]T + \rho T_s + q + (1+\rho+\omega)T_o, \\ \frac{dT_s}{d\vartheta} &= \xi T - \xi(1+\omega)T_s + \xi\omega T_o, \\ T_i &= \frac{1}{1+\rho+\omega}(q + T + \rho T_s + \omega T_o), \end{aligned} \tag{18}$$

where $\rho = \frac{R}{R_s}$, $\omega = \frac{R}{R_0}$, $q = QR$, $\vartheta = \frac{t}{\tau(1+\rho+\omega)}$, $\xi = \frac{\tau}{\tau_s}$, $\tau = RC$, $\tau_s = R_s C_s$.

The resulting mathematical model consists of two differential equations of state with respect to T and T_s, output equation with respect to T_i. System (18) contains a control q and a disturbance T_o.

The thermal process is determined by three parameters of similarity ξ, ρ, ω, which are calculated based on the structural features of the building.

7 Conclusions

The two-tier systems approach to mathematical modeling of thermal processes is an effective tool for the synthesis of mathematical models of heat supply of buildings of arbitrary structures.

The resulting reduced mathematical model in the form of a system of ordinary linear differential equations sufficiently precise simulates the distributed thermal processes in the external and partition walls and opens the way to the effective adaptation of the whole range of modern methods of analysis and synthesis of automatic control systems applied to the processes of heat supply in residential and office buildings.

References

1. Panferov, S.V.: Some problems of energy saving and automation in heating buildings, Herald SUSU. Series Computer technology, management, electronics (2010)
2. Tabunshchikov, Yu.A., Borodach, M.M.: Mathematical Modeling and Optimization of the Thermal Performance of Buildings. AVOK-PRESS, Moscow (2002)
3. Sokolov, E.Ya.: District Heating and Heat Networks. Publishing MPEI, Moscow (1999)
4. Medina, M.A.: Validation and simulations of a quasi-steady state heat balance model of residential walls. Math. Comput. Model. **30**(7–8), 93–102 (1999)
5. Malyarenko, V.A.: Basics Thermal Physics and Energy Efficiency of Buildings. Publishing SAGA, Kharkiv (2006)
6. Vasilyev, G.P., Lichman, V.A., Peskov, N.V.: A numerical optimization method for intermittent heating. Math. Model. **11**, 123–130 (2010)
7. Tejeda, G.: Hybrid predictive control for building climate control and energy optimization. In: Preprints of 2013 IFAC Conference, SAIT Petersburg, pp. 337–342 (2013)
8. Kutsenko, A.S., Kovalenko, S.V., Tovazhnyanskyy, V.I.: Mathematical modeling of controlled process of buildings heat supply. Integr. Technol. Energy Saving **1**, 36–43 (2013)
9. Kutsenko, A.S., Kovalenko, S.V., Tovazhnyanskyy, V.I.: System approach to mathematical modeling of thermal processes of buildings. East-Eur. J. Enterp. Technol. **4/4**(70), 9–12 (2014)
10. Atam, E., Helsen, L: Control oriented thermal modeling of multizone buildings: methods and issues. IEEE Control Syst. Mag., 86–111 (2016)
11. Florez, F., Fernandez de Cordoba, P., Higon, J.L., Olivar, G., Taborda, J.: Modeling, simulation, and temperature control of a thermal zone with sliding modes strategy. Mathematics **7**, 503 (2019)
12. Shorin, S.N.: Heat Transfer, 339 p. State ed. for Construction and Architecture, Moscow (1952)
13. Kutsenko, A., Kovalenko, S., Tovazhnyanskyy, V., Kovalenko, S.: Synthesis of a mathematical model of thermal processes of buildings. Systems approach. In: 2020 IEEE 2nd International Conference on System Analysis & Intelligent Computing (SAIC), pp. 276–281 (2020)

Chapter 4
Toward the Mentality Accounting in Large Social Systems Models

Alexander Makarenko

Abstract Modeling and simulation of social phenomena are very important for understanding such phenomena. This paper proposes the consideration of such human qualities as mentality, anticipation, the acquisition and generation of knowledge, especially when considering interacting individuals. Considerations are based on models of society in the form of networks with associative memory with elements that also take into account the internal mental representations of individuals. An example of such model is proposed.

Keywords Social systems · Modeling · Mentality · Sustainable development · External variables · Internal variables · Networks · Anticipation

1 Introduction

Sustainable development is one of the general concepts, becoming one of the basic components of the future world order. So far, the situation with the SD is developing in several directions. (The Club of Rome activities initially, there were Forrester's works on modeling (J. Forrester)). Later, especially after the activities of the Club of Rome (Club of Rome), the idea of sustainable development became widespread and has some practical applications. First of all, we remark humanitarian approaches at world summits Rio-de-Janeyro (1992, 2012), Johannesburg (2012) and many others. As examples remark general economies, or sustainable development considerations, shows a desirable direction for development. However, in view of the lack of mathematical formalization and sufficiently developed global mathematical models, the practical study of modeling problems is non-complete. That is why so far the success of developments refers to local tasks, regional or individual problems: global warming; climate change; sustainable development of regions; energy; infrastructure of firms; corporations, settlements. The success of such applications substantially depends on the capabilities of operational research methods, mathematical modeling, system analysis, etc.

A. Makarenko (✉)
ESC "Institute for Applied Systems Analysis" NTUU "KPI", Kyiv 03056, Ukraine

M. Zgurovsky and N. Pankratova (eds.), *System Analysis & Intelligent Computing*, Studies in Computational Intelligence 1022, https://doi.org/10.1007/978-3-030-94910-5_4

However, the further development of global ideas related to sustainable development of society as a whole is still an important issue for reflection and research, the success of which will depend on further mathematical formalization. Many specialists in operational research and mathematical modeling are working in this direction. However, the problem as a whole still remains unsolved. The author's work [1–5] proposes a formalization of the sustainable development issue that seems to be of great help in understanding sustainable development. In the proposed formalism the verbal description of sustainable development problem is formulated as evolutionary problems with certain limitations that take into account the specificity of limitations in the problem of sustainable development—resources and presence in the problem of new generations of population. As already mentioned, the approach [1–5] allows to take a new look at these issues and expand many areas of sustainable development research.

Usually the three interacting areas are considered in the overall sustainable development issue: ecology, economy and society. Currently, the science and practice of modeling environmental and economic blocks of sustainable development are more developed.

However, as stated in many works, a very important component of the sustainable development process is issues related to behavior, mind world of individual, personal history. Let us call all this components (and many others) as the mentality of individuals. It should be noted that, as indicated in [1, 2], the main resource in the problem of sustainable development is knowledge and the means to obtain it. In view of the relatively small amount of research on the mental factors, further research in modeling society with accounting mentality aspects.

1.1 Setting the Problem

The question of conscious transformation of large socio-economically-political systems is becoming increasingly important in modern conditions, both in theoretical and practical terms. It should be noted that in the Ukrainian context they are especially important in terms of the country's governance in the context of great challenges and internal and external uncertainty. When considering these issues, there is a need for an adequate understanding and consideration of the current stage of the evolution of society—namely, the postmodern state. Roughly speaking, this situation is characterized by the coexistence of different types of subsystems of society, pluralism of opinions, norms, moral teachings, stages of development, etc.

As one example, one can consider different concepts, approaches, definitions, methods to the problem of sustainable development, especially in the global aspect. Of course, a great role in the consideration of social systems is played by material factors: resources, the influence of the environment, technological structure, and infrastructure.

But it is obvious that purely human properties as a thinking being are very important (and maybe even outstanding). Very conventionally, this can be called the mentality of a person. In the context of an individual, these issues are dealt with by psychology, neurophysiology, computational neuroscience, and philosophy. The next very important stage is understanding of social systems as collectives of interacting subjects. In this case, one can conditionally speak about systems from a large number of thinking agents with different mental properties. Of course, many questions have already been dealt with by various disciplines related to society and have already been quite successfully dealt with by well-known scientific disciplines: sociology, political science, economics, public administration theory, social psychology, culturology, management theory, and many others. However, now it is becoming increasingly clear that the number, quality and depth of problems associated with the understanding of the mentality of properties is becoming increasingly necessary, even to solve current problems of management for post-industrial society in post-modernism. One example where mentality is important is the global sustainable development.

Despite the large amount of attention to the problem of sustainable development at all levels—from the world's leaders to the population of different countries—it is recognized that significant changes from the economic to the environmental are still ahead. But we can assume that the main thing in such changes and transformation processes will be carried out in the future. The main obstacle to sustainable development is the change of norms, preferences and attitudes of the society. Namely, these concepts are related to and based on the understanding of human mental properties. Therefore, the problem of sustainable development requires an adequate understanding of the impact of the properties of mentality, including archetypes.

1.2 Purpose of the Article

So far, studies of the influence of the mentality of individuals on the processes in society have been conducted to a large extent by methods of humanities, that is, intuitively and qualitatively. At the same time, it is well known that the mainstream of development of various sciences is the increasing use of methods of exact sciences, especially mathematics and physics. It should be noted that the author suggests the aspects of mathematical modeling of society, which allow formalizing and including the question of mentality and conduct modeling, including the formulation of real management plans. This is exactly what the proposed article is devoted to.

The author's models view society as a large complex entity, made up of many elements with connections. The consideration of the properties of society allows us to choose some interesting properties and then propose models that can imitate the regimes of society. The models resemble models of brain activity—neural network. Such models have been investigated by the author since 1992 and have already had some interesting applications.

2 Models with an Internal Structure and Mentality

Some new models for large social systems had been proposed by author since 1992 (see for example [1–4]). The structure of such models is familiar with classical artificial neural networks (for textbook on neural networks see [6]), especially Hopfield's neural networks [7, 8]. There were some useful applications: for geopolitics, stock market etc. But for further development it necessary to take into accounts the internal (mental) properties of individuals. So below we describe one presumable way for such accounting.

Consideration of mentality requires consideration of internal structures and their inclusion in global hierarchical models. There are many approaches for taking into account the mentality. The most natural way to accomplish this task is to consider the model for the internal structure also in the class of neuronet models. The easiest way is to represent the image of the World in the brain or the individual in the model as a collection of elements and connections between elements. In this image of the world, there is a place to represent the individual directly with personal faith, skills, knowledge, and benefits. We have presented some individual; he or she has some idea about the world structure. This representation is similar with the "pattern" above. But the essential new effect is that an individual can imagine himself as one of the elements of the "pattern". The mental structures of other personalities are also presented in the same way. Thus, society as a complex system is essentially a new representation. At the first level of description we are a collection of elements connected by connections. On the second level of description, we have added structure (some image of the world) to all elements.

2.1 *One Possible Way to Take into Account the Mentality*

The laws for elemental dynamics should depend on this view. To represent the image of the outside world in the individual brain: it is very important that each individual has his or her own personal image of the world. Some simple variants will be presented in the next section, in parallel with the description of the property of foresight. Of course, there can also be recursion with many levels of recursion, as in the theory of reflexive systems by N. Luman, J. Soros, S. Lefebvre and so on. In our scheme it can be represented as a mutual representation of all personalities in the internal representation of an individual.

2.2 *Internal Representation of the External World*

Recall that initially neural network models were introduced in brain research. First, we can change the basic laws of neural networks. At the phenomenological level, this can be implemented by subdividing the parameters of the elements into external (usual visible) and internal variables (sometimes externally visible and sometimes externally invisible) and establishing separate laws for two blocks of parameters—external and internal output and input parameters. Dynamical laws for such variables can have absolutely different forms. For example, the equation for external variables may take the form of neural network combined with differential equations for internal variables. Let's result here one very important remark which allows basically essential generalization of the offered technique and models, including to problems of archetypes, sustainable development, transformation of society and other similar problems. The internal variables should be divided into two classes. The first class includes variables that change relatively quickly in dynamics under the influence of the environment and the inner state of the individual. Actually, the majority of current tasks of the economy deal with such variables (and external factors). But the second class includes variables that are relatively stable, such as perceptions, archetypes, development patterns, etc. These constructs can also change, but much more slowly (for example, when several generations change).

Parameters of the first and second classes refer to what should be considered as components of the mentality. One of the most promising ways to take into account the mentality is to look for an equation in the neural network class. The simplest way is to represent the image of the World in the brain of an individual or in the model as a collection of elements and connections between elements. In this image of the world, there is a place for representing the individual with personal skills, knowledge, preferences.

2.3 An Important Representation of Some Individual

Each individual has a certain idea about the structure of the world. This representation is presented in a "model" as a network. But essentially new effect is that an individual can imagine himself as one of the elements of the "pattern". Mental structures of other personalities are also presented in the same way. Thus, society as a complex system is a new spectacle. On the first level of description there should be the collections of elements, which are connected by bonds. On the second level of description we have connected the structure (some image of the world) to all elements.

The laws for elemental behavior should depend on this view. Formally, we can introduce the operators of projection P to represent the image of the outside world in the individual brain: it is very important that each individual has his or her own personal image of the world. Some simple variants will be presented below, in parallel with the description of some society property.

3 Mentality Accounting in Models

The mentality accounting requires considerations internal structures and incorporating them in global hierarchical models. There are many approaches for mentality accounting (see review of some aspects in [3]). The most natural way for implementing this task is to consider as model for internal structure also neuronet models. Remember that originally neuronet models were introduced in the investigation of brain. Firstly we should change the basic laws of models for the goal of mentality accounting. On phenomenological level it may be implemented by introducing subdivision of elements variable which describe the state of each of each of the elements on external (externally visible variables) $Q_{i\ ext}$ and internal (mental) $Q_{i\ \text{int}}$ variables of *ith* element, where N—number of elements and establishing laws for two blocks of parameters.

Here we for simplicity consider the description of element's variables as vectors. But in more complex cases it may be patterns, mathematical structures, hierarchies of variables. In proposals below we should also use the improvements of dynamical laws for each of elements. Following the approaches for neural network models the *ith* element has output and input variables and some dynamical laws for solution calculations. In general case external (visible) variables of other elements may serve as the input parameters. We mark such external outputs for *ith* as a vector $\vec{X}^i_{ext}$. An example is the economic variables. But sometimes internal variables also may serve as the output. Mark such variables as a vector $\vec{X}^i_{\text{int}}$. Examples are stereotypes of individuals for decision-making. We should also introduce the dynamical laws for values of each of outputs $\vec{Y}^i_{ext}$ and $\vec{Y}^i_{\text{int}}$, $i = 1, 2, ..., N$:

$$\vec{Y}^i_{ext} = \vec{f}_{ext}(\vec{X}^i_{ext}, \vec{X}^i_{\text{int}}, \vec{P}^i_{ext}, \vec{E}^i_{ext}) \tag{1}$$

$$\vec{Y}^i_{\text{int}} = \vec{f}_{\text{int}}(\vec{X}^i_{ext}, \vec{X}^i_{\text{int}}, \vec{P}^i_{\text{int}}, \vec{E}^i_{\text{int}}) \tag{2}$$

where $\vec{f}^i_{ext}, \vec{f}^i_{\text{int}}$ are the dynamical functions for external and internal output and input variables. $\vec{P}_{ext}, \vec{P}_{\text{int}}, \vec{E}_{ext} \vec{E}_{\text{int}}$ are some parameters and biases for *ith* element. Remark that dynamical laws may take the forms useful for case of continual time or discrete time. The functions $\vec{f}^i_{ext}, \vec{f}^i_{\text{int}}$ may have absolutely different forms. For example equations for external variables may have neuronet form with ordinary differential equations for internal variables. But one of the most prospective ways for mentality account lies in searching Eq. (1), (2) also in neuronet class. Here proposed to introduce the intrinsic mental models of World in elements, which represent the individuals or decision-making organizations with human participation. The simplest way consists in representing image of World in the individual's brain or in model as collection of elements and bonds between elements. In such World pattern there exist place for representing individual himself with personal beliefs, skills, knowledge, preferences. The mental structures on other individuals are also represented.

3.1 One Approach to Modeling Mentality

At first we should take into account different representation of word. Namely we take into account 3 aspects (three images of the world) related to real world. The first image is the precise image (representation, knowledge on the real world) of real world. Such image exists in principle but nobody knows precisely such image. The second image of the world is knowledge of the individual about real state of the world in the internal representation of each of *i*th, $i = 1, 2, \ldots, N$ individuals (N is the number of elements of the society). Such representations may be different for different individuals. The third image of the world is "ideal" (desired) image of the world (including internal states of individual) in the internal representation of each of the individuals. Such "ideal" images also may be different for each of N individuals. It is very interesting that these 3 images of the world may be related to the concept of 3 worlds by K. Popper.

We call the set $Q(t)\}$ "the image of the real world" at a discrete time moment t. We also introduce—$Q_{wish}(t)$ "the desired image of the world at moment t for the first individual" as a set of elemental states and desired connections for the first individual at a time.

$$Q^{(1)}_{wish}(t) = (\{s_i^{wish}(t)\}, \{J_{ij}^{wish}(t)\}) \tag{3}$$

Then we assume that in the case of an isolated dynamic law, changes in the state of the first individual depend on the difference between the real and desirable image of the world:

$$D^{(1)}(t) = [[Q^{(1)}_{\text{wish}}(t) - Q^{(1)}(t)]]. \tag{4}$$

Here [[*]] is some norm. Equation (4) includes "deformed" vision of the World by the individual. Exactly the same type of representation has the desired ("ideal") image of the World. Let us note that only the state of the real World is one for all personalities (in such models), but perceptions and "ideals" are different for different personalities. If we want to represent the "inner" part of the dynamic law in the same way as the "outer" part in Sect. 3, then we can adopt a dynamic law. Then the laws for element evolution should depend on such representation. Formally we can introduce projection operators PR for representing image of outer World in the brain:

$$\begin{aligned} PR_i : W_{ext} &= \left(\left\{\vec{X}_{ext}\right\}, \left\{\vec{X}_{\text{int}}\right\}, \left\{\vec{f}_{ext}\right\}, \left\{\vec{f}_{\text{int}}\right\}, \left\{J_{ij}^{ext}\right\}, \vec{P}, \vec{E}\right) \\ \Rightarrow W_{\text{int}} &= \left(\left\{\vec{X}_{ext}^{\text{int}}\right\}, \left\{\vec{X}_{\text{int}}^{\text{int}}\right\}, \left\{\vec{f}_{ext}^{\text{int}}\right\}, \left\{\vec{f}_{\text{int}}^{\text{int}}\right\}, \left\{J_{ij}^{\text{int}}\right\}, \vec{P}^{\text{int}}, \vec{E}^{\text{int}}\right) \end{aligned} \tag{5}$$

where the last index i in right hand of (5) indicates that variables is the internal representation of original parameters in *ith* element. It is very important that each individual has own personal image of the World. Remark that the action of operator PR_i may be subdivides on the many local projection operators. The Eq. (5) may be replaced by more complicated by inserting self- representation of him in right hand of (5). This may lead to equation of type

$$PR_i(recursive) : W_{ext} \Rightarrow W_{\text{int}}(recursive) \quad (6)$$

Equation (6) remembers the Eq. (5) but right hand $W_{\text{int}}(recursive)$ now depends on self-representing by individual. Of course there may exist recursion with many levels of recursion as in the theory of reflexive systems by N. Luhman, D. Soros, S. Lefever and so on.

4 Models with Evolutionary Structure

4.1 Types of Mentality Variables in an Individual Representation

In previous subsections we proposed some general considerations allowing mentality aspects accounting. Such aspects should be implemented in mathematical modeling. But full general model of society may be very complicated. For example it is well known in psychology that each individual has about 30,000 different mental plans (different internal aspects, cognitive, behavioral etc.). It is impossible now to take into account all this aspects. So it is necessary to take more simple models with small number of mental plans. Just the models with accounting single aspects are very complex and interesting.

Here we present further opportunities for the development of an approach to mentality and the possibility of practical application of such models.

Here are the basic principles of taking into account the mentality in the proposed approach [1–4, 9, 10]. The internal representations of the individual from some slice of reality (in psychology, for example, this is called "internal plans") are represented as a network of objects, concepts and the like, depending on the "object" and purpose of modeling. There are three networks (patterns, drawings) of descriptions: the real state of affairs, presented in the form of a network; the state of things in the imagination of a particular individual and the desired (ideal) state of affairs in the imagination of a particular individual. The dynamics of changes in the parameters that characterize the individual depends on the state of the environment "externally" (similar to the model representations similar to those described in the previous section) and on the "internal" mental variables and the above three ideas about the external environment. Adequate consideration of the external environment by representing and internal representations of surrounding individuals is possible. Perhaps further clarification of these representations by taking into account the reflection of different levels of depth (the individual's ideas about his own representation in the eyes of other individuals, etc.). For internal variables of individuals, other model equations can be used than for externally visible variables. For example, for economic agents, external variables may be intentions to buy/sell some goods, and internal variables may be ideas about the intentions of others to buy goods.

Also, equations can depend on many other parameters that characterize the mental qualities of individuals. For example, the characteristics (indices) of individuals, important for the dynamics of mental variables can be the results of color tests in psychology (see [11]).

4.2 Fast and Slow Variables

In principle, when modeling specific socio-economic problems, the laws (equations) for modeling can be found in the processing of big data by means of artificial intelligence and machine learning. But when building modeling systems and interpreting modeling results, the issues of the velocity of internal variables are important. The corresponding equations are given in [1, 9, 10], and in this section we will only illustrate these aspects and some of their consequences, which are important for transformation issues. Figure 1 gives some illustration of the issues of mental variables.

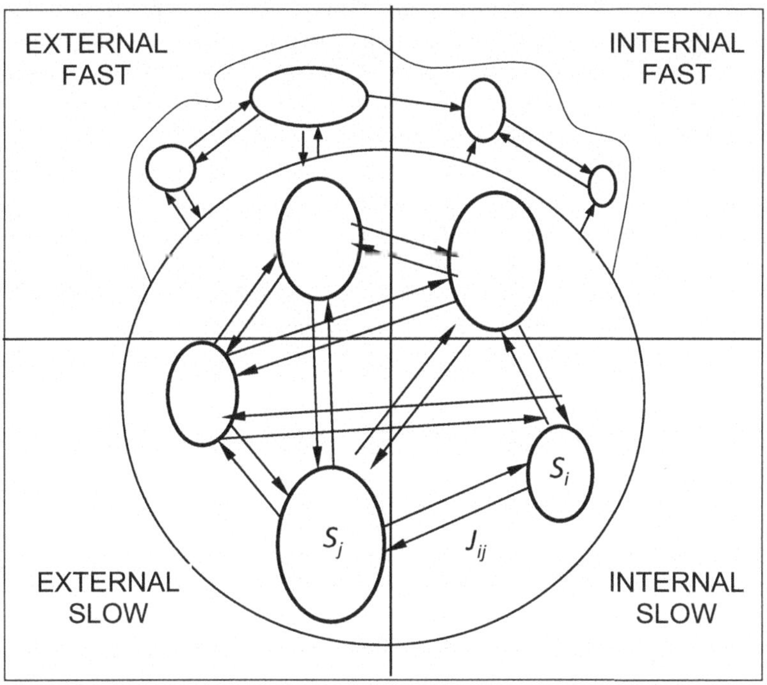

Fig. 1 Variables subdivision on external/internal, slow/fast

The simplest case is in accounting of each individuals (elements) states. Such models directly related to the classical Hopfield neural network models. We propose to use the same description of mentality properties of individual just as to the description of external world that is the network picture. More interesting case corresponds to accounting two type of variables as the description of elements—one eternally visible and one internal. The particularly interesting case is the case of dividing each of such type variables on two classes with different rate of such variables change (slow and fast). Such division of variables conditionally is illustrated on Fig. 1.

The figure on the left and right shows "visible from the outside" mental variables (left) and "internal" invisible from the outside (right). Such a division of variables is already important for understanding socio-economic systems and their transformation. But the figure shows another aspect – the division of the first and second categories of variables into two classes: fast and slow (upper and lower halves in the figure). In [1–4] the equations for modeling in this case are given. For example, the aforementioned "external" economic variables (intentions to buy or sell goods "now") can change quickly, and perceptions of the ideal state of the buying and selling system can change much more slowly. For example, fashion for some goods, which can change relatively slowly. If we look more broadly, the state of affairs in the country may change depending on the perception of the population of the country about the acceptable state of affairs in the country. In turn, changes in perceptions of the correct state of affairs can change very slowly, even in the process of changing generations. It should also be emphasized that such changes significantly depend on the growth of knowledge about the "optimal" state of affairs and their dissemination in the population through the education system in the country. An example is the spread of the idea of sustainable development in the world.

These, even qualitative, ideas about fast and slow variables can be useful for planning the processes of transformation of social systems. That is, first with the use of systems analysis it is necessary to distinguish the structure of fast and slow variables. It is also necessary to model, if possible, the influence of different parameters and controls on the behavior of systems and on the possibilities of different scenarios. You can then try to find appropriate controls for the different scenarios.

4.3 Bonds Between Different Types of Internal Variables of Elements

Also for this paper approach and models very important are bounds between different types of variables (external and internal). Remark that such bounds can changes with time and some external bias. Below (Fig. 2) we give for illustration the description of one of the simplest models of such kind with two variables for each of elements.

New aspects in proposed models appear with taking into account the variable with time structure of models. The simplest variant includes allowing bonds dependence

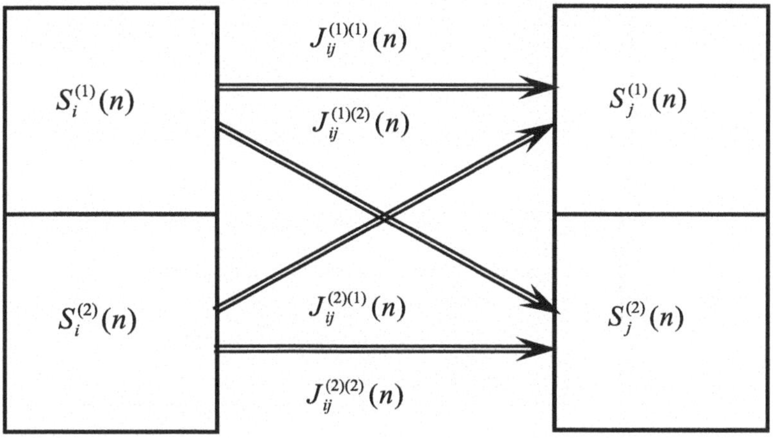

Fig. 2 Relationship of two two-state elements (i and j) at moment n, upper (1), (2) are the index for slow and fast variables

on time. Many variants of time dependence accounting in neuronal network architecture exist since the works by Hebb, McCallok and Pitts, T. Kohonenn, T. Hopfield and others. One simple example of proposed by author models with continuous time is (the index f corresponds to 'fast' variables and parameters and indexes s correspond to 'slow' variables and parameters'):

$$\begin{cases} C_j^f \frac{dv_j^f(t)}{dt} = -\frac{v_j^f(t)}{R_j^f} \\ \qquad + f^{ff}\left(\sum_{i=1}^{N} w_{ji}^f(t)\phi_i^f(v_i^f(t))\right) \\ \qquad + f^{fs}\left(\sum_{i=1}^{N} w_{ji}^s(t)\phi_i^s(v_i^s(t))\right) \\ \qquad + I_j^f + I_j^{fs}, \\ j = 1, \ldots, N \end{cases} \tag{7}$$

$$\begin{cases} C_j^s \frac{dv_j^s(t)}{dt} = -\frac{v_j^s(t)}{R_j^s} \\ \qquad + f^{ss}\left(\sum_{i=1}^{N} w_{ji}^s(t)\phi_i^s(v_i^s(t))\right) \\ \qquad + f^{sf}\left(\sum_{i=1}^{N} w_{ji}^f(t)\phi_i^s(v_i^f(t))\right) \\ \qquad + I_j^s + I_j^{sf}, \\ j = 1, \ldots, N \end{cases} \tag{8}$$

$$\begin{cases} \tau_{lk}^{f} \frac{dw_{lk}^{f}(t)}{dt} = \\ \quad \lambda_{lk}^{ff} F^{ff}(\phi_{f\ \ l}^{fs}(v_l^f(t))\phi_{f\ \ k}^{fs}(v_k^f(t))) \\ \quad -\gamma_{lk}^{fs} w_{lk}^{f}(t) \\ \quad +\lambda_{lk}^{fs} F^{sf}(\phi_{s\ \ l}^{fs}(v_l^s(t))\phi_{s\ \ k}^{fs}(v_k^s(t))) \\ \quad -\gamma_{lk}^{fs} w_{lk}^{s}(t) \end{cases} \tag{9}$$

$$\begin{cases} \tau_{lk}^{f} \frac{dw_{lk}^{f}(t)}{dt} = \\ \quad \lambda_{lk}^{ff} F^{ff}(\phi_{f\ \ l}^{fs}(v_l^f(t))\phi_{f\ \ k}^{fs}(v_k^f(t))) \\ \quad -\gamma_{lk}^{fs} w_{lk}^{f}(t) \\ \quad +\lambda_{lk}^{fs} F^{sf}(\phi_{s\ \ l}^{fs}(v_l^s(t))\phi_{s\ \ k}^{fs}(v_k^s(t))) \\ \quad -\gamma_{lk}^{fs} w_{lk}^{s}(t) \end{cases} \tag{10}$$

where $\{w_{lk}^f\}$, $\{w_{lk}^s\}$ are bonds between elements, $\{v_j^f\}$, $\{v_j^s\}$ are states of elements, $\{\lambda_{lk}^{ff}\}$, $\{\lambda_{lk}^{ss}\}$, $\{\gamma_{lk}^{fs}\}$, $\{\gamma_{lk}^{sf}\}$, $\{C_j^f\}$, $\{C_j^s\}$, $\{\tau_{lk}^f\}$, $\{\tau_{lk}^s\}$, $\{R_j^f\}$, $\{R_j^s\}$ are some parameters responsible on the learning rate and on the retrieval of system past patterns. $\{f^{ff}\}$, $\{f^{fs}\}$, $\{f^{sf}\}$, $\{f^{ss}\}$, $\{F^{ff}\}$, $\{F^{fs}\}$, $\{F^{sf}\}$, $\{F^{ss}\}$, $\{\phi_i^f\}$, $\{\phi_i^s\}$, $\{\phi_f^{fs}\}$, $\{\phi_s^{fs}\}$ are nonlinear functions.

Introduced models with bond time dependence lead to possibility of new interesting phenomena. So in preliminary numerical investigations of simple models derived from (10) we found abrupt transition from one quasi-state pattern to another during operating. Remark that such transition may have evident counterparts in the many process in real large social systems. This phenomenon is one of examples of punctuated equilibrium founded recently in large systems.

5 One Simple Example in Modeling with Mentality and with Time-Dependent Bonds

Proposed approach allows developing the software and trying to understand some properties of society and particularly eGovernment.

One simple example of such dependence received for continuous time model for stock market. The same model useful for investigation of opinion formation and others (11):

$$\begin{cases} C_j \frac{dv_j(t)}{dt} = -\frac{v_j(t)}{R_j} + \sum_{i=1}^{N} w_{ji}(t)\phi_i(v_i(t)) + I_j, \\ j = 1, \ldots, N \\ \frac{dw_{lk}(t)}{dt} = \lambda\phi_l(v_l(t))\phi_k(v_k(t)) - \gamma w_{lk}(t), \\ k \neq l;\ k, l = 1, \ldots, N \end{cases} \tag{11}$$

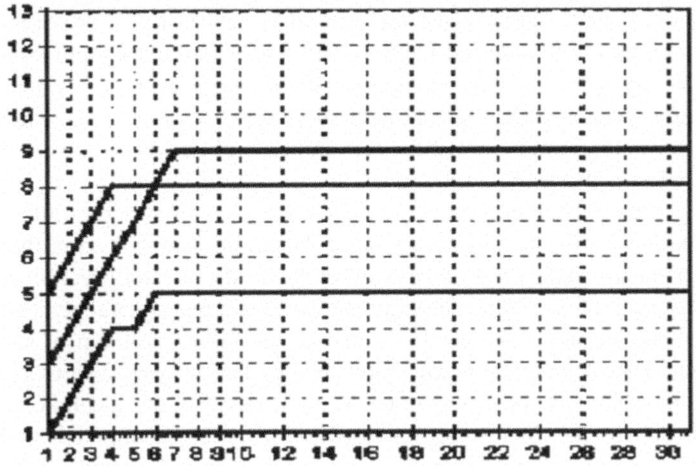

Fig. 3 Example of quasi-stable solution (Discrete time steps are represented on horizontal axe and values for three elements on the vertical)

Counterpart with discrete time also had been investigated.

Here we describe some examples of computer experiments with discrete-time models corresponds to models (10) which accounting the internal structure of participants and non-constant in time reputation of participants (Fig. 3). The variables $\{\nu_i\}$ have interpretation as opinion of individuals and $\{w_{lk}^{f}\}$, $\{w_{lk}^{s}\}$ have interpretation as reputation of *kth* agent's in opinion of *lth* agent. It is interesting that for time interval between 4 and 5 we can see the quasi-stable solution with further evolution.

First of all proposed internal representation may be considered as some correlate to ontology of participant. Also it may be interesting for considering classical problem of reputation. At second the approach reminiscent usual multi-agent approach. The description of participant remember participant with special representation of the internal and external worlds by network structure. Also the prospective feature in the approach is the associative memory in proposed models. Remark that recently we had found the possibility of multi-valued solution existing in case of individuals which can anticipate the future.

6 Some Qualitative Consequences of the Proposed Methodology of Modeling Large Social Systems

Here is one very important remark that allows us to generalize the proposed methodology and models in principle, including the problems of archetypes, sustainable development, transformation and other similar problems. The internal variables should be divided into two classes. The first class includes variables that change relatively quickly in dynamics under the influence of the environment and the inner state of the individual. Actually, the majority of current tasks of the economy deal with such variables (and external factors). But the second class includes variables that are relatively stable, such as perceptions, archetypes, development patterns, etc. These constructs can also change, but much more slowly (for example, when several generations change). Parameters of the first and second classes refer to what should be considered as components of the mentality.

As it has already been emphasized, the second class of variables allows taking into account aspects of archetypes. In particular, in the simplest case, they can be displayed for the use in the proposed models of the results of color psychological tests through the introduction of special parameters (or even one generalized parameter).

The methods are also suggested to consider the global problem of sustainable development. The idea of "economic" and "ecological" way of society's evolution can also be presented as constructs in terms of variables of the second class, i.e. as quasi-stable constructs. Therefore, the transition from the "economic" to the "ecological" path depends on the change of leading constructs of individuals. Sooner or later this will happen through education, media influence, etc.

It is also possible to assume that in the future the concept will also be useful for practical tasks of public administration. First, the concept can provide a qualitative understanding of the impact of various factors (including archival ones) on the processes in society. Also, if the proposed models are further developed and detailed, they may become part of the government's decision support systems.

Obviously, decision makers have forecasts for the future. In this case, the state of the elements in the model should depend on the images of the future, described in the internal representation. We call this case hyperincursion. Another important part of advance is the selection procedure.

Sustainable development is one of the most important problem which can be considered with propose concepts and models. As had been proposed earlier (see [4] the 'economical' and 'ecological' ways of society evolution corresponds to different attractors in the models. So the problem of SD corresponds to existence of different attractors in the models. Thus the problems of society transformation (including transition to sustainable development) follow to the problems of transition between attractors and design of new systems of attractors. Proposed models lead to many ways for such transformations. With fixed connections between elements it is possible with special external biases. But the case of flexible connections the attractors may changes, Following humanitarian sciences it may be accepted that positions of individuals in the problem of sustainable development mostly depends on the 'slow' components of internal variables and bounds between such slow changeable internal variables. Thus such objects are the task of change in transition to other ways of evolution. Such slow changeable also have a long time processes. The patterns and other parameters can be formed in schools for children, universities for young students, by mass-media for populations etc. The science and knowledge is the base of such transformations of 'slow' connection of elements. The network structures of internal variables of elements correspond to existing knowledge. Thus namely the knowledge and their representations into individuals are one of the very important components of the SD problem.

The system of equations and its modifications can form the basis for the study of many problems with internal and external images of the world. We should emphasize that the right side of the equation depends on the future values of the element's state. This form is oppositely constructed with respect to the form of delayed equations. It is very promising that the structure of such a system coincides in structure with the systems studied by Dubois [12]. This entails a possible similarity in properties.

7 Conclusions

In the proposed article we have outlined a part of the approach to modeling processes in large social systems. It has been suggested to include the properties of the mentality of individuals in society, as well as the properties of predicting individuals in the framework of a strict approach. As a result, we have obtained some new models that also take into account the properties of the individual's mentality. The possibility of including archetypical problems in mathematical models is also described. The possibilities of applying the proposed concepts to the problems of society management are also proposed. The approach is also useful for application in economic models.

Acknowledgements Author thanks for S. Levkov, V. Solia for discussions and V. Solia for the help in computer calculations.

References

1. Makarenko, A.: Sustainable Development and Principles of Social Systems Modeling. Generis Publisher (2020)
2. Makarenko, A.: New neuronet models of global socio-economical processes. In: Geurts, J., Joldersma, C., Roelofs, E. (eds.) Gaming Simulation for Policy Development and Organizational Change, pp. 128–132. Tillburg University Press, Tillburg (1998)
3. Makarenko, A.: Neuronet models of global processes with intellectual elements. Int. Bus. Innov. Psychol. Econ. **4**(1(6)), 65–83 (2013)
4. Makarenko, A.: Formalization, modeling and anticipatory properties in computational science for sustainable development. In: Electronic Preprint of EWG-ORD 2018 Workshop OR for Sustainable Development, Madrid, June 2018
5. Makarenko, A.S.: Mentality issues in the transformation processes of the postmodernity society. Public Manage. **1**(21), 154–168 (2020)
6. Haykin, S.: Neural Networks: Comprehensive Foundations. MacMillan, NY (1994)
7. Hopfield, J.J.: Neural networks and physical systems with emergent collective computational properties. Proc. Nat. Acad. Sci. (USA) **79**, 2554–2558 (1982)
8. Hopfield, J.J.: Hopfield network. Scholarpedia **2**(5) (1977)
9. Makarenko, A.: Toward the mentality accounting in social systems modeling. In: Proceedings International Conference on SAIC-2020, October 08–10, Kyiv (2020)
10. Makarenko, A.: Toward the methodology for considering mentality properties in egovernment problems. In: Gunchenko, Yu. (ed.) Intellectual Systems and Informational Technologies, pp. 155–168. Premier Publishing s.r.o., Vienna (2021)

11. Afonin, A., Martinov, A.: Archetypal principles of modeling social processes. Public Manag. **2**(3), 34–47 (2016)
12. Dubois, D.: Incursive and hyperincursive systems, fractal machine and anticipatory logic. In: AIP Conference Proceedings 573, American Institute of Physics, pp. 437–451 (2001)

Chapter 5
The Strategy of Underground Construction Objects Planning Based on Foresight and Cognitive Modelling Methodologies

Nataliya Pankratova, Galina Gorelova, and Vladimir Pankratov

Abstract The strategy of underground construction objects planning based on the mathematical support of foresight methodology with the aim of scenarios alternatives creating and cognitive modelling to build scenarios for the development of the desired future and ways of their implementation is proposed. These methodologies are suggested to be used together: the obtained results at the stage of the foresight methodology are used as initial data for cognitive modelling. Using the foresight process at the first stage of modelling allows to identify critical technologies and build alternative scenarios with quantitative characteristics applying expert assessment procedures. For the justified implementation of a particular scenario, cognitive modelling is used, which allows building causal relationships considering a large number of interconnections and interdependencies. The suggested strategy is applied to the study of underground communications in order to select reasonable scenarios for justification of their creation priority.

Keywords Foresight · Cognitive modelling · Pulse modelling · Scenarios · Underground communications

1 Introduction

The goal of the strategy presented in this article is to study some of the problems of the underground facility viability in extreme situations and emergencies. The proposed strategy of underground construction objects planning is based on the mathematical support of foresight and cognitive modelling methodologies. At the first stage the foresight methodology is applied based on the qualitative analysis methods: SWOT analysis, Topsis, Vikor, Delphi, analytic hierarchy, Delphi methods, and morphological analysis [1–8]. In some cases, when the output information for cognitive

N. Pankratova (✉) · V. Pankratov
Institute for Applied System Analysis, Igor Sikorsky Kyiv Polytechnic Institute, Kyiv, Ukraine

G. Gorelova
Engineering and Technology Academy of the Southern Federal University, "ITA SFEDU", Taganrog, Russia

M. Zgurovsky and N. Pankratova (eds.), *System Analysis & Intelligent Computing*, Studies in Computational Intelligence 1022, https://doi.org/10.1007/978-3-030-94910-5_5

modelling is given in statistical form as separate logical groups, the method of constructing an integrated indicator data is proposed [9]. This approach enables the construction of cognitive maps reasonably add or remove vertex, or break a sequence of interconnected nodes. As a result, alternatives of scenarios with quantitative characteristics should be created, which will be the initial parameters for cognitive modelling. Involving the cognitive modelling methodology at the next stage of the proposed strategy will make it possible to build scenarios of the possible development of a complex system that may arise under the influence of changes in an underground structure's internal and external environment.

In the cognitive modelling stage problem study, the cognitive map is created as a sign-oriented graph and a functional graph as a weighted sign digraph. Methods of analysis of structural stability and resistance to disturbances, methods of analysis of model connectivity (simplicial analysis), and graph theory methods are used to investigate the properties of the cognitive model [10]. The proofs of numerical stability of cognitive maps based on representing values and perturbations at the graph's vertices in matrix form are presented [11]. An impulse process model (simulation of disturbance propagation on cognitive models) is used to determine the possible development of processes in a complex system and create the scenario development. This allows to create scenarios of development in the process of dynamics and propose a scientifically based strategy for implementing the priority scenario [10].

Regulation of urban development to increase ecological standards and life safety in constantly growing metropolises is one of the most urgent, though insufficiently researched, and complex global problems [12]. It leads to the search for new places for production facilities, social and other objects of human activity.

Recently within the framework of mining sciences, a new field of research named construction geotechnology has been developed. The subject of the field is the technologically transformed bowels of the earth [13]. A significant contribution to it subject was made by the Russian scientist B.A. Kartosia [14–16]. The main direction is "... scientific direction—the integrated development of the underground space of the subsoil, a distinctive feature of which is the principle of priority of the work and rest level for comfort during the construction and operation of underground structures for various purposes, guaranteeing improvement of environmental and social living conditions in large cities and industrial areas ..." [13]. Many researchers in geoengineering concentrate different methodological approaches to design problems [17–19]. Risks in underground space development are considered in [20].

The purpose of the study is to substantiate the strategies and methods for developing the underground space, including the utilization and reuse of underground facilities. The development of construction geotechnology and its practical applications is the development of underground space, which requires the advancement of a methodology for studying underground construction problems. The existing methodology of underground structures study, design, and construction, both in terms of architectural and planning decisions as well as the safety of mining operations, needs to be improved. In any case, this methodology should integrate various scientific sections and directions, and it should be interdisciplinary.

Underground urban planning is a complex system in many aspects. Firstly, this system consists of many interconnected subsystems and objects. Secondly, the processes occurring in this system during construction and operation are also complex and occasionally poorly predictable, because they are associated mainly with various geological processes. The problems that accompany underground urban development can be attributed to poorly structured problems.

A system approach to studying underground urban planning based on the methodology of foresight, as a tool for the concept of sustainable development of megacities, was proposed in [21]. The strategy of planning underground space as a component of large cities' sustainable development was developed [22].

Models and methods of scientific foresight and cognitive modelling of complex systems allow developing a cognitive model for analysis of the system's structural properties, stability, and, most importantly, the possible ways of developing the system considering changes in its internal and external environment parameters and under various control actions. This is especially important for comprehending and preventing negative consequences, minimizing damage under the influence of the most unfavorable combination of negative factors: external and internal static and dynamic loads, all kinds of artificial influences inside an underground structure, harmful natural manifestations from a mountain range, etc.

2 Description of the Underground Construction Problem

Let us consider how studying the construction geotechnology problems can use scientific foresight and cognitive modelling of complex systems.

Currently, building geotechnology consists of four main sections: underground urbanism, mechanics of underground structures, geonics, and rock management during construction. All sections are of importance for research using the proposed cognitive methodology. Thus, the content of one of the assignments of the first section is the substantiation of strategies and methods for developing underground space, which is also one of the tasks of cognitive modelling complex systems.

At the design stage of underground structures, it is necessary to consider and justify the practical necessity, socio-economic expediency, and technical feasibility of constructing underground structures in mining and geological conditions and under the influence of construction technology, as well as the functional purpose of construction objects. The tasks of substantiating the necessity, feasibility, expediency, effectiveness of actions in complex systems are also included in the field of cognitive modelling tasks. The knowledge and data from the other scientific sections of construction geotechnology can also be helpful in cognitive modelling of the structure and behavior of a system that imitates a natural system from the perspective of the stated goal of research and functioning.

So, "assessment of the stability of mine workings; the study of the processes of engineering structures interaction with rock masses and the establishment of qualitative and quantitative characteristics of their stress–strain state" and others (second section) can provide data and gain new knowledge about them after cognitive modelling when analyzing of the cognitive model properties; the third section, geonics, including research on the "interconnections of the elements of mining construction technology, the establishment of qualitative and quantitative parameters that determine the choice of methods, engineering and construction technology; effective methods of organizing labor and managing construction works..." turn out to be necessary when developing a cognitive model, establishing relationships between its objects (or "concepts", "factors", "entities"), which are also concepts associated with underground urban planning.

A significant advantage of cognitive models is that their composition at different stages and levels of study and description can include both quantitative (for example, "hydrostatic pressure") and qualitative (for example, "socio-economic feasibility") characteristics.

In our opinion, imitation of cognitive modelling, especially at the stage of designing the development of the underground space, is extremely necessary. A serious reason for this may be the fact that it is necessary to anticipate and eliminate or reduce risks in advance, that are inevitably inherent in underground urban planning in advance. As shown in [13]: "Urban underground construction is characterized by dynamism and a high degree of uncertainty, so the risk factor is an integral attribute of the development of underground space." Therefore, risks under certain conditions are manifested and may have negative consequences for the entire underground infrastructure in the system "man—underground construction" [20, 23].

Knowledge of risks is necessary for everyone: including designers, builders, and operators. In [23], a risk classification was introduced. That classification consists of 8 groups: constructional, environmental, managerial and executive, commercial, economic, contractual, social, and operational. The classification provides the basis for the further development of environmentally friendly technologies and construction methods. Construction risk is dominant in terms of the impact on the entire life cycle of an underground structure, and the higher it is, the lower the requirements for personnel qualifications, construction quality, and terms, reliability of mining equipment, etc.

Wrong construction decisions are the basis for the occurrence of environmental, economic, operational, and other risks. Therefore, the basic principle laid down in research to improve the design and construction methods of underground facilities is the principle of minimizing damage from the consequences of negative manifestations of risks, taking into account the interaction and mutual influence of all natural, technical, technological and other factors. The probability of disasters caused not only by unpredictable natural accidents, but also by design errors, imperfection of existing technologies, requires the formulation and solution of the problem of the viability of the object in extreme and emergency situations.

One of the most challenging problems is underground communications, which provide the vital activity of both surface and underground urban planning. The sewer network intersection, ravines, shipping and drainage canals, railways, highways, dükers, overpasses and crossings is created. Significant advantages of water bodies underground crossings and coastal underground infrastructure put on the agenda large-scale agenda of underground construction in the area of water bodies influence [24, 25]. This paper examines the issue of the construction of pipe and tunnel underwater dükers and justifies the priority of their creation.

3 The System Approach to the Study of Underground Development Planning Based on Foresight and Cognitive Modelling Methodologies

3.1 Methodology of Foresight

The strategy of underground construction objects planning based on the mathematical support of foresight methodology is introduced. Such methodologies aim to create scenarios alternatives and cognitive modelling to build scenarios for the development of the desired future and ways of these scenarios' implementation. For the realization of this strategy, the system approach to solving of the underground construction problems is developed. This approach considers the totality of the properties and characteristics of the studied systems, as well as the features of the methods and procedures used to create them. Based on a comparison of the characteristics of the qualitative analysis methods, the requirements for their application, the disadvantages and advantages of each of them, researchers of foresight problems should choose the rational combination of methods, establish the correct sequence for their use, taking into account the totality of requirements for systems and the features of the tasks to be solved.

The methodological and mathematical support of a systematic approach to solving the problems of developing complex system as a two-stage model based on a combination of foresight and cognitive modelling methodologies is developed [10, 11]. The involvement of scanning methods, STEEP analysis, brainstorming, SWOT analysis at the first stage level allows using expert assessment to identify critical technologies in economic, social, environmental, technical, technological, information and other directions [1]. The basis of this level is the analysis subsystems, which are connected by direct and feedback links to the monitoring system. The quantitative data obtained after analysis and processing are initial for solving of foresight tasks. The SWOT analysis method is used to identify critical technologies. The TOPSIS (Technique for Order Preference by Similarity to an Ideal Solution) method is applied to rank the obtained critical technologies and identify the most topical ones [5]. According to the VIKOR method, a compromise solution to the problem should be an alternative that is closest to the ideal solution. Moreover, a multicriteria measure is used to assess the degree of the alternative proximity to the ideal solution [2]. As soon as the critical technologies are identified, we switch to the second-level system approach, using qualitative methods to create alternative underground systems scenarios [1, 10]. In this paper, the morphological analysis method is used to create alternatives.

The method of constructing an integrated indicator data is proposed when the output information for cognitive modelling is given in statistical form [9]. It allows the construction of cognitive maps to add or remove its vertex reasonably, to break a sequence of interconnected nodes. Formalization of this method includes the following procedures:

- the selection of indicator which will characterize the specific area of one of the directions of sustainable development (economic, environmental, social);
- grouping by specific characteristics of the data sets, which influence the dynamics, selected at the stage 1 of the indicator formation;
- forming a database for a specific period on the basis of the discrete samples;
- recovery of functional dependencies by the discrete samples [26];
- analysis of the results based on the recovered dependence.

The proposed strategy of underground construction objects planning based on the synthesis foresight and cognitive modelling methodologies allows building a science-based strategy to implement priority alternative scenario of different nature complex systems.

3.2 Methodology of Cognitive Modelling

According to the developed methodology of scientific foresight and simulation of cognitive modelling of complex systems [10], modelling is carried out in some stages. At the first stage the methodology of foresight is used. Then taking the theoretical and practical data on underground urban planning a cognitive model is developed.

At the first step cognitive models a sign oriented graph (1) and a functional graph in the form of a weighted sign digraph are used [10, 27–31]

$$G = \langle V, E \rangle, \tag{1}$$

where G is a cognitive map in which V are concepts, a finite set of vertices of the cognitive map $V_i \in V, i = 1, 2, ...k$; $E = \{e_{ij}\}$—the set of arcs e_{ij} of the graph, $i, j = 1, 2, \ldots m$, reflect the relationship between the vertices V_i and V_j. The cognitive map G corresponds to the square matrix of relations A_G

$$A_G = \{a_{ij}\} = \begin{cases} 1, \; if \; V_i \; is \; connected \; with \; V_j, \\ 0, \; otherwise. \end{cases} \tag{2}$$

Vector Functional Graph

$$\Phi = \langle G, X, F(X, E), \theta \rangle, \tag{3}$$

where G is a cognitive map; X is the set of vertex parameters; $F(X, E)$ is the arc transformation functional θ is the space of vertex parameters. If

$$F(X, E) = F(x_i, x_j, e_{ij}) = \begin{cases} +\omega_{ij}, if rising/falling\, X_i, \\ entails\, rising/\, falling\, X_j, \\ -\omega_{ij},\ if rising/falling\, X_i, \\ entails\ falling/rising\, X_j \end{cases} \tag{4}$$

then there is a weighted sign digraph, in which ω_{ij} is the weight coefficient.

In the case of studying the hierarchy systems, cognitive maps of individual levels can be combined into a hierarchical map [10, 27]

$$I_G = \langle G_k, G_{k+1}, E_k \rangle, k = 1, 2, 3, \ldots K, \tag{5}$$

where G_k and G_{k+1} are cognitive maps of k and $k + 1$ levels, respectively, whose vertices are connected by arcs G_k.

At the second step of cognitive modelling, to study the properties of the cognitive model the methods of structural stability and perturbation resistance analysis are applied [10, 32–34], methods for analyzing model connectivity (simplicial analysis [35, 36]), and graph theory methods [37]. The results of the analysis were compared with the available information on underground construction.

At the third step of cognitive modelling, to determine the possible development of processes in a complex system and develop development scenarios, the impulse process model (modelling the propagation of disturbances in cognitive models) is used [32, 37]:

$$x_{v_i}(n+1) = x_{v_i}(n) + \sum_{v_j:e=e_{ij}\in E}^{k-1} f(x_i, x_j, e_{ij}) P_j(n) + Q_{v_i}(n), \tag{6}$$

where $x(n)$, $x(n + 1)$ are the values of the indicator at the vertex V_i at the simulation steps at time $t = n$; and the next $t = n + 1$; $P_j(n)$ is the momentum that existed at the vertex V at the moment $t = n$; $Q_{V_i}(n) = \{q_1, q_2, \ldots, q_k\}$ is the vector of external pulses (disturbing or controlling actions) introduced to the vertices V_i at time moment n.

4 Modelling of Underground Communications

Data on the vertices (concepts) of the hierarchical cognitive model are presented without reference to a specific territory in a generalized form in Table 1. We have

Table 1 Vertices of the cognitive map G_1 "Pipe düker"

Code	Name of the vertex			Assignment of the vertex
V1	System "Pipe dükers"			Indicative
V2	Events of anthropogenic origin	V2–1	With malicious intent: fighting, sabotage, terrorism	Disturbing
		V2–2	Without malice: errors, negligence, non-compliance with construction or operation standards	
V3	Events of technogenic origin	Technical and technological (technical failures, breakdowns, destruction due to technological factors, corrosion, etc.)		Disturbing
V4	Natural disasters, weather cataclysms	V4–1	Natural disasters, weather cataclysms (atmospheric, hydrosphere and lithosphere disturbances)	Disturbing
		V4–1.1	Landslides, dips, subsidence of soil	
V5	Object protection	Organization of object protection from anthropogenic and natural hazards		Manager
V6	Emergency condition of dükers	Abrasive wear, breakage due to holes under pipes and external stresses of worn material		Regulatory
V7	The scale of the impact of the adverse event	V7–1	Structural, or a separate section	Regulatory
		V7–2	Functional element of an object, or several sections	
		V7–3	The object is completely destroyed	
		V7–4	City area and more	
V8	Ability to function	V8–1	The object can perform all the functions	Regulatory
		V8–2	The object stops working	
V9	Resumption time	How long will it take for the object to recover from an adverse event		Regulatory
V10	Environmental consequences	Potential risks to the environmental situation		Regulatory
V11	Economic consequences	Expected Economic Consequences in the Event of an Adverse Event		Regulatory

(continued)

Table 1 (continued)

Code	Name of the vertex			Assignment of the vertex
V12	Consequences for life	V12–1	Approximate number of persons whose living conditions may be violated in case of an adverse event	Regulatory
V13	Economic resources	Expected economic losses in the event of an adverse event		Manager
V14	Organizational, technical, etc. resources	Expected resources in the event of an adverse event		Manager
V15	Investor	Additional financing for the repair of pipe siphons		Basic
V16	Integrity of the system	How badly the object was damaged as a result of the impact of the undesirable event		Regulatory
V17	Material damage	Expected amount of losses in the event of an adverse event at the facility		Regulatory

used generalizing concepts (indicators, factors), independent of the specifics, which can be disclosed and taken into account in the future when developing the real object.

First Step. Cognitive Model Development. Based on the expert and statistical analysis of the düker system problems the following models have been developed: cognitive map G1 «Pipe düker» and G2 «Tunnel düker». The purpose of the these two models development is to compare the options of pipe and tunnel underwater dükers construction and justification of the priority of their creation.

When developing a cognitive model, it is convenient to present the analyzed and systematized information obtained with the method of morphological analysis at the stage of foresight in the form of cognitive map vertices. Table 1 shows the vertices (concepts) of the model of the cognitive map for the system "Pipe düker".

The data about the system, grouped in Table 1, are visualized using the CMSS [38] software system in the form of a cognitive map G_1—model 1 (Fig. 1).

The solid lines of arcs in Fig. 1 mean that with an increase (or decrease) in the signal at the vertex V_i, the same changes occur at the vertex V_j—an increase (or decrease). The dashed lines of arcs in Fig. 1 mean: an increase (or decrease) in the pulse at the vertex V_i leads to a decrease (or increase) in the pulse at the vertex V_j.

The cognitive map G_1 corresponds to its connectivity matrix R_{G_1} [10]. Various operations with the R_{G_1} matrix make it possible to investigate the multifaceted properties of cognitive maps. This is necessary both to verify that model G does not contradict a real complex system, and to use a cognitive map as a structure on which various scenarios for the development of situations in a real system are simulated.

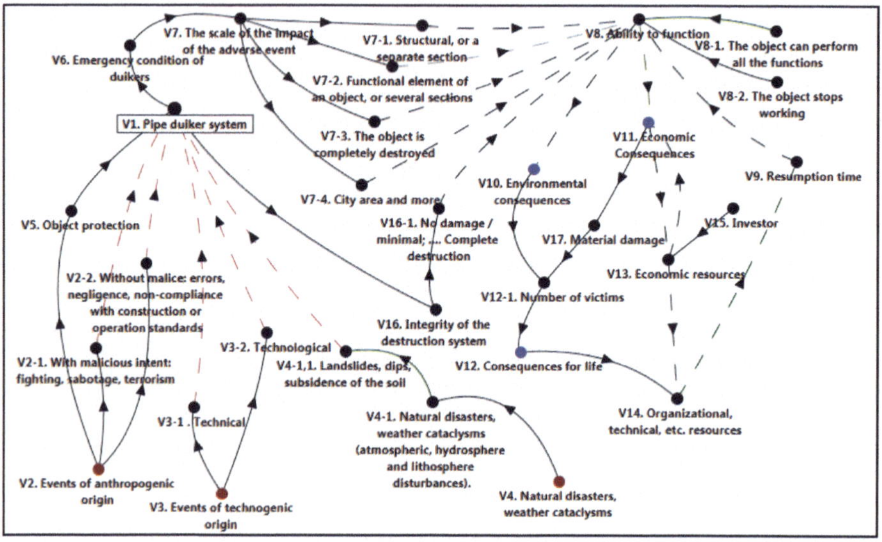

Fig. 1 Cognitive map G_1 "Pipe düker system"

The cognitive model is a simulation model that makes it possible not carry out an experiment on a "living" system, but to simulate its behavior and possible future development under the influence of various factors, generating new knowledge about the system. This allows to justify management decisions in certain situation.

The Second Step of Modelling. The second step of modelling analyzes the various properties of the model is realized before using the cognitive model to determine its possible behavior. In this case, the stability properties of the model must be analyzed.

Determination of the Degrees of Vertices. An analysis of the degree of vertices is necessary to identify the vertices with the highest and lowest degree and determine their significance for the entire system. Figure 2 shows the results of determining the number (degree P) of all arcs, as well as incoming P+ and outgoing P- arcs for each vertex.

Now analyze the model 2 connected with cognitive map G_2 "Tunnel düker". Vertices V2-V17 of the cognitive map G_2 correspond to the vertices of the cognitive map G_1 shown in Table 1. Vertices V18, V19 for the cognitive map G_2 "Tunnel düker system" are added to the model 2 (Table 2).

Based on the data of Tables 1 and 2, a cognitive map G_2—model 2 was built (Fig. 3).

Various operations with the connectivity matrix R_{G_2} make it possible to investigate the multifaceted properties of cognitive maps. This is necessary both to verify that model G_2 does not contradict a real complex system, and to use a cognitive map as a structure on which various scenarios for the development of situations in a real system are simulated.

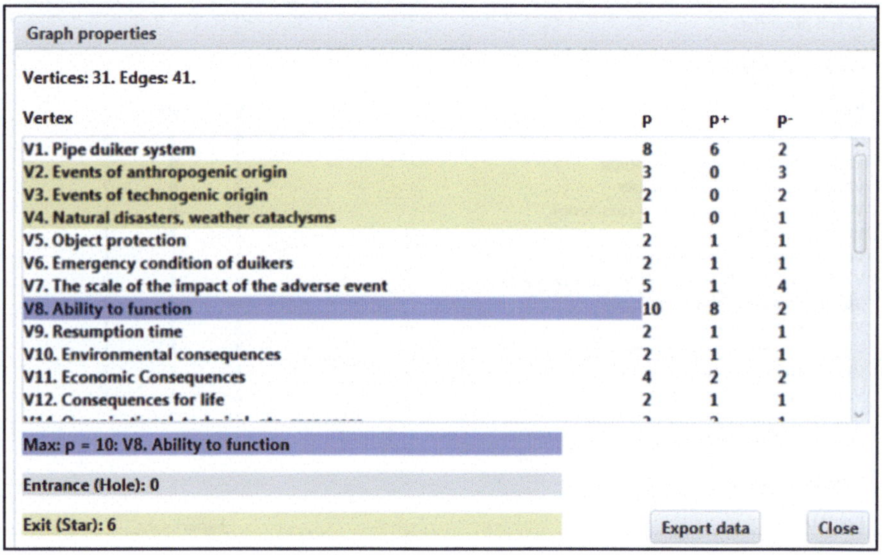

Fig. 2 The degrees of the vertices of the cognitive map G_1

Table 2 Vertices V18, V19 of the cognitive map G_2 "Tunnel düker system"

Code	Name of the vertex		Assignment of the vertex
V1	Tunnel düker system		Indicative
V18	Geotechnology of construction		Basic

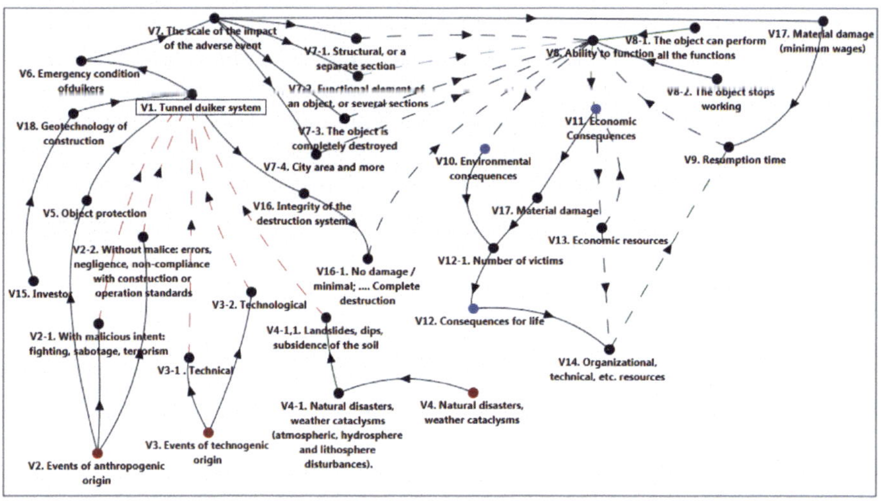

Fig. 3 Cognitive map G_2 "Tunnel düker system"

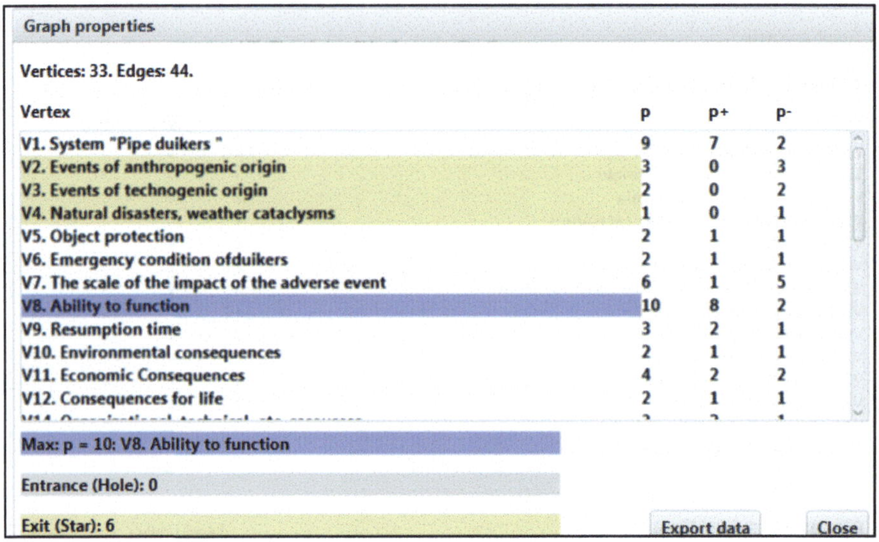

Fig. 4 Fragment of the analysis the degree of vertices the cognitive map G_2

Determination of the Degrees of Vertices. Figure 4 shows the results of the analysis of the number (degree P) of all arcs, as well as incoming P+ and outgoing P– arcs for each vertex.

The Third Step of Modelling. Scenario analysis is designed to anticipate possible trends in the development of situations on the model. To generate scenarios of the development of the system, impacts are introduced into the vertices of the cognitive map in the form of a set of impulses. The impulse process formula has the form (6).

Let us introduce perturbations of different sizes (normalized) to any of the vertices, as well as to their combination. In connection with a large number of theoretically possible variants of introduced disturbances, it is expedient to develop a plan for a computational experiment eliminating at least almost impossible variants. Introducing disturbances to the vertices, the decision-maker is looking for the answer to the question: “What will happen if …?”.

The CMLS [38] software system allows, in the process of impulse modelling and analysis of the obtained results, to introduce control or disturbing influences at any modelling step. It is allows to change (correct) scenarios in model dynamics, to determine the effects that bring the processes closer to the desired.

In order to substantiate the choice of a tunnel düker in comparison with a pipe düker the impulse modeling is carried out. It allows to consider of the process dynamics. During the computational experiment, numerous scenarios were considered. The results of the most representative four scenarios for pipe and tunnel dükers under the influence of anthropogenic and natural factors are presented in Fig. 5, 6, 7, 8, 9, 10, 11and12. The results (see Fig. 5, 7, 9, 11) of a computational experiment for Scenarios No 1- No 4 at 10 simulation steps are presented in the form of distributions: Ability to function V8, Environmental consequences V10, Economic Consequences V11, Consequences for life V12, With malicious intent (fighting, sabotage, terrorism) V2–1 and Natural disasters (Landslides, dips, subsidence of the soil) V4–1. Their representation in the form of histograms shown in Fig. 6, 8, 10, 12.

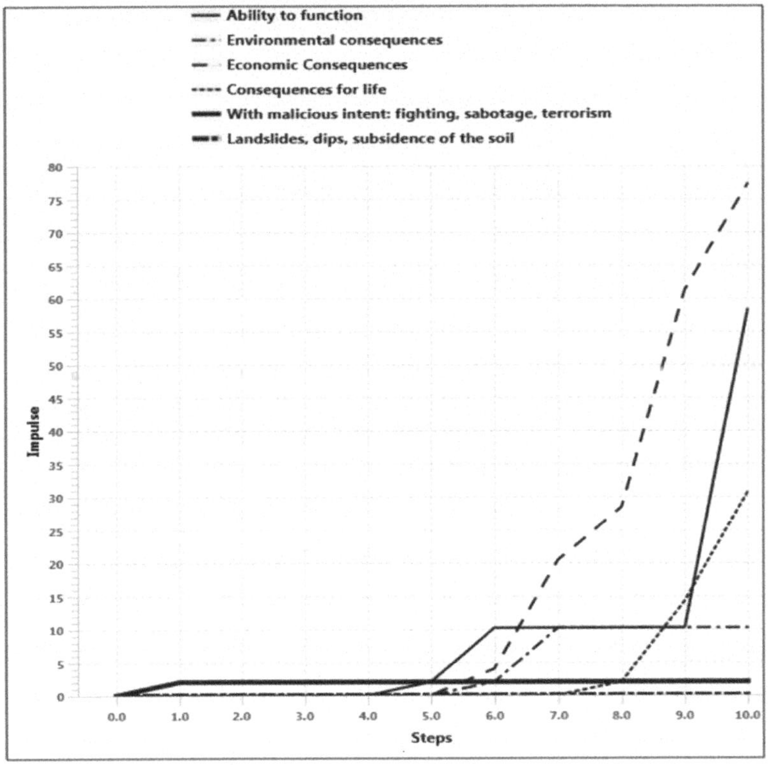

Fig. 5 Graphs of pulsed processes, from the first to the tenth step of modelling. Scenario No. 1

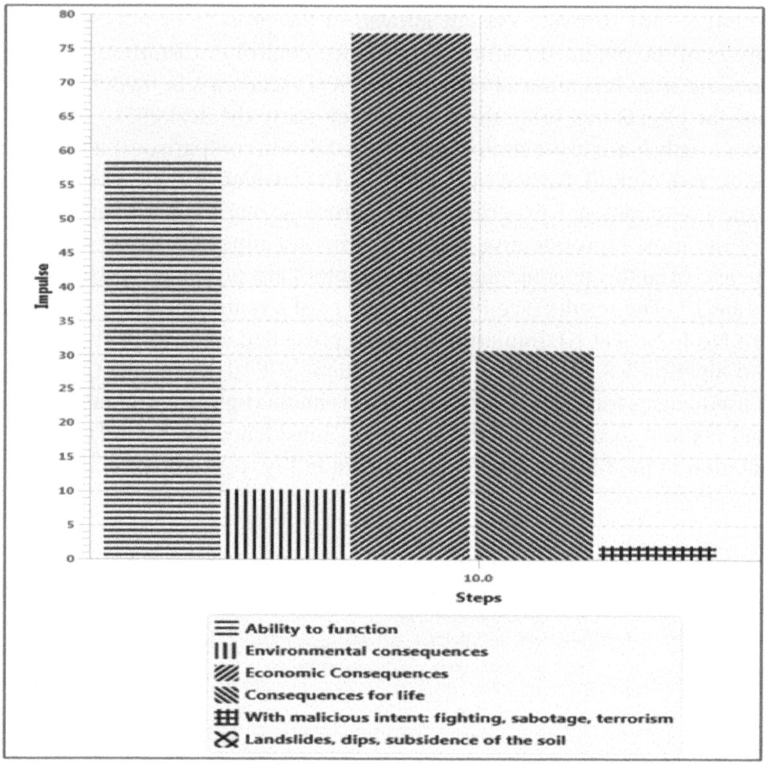

Fig. 6 Histograms of pulsed values at the 10th step of modelling. Scenario No. 1

The following scenarios were investigated.

Scenario No. 1. The "pipe düker" system is affected by anthropogenic factors " With malicious intent: fighting, sabotage, terrorism ". A disturbing impulse $q_{2.1} = +1$ is introduced at the vertex V2-1 (Fig. 5, 6).

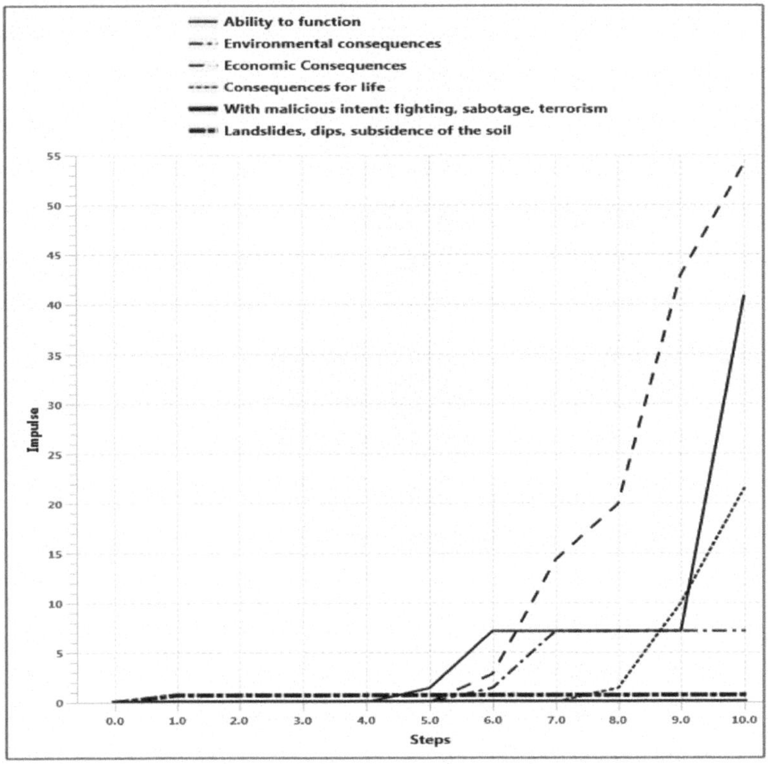

Fig. 7 Graphs of pulsed processes, from the first to the tenth step of modelling. Scenario No. 2

Scenario No. 2. The system "pipe düker" is influenced by Natural disasters, weather cataclysms (landslides). The disturbing impulse $q_{4-1.1} = +1$ is introduced to the vertex V4–1.1 (Fig. 7, 8).

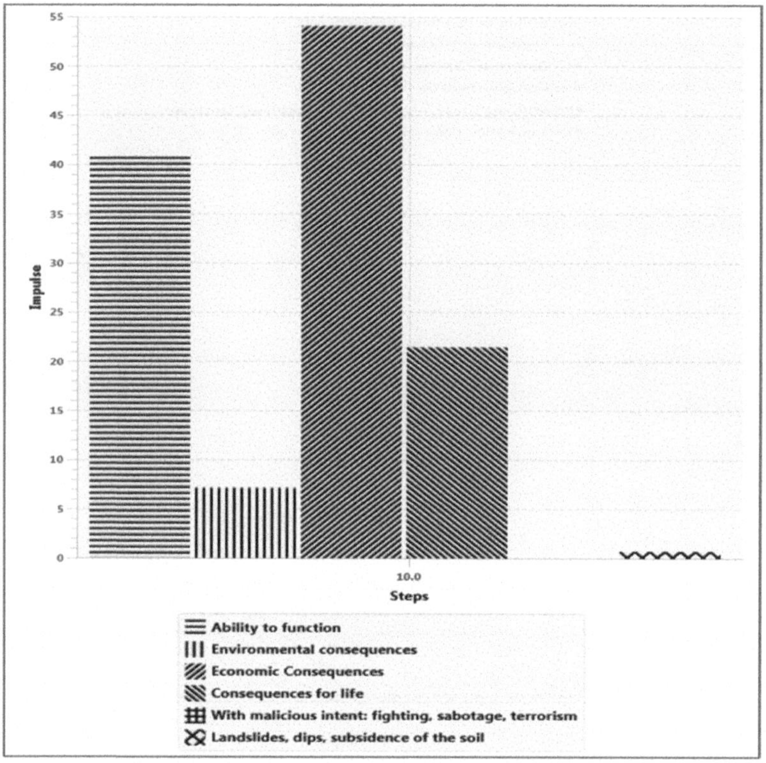

Fig. 8 Histograms of pulsed values at the 10th step of modelling. Scenario No. 2

Now consider the similar scenarios for tunnel dükers.

Scenario No. 3. The system "tunnel düker" is influenced by anthropogenic factors "With malicious intent: fighting, sabotage, terrorism". The disturbing impulse $q_{2.1} = +1$ is introduced to the vertex V2.1 (Fig. 9, 10).

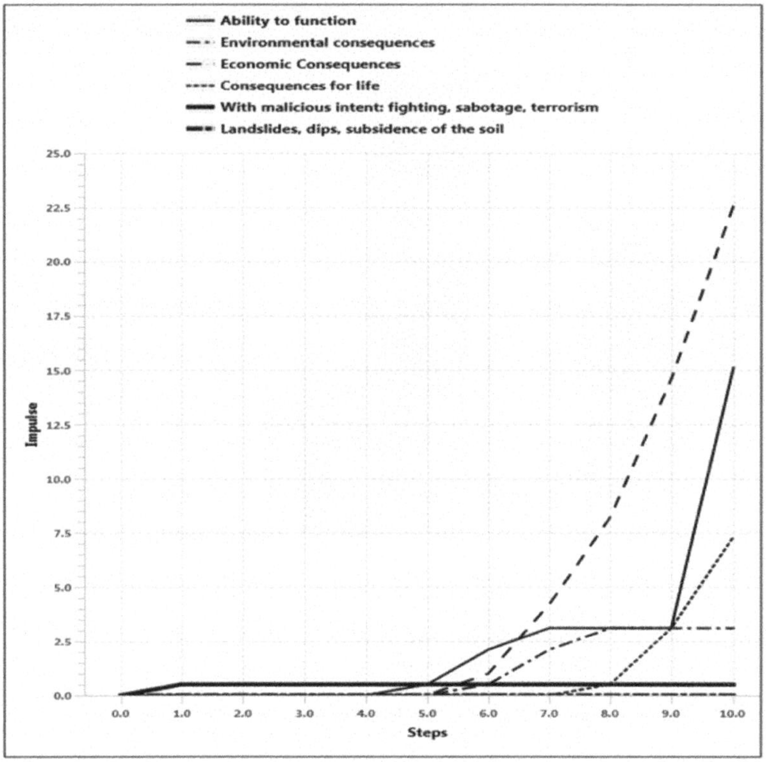

Fig. 9 Graphs of pulsed processes, from the first to the tenth step of modelling. Scenario No. 3

Scenario No. 4. The system "tunnel düker" is influenced by environmental factors (landslides). The disturbing impulse $q_{4-1.1} = +1$ is introduced to the vertex V4.1–1 (Fig. 11, 12).

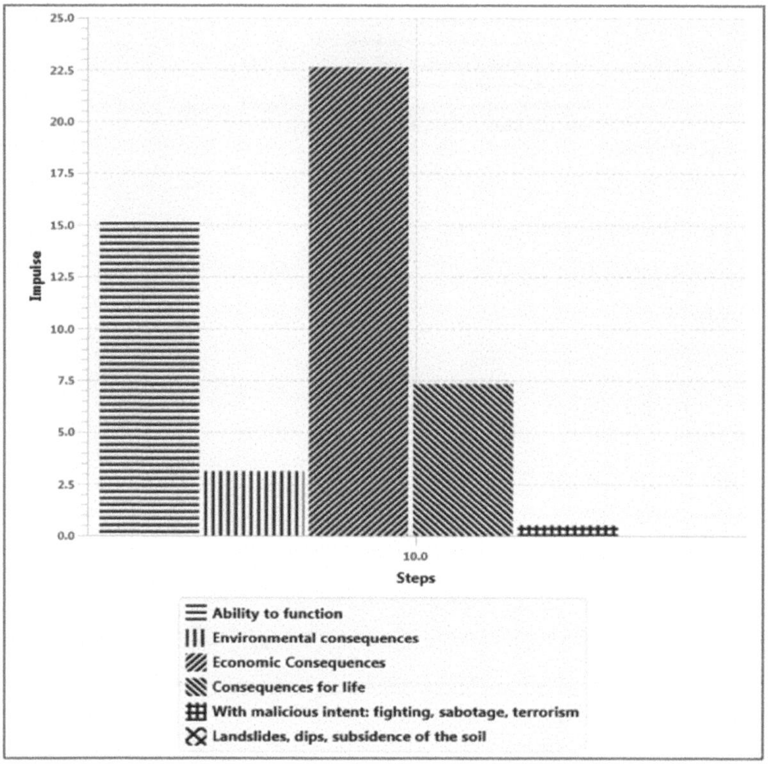

Fig. 10 Histograms of pulsed values at the 10th step of modelling. Scenario No. 3

The results (see Fig. 5, 6, 7,8, 9, 10, 11 and 12) a numerical experiment of impulse cognitive modeling for pipe and tunnel dükers for anthropogenic events (hostilities, sabotage, terrorism) and natural origin (landslides) are integrated in the form of Table 3. The given results of malicious intent impact (military operations, terrorism) (Fig. 5, 6; Scenario No. 1) for a pipe dükers and for a tunnel dükers (Fig. 9, 10; Scenario No. 3), show a significant, by 74%, decrease the ability to functioning (V8) and the environmental consequences (V10) of pipe dükers compared to tunnel dükers; by 75% decreasing the consequences of life (V12) and by 71.2% the environmental consequences (V10) of the pipe dükers using. Under the influence of natural origin events (landslides) the ability to functioning (V8) of tunnel dükers (Fig. 11, 12; Scenario No. 4) is 57% more reliable than pipe dükers (Fig. 7, 8; Scenario No. 2). As follows from the results shown in Table 3 the expediency of using tunnel dükers in comparison with pipe dükers for factors environmental consequences (v10), economic consequences (V11) and consequences of life (V12) is 61.5%, 50%, 62% respectively.

The carried out studies confirm the priority of the underwater construction of tunnel dükers in comparison with pipe dükers.

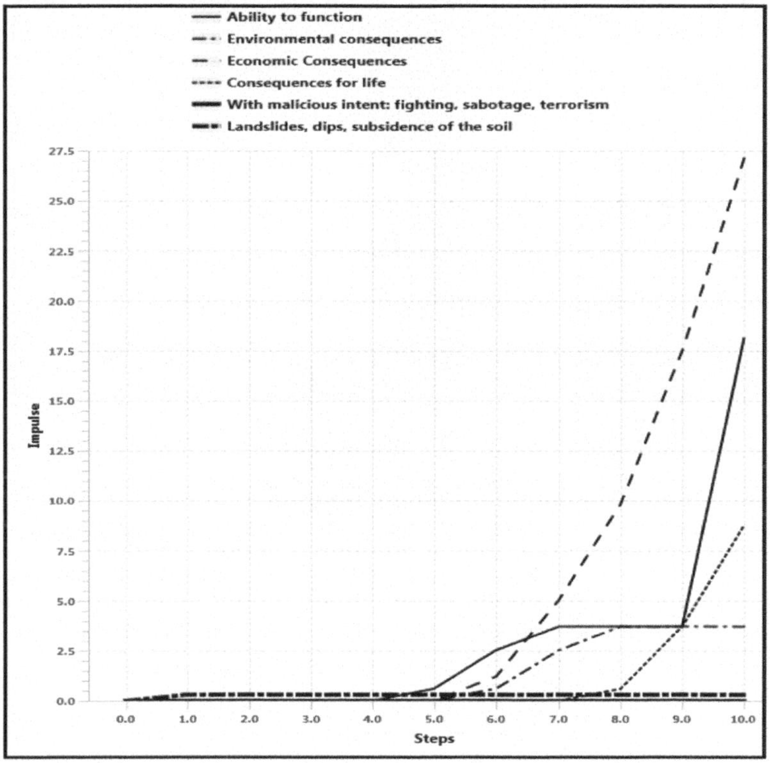

Fig. 11 Graphs of pulsed processes, from the first to the tenth step of modelling. Scenario No. 4

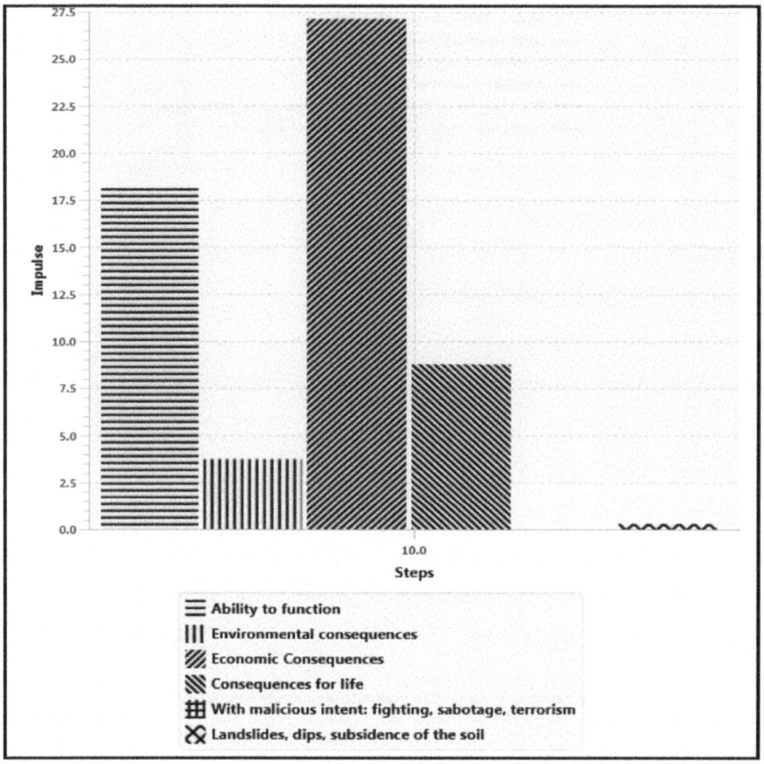

Fig. 12 Histograms of pulsed values at the 10th step of modelling. Scenario No. 4

Table 3 Comparing values at vertices $V_i (i = 8,\ 10 - 12)$ for tunnel and pipe dükers at the 10th step of modelling

Vertices	Events of anthropogenic origin. With malicious intent: fighting, sabotage, terrorism $q_{2.1} = +1$		Expediency of using tunnel dükers in relation to pipe dükers (%)	Natural disasters, weather cataclysms: Landslides, dips, subsidence of the soil $q_{4-1.1} = +1$		Expediency of using tunnel dükers in relation to pipe dükers (%)
	Pipe dükers	Tunnel dükers		Pipe dükers	Tunnel dükers	
V8	56	15	74	41	17.6	57
V10	10	2.6	74	7	2.7	61.5
V11	78	22.5	71.2	54	27	50
V12	30	7.5	75	21	8	62

5 Conclusion

The space of megalopolises, created by man in the process of underground construction, becomes a new, underground habitat, which should be comfortable and safe for humans. One of the most difficult problems is underground water sewer communications that support both surface and underground urban planning. Significant advantages of underground water crossings and coastal underground infrastructure have put on the agenda an ambitious agenda for underground construction in the zone of influence of water bodies. This paper examines the issue of the construction of pipe and tunnel underground water dükers and justifies the priority of their creation.

Solving of this problem is implemented on the basis of the developed strategy of underground construction objects planning based on the methodologies foresight and cognitive modelling. The suggested strategy is applied to the study of underwater communications in order to select reasonable scenarios for justification of their creation priority. In order to substantiate the choice of a tunnel düker in comparison with a pipe düker the impulse modeling is carried out. It allows investigating of the process dynamics. During the computational experiment, numerous scenarios were considered. The results of the most representative four scenarios for pipe and tunnel dükers under the influence of anthropogenic and natural factors are presented. The studies for carried out have shown the expediency of using tunnel dükers in comparison with pipe dükers for factors environmental, economic and life consequences.

Applying this strategy is especially important for comprehending and preventing negative consequences, minimizing damage under the influence of the most unfavorable combination of negative factors: external and internal static and dynamic loads, all kinds of artificial influences inside an underground structure, harmful natural manifestations from a mountain range, etc.

Combination of the underground planning, geoinformation, experience, equipment, production and supply of construction materials allows dramatically reduce costs for construction of underwater tunnel dükers and will increase the quality and safety of people's lives.

References

1. Zgurovsky, M.Z., Pankratova, N.D.: System Analysis: Theory and Applications. Springer, Berlin Heidelberg New York (2007)
2. Mardani, A., Zavadskas, E., Govindan, K., Senin, A., Jusoh, A.: VIKOR technique: a systematic review of the state of the art literature on methodologies and applications. Sustainability **8**(37), 1–38 (2016)
3. Mikhnenko, P.A.: Dynamic modification of SWOT analysis. Econ. Anal.: Theory Pract. **18**(417), 60–68 (2015)
4. Gopalakrishnan, K., Vijayalakshmi, V.: Using morphological analysis for innovation and resource and development: an invaluable tool for entrepreneurship. Ann. Res. J. Symbiosis Centre Manage. Stud., Pune **2**(1), 28–36 (2014).

5. García-Cascale, M.S., Lamata, M.T.: On rank reversal and TOPSIS method. Math. Comput. Model. **56**(5–6), 123–132 (2012)
6. Ritchey, T.: Futures Studies using Morphological Analysis. Adapted from an article for the UN University Millennium Project, Futures Research Methodology Series (2005)
7. Weimer-Jehle, W.: Cross-impact balances: a system-theoretical approach to cross-impact. Technol. Forecast. Soc. Change **73**, 334–361 (2006)
8. Alptekin, N.: Integration of SWOT analysis and TOPSIS method in strategic decision making process. Macrotheme Rev. **2**(7), Winter (2013)
9. Pankratov, V.: Development of the approach to formalization of vector's indicators of sustainable development. J. Inf. Technol. Knowl. ITHEA. SOFIA, **8**(3), 203–211 (2014)
10. Innovative development of socio-economic systems based on foresight and cognitive modelling methodologies. In: Gorelova G.V., Pankratova N.D. (eds.) (in Russian), Kiev, Nauk.Dumka (2015)
11. Zgurovsky, M.Z., Pankratov, V.A.: Strategy of innovative development of the region based on the synthesis of foresight methodology and cognitive modelling. Syst. Res. Inf. Technol. **2**, 7–17 (in Russian) (2014)
12. World Urbanization Prospects: Highlights. United Nations, New York (2019)
13. Levchenko, A.N.: About a new direction of scientific research in construction geotechnology. Mining Inf. Anal. Bull. (Sci. Tech. J.) **2**, 15–21 (2007)
14. Kartosia, B.A.: Mastering the underground space of large cities. New Trends. Mining Information and Analytical Bulletin (scientific and technical magazine), "Construction and Architecture", pp. 615–628 (2015)
15. Kartozia, B.: Fundamentals of Underground Space Development (in Russian). M, Press Department of Moscow State University for the Humanities (2009)
16. Kartosia, B.A.: The development of the underground space is a global problem of science, production and higher mountain education. Materials conf. Prospects for the development of underground space, pp. 12–26 (2010)
17. Bondarik, G.K.: General Theory Of Engineering (Physical) Geology. Nedra, Moscow (1981).(in Russian)
18. Tajdus, A., Cala, M., Tajdus, K.: Geomechanika w budownictwie podziemnym. Projektowaniei budowa tuneli. Krakow: AGH (2012)
19. Owen, C.L., Bezerra, C.: Evolutionary structured planning. a computer-supported methodology for the conceptual planning process. In: Gero J.S. (ed.). Artificial Intelligence in Design'00. Kluwer Academic Publishers, Dordrecht, pp. 287–307(2000)
20. Saługa, P.: Ocena ekonomiczna projektów i analiza ryzyka w górnictwie. Economic Evaluation and Risk Analysis of Mineral Projects. Studia, Rozprawy, Monografie, nr 152, Wyd. IGSMiE PAN, Kraków (2009)
21. Pankratova, N., Savchenko, I., Haiko, H., Kravets, V.: System approach to planning urban underground development. J. Inf. Content Process. **6**(1), 3–17 (2019)
22. Pankratova, N.D., Gaiko, G.I., Savchenko, I.A.: Development of underground urban studies as a system of alternative design configurations. M.: Naukova Dumka, In Russian (2020)
23. Kulikova, E.Y., Korchak, A.B., Levchenko, A.N.: Strategy of Risk Management in Urban Underground Construction (in Russian), M.: Moscow State University State University Publishing House (2005)
24. Sakellariou, M.: Tunnel Engineering—Selected Topics. Athens, National Technical University of Athens, Books, IntechOpen, number 6201 (2020)
25. Hong, K.: Typical underwater tunnels in the mainland of China and related tunneling technologies. Engineering. **3**(6), 871–879 (2017)
26. Pankratova, N.D.: A rational compromise in the system problem of disclosure of conceptual uncertainty. J. Cybernetics Syst. Anal. **38**(4), 618–631 (2002)
27. Pankratova, N.D., Gorelova, G.V., Pankratov, V.A.: Strategy for the study of interregional economic and social exchange based on foresight and cognitive modelling methodologies. In: Proceedings of the 8th Int.Conf. on "Mathematics. Information Technologies. Education, Shatsk, Ukraine, June 2–4, (2019) pp. 136–141

28. Langley, P., Laird, J.E., Rogers, S.: Cognitive architectures: research issues and challenges. Cognitive Syst. Res. **10**(2), 141–160 (2009)
29. Abramova, N.A., Avdeeva, Z.K.: Cognitive analysis and management of the development of situations: problems of methodology, theory and practice. J. Problems Control **3**, 85–87 (2008)
30. Avdeeva, Z.K., Kovriga, S.V.: On governance decision support in the area of political stability using cognitive maps. In 18-th IFAC Conference on Technology, Culture and International Stability (TECIS2018), **51**(30), 498–503 (2018)
31. Kovriga, S.V., Maksimov, V.I.: Cognitive technology of strategic management of the development of complex socio-economic objects in an unstable external environment, 1st issue of Cognitive Analysis and Situational Management (in Russian) (2001)
32. Kulba, V., Kononov, D.A., Kovalevsk, S.S., Kosyachenko, S.A., Nizhegorodtsev, R.M., Chernov, I.V.: Scenario analysis of the dynamics of behavior of socio-economic systems (in Russian). M.: IPU RAS (2002)
33. Maksimov, V.I.: Cognitive technology—from ignorance to understanding. 1st work "Cognitive analysis and management of the development of situations", (CASC'2001),**1**, 4–18 (2001)
34. Atkin, R.H.: Combinatorial connectivies in social systems. An Application of Simplicial Complex Structures to the Study of Large Organisations', Interdisciplinary Systems Research (1997)
35. Atkin, R.H., Casti, J.: Polyhedral dynamics and the geometry of systems, RR-77-International Institute for Applied Systems Analysis, Laxenburg, Austria (1977)
36. Casti, J.: Connectivity, complexity, and catastrophe in large-scale systems. A Wiley—Interscience Publication International Institute for Applied Systems Analysis. JOHN WILEY and SONS. Chichester, New York, Brisbane, Toronto (1979)
37. Roberts, F.: Graph Theory and its Applications to Problems of Society. Society for Industrial and Applied Mathematics, Philadelphia (1978)
38. Program for cognitive modeling and analysis of socio-economic systems at the regional level. Certificate of state registration of computer programs N2018661 (2018)

Chapter 6
Assessing Territories for Urban Underground Objects Using Morphological Analysis-Based Model

Hennadii Haiko and Illia Savchenko

Abstract The paper is devoted to the development and testing of a model that formalizes and supports decision-making process regarding the appropriateness of using territory (geological environment) for urban underground construction of different object types. It was shown how the previously designed general morphological model of territorial development for underground city planning that describes geological environment, can be complemented by additional modules aimed at taking into account the features of specific underground objects, namely parking lots and car tunnels. These modules address the structural and functional factors to assess the suitability of a site for a given purpose, taking into account the relied ecological and technogenic risks, and the potential object's capability of mitigating these risks. The purpose of the designed model is to provide decision-making support and set priorities for urban underground development. The application of this model is shown for different types of underground objects and tested on the real existing and planned construction sites in Kyiv. The analysis based on the results of the model is given, demonstrating the model's capabilities of determining priorities of underground object construction at the pre-project stage, and the opportunity to consider the ecological and technogenic risks of urban space.

Keywords Underground infrastructure · Morphological analysis · System analysis · Geological environment · Parking lots · Car tunnels

H. Haiko
Institute for Energy Saving and Energy Management, National Technical University of Ukraine "Igor Sikorsky Kyiv Polytechnic Institute", Peremohy ave., 37, Kyiv 03056, Ukraine

I. Savchenko (✉)
Institute for Applied System Analysis, National Technical University of Ukraine "Igor Sikorsky Kyiv Polytechnic Institute", Peremohy Ave, 37, Kyiv 03056, Ukraine

M. Zgurovsky and N. Pankratova (eds.), *System Analysis & Intelligent Computing*, Studies in Computational Intelligence 1022, https://doi.org/10.1007/978-3-030-94910-5_6

1 Introduction

1.1 System Approach to Urban Underground Construction

The regulation of urban development aimed at increasing ecological standards and life safety in ever-growing metropolises is one of the most urgent yet complex and insufficiently researched world problems [1]. Modern approaches to urban development pay significant heed to the capacity of underground space of taking over the most hazardous and risk-bearing functions from surface objects and communications, providing the minimization of ecological and technogenic risks of large cities [2–5]. The sustainable development concept changes the past trend of constructing separate underground facilities as local objects to designing large scale projects for systemic organization of underground space as an integral part of the whole urbanized space. This allows to solve comprehensively the actual metropolises problems, including territorial, transport, power supply, ecological, social problems.

Underground urban territory is a complex system in many aspects. It comprises numerous interconnected subsystems and objects, and the processes that flow in this system both during construction and during operation, are also complex and often hard to predict, as they are related to the variable geological environment and a multitude of structural and functional factors of urban space. That is why the problems accompanying the facilitation of underground space in large cities can be defined as weakly structured problems with a bunch of uncertainties, and the approaches to solving these problems should be based on system methodology [6].

It should be noted that even individual examples of successful plans of territorial development of underground urban space (e.g. more than 400 interconnected underground objects in Helsinki [7, 8]), are hard to apply in the conditions of a different urban space, other geological environment, and the existing specific underground infrastructure that should be taken into account. This calls for creating a universal systemic tool set for underground space planning in metropolises that would take into account the ideology, structural, natural and technogenic features of development planning for different cities.

Individual studies, aimed at the overview of system approaches to the underground construction development, reached only general recommendations and statements of the system research problems, and were not concentrated on the ecology and social security in metropolises [9]. The system approaches to surface urban planning [10] cannot be directly carried over to underground space, due to the sharp influence of geological environment and the peculiarities of geoconstruction technologies. The modern development of applied system analysis provides new prospects for planning underground infrastructure in large cities [6, 11, 12]. Adaptation of these methods for urban problems, including the estimation of favorability of city areas for underground construction [13–15], allowed to obtain the efficient tools for underground space planning, considering the interaction with engineering geological environment; however the ecological and safety components were not paid due attention. The construction priorities of the underground objects and complexes, considering the impact on ecological and technical risks, remained an important open issue of the system approach.

1.2 Objects of Study

One of the most urgent problems of urban metropolises is their limited traffic capabilities. This issue can often be mitigated by utilizing underground infrastructure, which is often underdeveloped. Thus, underground parking lots and car tunnels were selected as the main targets for application of the model.

Among the large diversity of urban underground objects the underground parking lots stand out especially due to the significant urgency of the problem of parking in central (business) regions of a metropolis, as well as its periphery, where the dwellers of satellite cities and localities commonly switch to the city transport (mainly subway) [10, 16, 17]. As shown in the paper [18], the cluttering of traffic passageways by temporary vehicle parking is an important issue in the goal of eliminating traffic jams and increasing average move speed in large cities. Thus the role of parking lots becomes even more important. On the other hand, spare sites for constructing open air parking lots are practically non-existent in the central regions of a metropolis, as former city planning did not foresee the present quantity of cars. Additionally, any "spare" sites were occupied, often using questionable methods, by construction companies for erecting apartment houses, office buildings or malls. This leads to the conclusion that the parking problem downtown and near terminal subway stations has only one profound solution—constructing underground parking lots. Among the examples of the system approach to this problem one can note nearly simultaneous construction of 41 parking lots in Paris, moreover the Scientific Coordination Council of Urban Underground Construction, directed by the famous organizer of underground development in European cities E. Utudzhyan, conducted complex and controversial discussions regarding the placement of these underground objects [19].

The General Plan of Kyiv city development up to 2025 envisions the construction of eight car tunnels, three of which will pass under Dnipro river. The "optimistic scenario" includes laying over 20 km of car tunnels in the nearest 20 years.

It should be noted that the city planning history never foresaw the current amounts of automobiles, which is why the organization of city space acutely discords with the modern traffic flows. Numerous traffic jams in downtown, especially at peak hours caused the average move speed of 15 km/h with constantly increasing accident rates. Another urgent problem is air pollution by exhaust fumes: carbon monoxide, nitrogen oxide, carcinogenic hydrocarbons, with CO being the largest portion of them. The concentration of harmful substances in the air of main traffic routes reaches the levels of large chemical, mining and metallurgical plants. Ecological organizations estimate the contribution of car exhaust fumes in the air of metropolises at 70%, power plants at 24%, industrial plants at 6% [20, 21]. Thus, building underground car tunnels is not only aimed at solving logistical problems, but should also improve the ecological situation, as ecologized tunnels provide the opportunity for purposeful redirection and utilization of noxious exhaust fumes from transport [22, 23]. It is important to analyze the planned tunnel tracks and determine the priorities of tunnel construction, using the ecological safety criterion, to select those that provide the maximum mitigation of ecological risks.

2 Methods

This study employs the modified morphological analysis method (MMAM) [12] as the modeling technique. This method excellently performs in conditions of situational uncertainty, which is common in most problems of system analysis and modeling of complex systems. The original morphological analysis method is flexible and universal, it has shown good results in modeling problems with large numbers of alternative configurations [24, 25]. The idea of the modified method lies in forming the multitude of configurations for an object (system, event, phenomenon etc.) and processing all of these configurations by applying Bayesian probability apparatus.

The task of evaluating a construction site regarding its suitability for underground construction has several uncertainty factors. This uncertainty is caused by a couple of peculiarities:

(1) the exact assessment of all influence factors requires conducting engineering and/or geological works and thorough measurements that consume time and resources and are often economically inexpedient at the pre-project stage;
(2) most of the sites are heterogeneous and, consequently, have spatially varied characteristics.

Thus the evaluation requires input from experts that are able to make decisions based on their experience, intuition and relatively scarce information about the construction site.

Constructing a model in MMAM bears the following steps:

(1) defining the objects (entities) which will be described by morphological tables, and relations between them;
(2) constructing morphological tables for each of the objects;
(3) estimating the dependencies between morphological table parameters.

The resulting model can be applied for calculating the performance values for each alternative of the specified object, using expert input data about the object.

A special two-stage MMAM procedure [12] was applied for the problems of study. It includes the estimation of the uncontrolled, uncertain parameters at the first stage, and the inference or decision alternatives at the second stage.

The MMAM procedures in this study were implemented using the SAS Studio software with user C# modules, compiled in Microsoft Visual Studio 2017 environment. The modules correspond to the main steps of MMAM: morphological table (MT) construction; MT estimation; cross-consistency matrix estimation; weight calculation for one-stage and two-stage MMAM procedures. This tool set allows to design and perform morphological studies for complex tasks, including the task of this research. Working with this tool set requires inputting data obtained from expert estimation on the tunnel site, after which the following MMAM procedure performs automatically, generating the results from the two stages of the method.

3 Models

A common trait for analysis of sites for underground construction is that their geological environment should be taken into account. A general two-stage morphological model was designed in previous studies [15] that analyzes the geological environment considering 10 parameters:

1. level of dynamic load;
2. static load from surface buildings;
3. static load from soil;
4. influence of existing underground objects;
5. genetic type and lithologic composure of soil;
6. effective soil strength;
7. influence of aquifers and perched groundwater;
8. landscape type and morphometrics;
9. geological engineering processes;
10. geotechnologies of underground construction.

Each parameter has several possible alternative values or ranges. The interrelations between them were assessed to construct the general model, so this model could be used to assess a given site using expert estimation of the initial probabilities of alternatives. The evaluation using MMAM algorithm is based on solving the system of equations, the size of which is determined by the total number of alternatives in a MT (38 for this study). This procedure allows to assess more precisely the probabilities for each alternative, taking into account their interdependence, under conditions of uncertainty imposed by a studied area.

These results are then used as input data for the second stage, which evaluates the decision alternatives regarding the area. The MT for the second stage comprises of six parameters, necessary for the thorough analysis of the area:

- Site suitability;
- Recommended object scale;
- Recommended construction depth;
- Dominant risk factors;
- Risk degree;
- Risk level.

Fig. 1 The network of morphological tables for evaluating suitability of construction sites for parking lots. Arrows indicate influence

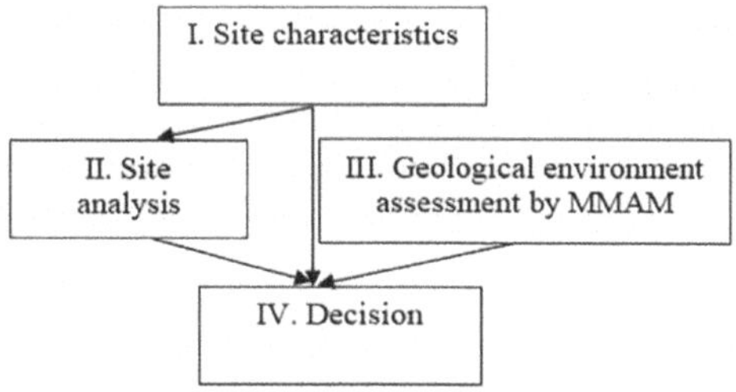

This model provides a fine general analysis of the geological environment for a site itself. However, specific types of underground objects may call for slight adjustments of this model, which will be addressed when appropriate.

Parking lots. The main questions that the model for assessing sites of potential parking lots should answer, are: will this site generate high demand for parking places? How big should a parking lot be if decided to be constructed here? As was mentioned in the introduction, most of the places in metropolises will have high demand for parking places, so additionally the model should be able to compare different sites to set construction priorities.

To answer these questions, the structural and functional factors of a site are involved, which describe the neighboring objects and present parking lots to infer the approximate demand for parking places at a site, and this numbers, along with the geological assessment, are then used to make conclusions about favorability of a site for an underground parking lot.

This concept represented by a network of morphological tables (Fig. 1):

The parameters of objects in morphological analysis method were grouped in four separate morphological tables. The first table "I. Site characteristics" contains six parameters with respective alternatives that correspond to the construction site:

1. nearby urban object types (residential buildings; office and administrative buildings; shopping and entertainment centers; stadiums, concert halls, theaters; schools and universities; architectural and tourist attractions; industrial objects);
2. number of residents in vicinity (up to 1 000; 1 000–3 000; 3 000–5 000; 5 000–10 000; over 10 000);
3. number of workplaces in vicinity (up to 500; 500–1 000; 1 000–3 000; 3 000–5 000; over 5 000);
4. traffic speed (high—over 60 km/h; average—30–60 km/h; low—15–30 km/h; very low—less than 15 km/h);
5. existing open air and underground parking places (up to 50; 50–200; 200–400; over 400);
6. accessibility of the territory for construction (no complications; slight complications; severe complications).

As can be seen, the parameters are characterized by different types of uncertainty—spatial distribution uncertainty (parameter 1—a single site may contain different types of buildings), temporal distribution uncertainty (parameter 4—the traffic speed may change over time), informational uncertainty (the exact measurements for parameters 2, 3 require additional cumbersome research, which is why the expert opinions are deemed sufficient).

For the purposes of the study, "nearby" was defined as a radius of 300 m as a rational assumption for the longest distance that a person would be content to walk from a parked car.

As most of these parameters are independent, the cross-consistency matrix for the table was not introduced. It is implied that the direct estimates for the alternatives would be sufficiently precise and would not require re-adjustment.

The next table "II. Site analysis" contains two parameters that describe the demand for establishing a parking lot on the site—the approximate number of required parking places (parameter "7. Parking place demand", with the same set of alternatives as parameter 5 in the previous table), and the type of demand (parameter "8. Parking demand type"):

- constant—the demand for parking is close to identical at different time;
- pendular—the demand for parking varies at different daytimes (e.g. office buildings or malls);
- peak—the demand emerges in case of some singular mass events (e.g. stadiums, concert halls).

This table is linked to the previous table via a dependency matrix, and the values in this table are calculated by the MMAM procedure using the estimates in the previous table and the dependency matrix.

The table "III. Geological environment assessment by MMAM" contains the parameters from the research of geological environment by MMAM, taken from papers [14, 15]. Only the parameters that directly influence the decision about parking lots, are selected, i.e. "A. Site suitability", "B. Object scale", and "C. Construction depth". In respect to the morphological table network in Fig. 2 these parameters were numbered 9–11.

The table "IV. Decision" contains parameters that summarize the decision regarding the considered construction site—the weights for suitability or unsuitability of a site regarding the parking lot construction (parameter "A. Suitability for a parking lot"), and the most advisable sizes of the potential parking lot, taking into account both the demand, and the surrounding geological environment (parameter "B. Advisable parking lot size").

The table IV is linked with all the previous tables by the dependence matrix. After the filling of the dependency matrices by expert estimation, the model for evaluating construction sites for a parking lot was obtained.

Car tunnels. The analysis of a potential track for a car tunnel was decomposed into two separate tasks for the two-stage MMAM: the analysis of the geological environment, and the analysis of the structural–functional factors for a studied area. The analysis of geological factors was conducted using a network of two previously described MTs, with certain changes aimed at considering the specifics of tunnels as underground objects:

Fig. 2 The scheme of car tunnel tracks (the General Plan of Kyiv city development up to 2025)

- the impact of the alternative "7.4. Flooded areas/quicksand are present" on the unsuitability of a construction site was significantly increased;
- the ranges of parameters "B. Object scale" and "C. Construction depth" were slightly revised to better reflect the type of analyzed underground objects (i.e. tunnels);
- the set of alternatives for parameter "D. Risk factor" was altered, as a portion of risk factors were transferred to structural and functional factor analysis.

The analysis of structural and functional factors for car tunnels was viewed in this study as a separate task. Constructing a car tunnel is generally a more cumbersome work than constructing an underground parking lot, but it also solves more issues, and this should be reflected in the model. As such, the purpose of this analysis is to answer the following question: will there be a demand for a tunnel in this area? Which problems will it solve? How big is its potential to mitigate different risk factors?

To construct a model that provides the answers, firstly a group of considered technogenic and ecological risks was formed, that could be mitigated by construction of a tunnel:

- R1. Air pollution (exhaust emission);
- R2. Noise and dynamic impact (engine hum, vehicle clanking, vibration etc.);

- R3. Traffic jams (lowered average traffic speed, disrupting transport communications, increased exhaust emission);
- R4. Traffic accidents (injuries, traffic block).

The chosen set of factors conditioned the MMAM application for the task of analyzing structural and functional factors. The MT for the first MMAM stage comprised of 8 parameters that were important in regard to the favorability of a planned tunnel and its capacity to mitigate one or more of the risk factors:

1. neighboring urban objects in the planned tunnel's area—this parameter influences the weights of different risk factors. For example, residential buildings, tourist attractions, parks make mitigating pollution and noise risk factors much more important, compared to industrial objects or undeveloped areas;
2. residential building density—this parameter complements the previous one, influencing risk factor weights;
3. downtown factor—describes the proximity of a studied area to the center of the city, or its influence on the traffic in the city center;
4. crowd density in the planned tunnel's area—this parameter influences the risk factor weights, primarily traffic accidents (more pedestrians means higher accident risk), and pollution and noise factors as well. This parameter also impacts the tunnel's capacity to mitigate some of the risk factors, as its construction leads to less opportunities for accidents and traffic jams caused by crosswalks;
5. intensity of traffic in the planned tunnel's area—this parameter is the most critical for the advisability of tunnel construction. Additionally, it influences the risk factor importance—more intensive traffic demands smooth, unobstructed movement, meaning higher weight of the traffic jam risk factor. High traffic intensity also increases the tunnel's capacity of dealing with all the risk factor groups: rerouting more intensive traffic underground means stronger mitigation of all the considered risk factors;
6. average traffic speed on the busiest road sections at peak hours in the planned tunnel's area—this parameter varies the tunnel's influence on risk factors: low movement speed means that the tunnel's construction will stronger impact the pollution and traffic jam factors, while high movement speed means the same for noise and traffic accident factors. Higher movement speed also increases the weight of traffic accident factor, as the accidents become potentially more dangerous;
7. surface connectivity of the planned tunnel ends by existing roads—the parameter obviously determines the advisability of the tunnel's construction. Also this parameter provides weight to the traffic jam factor, as under bad connectivity the presence or absence of traffic jams becomes critical. Accordingly, there is an opposite effect—if the connectivity was poor, the tunnel's construction creates a positive effect on traffic by creating an alternative route;
8. surface road throughput in the planned tunnel's area (road width, presence of crossings, especially unregulated crossing)—the parameter influences the advisability of the tunnel, and the traffic jam factor weight. It also contributes to tunnel's capacity of mitigating traffic jam and accident risk factors.

Table 1 Presence (+) or absence (–) of direct relation between structural–functional factors

	1	2	3	4	5	6	7	8
1		+	+	+	+	–	–	–
2	+		+	+	+	–	–	–
3	+	+		–	+	+	+	+
4	+	+	–		–	–	–	–
5	+	+	+	–		+	+	–
6	–	–	+	–	+		+	+
7	–	–	+	–	+	+		–
8	–	–	+	–	–	+	–	

Some of the parameters 1–8 are, obviously, interrelated. To account this relations, an expert estimation of the dependences between alternatives was made according to the MMAM procedure. Only the relevant assessments were made, which are shown in Table 1.

At the second MMAM stage a MT was constructed that describes the advisability of tunnel's construction from the structural–functional point of view, as well as the profile of the planned tunnel's area from the ecological and safety risk factors point of view. Five parameters in the table were assigned to risk analysis: one parameter compares the weights of the different risk factors, describing a "profile" of the area in regard to the risk factor importance; the other four parameters indicate the tunnel's capacity of mitigating each of the considered risk factor groups R1–R4. The MT consists of these parameters:

- Tunnel construction advisability (advisable, not advisable);
- Risk factor importance in the planned tunnel's area (air pollution; noise and dynamic impact; traffic jams; traffic accidents);
- Impact of tunnel construction at the pollution risk factor (no impact; slightly mitigates; moderately mitigates; significantly mitigates);
- Impact of tunnel construction at the noise and dynamic impact risk factor (no impact; slightly mitigates; moderately mitigates; significantly mitigates);
- Impact of tunnel construction at the traffic jams risk factor (no impact; slightly mitigates; moderately mitigates; significantly mitigates);
- Impact of tunnel construction at the traffic accidents risk factor (no impact; slightly mitigates; moderately mitigates; significantly mitigates).

Again, not all pairs of parameters from both MTs are related, an expert assessment of dependence between alternatives was conducted only for evidently present relations, shown in Table 2.

These two MTs, along with the estimates of relation between pairs of alternatives in related parameters, comprise the model for assessing the suitability of a site for a car tunnel.

Table 2 Presence (+) or absence (–) of impact of parameters 1–8 on parameters A–F

	A	B	C	D	E	F
1	–	+	–	–	–	–
2	–	+	–	–	–	–
3	+	+	–	–	–	–
4	+	+	–	–	+	+
5	+	+	+	+	+	+
6	+	+	+	+	+	+
7	+	+	–	–	+	–
8	+	+	–	–	+	+

4 Results

Parking lots. The model was tested on the same construction sites as in the study [15]:

- site 1 at the Shevchenkivsky district at the Peremohy avenue;
- site 2 at the Shevchenkivsky district between the Bulvarno-Kudriavska and Honchara streets.

The evaluation of the sites was made by expert estimation, in which the expert was tasked with estimation of each alternative of each parameter in the table "I. Site characteristics" (totaling 28 questions for a single site). The resulting input data for both sites is given in Table 3. For each parameter, the weights were normalized so that the sum equaled 1.

The first stage of calculations processed the dependency between the morphological tables "I. Site characteristics" and "II. Site analysis", which resulted in the following estimates for the potential demand in parking places (Table 4).

As can be seen, the distribution of demand types (parameter 8) is nearly identical, as for the demand values (parameter 7), the site 1 is skewed towards bigger sizes of parking lots compared to site 2.

Next, taking the data from Tables 1, 2 as input, as well as the results from [15] that describe the geological environment, the decision table was evaluated (Table 5).

The obtained results allow to make several conclusions. Both of the sites are quite favorable for construction of parking lots, which can be seen from the estimates of parameter "A. Suitability for a parking lot". This is stipulated by the location of both sites in places with high functional urban activity, close to office and administrative buildings, shopping and entertainment centers, educational facilities etc. However, the second site appeared more favorable, which is explained by its better geological environment assessments, obtained in [15]. The most advisable size for potential parking lots was described by the alternative "50–200 parking places" (with weights of 0.444 and 0.456 respectively). High weights of relatively small parking lots can be explained by the quantitative characteristics of the considered sites (bound by

Table 3 The input data for the study, obtained by expert estimation

Parameter	Alternative	Raw input		Normalized input	
		Site 1	Site 2	Site 1	Site 2
1	1.1. Residential buildings	0.8	0.8	0.242	0.184
	1.2. Office and administrative buildings	0.8	0.8	0.242	0.184
	1.3. Shopping and entertainment centers	0.65	0.8	0.197	0.184
	1.4. Stadiums, concert halls, theaters	0.65	0.65	0.197	0.149
	1.5. Schools and universities	0.2	0.65	0.061	0.149
	1.6. Architectural and tourist attractions	0	0.65	0.000	0.149
	1.7. Industrial objects	0.2	0	0.061	0.000
2	2.1. Up to 1 000	0	0.2	0.000	0.091
	2.2. 1 000–3 000	0.35	0.65	0.206	0.295
	2.3. 3 000–6 000	0.8	0.8	0.471	0.364
	2.4. 6 000–10 000	0.35	0.35	0.206	0.159
	2.5. Over 10 000	0.2	0.2	0.118	0.091
3	3.1. Up to 500	0.35	0.2	0.175	0.100
	3.2. 500–1 000	0.8	0.65	0.400	0.325
	3.3. 1 000–3 000	0.65	0.8	0.325	0.400
	3.4. 3 000–5 000	0.2	0.35	0.100	0.175
	3.5. Over 5 000	0	0	0.000	0.000
4	4.1. High (over 60 km/h)	0.35	0.35	0.206	0.163
	4.2. Average (30…60 km/h)	0.8	0.65	0.471	0.302
	4.3. Low (15…30 km/h)	0.35	0.8	0.206	0.372
	4.4. Very low (less than 15 km/h)	0.2	0.35	0.118	0.163
5	5.1. Up to 50 parking places	0.8	0.8	0.593	0.400
	5.2. 50–200 parking places	0.35	0.65	0.259	0.325
	5.3. 200–400 parking places	0.2	0.35	0.148	0.175
	5.4. Over 400 parking places	0	0.2	0.000	0.100
6	6.1. No complications	0.8	0.65	0.485	0.361
	6.2. Slight complications	0.65	0.8	0.394	0.444
	6.3. Severe complications	0.2	0.35	0.121	0.194

the 300 m restriction), as well as the assessments of geological environment which favored lesser scale structures due to their stability (though the impact of this factor can be variable depending on the type of a chosen parking lot).

Car tunnels. For the purposes of testing, two potential car tunnel tracks of right bank Kyiv (Fig. 2) were taken: Tunnel 1 (M. Gryshko National Botanical garden—Darnytskyi bridge), and Tunnel 5 (Peremohy square—Dnipro river; only the section before Dnipro was considered, without the underwater tunnel section).

Table 4 Site analysis for parking lots

Parameter	Alternative	Site 1	Site 2
7	7.1. Up to 50 parking places	0.306	0.379
	7.2. 50–200 parking places	0.407	0.390
	7.3. 200–400 parking places	0.206	0.179
	7.4. Over 400 parking places	0.080	0.053
8	8.1. Constant	0.272	0.275
	8.2. Pendular	0.375	0.372
	8.3. Peak	0.352	0.354

Table 5 Decision for parking lots

Parameter	Alternative	Site 1	Site 2
A	A.1. Construction suitable	0.853	0.979
	A.2. Construction unsuitable	0.147	0.021
B	B.1. Up to 50 parking places	0.429	0.355
	B.2. 50–200 parking places	0.444	0.456
	B.3. 200–400 parking places	0.121	0.160
	B.4. Over 400 parking places	0.007	0.028

The results of expert assessment of the geological environment, and the values calculated by the first MMAM stage are presented in Table 6.

The column "Input" in Table 6 shows the normalized expert assessments, and the column "Result" shows the processed by MMAM values which take into account the mutual dependencies between parameters. As the table shows, the influence of interdependence allowed to obtain a more precise representation, however significant differences in distributions between alternatives were not observed, thus affirming that the expert assessment was conducted quite accurately, and the parameter estimates are consistent.

Using the calculated values, the second MMAM stage was implemented to evaluate decisions regarding the geological engineering factors (Table 7).

Table 2 implies that the geological environments around the both tunnel tracks are favorable enough for underground construction (the weight of "A.1. Suitable" alternative is considerably higher than the weight of "A.2. Not suitable" alternative), and the difference between this parameter for both tunnels is negligible. According to the characteristics of Tunnel 5 (shown in Table 1), the model proposes smaller cross-section and deeper construction depth compared to Tunnel 1 ("C.4. beneath 60 m" weight is 0.518 for Tunnel 5 and 0.276 for Tunnel 1), which can be explained by land relief. The two considered sites also have slightly different profiles regarding risk factors—the most tangible risk factor for Tunnel 1 is "D.4. Initiating displacements and other unwanted geological processes" with weight 0.547, which is significantly higher than the next significant risk factor "D.2. Increasing construction and operation cost" with weight 0.348, whereas for Tunnel 5 both of these factors are nearly

Table 6 The geological environment assessments for two tunnel tracks

Parameter	Alternative	Tunnel 1		Tunnel 5	
		Input	Result	Input	Result
1. Level of dynamic load	1.1. Low (46–53 dB)	0.143	0.137	0.093	0.070
	1.2. Medium (53–73 dB)	0.327	0.489	0.302	0.448
	1.3. Increased (73–96 dB)	0.327	0.238	0.372	0.299
	1.4. High (over 96 dB)	0.204	0.136	0.233	0.182
2. Static load from surface buildings	2.1. Insignificant (Ksl <1)	0.206	0.101	0.000	0.000
	2.2. Medium (1 < Ksl <2)	0.471	0.584	0.194	0.164
	2.3. Increased (2 < Ksl <3.5)	0.206	0.215	0.444	0.451
	2.4. High (Ksl >3.5)	0.118	0.100	0.361	0.385
3. Static load from soil	3.1. Insignificant (Kmas < 0.05, MPa)	0.175	0.063	0.093	0.015
	3.2. Medium (0.05< Kmas < 0.3, MPa)	0.400	0.703	0.302	0.485
	3.3. High (0.3< Kmas < 0.5, MPa)	0.325	0.200	0.372	0.362
	3.4. Very high (Kmas > 5, MPa)	0.100	0.034	0.233	0.138
4. Influence of existing underground objects	4.1. Absent (distance over 50 m)	0.129	0.055	0.108	0.032
	4.2. Slight (distance 20–50 m)	0.516	0.422	0.432	0.336
	4.3 Significant (distance 10–20 m)	0.226	0.437	0.351	0.584
	4.4 Hazardous (distance less than 10 m)	0.129	0.087	0.108	0.048
5. Genetic type and lithologic composure of soil	5.1. Unweathered clays and average density sands	0.175	0.269	0.265	0.320
	5.2. Technogenic deposits (alluvial and bulk types)	0.100	0.127	0.143	0.157
	5.3. Deluvial clay soils (water-saturated), water-saturated overfloodplain sands	0.400	0.402	0.327	0.354
	5.4. Sedentary soils, soils with special properties (loess, peat, silt)	0.325	0.203	0.265	0.169
6. Effective soil strength	6.1. Very strong soils > 300 kPa	0.000	0.000	0.093	0.091
	6.2. Strong soils 200–300 kPa	0.121	0.113	0.233	0.328
	6.3. Average strength soils 150–200 kPa	0.485	0.575	0.372	0.402

(continued)

Table 6 (continued)

Parameter	Alternative	Tunnel 1		Tunnel 5	
		Input	Result	Input	Result
	6.4. Relatively strong soils < 150 kPa	0.394	0.313	0.302	0.179
7. Influence of aquifers and perched groundwater	7.1. Water-bearing horizons at P-N1np	0.217	0.193	0.372	0.383
	7.2. Groundwater depth > 3 m, pressurized groundwater > 10 m	0.283	0.365	0.372	0.474
	7.3. Groundwater depth < 3 m, pressurized groundwater < 10 m	0.348	0.358	0.093	0.089
	7.4. Flooded areas/quicksand are present	0.152	0.084	0.163	0.054
8. Landscape type and morphometrics	8.1. Flat areas of overfloodplain terraces, morainic-glacial plains	0.143	0.159	0.000	0.000
	8.2. Slightly tilted overfloodplain terraces, watershed ares	0.265	0.340	0.194	0.353
	8.3. Small river valleys, slightly irregular slopes, high floodplain	0.327	0.416	0.361	0.536
	8.4. Slope areas with ravines and steep banks, low floodplain	0.265	0.086	0.444	0.111
9. Geological engineering processes	9.1. Absent	0.082	0.043	0.093	0.027
	9.2. Stabilized	0.265	0.320	0.302	0,303
	9.3. Low displacement processes	0.327	0.527	0.372	0.586
	9.4. Active manifestations of subsidence, underflooding, gravitational processes	0.327	0.109	0.233	0.084
10. Geotechnologies of underground construction	10.1. Open	0.200	0.364	0.200	0.119
	10.2. Underground	0.800	0.636	0.800	0.881

equivalent (their weights are 0.426 and 0.417 respectively). The profiles for risk degree (probability) and risk level (cost of consequences) are close for both tunnels, with a slight skew towards higher values for Tunnel 5, which means that the cost of support and potential renovations for Tunnel 5 will be 5–7% of total construction value more compared to Tunnel 1. Therefore, the geological environments, that are favorable enough for both of the tunnels, cannot be the defining factor for construction priority.

Table 7 Estimates of the decision alternatives weights obtained at second MMAM stage

Parameter	Alternative	Tunnel 1	Tunnel 5
A. Track suitability	A.1. Suitable	0.777	0.799
	A.2. Not suitable	0.223	0.201
B. Object scale	B.1. Cross-section up to 10 m^2	0.643	0.712
	B.2. Cross-section up to 25 m^2	0.255	0.232
	B.3. Cross-section up to 40 m^2	0.084	0.049
	B.4. Cross-section up to and over 40 m^2	0.018	0.007
C. Construction depth	C.1. 0–10 m	0.096	0.021
	C.2. 10–20 m	0.183	0.100
	C.3. 20–50 m	0.445	0.360
	C.4. beneath 60 m	0.276	0.518
D. Risk factor	D.1. Construction failure, malfunction	0.039	0.046
	D.2. Increasing construction and operation cost	0.348	0.417
	D.3. Dangerous influence on surface or neighboring underground objects	0.067	0.111
	D.4. Initiating displacements and other unwanted geological processes	0.547	0.426
E. Risk degree	E.1. < 3%	0.089	0.050
	E.2. 3–10%	0.598	0.584
	E.3. 10–20%	0.274	0.308
	E.4. 20–50%	0.031	0.044
	E.5. > 50%	0.009	0.014
F. Risk level	F.1. 0,1–5% Q	0.212	0.137
	F.2. 5–20% Q	0.701	0.768
	F.3. 20–50% Q	0.079	0.085
	F.4. > 50% Q	0.008	0.010

The next step was to assess the structural–functional factors by the second morphological table (Table 8). Here the input estimates are given (the "Input" column), as well as the values calculated by MMAM (the "Result" column).

The results of this estimation allow to make a comparison of the two areas:

- the area of Tunnel 5 is mostly covered in residential, commercial, administrative buildings, while a large portion of Tunnel 1 area is taken by parks and undeveloped areas. Accordingly, Tunnel 5 has much higher density of residential buildings (Parameter 2): the highest-ranking alternatives are "High" and "Average", whereas for Tunnel 1 they are "Average" and "Low";
- the area of Tunnel 5 can undoubtedly be labeled as historical city center zone, and it definitely influences the traffic in the city center; not so much for the Tunnel 1 area, which is mostly outside the historical city center;

Table 8 Input data for assessing structural–functional factors, and the results of the first MMAM stage

Parameter	Alternative	Tunnel 1		Tunnel 5	
		Input	Result	Input	Result
1. Neighboring urban objects	1.1. Residential buildings	0.197	0.289	0.258	0.393
	1.2. Administrative, commercial buildings	0.197	0.191	0.210	0.285
	1.3. Tourist attractions, sights	0.061	0.056	0.258	0.219
	1.4. Parks	0.242	0.270	0.210	0.103
	1.5. Industrial buildings	0.061	0.062	0.000	0.000
	1.6. Undeveloped areas	0.242	0.134	0.065	0.001
2. Residential building density	2.1. Very low	0.175	0.088	0.000	0.000
	2.2. Low	0.400	0.275	0.121	0.020
	2.3. Average	0.325	0.518	0.485	0.532
	2.4. High	0.100	0.119	0.394	0.449
3. Downtown factor	3.1. Area is in the city center	0.121	0.187	0.500	0.650
	3.2. Area influences traffic in the city center	0.394	0.283	0.500	0.350
	3.3. Area is far from city center	0.485	0.531	0.000	0.000
4. Crowd density	4.1. Very low	0.233	0.103	0.100	0.004
	4.2. Low	0.372	0.255	0.100	0.019
	4.3. Average	0.302	0.490	0.400	0.421
	4.4. High	0.093	0.153	0.400	0.558
5. Intensity of traffic	5.1. Low	0.093	0.106	0.000	0.000
	5.2. Average	0.302	0.432	0.289	0.201
	5.3. High	0.372	0.368	0.356	0.507
	5.4. Very high	0.233	0.096	0.356	0.293
6. Average traffic speed at peak hours	6.1. Lower than 15 km/h	0.148	0.095	0.194	0.228
	6.2. 15–30 km/h	0.259	0.307	0.444	0.637
	6.3. 30–60 km/h	0.593	0.599	0.361	0.136
7. Surface connectivity	7.1. Very poor	0.129	0.073	0.325	0.230
	7.2. Poor	0.516	0.410	0.400	0.432
	7.3. Average	0.226	0.286	0.175	0.193
	7.4. Good	0.129	0.232	0.100	0.146
8. Surface road throughput	8.1. Low	0.233	0.165	0.361	0.452
	8.2. Average	0.533	0.594	0.444	0.473
	8.3. High	0.233	0.242	0.194	0.077

- the crowd density in the Tunnel 5 area is average to high (due to few high-rise residential buildings), and in the Tunnel 1 area the crowd density is low to average (due to average building density, presence of park areas);
- intensity of traffic is higher in the Tunnel 5 area, however at peak hours the traffic speed is lower in comparison with the Tunnel 1 area;
- the surface road connectivity and throughput is better for the Tunnel 1 area.

Using the estimates obtained in Table 8, and the constructed dependency matrix, the second MMAM stage was conducted for assessing the impact of structural–functional factors. The results are given in Table 9.

The results of evaluation allow to make several conclusions regarding the planned tunnels' sites:

(1) The construction of both tunnels is highly advisable: the weight of alternative "A.1. Advisable" significantly overcomes the weight of "A.2. Not advisable" (accordingly, 0.895 versus 0.105 for Tunnel 1, and 0.994 versus 0.006 for

Table 9 Assessing impact of structural–functional factors at tunnel construction areas on the ecological and safety factors

Parameter	Alternative	Tunnel 1	Tunnel 5
A. Tunnel construction advisability	A.1. Advisable	0.895	0.994
	A.2. Not advisable	0.105	0.007
B. Risk factor importance in the planned tunnel's area	B.1. Air pollution	0.143	0.173
	B.2. Noise and dynamic impact	0.166	0.151
	B.3. Traffic jams	0.280	0.454
	B.4. Traffic accidents	0.412	0.223
C. Impact of tunnel construction at the pollution risk factor	C.1. No impact	0.088	0.022
	C.2. Slightly mitigates	0.404	0.289
	C.3. Moderately mitigates	0.311	0.358
	C.4. Significantly mitigates	0.199	0.332
D. Impact of tunnel construction at the noise and dynamic impact risk factor	D.1. No impact	0.086	0.056
	D.2. Slightly mitigates	0.312	0.270
	D.3. Moderately mitigates	0.362	0.371
	D.4. Significantly mitigates	0.241	0.304
E. Impact of tunnel construction at the traffic jams risk factor	E.1. No impact	0.108	0.010
	E.2. Slightly mitigates	0.442	0.237
	E.3. Moderately mitigates	0.285	0.402
	E.4. Significantly mitigates	0.166	0.352
F. Impact of tunnel construction at the traffic accidents risk factor	F.1. No impact	0.085	0.023
	F.2. Slightly mitigates	0.341	0.377
	F.3. Moderately mitigates	0.348	0.359
	F.4. Significantly mitigates	0.226	0.242

Tunnel 5). This is a very reasonable result, as the test dealt with the real, previously rationaled objects from the General Kyiv city plan. The integral advisability factor for Tunnel 5 emerged 10% higher than for Tunnel 1.

(2). The structure of the most influential risk factors has slight differences for the considered tunnel areas. The biggest difference is that the traffic jams risk factor is more important in the Tunnel 5 area, while the same is true for the traffic accident risk factor in the Tunnel 1 area, which can be explained by higher traffic speed. Also, the air pollution risk factor has more weight for Tunnel 5 area.

(3) The construction of each of the studied tunnels will definitely provide a certain mitigation of the considered risk factors, and the extent of this mitigation varies for each specific tunnel and risk factor type.

The weight of alternative "No impact" is notably small, especially for Tunnel 5, proving the high potential of underground construction for ecologization of transport infrastructure. This fact is also confirmed by high values of the alternative "A.1. Tunnel advisability", as was noted earlier.

In general Tunnel 5 provides higher mitigation of ecological and safety risks of existing surface infrastructure. For the factors "R2. Noise and dynamic impact" and "R4. Traffic accidents" its advantage is negligible (within the estimation error margin), however for the more important factors "R1. Air pollution" and "R3. Traffic jams" the advantage of Tunnel 5 is more evident. While the impact of Tunnel 1 can be characterized as slight to moderate, Tunnel 5 shifts the impact to the moderate to significant zone.

The obtained data allows to make conclusions that the construction of Tunnel 5 is the first priority task for quick improvement of transport and ecological situation in Kyiv city center.

5 Discussion

The obtained results show how a general morphological model for assessment of geological environment at potential construction sites can be incorporated as a powerful decision support tool for identifying construction priorities of various types of underground objects.

The adjustments needed to make in the model depend on the complexity of the studied type of objects: structures that have simpler functions, like parking lots, need less input and smaller supplements to the core model; while more complex structures like car tunnels that cause more significant impact on the urban functioning, require multi-criteria additions.

The developed technique and tool set allow to incorporate the assessment of impacts and relations of geologic, technogenic, and structural–functional factors for analysis of favorability of urban territories for different types of underground objects, taking into account economic factors, impact on ecological and technogenic risks.

6 Conclusions

The developed tool set was based on the modified morphological analysis method, which performed well for modeling situations comprising of objects and entities that are characterized by a very large multitude of possible configurations spawned from combining different alternatives of their innate parameters. The method allowed to use the selected groups of geological, technogenic factors, as well as the functional site characteristics in the thorough study of decisions and risks related to the underground development of chosen territories. The applied technique is suitable for evaluating the prospect of underground construction at the pre-project stage, capabilities for risk management of urban underground city space development, diminishing the potential for project flaws caused by neglecting certain factors or specifics of a geological environment and technogenic impacts, convenient form of information generation as tables, charts or graphs. This modeling tool set is a valuable instrument for investors and city state administrations to manage risks and investments in mastering the underground space of metropolises. The proposed technique and tool set can be utilized in developing strategic master plans for large cities.

Acknowledgements The presented results were obtained in the National Research Fund of Ukraine project 2020.01/0247 «System methodology-based tool set for planning underground infrastructure of large cities providing minimization of ecological and technogenic risks of urban space».

References

1. World Urbanization Prospects 2018: Highlights. United Nations, New York (2019)
2. Berkowitz, A.R., Nilon, C.H., Hollweg, K.S.: Understanding Urban Ecosystems. Springer, New York (2003)
3. Korendiaseva, E.V.: Ecological Aspects of City Management. MGUU of Moscow Government, Moscow (2017)
4. Gilbert, P.H., et al.: Underground Engineering for Sustainable Urban Development. The National Academies Press, Washington (2013)
5. Golubev, G.E.: Underground Urbanistics and City. MIKHiS, Moscow (2005)
6. Pankratova, N.D., Haiko, H.I., Savchenko, I.O.: Development of Urban Underground Studies as a System of Alternative Project Configurations. Naukova Dumka, Kyiv (2020)
7. Vähäaho, I.: Underground space planning in Helsinki. J. Rock Mech. Geotech. Eng. **6**(5), 387–398 (2014)
8. Sterling, R., Admiraal, H., Bobylev, N., Parker, H., Godard, J.P., Vähäaho, I., Shi, X., Hanamura, T.: Sustainability issues for underground spaces in urban areas. Proc. ICE—Urban Des. Plan. **165**(4), 241–254 (2012)
9. Kartosiya, B.A.: Developing underground space of large cities new tendencies. Min. Inf.-Anal. Bull. (Sci. Tech. J.) **1**, 615–629 (2015)
10. Resin, V.I., Popkov, Y.S.: Development of Large Cities under Conditions of Transitional Economy. System Approach. Bookhouse «LIBROKOM», Moscow (2013)
11. Zgurovsky, M.Z., Pankratova, N.D.: System Analysis: Theory and Applications. Springer, New York (2007)
12. Pankratova, N.D., Savchenko, I.O.: Morphological analysis. In: Problems, Theory, Applications. Naukova Dumka, Kyiv (2015)

13. Pankratova, N., Gayko, G., Kravets, V., Savchenko, I.: Problems of megapolises underground space system planning. J. Autom. Inf. Sci. **48**(4), 32–38 (2016)
14. Pankratova, N.D., Savchenko, I.O., Haiko, H.I., Kravets, V.G.: System approach to mastering underground space of metropolises under conditions of uncertainty and multi-factor risks. Rep. Natl. Acad. Sci. Ukraine **10**, 18–25 (2018)
15. Haiko, H.I., Savchenko, I.O., Matviichuk, I.O.: Development of a morphological model for territorial development of underground city space. Naukovyi Visnyk Natsionalnoho Hirnychoho Universytetu **3**, 92–98 (2019)
16. Haiko, H.: Problems of system planning of underground space in large cities. Visnyk NTUU "KPI", Mining Series **25**, 35–40 (2014)
17. Samedov, A., Kravets, V.: Construction of Urban Underground Structures. NTUU "KPI", Kyiv (2011)
18. Pankratova, N., Savchenko, I.: Strategy of application of morphological analysis methods in technology foresight process. Naukovi visti NTUU "KPI" **2**, 35–44 (2009)
19. Kelemen, Y., Vayda, Z.: City Underground. Stroyizdat, Moscow (1985)
20. Regional environment state report for Kyiv city in 2017. Kyiv city state administration, https://ecodep.kyivcity.gov.ua/files/2019/1/22/REG_DOP_2017.pdf. Last accessed 05 Jul 2021
21. Eliseev, D.O.: Risks in modern metropolises: types, features, specifics. Regional Econ. Manage. **4**(52), 5211 (2017)
22. Haiko, H.I., Bulhakov, V.P., Siveryn, M.O.: Construction of a system of road tunnels as a way to solve transport and environmental challenges of a metropolis. Visnyk NTUU "KPI", Mining Series **30**, 196–206 (2016)
23. Cuia, J., Nelson, J.: Underground transport: an overview. Tunn. Undergr. Space Technol. **87**, 122–126 (2019)
24. Duczynski, G.: Morphological analysis as an aid to organisational design and transformation. Futures **86**, 36–43 (2017)
25. Ji, S.H., Ahn, J.: Scenario-planning method for cost estimation using morphological analysis. Adv. Civil Eng. **4**, 1–10 (2019)

Chapter 7
Application of Impulse Process Models with Multirate Sampling in Cognitive Maps of Cryptocurrency for Dynamic Decision Making

Viktor Romanenko, Yurii Miliavskyi, and Heorhii Kantsedal

Abstract In this chapter a cognitive map (CM) of cryptocurrency application in the financial market is developed, based on which the dynamic model of impulse processes of CM in the form of a system of difference equations (Roberts equations) with multirate sampling is described. The original CM model is decomposed into subsystems with fast- and slow-measured node coordinates. The subsystems are interconnected with each other and are represented with multirate sampling of coordinates. The realization of external control vectors for the subsystem with fast-measured and slow-measured node coordinates by varying some coordinates that can be measured by a decision maker is performed. Closed-loop subsystems of CM impulse process control are implemented, which include multidimensional discrete controllers designed based on the method of invariant ellipsoids. The controllers in the subsystems generate external controls with multirate sampling and affect directly the CM nodes by varying their coordinates. The problem of designing the above discrete-time controllers for suppression of constrained internal and external disturbances is solved. External disturbances include a variety of information disturbances acting on the system. Internal disturbances include changes in the influence of CM nodes on each other, for instance, fluctuations of weight coefficients of CM with respect to their basic values. The basic values are determined by an expert based on cause-effect relations or by the preliminarily identification of the CM model parameters. The mutual influence of interconnected CM subsystems on each other is also taken into account as internal disturbances. The designed control is implemented by a decision maker in the corresponding sections of the CM. By numerical simulation, the efficiency of the designed discrete controllers was investigated and the system performance was compared in the presence and absence of controls.

Keywords Cognitive map · Multirate sampling · Linear matrix inequalities · Cryptocurrency · State controller · Invariant ellipsoid

V. Romanenko (✉) · Y. Miliavskyi · H. Kantsedal
Institute of Applied System Analysis, National Technical University of Ukraine "Igor Sikorsky Kyiv Polytechnic Institute", 37a Peremohy av., Kyiv 03056, Ukraine

H. Kantsedal
e-mail: g.kantsedal@protonmail.com

M. Zgurovsky and N. Pankratova (eds.), *System Analysis & Intelligent Computing*, Studies in Computational Intelligence 1022, https://doi.org/10.1007/978-3-030-94910-5_7

1 Introduction

This chapter applies cognitive modeling, which is one of the most relevant directions of scientific and practical research of complex systems of different nature, to study dynamic processes in the application of cryptocurrency. It is based on the concept of a cognitive map (CM) [1–10], which is a weighted directed graph, the nodes of which display a complex system, and the edges (arcs) of the graph with weight coefficients describe the cause-effect relationship between the nodes of CM. When disturbances affect the CM nodes, an impulse process arises, the dynamics of which is described by the difference equation [1].

$$\Delta y_i(k+1) = \sum_{j=1}^{n} a_{ij} \Delta y_j(k) \tag{1}$$

where $\Delta y_i(k+1) = y_i(k) - y_i(k-1), i = 1, \ldots, n$; a_{ij}—the weight of the edge that connects the node j with node i. Equation (1), describing the free motion of the CM coordinate i without the application of external control actions, can be written in vector–matrix form:

$$\Delta \overline{Y}(k+1) = A \Delta \overline{Y}(k) \tag{2}$$

where A—weighted transposed adjacency matrix of the CM.

In order to implement CM impulse process control based on modern control theory, it is necessary to be able to physically vary some of the CM node coordinates when generating control actions. Then the forced motion of the CM impulse process under external control can be represented in the following form:

$$\Delta \overline{Y}(k+1) = A \Delta \overline{Y}(k) + B \Delta \overline{U}(k) \tag{3}$$

where $\Delta \overline{U}(k) = \overline{U}(k) - \overline{U}(k-1)$—vector of controls increments with size $m \leq n$. The operator (human) fills the control matrix B and usually uses ones and zeros.

In [11] we considered a model of impulse CM process for the application of cryptocurrency with unirate sampling, that is, with one sampling period T_0. The stabilization of a CM impulse process by designing a modal state controller based on the method of modal control is described in [12] with $m \leq n$. However, [11] does not account for a very important fact that the sampling period T_0 is chosen according to the Shannon theorem in the model (3). According to the theorem, the sampling period is bound to the fastest varying node y_i of CM. Thus, in practice, there are slowly changing nodes in a CM, the coordinates of which are impossible or very difficult to measure at discrete moments of time with a small period T_0. Therefore, for part of the CM nodes of the cryptocurrency application it is necessary to apply an increased sampling period h when measuring their coordinates. It is assumed that n

coordinates in model (2) are divided into p coordinates with a sampling period T_0, and other $(n - p)$ coordinates should be measured with an increased period $h = mT_0$, where m is an integer greater than one. A similar problem has already been solved in [13] considering the principles of CM impulse processes control, where a preliminary decomposition of the initial CM model (2) into two parts is recommended. The first part of the CM consists of p coordinates y_{fi} that are fast changing and need to be represented in discrete form with a period T_0. The second part consists of $(n - p)$ coordinates y_{si} which need to be further converted to discrete form with an increased period h.

Then, the initial model (1) for the first part of the p coordinates can be written in the following form:

$$\Delta y_{f_i}(k+1) = \sum_{j=1}^{p} a_{i_j} \Delta y_j(k) + \sum_{\mu=p+1}^{n} a_{i_\mu} \Delta y_\mu(k), i = 1, \ldots, p \tag{4}$$

The second part of the model (1) for the $(n - p)$ coordinates will have the form:

$$\Delta y_{s_l}(k+1) = \sum_{r=1}^{p} a_{l_r} \Delta y_{f_r}(k) + \sum_{j=p+1}^{n} a_{l_j} \Delta y_{s_j}(k), l = p+1, \ldots, n \tag{5}$$

Expressions (4) and (5) can be written in vector–matrix form:

$$\Delta \overline{Y}_f(k+1) = A_{11} \Delta \overline{Y}_f(k) + A_{12} \Delta \overline{Y}_s(k) \tag{6}$$

$$\Delta \overline{Y}_s(k+1) = A_{21} \Delta \overline{Y}_f(k) + A_{22} \Delta \overline{Y}_s(k) \tag{7}$$

where the matrices A_{11}, A_{12}, A_{21}, A_{22} have the following dimensions $p \times p$, $p \times (n - p)$, $(n - p) \times p$, $(n - p) \times (n - p)$. Subsystems (6) and (7) are interconnected.

Problem statement. The first problem, which is solved in this chapter, is to develop a CM of the application of cryptocurrency in the financial market, compared with the CM presented in [11]. This problem considers the allocation of CM nodes that change in discrete time with period T_0 and period h. The second problem is to represent models (6) and (7) with multirate sampling of coordinates. The third problem is to develop closed-loop control subsystems for fast and slow measured coordinates with multirate sampling to suppress constrained external and internal disturbances affecting the CM nodes coordinates.

2 Materials and Methods

2.1 Building an Improved Cognitive Map of Cryptocurrency Applications in the Financial Market

Figure 1 shows the improved CM of cryptocurrency application compared to the one given in [11]. The subsystem with fast-measured coordinates includes nodes:

1. Cryptocurrency exchange rate (bitcoin value)
2. Cryptocurrency trading volume
3. Supply of cryptocurrency
4. Demand of cryptocurrency
5. Volume of speculation in cryptocurrency
6. Risk of cryptocurrency collapse.

The CM subsystem with slowly measured coordinates includes nodes:

7. Number of cryptocurrency users
8. The volume of investment (interest in bitcoin from institutional investors)
9. Volume of capitalization
10. Indirect profits
11. Level of confidence in cryptocurrency
12. The risk of losing the number of users.

A separate node 13 represents various information disturbances, which consist mostly of fast changing processes.

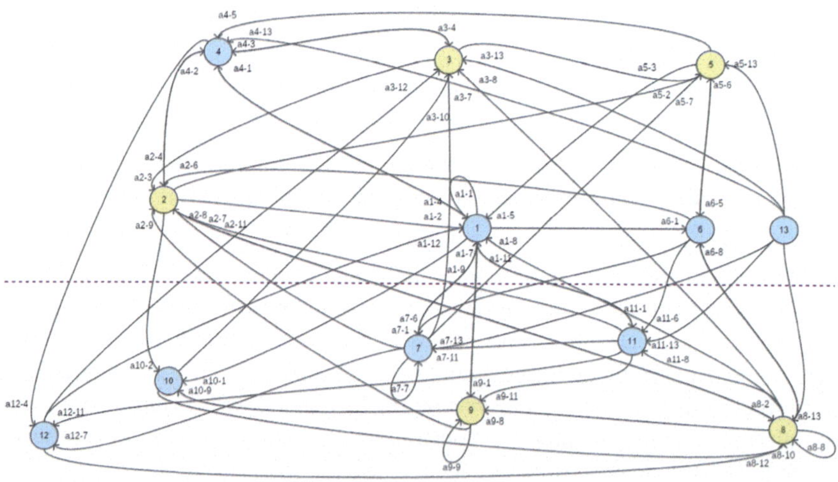

Fig. 1 Interactions of the CM nodes. The dashed line separates the fast-measured and slow-measured systems. Yellow nodes can be controlled

As control actions based on varying the resources of CM coordinates the following nodes can be used:

- For subsystems with fast-measured node coordinates:
  - Varying the volume of cryptocurrency trading $\Delta u_2(k)$
  - Varying the offer of cryptocurrency $\Delta u_3(k)$
  - Varying the volume of speculation $\Delta u_5(k)$
- For a subsystem with slowly-measured node coordinates:
  - Varying the amount of investment $\Delta u_8(k)$
  - Varying the volume of capitalization $\Delta u_9(k)$

To determine (measure) the risk of collapse of the cryptocurrency rate (node 6) the method of financial risk assessment Value-at-risk (VAR) is used, which is based on the analysis of the statistical nature of the market and involves a universal methodology for assessing various types of risks (price, currency, credit and liquidity risk). In fact, the VaR methodology is currently used as a standard for risk assessment [14, 15]. Parametric VaR is calculated as follows [16]:

$$VaR = \alpha\sigma P\sqrt{M} \tag{8}$$

where α—the quantile of the confidence interval; σ—volatility (measure of variability); P—open position value; M—forecast period. Volatility is determined based on the preliminary calculation of the variance of the cryptocurrency rate (node 1), which is calculated on the interval NT_0 according to the following algorithm:

$$\sigma_{y_1}(kT_0) = \frac{1}{N}\sum_{i=1}^{N}\left\{y_1[(k-i)T_0] - \frac{1}{N}\sum_{i=1}^{N} y_1[(k-i)T_0]\right\}^2 \tag{9}$$

where the sampling time interval NT_0 is chosen experimentally.

To determine the risk of losing the number of users (node 12) the VaR method (8) is also used. In which the variance of the coordinate Y_7 (number of users) is calculated as follows:

$$\sigma_{y_7}\left(\left\lfloor\frac{k}{m}\right\rfloor h\right) = \frac{1}{N}\sum_{i=1}^{N}\left\{y_7\left[\left(\left\lfloor\frac{k}{m}\right\rfloor - i\right)h\right] - \frac{1}{N}\sum_{i=1}^{N} y_7\left[\left(\left\lfloor\frac{k}{m}\right\rfloor - i\right)h\right]\right\}^2 \tag{10}$$

where $h = mT_0$ and $\left\lfloor\frac{k}{m}\right\rfloor$—integer part of division.

Consider the nature of the main disturbances (node 13), which affect the CM nodes and are called "information disturbances". These influences have a significant impact on rate fluctuations, as a consequence of a rapid increase in demand depending on positive or negative news. Because the cryptocurrency environment is quite open and a large number of cryptocurrency users have no professional education in this field. A vivid example of informational impact was Elon Musk's recommendation to use the messenger "Signal". In this case, the company's shares rose by 1300% in an hour. This type of disturbance can affect the nodes: 3, 4, 5, 7, 8, 11. Information disturbances are also generated by fluctuations in the global economy, causing the uncertainty of institutional investors (e.g., large banks) and forcing them to look for new areas for investment. At the same time, the entry into the market of a significant number of banks or stock market players leads to a sharp fluctuation of the bitcoin hurricane. Their behavior in the market is similar to long-term investing and affects CM nodes 5, 8, which leads to an increase in demand for cryptocurrency (node 4). The opposite effect is also possible. If the global economy stabilizes, some investors will move to the stock market, which will lead to a sharp increase in the supply of cryptocurrencies (node 3).

Information disturbances are also caused by fluctuations in energy prices, which are one of the factors of instability in the world economy. Thus, a decrease in energy prices attracts new users to the process of mining and a significant number of people have cryptocurrencies. This leads to the growth of the number of speculations (node 5) and naturally to the growth of the rate (node 1).

Legislative fluctuations also generate information outrage. For example, the news about the introduction in Denmark of cryptocurrency as a unit of payment fueled interest in cryptocurrency and sharply increased the number of users (node 7) and the growth of speculation (node 5).

Information disturbances are also generated by the mismatch between the level of trust (node 11) and the volume of investments (node 8), which can be caused by the confrontation between small and large investors. This can be explained by the fact that small investors often use borrowed money and are vulnerable to significant fluctuations of the rate of cryptocurrency (node 1), which can be easily caused by institutional investors. This leads to a decrease in the number of cryptocurrency users (node 7).

2.2 *Development of Subsystem Models with Slow-Measured and Fast-Measured CM Coordinates with Multirate Sampling*

The development of CM models with multirate sampling of coordinates is based on CM subsystem models (6) and (7) with unirate sampling. Model (6) is the basic one for developing the model of CM subsystem with fast-measured coordinates. If we assume that the vector of coordinates $\overline{Y}_f$ of CM nodes is measured at discrete time moments with period T_0, and the vector $\overline{Y}_s$ with period $h = mT_0$, then model (6) with multirate sampling can be written in the following form:

$$\Delta\overline{Y}_f[rh + (l + 1)T_0] = A_{11}\Delta\overline{Y}_f(rh + lT_0) + A_{12}\Delta\overline{Y}_s(rh) \tag{11}$$

where $l = 0, 1, \ldots, (m - 1)$, $\Delta\overline{Y}_s(rh)$ is non-zero only when $l = 0$. According to the Theorem 2 in [13], the components of the vector $\Delta\overline{Y}_s(rh)$ are calculated based on the following equation:

$$\Delta y_{s_l}(rh) = \sum_{\mu=1}^{m} \Delta y_{s_l}[(r - 1)h + \mu T_0]$$
$$= y_{s_l}[(r - 1)h + mT_0] - y_{s_l}[(r - 1)h], i = 1, \ldots, p$$

Under the influence of information disturbances (node 13) the model (11) will take the form:

$$\Delta\overline{Y}_f[rh + (l + 1)T_0] = A_{11}\Delta\overline{Y}_f(rh + lT_0) + A_{12}\Delta\overline{Y}_s(rh) + \overline{\Psi}_f\Delta\overline{\xi}_f(rh + lT_0), l = 0, 1, \ldots, (m - 1) \tag{12}$$

where $\Delta\overline{\xi}_f$—increments (differences) of the coordinates of the information disturbances that affect the vector $\Delta\overline{Y}_f$.

The initial model (7) for the CM subsystem with slow-measured coordinates can be written in an intermediate form with unirate sampling:

$$\Delta\overline{Y}_s[rh + (i + 1)T_0] = A_{21}\Delta\overline{Y}_f(rh + iT_0) + A_{22}\Delta\overline{Y}_s(rh + iT_0) + \overline{\Psi}_s\Delta\overline{\xi}_f(rh + iT_0), i = 0, 1, \ldots, (m - 1) \tag{13}$$

Consider the iterative procedure for model (13) to represent a vector $\overline{Y}_s$ in discrete form with a large sampling period $h = mT_0$:

1. For $i = 0$

$$\Delta\overline{Y}_s(rh + T_0) = A_{21}\Delta\overline{Y}_f(rh) + A_{22}\Delta\overline{Y}_s(rh) + \overline{\Psi}_s\Delta\overline{\xi}_f(rh)$$

2. For $i = 1$, taking into account the previous step, we get:

$$\Delta\overline{Y}_s(rh + 2T_0) = \sum_{i=0}^{1} A_{22}^{i}A_{21}\Delta\overline{Y}_f[rh + (1 - i)T_0]$$
$$+ A_{22}^{2}\Delta\overline{Y}_s(rh) + \sum_{j=0}^{1} A_{22}^{j}\overline{\Psi}_s\Delta\overline{\xi}_f[rh + (1 - j)T_0]$$

3. For $i = 2$, taking into account the previous step, we get:

$$\Delta\overline{Y}_s(rh+3T_0)=\sum_{i=0}^{2}A_{22}^{i}A_{21}\Delta\overline{Y}_f[rh+(2-i)T_0]+A_{22}^{3}\Delta\overline{Y}_s(rh)$$
$$+\sum_{j=0}^{2}A_{22}^{j}\overline{\Psi}_s\Delta\overline{\xi}_f[rh+(2-j)T_0]$$

Eventually for $i=(m-1)$, we obtain a CM subsystem model with slow-measured node coordinates with multirate sampling:

$$\Delta\overline{Y}_s[(r+1)h]=\sum_{i=0}^{m-1}A_{22}^{i}A_{21}\Delta\overline{Y}_f[rh+(m-1-i)T_0]+A_{22}^{m}\Delta\overline{Y}_s(rh)$$
$$+\sum_{j=0}^{m-1}A_{22}^{j}\overline{\Psi}_s\Delta\overline{\xi}_f[rh+(m-1-j)T_0] \quad (14)$$

where $\Delta\overline{\xi}_f$—external disturbance (node 13), $\sum_{i=0}^{m-1}A_{22}^{i}A_{21}\Delta\overline{Y}_f[rh+(m-1-i)T_0]$—internal disturbance for a subsystem with slow-measured coordinates.

It can be shown that for big sampling period $\Delta y_{s_l}(rh)=\sum_{\mu=1}^{m}\Delta y_{s_l}[(r-1)h+\mu T_0]$ the formula (14) still holds true, but fast components should be calculated according to the Theorem 1 in [13], i.e.,

$$\Delta\overline{Y}_f[rh+(m-1-i)T_0]=\overline{Y}_f[rh+(m-1-i)T_0]-\overline{Y}_f[(r-1)h+(m-1-i)T_0]$$

$$\Delta\overline{\xi}_f[rh+(m-1-i)T_0]=\overline{\xi}_f[rh+(m-1-i)T_0]-\overline{\xi}_f[(r-1)h+(m-1-i)T_0]$$

2.3 *Suppression of External and Internal Constrained Disturbances in CM Subsystems with Fast-Measured Coordinates with Multirate Sampling*

Suppression of external and internal constrained disturbances in a CM subsystem with fast-measured coordinates. In the process of application of cryptocurrency in the financial market the weight coefficients of the adjacency matrix A_{11} in the model (12) change relative to their basic values previously estimated by experts based on cause-effect relationships or determined based on identification methods [17]. These changes in the dynamic CM parameters can be represented by an increment $\Delta A_{11}(rh+lT_0)=\hat{A}_{11}-A_{11\text{var}}(rh+lT_0)$ and taken into account in the model (12) as follows:

$$\Delta\overline{Y}_f[rh+(l+1)T_0]=A_{11}\Delta\overline{Y}_f(rh+lT_0)$$

$$+ \Delta A_{11}(rh + lT_0)\Delta \overline{Y}_f(rh + lT_0)$$
$$+ A_{12}\Delta \overline{Y}_s(rh) + \overline{\Psi}_f \Delta \overline{\xi}_f(rh + lT_0), l = 0, 1, \ldots, (m-1) \tag{15}$$

To suppress external disturbances, $\Delta\overline{\xi}_f$ and internal disturbances: $\Delta\overline{Y}_s(rh)$, $\Delta A_{11}(rh + lT_0)\Delta\overline{Y}_f(rh + lT_0)$, let us perform a modification of the method of invariant ellipsoids described in [18] to suppress external disturbances.

Let us represent the internal disturbances in the following form:

$$\Delta A_{11}(rh + lT_0)\Delta \overline{Y}_f(rh + lT_0) = \Delta \overline{W}_f(rh + lT_0)$$

Then the norm l_∞ for all constrained disturbances in the model (15) will be written as

$$\left\| \begin{bmatrix} \Delta\overline{W}_f(rh + lT_0) \\ \Delta\overline{Y}_s(rh) \\ \Delta\overline{\xi}_f(rh + lT_0) \end{bmatrix} \right\|_\infty = \sup_{\substack{r \ge 0 \\ l \ge 0}} \left\{ \left[\Delta\overline{W}_f^T(rh + lT_0)\ \Delta\overline{Y}_s^T(rh)\ \Delta\overline{\xi}_f(rh + lT_0) \right] \right.$$
$$\left. \times \begin{bmatrix} \Delta\overline{W}_f(rh + lT_0) \\ \Delta\overline{Y}_s(rh) \\ \Delta\overline{\xi}_f(rh + lT_0) \end{bmatrix} \right\}^{\frac{1}{2}} \le 1 \tag{16}$$

As a characteristic of the influence of disturbances of the type (16) on the trajectory of subsystem (15) we apply an invariant ellipsoid on the coordinates $\Delta\overline{Y}_f$:

$$\varepsilon_{\Delta\overline{Y}_f} = \left\{ \Delta\overline{Y}_f(rh + lT_0) \in R^p;\ \Delta\overline{Y}_f^T(rh + lT_0)P_f^{-1}\Delta\overline{Y}_f(rh + lT_0) < 1 \right\},$$
$$P_f > 0; l = 0, 1, \ldots, (m-1) \tag{17}$$

If from condition $\Delta\overline{Y}_f(0) \in \varepsilon_{\Delta\overline{Y}_f}$ it follows that $\Delta\overline{Y}_f(rh + lT_0) \in \varepsilon_{\Delta\overline{Y}_f}$ for all discrete moments of time $r = 1, 2, 3, \ldots; l = 0, 1, \ldots, (m-1)$, then the matrix P_f is called the matrix of the ellipsoid $\varepsilon_{\Delta\overline{Y}_f}$.

Let us prove the condition of invariance of the ellipsoid (17) under disturbances (16). To do this, we introduce a quadratic Lyapunov function: $V_f\{\Delta\overline{Y}_f(rh + lT_0)\} = \Delta\overline{Y}_f^T(rh + lT_0)Q_f\Delta\overline{Y}_f(rh + lT_0)$ where $Q_f > 0$, that was built on the solutions of the system (15).

To keep the trajectories of the system (15) within the boundaries of the ellipsoid $\varepsilon_{\Delta\overline{Y}_f}$ (17) it is necessary to fulfill: $V_f\{\Delta\overline{Y}_f(rh + (l+1)T_0)\} \le 1$ and $V_f\{\Delta\overline{Y}_f(rh + lT_0)\} \le 1$. So, based on (15) we obtain:

$$\Delta\overline{Y}_f^T(rh + (l+1)T_0)Q_f\Delta\overline{Y}_f(rh + (l+1)T_0)$$

$$= \left\{ \Delta\overline{Y}_s^T(rh + lT_0)A_{11}^T + \left[\Delta\overline{W}_f^T(rh + lT_0)\ \Delta\overline{Y}_s^T(rh)\ \Delta\overline{\xi}_f(rh + lT_0) \right] \begin{bmatrix} I_f \\ A_{12}^T \\ \overline{\Psi}_f^T \end{bmatrix} \right\} Q_f$$

$$\times \left\{ A_{11}\Delta\overline{Y}_f(rh + lT_0) + \left[I_f\ A_{12}\ \overline{\Psi}_f \right] \begin{bmatrix} \Delta\overline{W}_f(rh + lT_0) \\ \Delta\overline{Y}_s(rh) \\ \Delta\overline{\xi}_f(rh + lT_0) \end{bmatrix} \right\} \le 1$$

After multiplication we get:

$$\Delta\overline{Y}_f^T(rh + (l+1)T_0)Q_f\Delta\overline{Y}_f(rh + (l+1)T_0)$$
$$= \left\{ \Delta\overline{Y}_s^T(rh + lT_0) \left[\Delta\overline{W}_f^T(rh + lT_0)\ \Delta\overline{Y}_s^T(rh)\ \Delta\overline{\xi}_f(rh + lT_0) \right] \right\}$$
$$\bullet \begin{bmatrix} A_{11}^T Q_f A_{11} & A_{11}^T Q_f \left[I_f\ A_{12}\ \overline{\Psi}_f \right] \\ \begin{bmatrix} I_f \\ A_{12}^T \\ \overline{\Psi}_f^T \end{bmatrix} Q_f A_{11} & \begin{bmatrix} I_f \\ A_{12}^T \\ \overline{\Psi}_f^T \end{bmatrix} Q_f \left[I_f\ A_{12}\ \overline{\Psi}_f \right] \end{bmatrix} \begin{bmatrix} \Delta\overline{Y}_f(rh + lT_0) \\ \begin{bmatrix} \Delta\overline{W}_f(rh + lT_0) \\ \Delta\overline{Y}_s(rh) \\ \Delta\overline{\xi}_f(rh + lT_0) \end{bmatrix} \end{bmatrix}$$

Let us apply the S-procedure [19, 20] where $\overline{S}_f = \begin{bmatrix} \Delta\overline{Y}_f(rh + lT_0) \\ \begin{bmatrix} \Delta\overline{W}_f(rh + lT_0) \\ \Delta\overline{Y}_s(rh) \\ \Delta\overline{\xi}_f(rh + lT_0) \end{bmatrix} \end{bmatrix}$, $\dim \Delta\overline{Y}_f = p$; $\dim \Delta\overline{W}_f = p$; $\dim \Delta\overline{Y}_s = n - p$; $\dim \Delta\overline{\xi}_f = 1$. Then $\dim \overline{S}_f = n + p + 1$.

Let us introduce the following quadratic formulas: $f_0(\overline{S}_f) = \overline{S}_f^T M_{f_0} \overline{S}_f = \left\{ \Delta\overline{Y}_s^T(rh + lT_0) \left[\Delta\overline{W}_f^T(rh + lT_0)\ \Delta\overline{Y}_s^T(rh)\ \Delta\overline{\xi}_f(rh + lT_0) \right] \right\}$

$$\bullet \begin{bmatrix} A_{11}^T Q_f A_{11} & A_{11}^T Q_f \left[I_f\ A_{12}\ \overline{\Psi}_f \right] \\ \begin{bmatrix} I_f \\ A_{12}^T \\ \overline{\Psi}_f^T \end{bmatrix} Q_f A_{11} & \begin{bmatrix} I_f \\ A_{12}^T \\ \overline{\Psi}_f^T \end{bmatrix} Q_f \left[I_f\ A_{12}\ \overline{\Psi}_f \right] \end{bmatrix} \begin{bmatrix} \Delta\overline{Y}_f(rh + lT_0) \\ \begin{bmatrix} \Delta\overline{W}_f(rh + lT_0) \\ \Delta\overline{Y}_s(rh) \\ \Delta\overline{\xi}_f(rh + lT_0) \end{bmatrix} \end{bmatrix} \le 1$$

$$f_1(\overline{S}_f) = \overline{S}_f^T M_{f_1} \overline{S}_f = \left\{ \Delta\overline{Y}_s^T(rh + lT_0) \left[\Delta\overline{W}_f^T(rh + lT_0)\ \Delta\overline{Y}_s^T(rh)\ \Delta\overline{\xi}_f(rh + lT_0) \right] \right\}$$
$$\bullet \begin{bmatrix} Q_f & \overline{O}_1 \\ 0 & \overline{O}_1 \end{bmatrix} \begin{bmatrix} \Delta\overline{Y}_f(rh + lT_0) \\ \begin{bmatrix} \Delta\overline{W}_f(rh + lT_0) \\ \Delta\overline{Y}_s(rh) \\ \Delta\overline{\xi}_f(rh + lT_0) \end{bmatrix} \end{bmatrix} \le 1$$

where $\overline{O}_1$—zero-dimensional matrix of dimension $(n + 1) \times (n + 1)$.

$$f_2(\overline{S}_f) = \overline{S}_f^T M_{f_2} \overline{S}_f$$
$$= \left\{ \Delta\overline{Y}_s^T(rh + lT_0) \left[\Delta\overline{W}_f^T(rh + lT_0)\ \Delta\overline{Y}_s^T(rh)\ \Delta\overline{\xi}_f(rh + lT_0) \right] \right\}$$

$$\bullet \begin{bmatrix} 0 & \overline{O}_1 \\ 0 & I_1 \end{bmatrix} \left[\begin{bmatrix} \Delta\overline{Y}_f(rh + lT_0) \\ \Delta\overline{W}_f(rh + lT_0) \\ \Delta\overline{Y}_s(rh) \\ \Delta\overline{\xi}_f(rh + lT_0) \end{bmatrix} \right] \le 1$$

where I_1—unit matrix $(n + 1) \times (n + 1)$.

According to the statement S-procedure [19, 20], the inequality must be satisfied $M_{f_0} \le \sum_{i=1}^{2} \tau_{f_i} M_{f_i}$. In another form:

$$\begin{bmatrix} A_{11}^T Q_f A_{11} & A_{11}^T Q_f [I_f \ A_{12} \ \overline{\Psi}_f] \\ \begin{bmatrix} I_f \\ A_{12}^T \\ \overline{\Psi}_f^T \end{bmatrix} Q_f A_{11} & \begin{bmatrix} I_f \\ A_{12}^T \\ \overline{\Psi}_f^T \end{bmatrix} Q_f [I_f \ A_{12} \ \overline{\Psi}_f] \end{bmatrix} \le \tau_{f_1} \begin{bmatrix} Q_f & \overline{O}_1 \\ 0 & \overline{O}_1 \end{bmatrix} + \tau_{f_2} \begin{bmatrix} 0 & \overline{O}_1 \\ 0 & I_1 \end{bmatrix}$$

Or in the following form:

$$\begin{bmatrix} A_{11}^T Q_f A_{11} - \tau_{f_1} Q_f & A_{11}^T Q_f [I_f \ A_{12} \ \overline{\Psi}_f] \\ \begin{bmatrix} I_f \\ A_{12}^T \\ \overline{\Psi}_f^T \end{bmatrix} Q_f A_{11} & \begin{bmatrix} I_f \\ A_{12}^T \\ \overline{\Psi}_f^T \end{bmatrix} Q_f [I_f \ A_{12} \ \overline{\Psi}_f] - \tau_{f_2} I_1 \end{bmatrix} \le 0$$

Using Schur's formula, this inequality can be reduced to the form:

$$A_{11}^T Q_f A_{11} - \tau_{f_1} Q_f \le A_{11}^T Q_f [I_f \ A_{12} \ \overline{\Psi}_f] \left[\begin{bmatrix} I_f \\ A_{12}^T \\ \overline{\Psi}_f^T \end{bmatrix} Q_f [I_f \ A_{12} \ \overline{\Psi}_f] - \tau_{f_2} I_1 \right]^{-1} \begin{bmatrix} I_f \\ A_{12}^T \\ \overline{\Psi}_f^T \end{bmatrix} Q_f A_{11}$$

This inequality can be reduced to the following form after performing elementary transformations:

$$\tau_{f_1} Q_f \ge A_{11}^T \left[Q_f - Q_f [I_f \ A_{12} \ \overline{\Psi}_f] \left[\begin{bmatrix} I_f \\ A_{12}^T \\ \overline{\Psi}_f^T \end{bmatrix} Q_f [I_f \ A_{12} \ \overline{\Psi}_f] - \tau_{f_2} I_1 \right]^{-1} \begin{bmatrix} I_f \\ A_{12}^T \\ \overline{\Psi}_f^T \end{bmatrix} Q_f \right] A_{11}$$

When $\tau_{f_2} = 1 - \tau_{f_1}$ we get:

$$\tau_{f_1} Q_f \ge A_{11}^T [Q_f + Q_f [I_f \ A_{12} \ \overline{\Psi}_f]$$

$$\left[(1-\tau_{f_1})I_1 - \begin{bmatrix} I_f \\ A_{12}^T \\ \overline{\Psi}_f^T \end{bmatrix} Q_f \left[I_f \; A_{12} \; \overline{\Psi}_f \right]\right]^{-1} \left[\begin{bmatrix} I_f \\ A_{12}^T \\ \overline{\Psi}_f^T \end{bmatrix} Q_f\right] A_{11} \tag{18}$$

According to the lemma of matrix inversion [21] we have:

$$Q_f + Q_f\left[I_f \; A_{12} \; \overline{\Psi}_f \right]\left[(1-\tau_{f_1})I_1 - \begin{bmatrix} I_f \\ A_{12}^T \\ \overline{\Psi}_f^T \end{bmatrix} Q_f \left[I_f \; A_{12} \; \overline{\Psi}_f \right]\right]^{-1}$$

$$\begin{bmatrix} I_f \\ A_{12}^T \\ \overline{\Psi}_f^T \end{bmatrix} Q_f = \left[Q_f^{-1} - (1-\tau_{f_1})^{-1}\left[I_f \; A_{12} \; \overline{\Psi}_f \right]\begin{bmatrix} I_f \\ A_{12}^T \\ \overline{\Psi}_f^T \end{bmatrix}\right]$$

Then expression (18) can be reduced to the form:

$$\tau_{f_1} Q_f \ge A_{11}^T\left[Q_f^{-1} - (1-\tau_{f_1})^{-1}\left[I_f \; A_{12} \; \overline{\Psi}_f \right]\begin{bmatrix} I_f \\ A_{12}^T \\ \overline{\Psi}_f^T \end{bmatrix}\right]^{-1} A_{11}$$

Let us perform the elementary transformation, where $P_f = Q_f^{-1}$:

$$\tau_{f_1} P_f^{-1} \ge \left[A_{11}^T\left[P_f - (1-\tau_{f_1})^{-1}\left[I_f \; A_{12} \; \overline{\Psi}_f \right]\begin{bmatrix} I_f \\ A_{12}^T \\ \overline{\Psi}_f^T \end{bmatrix}\right]\left(A_{11}^T\right)^{-1}\right]^{-1}$$

After inverting the left and right parts we get:

$$\frac{P_f}{\tau_{f_1}} \ge A_{11}^{-1}\left[P_f - (1-\tau_{f_1})^{-1}\left[I_f \; A_{12} \; \overline{\Psi}_f \right]\begin{bmatrix} I_f \\ A_{12}^T \\ \overline{\Psi}_f^T \end{bmatrix}\right]\left(A_{11}^T\right)^{-1}$$

Multiply from the left by A_{11}, and then from the right by A_{11}^T and replace with $\tau_{f_1} = \alpha_f$. Then the linear matrix inequality (LMI) will take the final form:

$$\frac{1}{\alpha_f} A_{11} P_f A_{11}^T - P_f + \frac{I_f + A_{12}A_{12}^T + \overline{\Psi}_f \overline{\Psi}_f^T}{\left(1-\alpha_f\right)} \le 0 \tag{19}$$

Let us consider an algorithm for the design of a state controller for a CM subsystem with fast-measured node coordinates:

$$\Delta\overline{U}_f(rh + lT_0) = -K_{P_f}\Delta\overline{Y}_f(rh + lT_0) \tag{20}$$

If $\Delta A_{11}(rh+lT_0)\Delta\overline{Y}_f(rh+lT_0)=\Delta\overline{W}_f(rh+lT_0)$ then the equation of state for the controlled impulse process (15) takes the form:

$$\Delta\overline{Y}_f[rh+(l+1)T_0]=A_{11}\Delta\overline{Y}_f(rh+lT_0)+B_f\Delta\overline{U}_f(rh+lT_0) + \begin{bmatrix} I_f & A_{12} & \overline{\Psi}_f \end{bmatrix}\begin{bmatrix} \Delta\overline{W}_f(rh+lT_0) \\ \Delta\overline{Y}_s(rh) \\ \Delta\overline{\xi}_f(rh+lT_0) \end{bmatrix} \tag{21}$$

Then the control of the closed-loop subsystem of equations based on (20), (21) we write in the form:

$$\Delta\overline{Y}_f[rh+(l+1)T_0]=\left(A_{11}-B_fK_{P_f}\right)\Delta\overline{Y}_f(rh+lT_0) + \begin{bmatrix} I_f & A_{12} & \overline{\Psi}_f \end{bmatrix}\begin{bmatrix} \Delta\overline{W}_f(rh+lT_0) \\ \Delta\overline{Y}_s(rh) \\ \Delta\overline{\xi}_f(rh+lT_0) \end{bmatrix} \tag{22}$$

It is assumed that the pair A_{11}, B_f in the model (21) is controllable. Then the LMI (19) for the closed-loop control subsystem (22) takes the form:

$$\frac{1}{\alpha_f}\left(A_{11}-B_fK_{P_f}\right)P_f\left(A_{11}-B_fK_{P_f}\right)^T-P_f+\frac{I_f+A_{12}A_{12}^T+\overline{\Psi}_f\overline{\Psi}_f^T}{\left(1-\alpha_f\right)}\le 0 \tag{23}$$

As an optimality criterion for the design of the controller (20) in this chapter we consider the minimization of the following matrix of the ellipsoid (17):

$$tr P_f\left(\alpha_f\right) \to \min, 0 \le \alpha_f < 1 \tag{24}$$

This minimizes the size of the invariant ellipsoid (17) with the largest suppression of disturbances $\begin{bmatrix} \Delta\overline{W}_f^T(rh+lT_0) & \Delta\overline{Y}_s^T(rh) & \Delta\overline{\xi}_f(rh+lT_0) \end{bmatrix}^T$, which are limited only by the maximum range (16). After multiplying the factors in the inequality (23), we obtain:

$$\frac{1}{\alpha_f}\left(A_{11}P_fA_{11}^T-B_fK_{P_f}P_fA_{11}^T-A_{11}P_fB_f^TK_{P_f}^T+B_fK_{P_f}P_fB_f^TK_{P_f}^T\right)-P_f + \frac{I_f+A_{12}A_{12}^T+\overline{\Psi}_f\overline{\Psi}_f^T}{\left(1-\alpha_f\right)}\le 0 \tag{25}$$

This inequality is nonlinear with respect to P_f and K_{P_f}, which needs to be optimized. In [18] a linearization by replacement $L_f=K_{P_f}P_f$ with the introduction of an additional constraint is proposed:

$$\begin{bmatrix} R_f & L_f \\ L_f^T & P_f \end{bmatrix} \geq 0 \tag{26}$$

Here $R_f = R_f^T$. Thus (26) is equivalent $R_f \geq L_f P_f^{-1} L_f^T = K_{P_f f} P_f^{-1} K_{P_f}^T$, according to the Schur's formula for $P_f > 0$. Then in order to satisfy inequality (25), it is sufficient that the following inequality is satisfied:

$$\begin{aligned} &\frac{1}{\alpha_f}\left(A_{11}P_f A_{11}^T - B_f L_f A_{11}^T - A_{11}L_f^T B_f^T + B_f L_f B_f^T\right) - P_f \\ &+ \frac{I_f + A_{12}A_{12}^T + \overline{\Psi}_f \overline{\Psi}_f^T}{\left(1-\alpha_f\right)} \leq 0 \end{aligned} \tag{27}$$

Minimization of criterion (24) under constraints (26), (27) is performed on variables P_f, L_f, R_f by means of solving the semi-definite programming problem by using SeDuMi Tolbox based on Matlab tools. Then the matrix $\hat{K}_{P_f}$ of the optimal state controller for the CM subsystem with fast-measured node coordinates (20) is defined as:

$$\hat{K}_{P_f} = \hat{L}_f \hat{P}_f^{-1} \tag{28}$$

With the estimated values of α_f, $\hat{L}_f$, $\hat{P}_f$, $\hat{R}_f$, we provide minimization of criterion (24) under constraints (26), (27).

Suppression of external and internal constrained disturbances in a CM subsystem with slow-measured coordinates. The adjacency matrix weights A_{22}^m in the CM subsystem with slow node coordinates (14) change over time relative to the known basic values during the application of cryptocurrency. These changes can be conventionally represented in the following increments: $\Delta A_{22}^m(rh) = \hat{A}_{22}^m - A_{22}^m(rh)$, which are unknowns. Taking into account these changes we propose to represent the model (14) in the form:

$$\begin{aligned} \Delta\overline{Y}_s[(r+1)h] = {} & A_{22}^m \Delta\overline{Y}_s(rh) + \Delta A_{22}^m(rh)\Delta\overline{Y}_s(rh) \\ &+ \sum_{i=0}^{m-1} A_{22}^i A_{21} \Delta\overline{Y}_f[rh + (m-1-i)T_0] \\ &+ \sum_{j=0}^{m-1} A_{22}^j \overline{\Psi}_s \Delta\overline{\xi}_f[rh + (m-1-j)T_0] \end{aligned} \tag{29}$$

Let us represent $\Delta A_{22}^m(rh)\Delta\overline{Y}_s(rh)$ as $\Delta\overline{W}_s(rh)$—internal disturbances of the impulse process. Then for all disturbances in the model (29) we can write the constrained norm l_∞:

$$\left\|\begin{bmatrix} \Delta\overline{W}_s(rh) \\ \sum_{i=0}^{m-1} A_{22}^i A_{21}\Delta\overline{Y}_f[rh+(m-1-i)T_0] \\ \sum_{j=0}^{m-1} A_{22}^j \overline{\Psi}_s \Delta\overline{\xi}_f[rh+(m-1-j)T_0] \end{bmatrix}\right\|_\infty$$

$$= \sup_{r\geq 0}\left\{\left[\Delta\overline{W}_s^T(rh)\ \left(\sum_{i=0}^{m-1} A_{22}^i A_{21}\Delta\overline{Y}_f[rh+(m-1-i)T_0]\right)^T\right.\right.$$

$$\left.\left(\sum_{j=0}^{m-1} A_{22}^j \overline{\Psi}_s \Delta\overline{\xi}_f[rh+(m-1-j)T_0]\right)^T\right]$$

$$\left.\times\begin{bmatrix} \Delta\overline{W}_s(rh) \\ \sum_{i=0}^{m-1} A_{22}^i A_{21}\Delta\overline{Y}_f[rh+(m-1-i)T_0] \\ \sum_{j=0}^{m-1} A_{22}^j \overline{\Psi}_s \Delta\overline{\xi}_f[rh+(m-1-j)T_0] \end{bmatrix}\right\}^{1/2} \leq 11 \tag{30}$$

To suppress external $\sum_{j=0}^{m-1} A_{22}^j \overline{\Psi}_s \Delta\overline{\xi}_f[rh+(m-1-j)T_0]$ and internal disturbances $\Delta\overline{W}_s(rh)$, $\sum_{i=0}^{m-1} A_{22}^i A_{21}\Delta\overline{Y}_f[rh+(m-1-i)T_0]$, we use the method of invariant ellipsoids according to the method described in the previous section of this chapter. To describe the characteristic of the influence of disturbances (30) on the motion trajectory of the subsystem (29), we propose to consider the invariant ellipsoids in the form:

$$\varepsilon_{\Delta\overline{Y}s} = \left\{\Delta\overline{Y}_s(rh) \in R^{n-p};\ \Delta\overline{Y}_s^T(rh)P_s^{-1}\Delta\overline{Y}_s(rh) \leq 1\right\},\ P_s > 0 \tag{31}$$

If from $\Delta\overline{Y}_s(0) \in \varepsilon_{\Delta\overline{Y}_s}$ the condition $\Delta\overline{Y}_s(rh) \in \varepsilon_{\Delta\overline{Y}_s}$ follows for all discrete moments of time $r = 1, 2, 3, \ldots$, the matrix P_s is called the matrix of the ellipsoid $\varepsilon_{\Delta\overline{Y}_s}$.

If we apply the method described in the previous section, the invariance ellipsoid $\varepsilon_{\Delta\overline{Y}_s}$ condition for the disturbance characteristic (30) will be represented as the following linear matrix inequality:

$$\frac{1}{\alpha_s} A_{22}^m P_f \left(A_{22}^m\right)^T - P_f + \frac{I_s + I_s + I_s}{(1-\alpha_s)} \leq 0 \tag{32}$$

In Eq. (32) I_s—identity matrix $(n-p)\times(n-p)$, α_s—scalar coefficient.

The equation of the controlled impulse process of the CM subsystem with slow-measurable coefficients (13) under the influence of a control vector $\Delta\overline{U}_s(rh+iT_0)$ of size m_s with unirate sampling takes the following form:

$$\begin{aligned}\Delta\overline{Y}_s[rh+(i+1)T_0] &= A_{21}\Delta\overline{Y}_f(rh+iT_0) + A_{22}\Delta\overline{Y}_s(rh+iT_0) \\ &\quad + B_s\Delta\overline{U}_s(rh+iT_0) + \overline{\Psi}_s\Delta\overline{\xi}_f(rh+iT_0), \\ &\quad i = 0, 1, \ldots, (m-1)\end{aligned} \tag{33}$$

If we apply the iterative procedure (13–14) to this equation, taking into account that $\Delta\overline{U}_s(rh + iT_0) = \Delta\overline{U}_s(rh)$, the model (33) with a multirate sampling (29) for internal disturbances $\Delta A_{22}^m(rh)\Delta\overline{Y}_s(rh) = \Delta\overline{W}_s(rh)$ will be written as follows:

$$\Delta\overline{Y}_s[(r+1)h] = A_{22}^m \Delta\overline{Y}_s(rh) + \sum_{i=0}^{m-1} A_{22}^i B_s \Delta\overline{U}_s(rh)$$
$$+ \begin{pmatrix} I_s & I_s & I_s \end{pmatrix} \begin{bmatrix} \Delta\overline{W}_s(rh) \\ \sum_{i=0}^{m-1} A_{22}^i A_{21} \Delta\overline{Y}_f[rh + (m-1-i)T_0] \\ \sum_{j=0}^{m-1} A_{22}^j \Psi_s \Delta\overline{\xi}_f[rh + (m-1-j)T_0] \end{bmatrix} \quad (34)$$

where matrix B_s with dimension $(n-p) \times m_s$ is filled by the operator-programmer and consists of ones and zeros.

The control vector $\Delta\overline{U}_s(rh)$ is formed by a discrete controller for slow-measured CM coordinates:

$$\Delta\overline{U}_s(rh) = -K_{P_s} \Delta\overline{Y}_s(rh) \quad (35)$$

Then, based on (34), (35), the equation of the closed-loop CM subsystem will have the form:

$$\Delta\overline{Y}_s[(r+1)h] = \left(A_{22}^m - \sum_{i=0}^{m-1} A_{22}^i B_s K_{P_s} \right) \Delta\overline{Y}_s(rh)$$
$$+ \begin{pmatrix} I_s & I_s & I_s \end{pmatrix} \begin{bmatrix} \Delta\overline{W}_s(rh) \\ \sum_{i=0}^{m-1} A_{22}^i A_{21} \Delta\overline{Y}_f[rh + (m-1-i)T_0] \\ \sum_{j=0}^{m-1} A_{22}^j \Psi_s \Delta\overline{\xi}_f[rh + (m-1-j)T_0] \end{bmatrix} \quad (36)$$

It is assumed that the pair A_{22}^m, $\sum_{i=0}^{m-1} A_{22}^i B_s$ in the model (34) is controllable. Then the LMI (32) for the closed-loop control subsystem (36) takes the form:

$$\frac{1}{\alpha_s} \left[A_{22}^m - \sum_{i=0}^{m-1} A_{22}^i B_s K_{P_s} \right] P_s \left[A_{22}^m - \sum_{i=0}^{m-1} A_{22}^i B_s K_{P_s} \right]^T - P_s + \frac{I_s + I_s + I_s}{(1-\alpha_s)} \leq 0 \quad (37)$$

The minimization of the trace of the ellipsoid matrix (31) is considered as the optimality criterion for the design of the controller (35):

$$\mathrm{tr} P_s(\alpha_s) \to \min, 0 \leq \alpha_s < 1 \quad (38)$$

which ensures that the size of the invariant ellipsoid (31) is minimized with the largest disturbance suppression:

$$\left[\Delta\overline{W}_s^T(rh) \left(\sum_{i=0}^{m-1} A_{22}^i A_{21} \Delta\overline{Y}_f[rh + (m-1-i)T_0]\right)^T\right.$$

$$\left.\left(\sum_{j=0}^{m-1} A_{22}^j \overline{\Psi}_s \Delta\overline{\xi}_f[rh + (m-1-j)T_0]\right)^T\right]^T$$

These disturbances are restricted to the maximum range only. After the multiplication of the factors in the inequality (37) we obtain:

$$\frac{1}{\alpha_s}\Big(\sum_{i=0}^{m-1} A_{22}^i B_s K_{P_s} P_s \left(\sum_{i=0}^{m-1} A_{22}^i B_s K_{P_s}\right)^T + A_{22}^m P_s (A_{22}^m)^T$$
$$-\sum_{i=0}^{m-1} A_{22}^i B_s K_{P_s} P_s (A_{22}^m)^T - A_{22}^m P \left(\sum_{i=0}^{m-1} A_{22}^i B_s K_{P_s}\right)^T\Big) - P_s + \frac{I_s + I_s + I_s}{(1-\alpha_s)} \le 0 \tag{39}$$

This inequality is nonlinear with respect to K_{P_s} and P_s. As in the case of (25), we perform a linearizing substitution $L_s = K_{P_s} P_s$ with the introduction of an additional restriction:

$$\begin{bmatrix} R_s & L_s \\ L_s^T & P_s \end{bmatrix} \ge 0 \tag{40}$$

In Eq. (40) we have $R_s = R_s^T$. Thus (26) is equivalent $R_s \ge L_s P_s^{-1} L_s^T = K_{P_s} P_s^{-1} K_{P_s}^T$ according to the Schur formula for $P_s > 0$. Then, to fulfill inequality (39), it is sufficient that the following inequality is satisfied:

$$\frac{1}{\alpha_s}\Big(\sum_{i=0}^{m-1} A_{22}^i B_s L_s \left(\sum_{i=0}^{m-1} A_{22}^i B_s\right)^T + A_{22}^m P_s (A_{22}^m)^T$$
$$-\sum_{i=0}^{m-1} A_{22}^i B_s L_s (A_{22}^m)^T - A_{22}^m L_s \left(\sum_{i=0}^{m-1} A_{22}^i B_s\right)^T\Big) - P_s + \frac{I_s + I_s + I_s}{(1-\alpha_s)} \le 0 \tag{41}$$

Minimization of criterion (38) under constraints (40), (41) is performed w.r.t. the variables P_f, L_f, R_f similarly with the fast-measured subsystem. Then the matrix $\hat{K}_{P_f}$ of the optimal controller (35) for the subsystem with slow-measured node coordinates is defined as:

$$\hat{K}_{Ps} = \hat{L}_s \hat{P}_s^{-1} \tag{42}$$

with the estimated values of α_s, $\hat{L}_s$, $\hat{P}_s$, $\hat{R}_s$, ensuring minimization of criterion (38) under the constraints (40), (41).

3 Results

3.1 *Experimental Studies of the System of Suppression of Constrained Internal and External Disturbances in a CM Subsystem with Fast-Measured Node Coordinates*

The experimental study uses the model of closed-loop subsystem (22) with multi-rate sampling, where the vector $\Delta\overline{Y}_f$ with fast-measured node coordinates includes: Δy_{f_1}—cryptocurrency rate, Δy_{f_2}—cryptocurrency trading volume, Δy_{f_3}—supply of cryptocurrency, Δy_{f_4}—demand for cryptocurrency, Δy_{f_5}—cryptocurrency speculation volume, Δy_{f_6}—risk of cryptocurrency rate collapse. For these coordinates a small sampling period is chosen T_0, which is defined with respect to the fastest changing coordinate y_{f_1}. The vector $\Delta\overline{Y}_s$ includes increments of slow-measured coordinates, which are internal disturbances in the model (22). This includes: Δy_{s_7}–the number of users of the cryptocurrency, Δy_{s_8}—the volume of investment, Δy_{s_9}—the volume of capitalization, $\Delta y_{s_{10}}$—indirect profit, $\Delta y_{s_{11}}$—the level of trust in the cryptocurrency, $\Delta y_{s_{12}}$—the risk of losing the number of users. For these coordinates the sampling period $h = mT_0$, $m = 5$ is applied. The control vector $\Delta\overline{U}_f$ includes increments of fast measurable coordinates: u_{f_2}—cryptocurrency trading volume, u_{f_3}—cryptocurrency supply, u_{f_5}—cryptocurrency speculation volume, which are generated by the controller (20) and fed to the corresponding CM nodes by varying them by the operator making decisions on trading exchanges.

The model (22) has the information disturbance increment (node 13) as an external disturbance $\Delta\overline{\xi}_f$. According to the cognitive map (Fig. 1), the matrices A_{11}, A_{12}, B_f and the vector $\overline{\Psi}_f$ have the following form:

$$A_{11} = \begin{bmatrix} 0.1 & 0.4 & 0 & 0.1 & 0.4 & 0 \\ 0 & 0 & 0.2 & 0.1 & 0 & -0.6 \\ 0 & 0 & 0 & 0.7 & 0 & 0 \\ -0.2 & -0.8 & -0.5 & 0 & 0.7 & 0 \\ 0 & 0.5 & 0.6 & 0 & 0 & -0.5 \\ 1 & 0 & 0 & 0 & -0.15 & 0 \end{bmatrix}; \; B_f = \begin{bmatrix} 0\,0\,0 \\ 1\,0\,0 \\ 0\,1\,0 \\ 0\,0\,0 \\ 0\,0\,1 \\ 0\,0\,0 \end{bmatrix};$$

$$A_{12} = \begin{bmatrix} 0.4 & 0.3 & 0.15 & 0 & 0.3 & -0.3 \\ 0.3 & 0.5 & 0.1 & 0 & 0.2 & 0 \\ 0.8 & 0.3 & 0 & 0.1 & 0 & -0.5 \\ 0 & 0 & 0 & 0 & 0.7 & 0 \\ 0.1 & 0 & 0 & 0 & 0 & 0 \\ 0 & -0.2 & 0 & 0 & 0 & 0 \end{bmatrix}; \overline{\Psi}_f = \begin{bmatrix} 0 \\ 0 \\ -0.1 \\ -0.2 \\ -0.5 \\ 0 \end{bmatrix}$$

The control matrix B_f is formed depending on the number of control actions in $\Delta\overline{U}_f$, which is generated based on of (20), (28).

When modeling the dynamics of the closed-loop control subsystem with fast-measurable coordinates of CM, the external disturbance $\overline{\xi}_f$ is a step change in the coordinate of node 13 with amplitude 1, acting at the initial moment of time on nodes 3, 4, 5 in the direction of reducing the supply and demand for cryptocurrency and the volume of speculation.

Internal disturbances $\Delta\overline{W}_f(rh + lT_0) = \Delta A_{11}(rh + lT_0)\Delta\overline{Y}_f(rh + lT_0)$, where $\Delta A_{11}(rh + lT_0) = \hat{A}_{11} - A_{11\text{var}}(rh + lT_0) = \hat{A}_{11} - \hat{A}_{11}\gamma(rh + lT_0)$, are results of varying the non-zero coefficients of the matrix A_{11} at each period T_0, where $\gamma(rh + lT_0)$ is a normally distributed random variable (Gaussian white noise). Only the matrix A_{11} is used to calculate the control, while $A_{11\text{var}}(rh + lT_0)$ remains unknown.

Initial levels of all CM nodes increments are taken to be zero for convenience. The graphs of transient processes of fast measured CM nodes and their increments are shown in Fig. 2, where the dotted line indicates transients without control, and the solid line indicates transients with control.

3.2 *Experimental Studies of Suppression of Constrained Internal and External Disturbances in a CM Subsystem with Slowly Measured Node Coordinates*

For the experimental study, the model of the closed-loop system (36) with multirate sampling of coordinates is used, where the vector $\Delta\overline{Y}_s$ with slow-measured coordinates is included. For these coordinates the sampling period $h = 5T_0 (m = 5)$ is chosen. The coordinates y_{f_1}, y_{f_2}, y_{f_3}, y_{f_4}, y_{f_5}, y_{f_6} are internal disturbances in the subsystem, and their increments $\Delta\overline{Y}_f$ are represented with a sampling period T_0.

The control vector $\Delta\overline{U}_s$ in the control law (35) includes the increments Δu_{s_8}, Δu_{s_9}, which are directly fed to the corresponding CM nodes. Just as in subsystem (22), a unit step increment of the external disturbance $\overline{\xi}_f$ is fed to the nodes y_{s_7}, y_{s_8}, $y_{s_{11}}$. The matrices A_{22}, A_{21}, B_s and the vector $\overline{\Psi}_s$ have the following form:

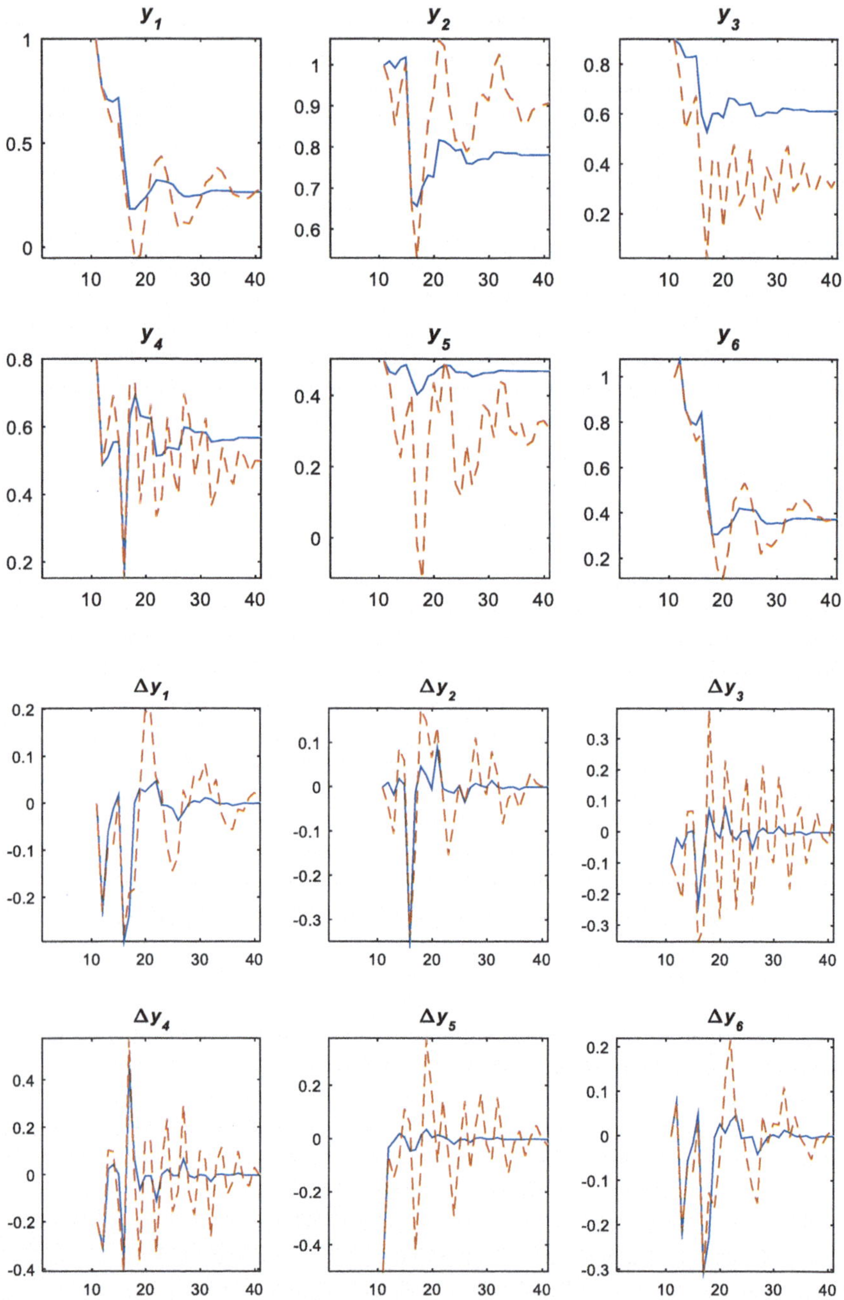

Fig. 2 Fast CM nodes and their increments

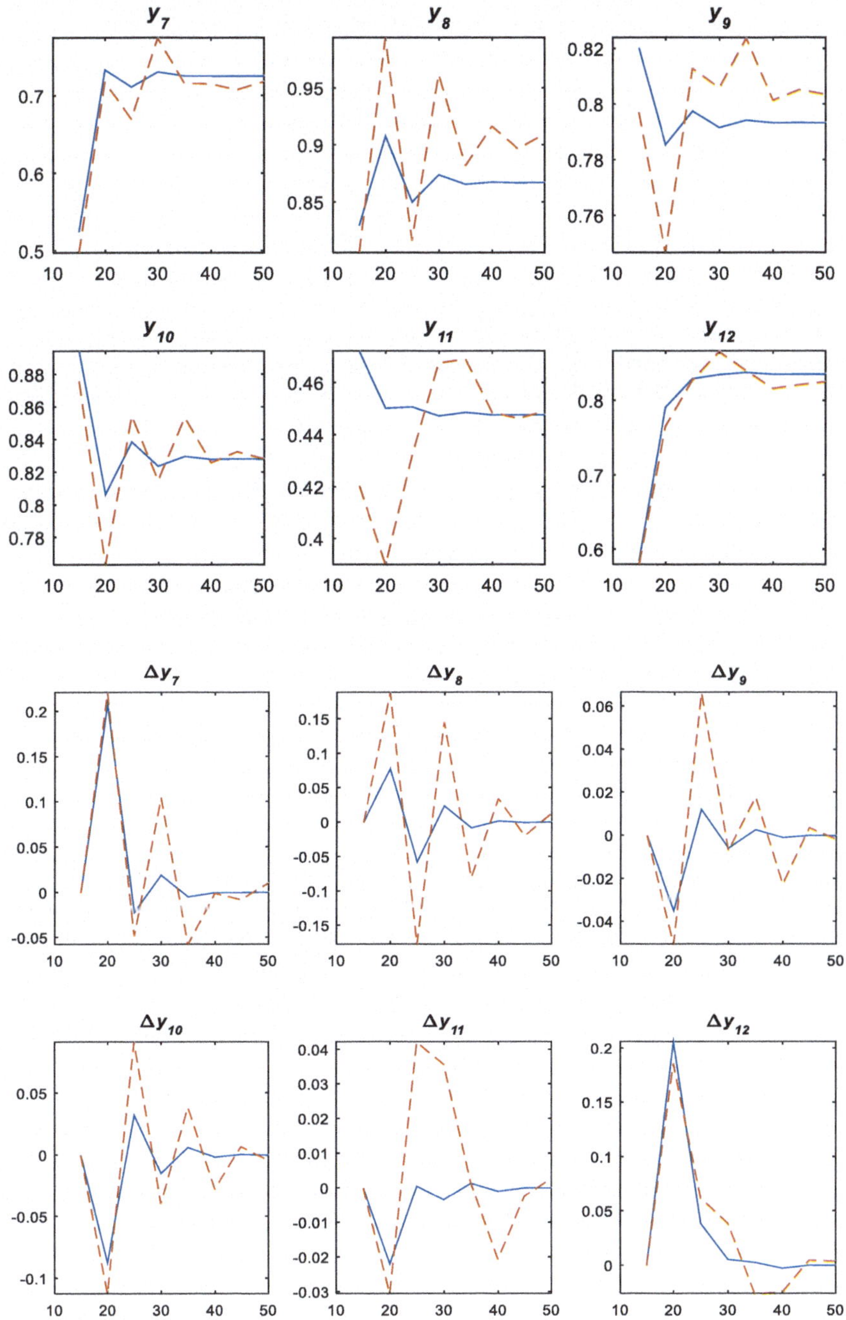

Fig. 3 Slow CM nodes and their increments

$$A_{22} = \begin{bmatrix} 0.1 & 0 & 0 & 0 & 0.5 & 0 \\ 0 & 0.1 & 0 & 0.2 & 0 & -0.2 \\ 0 & 0.2 & 0.1 & 0 & 0.15 & 0 \\ 0 & 0 & 0.3 & 0 & 0 & 0 \\ 0 & 0.2 & 0.1 & 0 & 0 & 0 \\ 1 & 0 & 0 & 0 & -0.2 & 0 \end{bmatrix};$$

$$A_{21} = \begin{bmatrix} 0.2 & 0 & 0 & 0 & 0 & -0.6 \\ 0 & 0.3 & 0 & 0 & 0 & -0.4 \\ 0.1 & 0 & 0 & 0 & 0 & 0 \\ 0.15 & 0 & 0 & 0 & 0 & 0 \\ 0.4 & 0 & 0 & 0 & 0 & -0.3 \\ 0 & 0 & 0 & 0 & 0 & 0 \end{bmatrix}; B_s = \begin{bmatrix} 0 & 0 \\ 1 & 0 \\ 0 & 1 \\ 0 & 0 \\ 0 & 0 \\ 0 & 0 \end{bmatrix}; \overline{\Psi}_s = \begin{bmatrix} -0.2 \\ -0.3 \\ 0 \\ 0 \\ -0.4 \\ 0 \end{bmatrix}$$

Internal disturbances $\Delta\overline{U}_s(rh)$ are formed at each sampling period h according to the formula $\Delta A_{22}^5(rh)\Delta\overline{Y}_s(rh) = \left[A_{22}^5 - A_{22}^5\gamma(rh)\right]\Delta\overline{Y}_s(rh)$, where is a normally distributed random variable (Gaussian white noise).

The external disturbance $\overline{\xi}_f$ is common for subsystems with fast and slow-measured coordinates. The graphs of transients of slow-measured coordinates $\Delta\overline{Y}_s(rh)$ are shown in Fig. 3, where the dotted line indicates transients without control and the solid line indicates transients with control.

4 Conclusion

This chapter develops a cognitive map of the application of cryptocurrency in the financial market. The model of impulse CM process, represented as a system of difference equations with unirate sampling with small period T_0, is decomposed into two interrelated subsystems with fast and slow-measured CM node coordinates, which are represented with multirate sampling. For each subsystem, the vectors of control actions, which can be implemented by varying the corresponding coordinates of the CM nodes in the closed-loop multidimensional system, are selected. Based on the method of invariant ellipsoids for each subsystem, the fast and slow state controllers with multirate sampling for suppression of external and internal disturbances are designed. Experimental studies of closed-loop control systems for the suppression of constrained internal and external disturbances in subsystems of CM were carried out. They showed high efficiency in reducing the risk of collapse of the cryptocurrency rate and the risk of losing the number of users.

References

1. Roberts, F.: Discrete Mathematical Models with Applications to Social, Biological and Environmental Problems. Prentice-Hall, Englewood Cliffs (1976)
2. Tolman, E.C.: Cognitive maps in rats and men. Psychol. Rev. **55**(4), 189–208 (1948)
3. Axelrod, R.: The Structure of Decision: Cognitive Maps of Political Elites. Princeton University Press (1976)
4. Kosko, B.: Fuzzy Cognitive Maps. Int. J. Man Mach. Stud. **24**, 65–75 (1986)
5. Papageorgiou, E., Salmeron, J.: A review of fuzzy cognitive maps research during the last decade. IEEE Trans. Fuzzy Syst. **21**(1), 66–79 (2013)
6. Aguilar, J.: A survey about fuzzy cognitive maps papers. Int. J. Comput. Cogn. **3**, 27–33 (2005)
7. Wang, Y., Zhang, W.: A brief survey on fuzzy cognitive maps research. In: Advanced Intelligent Computing Theories and Applications, ICIC 2015, Lecture Notes in Computer Science, vol. 9227, Springer, Cham (2015)
8. Stylios, C., Groumpos, P.P.: Modeling complex systems using fuzzy cognitive maps. IEEE Trans. Syst. Man Cybernetics-Part A: Syst. Humans **34**(1), 155–162 (2004)
9. Gorelova, G., Pankratova, N.: Innovationa Development of Socio-Economic Systems Based on Foresight Methodology and Cognitive Modeling. Kyiv, Naukova dumka (2015) (in Russian)
10. Gorelova, G., Pankratova, N.: Possibility of simulation modeling to identify the impact of a pandemic on the education. In: Proceedings of Automatics—2020. XXVI International conference on automated control, pp. 65–66 (2020)
11. Romanenko, V., Miliavskiy, Y., Kantsedal, H.: An Adaptive System for Stabilization Unstable Cryptocurrency rate based on a Cognitive Map Impulse Process Model. Problems Control Inf. **2**, 11–23 (2021). (in Russian)
12. Egupov, N.: Methods of classical and modern automatic control theory: textbook in 3 volumes. V2: Synthesis of Controllers and Optimization Theory of Automatic Control Systems. Publishing house of Bauman Moscow State Technical University, Moscow (2000) (in Russian)
13. Zgurowsky, M., Romanenko, V., Milyavskiy, Y.: Principles and methods of impulse processes control in cognitive maps of complex systems. Part 1. Autom. Inf. Sci. **48**, 36–45 (2016)
14. Perepelytsya, V.: Mathematical models and methods of risk assessment of economic, social and agrarian processes. Center for Educational Literature, Kyiv (2013) (in Ukrainian)
15. Bashkirov, O.: Comparative VaR method of assessing the risk of bank assets. Problems and Prospects of Development of the Banking System of Ukraine. UABS NBU, Sumy (2005) (in Ukrainian)
16. Kuznetsova, N., Bidyuk, P.: Theory and Practice of Financial Risk Analysis: Systemic Approach. Lira Publishing House, Kyiv (2020).(in Ukrainian)
17. Gubarev, V., Romanenko, V., Miliavskyi, Y.: Identification in cognitive maps in the impulse processes mode with full information. J. Autom. Inf. Sci. **50**, 1–15 (2018)
18. Nazin, S., Polyak, B., Topunov, M.: Suppression of Restricted External Disturbances by the Method of Invariant Ellipsoids. Autom. Telemech. 106–125 (2007) (in Russian)
19. Polyak, B.: Convexity of guadratic transformations and its use in control and optimization. Optim. Theory Appl. **99**, 553–583 (1998)
20. Gusev, S., Likhtarnikov, A.: Sketch of history of Kalman-Popov-Yakubovich lemma and S-procedure. Autom. Telemech. 77–121 (2006) (in Russian)
21. Hager, W.: Updating the inverse of a matrix. SIAM Rev. **31**, 221–239 (1989)

Chapter 8
Systemic Approach to Risk Estimation Using DSS

Vira Huskova, Petro Bidyuk, Oxana Tymoshchuk, and Oleksandr Meniailenko

Abstract A concept is proposed for solving the problem of adaptive risk estimation that is based on the system analysis methodology and combined use of preliminary data processing techniques, mathematical and statistical modeling, optimal state estimation of the nonlinear nonstationary processes widely met in risk analysis and forecasting. The cyclical adaptation of a model structure and its parameters on the basis of a set of statistical criteria of the process being studied provides a possibility for computing high quality estimates of conditional variance forecasts under condition that data available is informative. To identify and take into consideration possible stochastic, structural and parametric uncertainties it is proposed to use optimal and digital filtering and the methods of intellectual data analysis such as Bayesian networks, adaptive Bayesian networks, various approaches to Kalman filtering, particle filters and other necessary instruments. Possible parametric uncertainties of the model being developed are minimized with application of several alternative parameter estimation techniques such as LS, RLS, ML and Markov chains Monte Carlo sampling. The models constructed are used for short term volatility (conditional variance) forecasting to be further used for market risk estimation. Bayesian networks allow for the formal description of other risk types such as operational, credit and actuarial risks, etc. The system proposed has wide possibilities for risk analysis application and further enhancement of its functional possibilities with new methods and computational procedures directed towards refining risk estimation and forecasting procedures.

Keywords Risk estimation · Nonlinear nonstationary processes · Uncertainties · Filtering · Heteroscedastic processes · Volatility forecasting · Bayesian networks

V. Huskova (✉) · P. Bidyuk · O. Tymoshchuk
National Technical University of Ukraine "Igor Sikorsky Kyiv Polytechnic Institute", Kyiv 03056, Ukraine

O. Meniailenko
Luhansk Taras Shevchenko National Technical University, Luhansk 92703, Ukraine

M. Zgurovsky and N. Pankratova (eds.), *System Analysis & Intelligent Computing*, Studies in Computational Intelligence 1022, https://doi.org/10.1007/978-3-030-94910-5_8

1 Introduction

Modern decision support systems (DSS) create convenient instrumentation for solving many sophisticated problems in many areas of human activities. They find applications in mathematical modeling, forecasting future developments of complex processes, estimation and management of various risks, solving optimization problems, generating decision alternatives etc. [1–4]. Comparing DSS with other types of information systems reveals their following advantages: (1) DSS is designed and implemented using the same system analysis principles that are used by the experts hired for making decisions; (2) DSS should be developed and implemented in such a way that it should become an active member of the decision making process; (3) the system should contain data and knowledge base for keeping numerical procedures necessary for estimating mathematical model structure and computing its parameters; statistical criteria for estimating quality of data, adequacy of models, quality of forecast estimates and possible alternative decisions (alternatives), rules for selection best alternatives using appropriate criteria etc.; (4) active monitoring of computational procedures at all stages of data and knowledge processing by making use of formalized quality criteria; (5) DSS should have modular structure that provides for possibility of its operative modification directed towards enhancement and improvement of the system functionality; (6) availability of optimization procedures for quality estimation of a model structure and parameters as well as for generating optimal development trajectories for the processes under study and respective controls; (7) development of high level interface with intellectual functions that corresponds to requirements of users of different levels and human factor requirements [1].

The DSS designed and implemented with taking into consideration all the features and elements mentioned will create a highly effective instrument for making objective decisions by expert and actually will become a member of the decision making process. This study is directed towards development of systemic methodology directed towards solution of financial risks analysis problems using appropriately designed decision support system. The methodology is based upon model constructing approach for the nonlinear nonstationary processes (NNP) met in the systems being studied. For example, practically all the modern financial processes taking place at stocks, investment companies, banks and many other companies are related to NNP what requires development of new model structures, forecasting techniques and risk estimation procedures [5–8].

The existing methods of risk estimation and forecasting that are based upon analytic procedures, logical rules, and rational expert judging not always lead to desirable result regarding high quality risk estimates what supports the necessity for development of new models, risk estimation and forecasting approaches, and DSS based upon these new methods. An effective modeling and estimation of risks requires application of modern system analysis methodology to data analysis methods, constructing mathematical models for arbitrary NNP, estimation of forecasts on the basis of modern achievements in the area of statistical data analysis and estimation theory. Some possibilities for solving the problems of adaptive forecasting aiming improvement of the forecast estimates are considered in [7–10], in particular adaptive methods of exponential smoothing and filtering statistical data. However, the methods considered in these studies do not suppose application of systemic approach to solving the problems of modeling, forecasting and control, and do not give an answer to the basic question: how to organize the process of data processing in such a way so that to get the best (possible) estimates of risks, and respective forecasts of possible loss, in conditions of influence of various uncertainties, for example of structural, parametric and statistical origin?

The uncertainties mentioned above are provoked by substantial non-stationarity and nonlinearity of the processes under study, missing data values, stochastic external influences and measurement noise, availability of outliers, jumping transitions between regimes etc. Effective (from the point of view of quality of final result) methods of adaptive estimation and forecasting states of dynamic processes using filtering techniques are given in [8, 9, 11]. To perform adaptation of state estimation and forecasting algorithms, estimates of statistical parameters for external disturbances and measurement errors are usually required. The optimal filtering procedures have their own advantages and disadvantages. The advantages allow for taking into consideration in explicit form statistical characteristics of state disturbances and measurement noise (errors), and finding optimal state and forecast estimates, estimation of non-measurable state vector components as well as simultaneous estimation of states and some model parameters. The disadvantages include substantial decreasing of forecast estimates quality with growing number of forecasting steps (growing forecasting horizon), and possible divergence of the state and forecasts estimation procedures due to model inadequacy [11, 12], what requires correct application of linearization procedures in a case of processing nonlinear processes. Nevertheless, optimal state and parameter estimation procedures are often used to fight uncertainties of stochastic types mentioned.

Today necessity exists for development of new methods and procedures for model and forecasts estimation that would guarantee computing of high quality forecast in conditions of uncertainty, short samples, and insufficient information content in statistical data. Such problem is partially solved in the study presented here by the authors thanks to development of information decision support system (IDSS) on the principles of systemic approach based upon the usage of modern methods for preliminary data processing, identification and taking into consideration uncertainties, adaptive estimation of a model structure and parameters, estimation of forecast estimates as well as development of resulting decision alternatives. The approach also requires application of several sets of statistical quality criteria for analyzing quality of intermediate and final computational results [2, 3].

2 Problem Statement

The purpose of the study is as follows: 1—to develop the concept of adaptive modeling and forecasting development of arbitrary nature processes using statistical data; 2—the concept should be based upon system analysis methodology that supposes application of hierarchical analysis of the processes under study, identification and taking into consideration possible uncertainties, adaptation of model structure and its parameters, application of alternative techniques for computing risk and forecast estimates and alternative decisions; 3—to develop new computational procedures for constructing risk forecasting systems with appropriate feedback on the basis of several sets of quality analysis parameters for model adequacy and quality of the forecasts; 4—develop and implement the project of IDSS on the basis of systemic concept for adaptive modeling, forecasting, and risk estimation.

Concept of constructing adaptive system for modeling, forecasting and risk estimation. Figure 1 illustrates structure of the computer based system implementing the systemic approach to data processing, model constructing for the process under study, and selected risk estimation. First, the selected process data analyses is performed directed towards uncertainty identification and processing, estimation of correlation

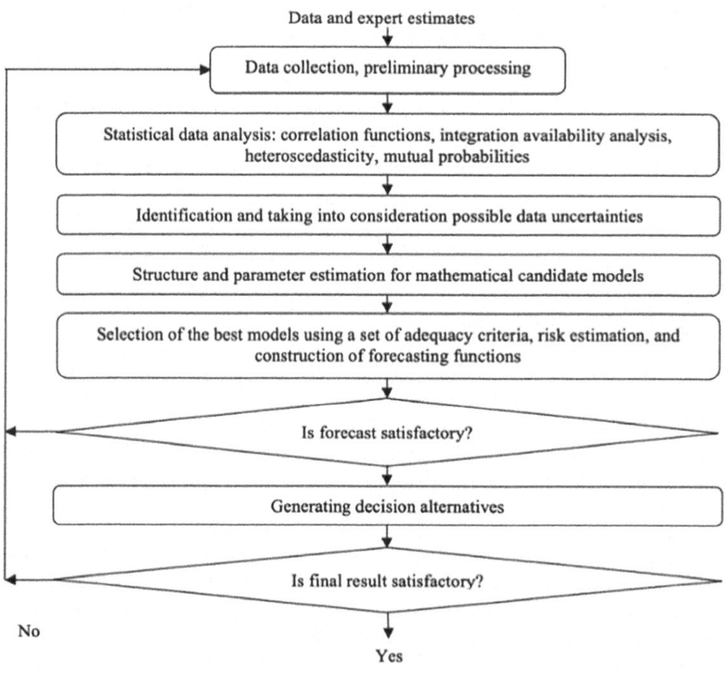

Fig. 1 Structure of the systemic approach implementing to data processing, model constructing for a process under study, and selected risk estimation

characteristics necessary for model constructing, model structure and parameter estimation as well forecast computing necessary for the subsequent risk estimation. All these tasks are solved by making use of IDSS.

Figure 2 shows some details of adaptive modeling and forecasting scheme implemented in the IDSS proposed. Now consider in some detail the stages of IDSS constructing. Functioning of the system starts from selecting of the type of risk for analysis and necessary data for processing, analysis of existing models and possible formal approaches to the problem description, and forecasting its further development. At this stage it is necessary to analyze special references that could provide for substantial help in the search for appropriate existing models describing behavior of the processes under consideration. The models can be hired in the form of systems of

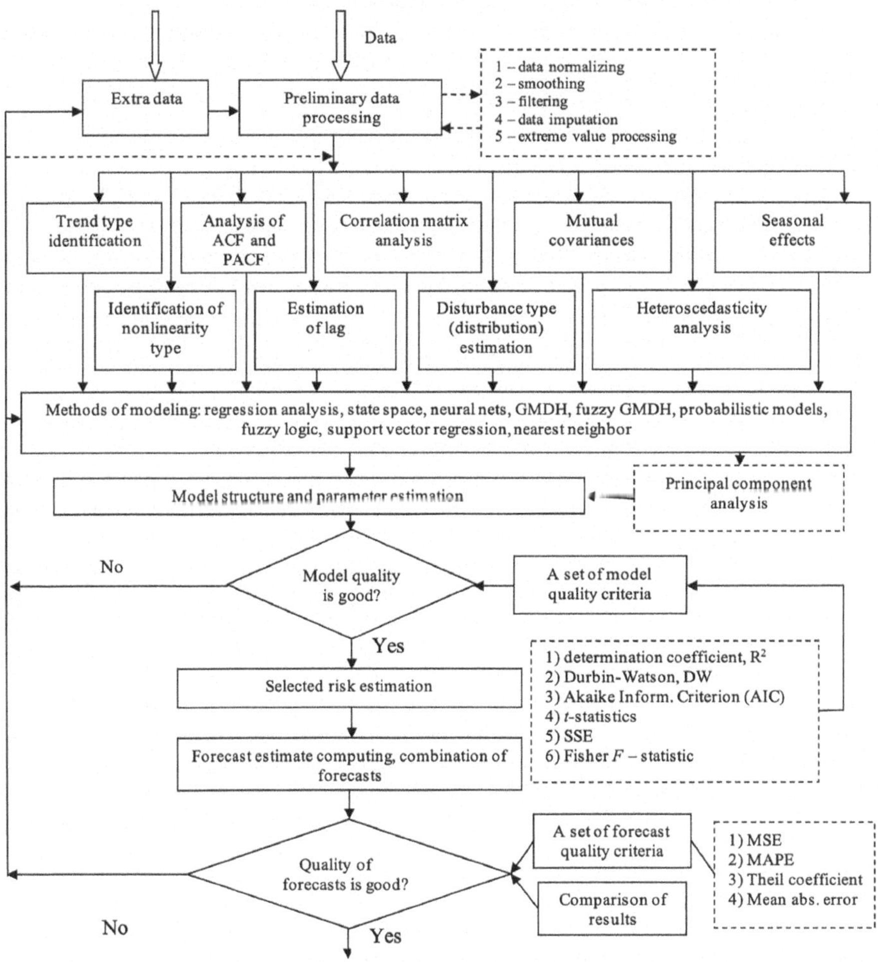

Fig. 2 Functional layout of the IDSS developed

differential, difference or algebraic equations, probability distributions of input and output variables (statistical models), sets of logical rules (e.g. fuzzy sets) that describe logic of interactions between inputs and outputs [13, 14]. To characterize risks are especially suitable probabilistic models of Bayesian type, such as generalized linear models (GLM), multivariate conditional distributions, Bayesian networks etc. [15–17]. Such models are very close to the expert way of thinking what is especially good for the cases of managing the situations with uncertainties. At the same time BN have substantial analytic support in the form of rather complicated computational procedures and logic.

Selection of the type and structure of a formal model plays substantial role for implementing further stages of the IDSS functioning. The model created on the basis of theoretical considerations and laws for a specific process may require only some refinement of its parameters using statistical data.

At the same time the model that is completely based upon statistical/experimental data may require much more information and time for its completion. A survey of special literature can also be useful regarding estimation of a model structure and its parameters. Each estimation method has its application frames that should be known for correct specific application.

IDSS that has been developed is based upon the following type of models: regression, state space and polynomial models including Kolmogorov-Gabor polynomials (constructed by the group method of data handling), Bayesian and neural networks [18–20]. The practice of model constructing and forecasting system development provides the ground for stating that ready-for-use models cannot be met often. Even existing tested models for specific processes require correction of their structure and/or parameters to adapt them to specific data and conditions of application. That is why practically it is always necessary to construct new models based on thorough study of a process under investigation and available data. It is known that quality of data plays crucial role in the process of model constructing. That is why in collecting database it is necessary to remember the basic requirements regarding their information quality, synchronization, and correctness.

Preliminary data processing procedures are required to bring data to the form that provides possibilities for the correct application of model structure and parameter estimation techniques aiming to computing of their statistically significant estimates. Thus, it is often necessary to apply appropriate data normalizing procedures, perform imputation of lost values, process extreme values, to perform digital or optimal filtering and solve multi-collineation problem. The data filtering can be digital or optimal dependently on the quality of data, specific problem statement, quality and volume of measurement information available [21–26].

The sequence of actions necessary for identification, processing and taking into consideration possible uncertainties is given in Fig. 3. This sequence of computational procedures differs from known data processing schemes due to explicit application of uncertainty identification techniques and uncertainty reduction procedures. Usually all the tasks mentioned in Fig. 3 are solved successfully with appropriately designed and implemented IDSS possessing special features for processing specific uncertainties mentioned in Table 1.

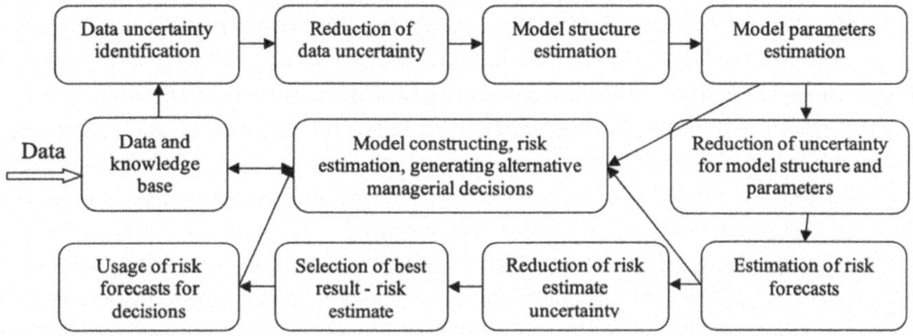

Fig. 3 Sequence of actions necessary for uncertainty identification, processing and taking them into consideration

Table 1 Some types of uncertainties met in the process of modeling

№	Type of uncertainty	Reason for uncertainty	Method of overcoming
1	Structural uncertainty of a model	– Problem with establishing causal relations – Approximate values of structural characteristics	– Expert estimates – Application of statistical tests – Testing of hypothesis
2	Statistical data uncertainties	– Measurement errors – Stochastic disturbances – Multi-collineation – Extreme values – Lost measurements	– Filtering techniques – Refining distribution type – Principal component method – Extreme value theory – Lost data imputation
3	Parametric uncertainties of a model	– Incorrect parameter estimation method – Short sample – Non informative data	– Alternative estimation techniques – Appropriate data generation – Application of simulation techniques
4	Probabilistic uncertainties	– Complicated mechanisms of causality – Absence of deterministic components in data – Complex distributions	– Static and dynamic Bayesian networks – Other probabilistic models and filters – Conditional distributions
5	Amplitude uncertainty	– Non-measurable variables	– Fuzzy logic methods – Bayesian techniques

We consider uncertainties as the factors that influence negatively the whole process of data processing, model structure and parameter estimation, risk estimating and alternative decisions generating. They are inherent to the process of modeling due to incompleteness or inexactness of our knowledge regarding the objects (systems) under study, incorrect selection or application of computational procedures etc. [26–29]. The uncertainties very often appear due to incompleteness of measurement data, noise contaminated measurements (measurement errors) or they are provoked by stochastic external disturbances with unknown probability distribution, poor estimates of model structure or by a wrong selection of parameter estimation method. The problem of uncertainty identification is solved with special statistical tests, appropriate computational procedures and visual data studying. Table 1 provides some information on the possibilities of processing uncertainties associated with preliminary data processing and model constructing.

Generally speaking the problem of identifying uncertainties and taking them into consideration is rather sophisticated due to the fact that it touches upon different aspects of various data structures, modeling, forecasting, risk estimation, and decision making [30]. Usually that's separate problem to be solved jointly by theoretical investigators and practitioners.

3 Data, Model and Forecasts

Correctly processed data are used for estimating structure and parameters for candidate models of the process being studied. Model structure estimation is non-trivial procedure that requires experience from a researcher. It is proposed to define formally the model structure in the following way:

$$S = \{r, p, m, n, d, z, l\}, \tag{1}$$

where r is model dimension determined by the number of equations creating a model; p is model order, i.e. maximum order of differential or difference equations in a model; m is a number of independent variables in the right side of specific equation; n is nonlinearity and its type (nonlinearity with respect to variables or model parameters); d is delay time (lag) of system output reaction with respect to the moment of applying input (or control) action; z is external disturbance and its type (stochastic or deterministic); l denotes possible constraints on variables (or parameters). Model structure estimation requires knowledge of functioning of the process under study and respective statistical/experimental data that describe its evolution in time. As a rule, for one process several candidate models are estimated from which the best one is selected using a set of statistical adequacy parameters. Such approach is enhancing substantially the probability of constructing adequate model for a specific case.

Time series data in technology, technical systems, economy and finances usually contain deterministic and stochastic components. The stochastic component is provoked by stochastic disturbances, availability of measurement noise, computational errors, approximate estimates of structure and parameters. That is why preliminary data processing is usually necessary to "clean" data of the undesirable components.

Under statistical model we will understand the model in the form of random values distribution. Substantiated selection of the distribution type and its parameters estimation with statistical/experimental data represents the process of statistical model constructing.

However, even high adequacy model still may not guaranty high quality forecasting because the main purpose of constructing forecasting model is in high quality approximation of basic statistical characteristics of the data available: mathematical expectation, variance and covariance. That is why after constructing the model it is necessary to test its forecasting possibility.

Today there exist wide spectrum of forecasting methods and models that can be used in economy, finances, ecology and many other areas. However, not all the methods provide for high quality forecasts in specific cases due to complexity of the process under examination and the reasons mentioned above. That is why selection of forecasting method may be complicated problem that requires simultaneous application of several alternative procedures, combination of forecasts generated by different methods etc.

The most popular today are the following methods of forecasting future development of various origin processes: regression model approach, neural networks, fuzzy logic, probabilistic and statistical models, group method for data handling (GMDH), the methods based on the soft computing (genetic and immune algorithms), support vector machine, method of similar trajectories and some others. Each of the methods mentioned can take (to some extent) into consideration uncertainties of structural, statistic and parametric type. According to the results presented by researchers usually high quality results of forecasting can be reached with GMDH, neural networks, probabilistic models and fuzzy logic. These methods and techniques are close by their inherent features to the approaches of modeling situations and decision making by human experts. That is why their application in IDSS as a rule provide for a significant positive effect. Modern directions of development probabilistic methods of modeling and forecasting rather complicated nonlinear nonstationary processes (NNP) are based upon generalized linear models (GLM), hierarchical and structural models, particle filters, static and dynamic Bayesian networks (BN) [30, 31]. Models of this type exhibit their own advantages and disadvantages in solving practical problems. Some possibilities for hiring linear and nonlinear models are shown in Table 2.

The models (No. 1–8) presented in Table 2 have known structure though it can be modified in the process of adaptation using specific statistical data. Model 1 was successfully applied for trend modeling of various orders together with short-term fluctuations around conditional mean. Models 2 and 4 can describe bilinear and exponential nonlinearities or nonlinearity with saturation (model 3). Models 5 and 6 are used for describing conditional variance dynamics while modeling heteroscedastic process. The last one turned out to be the best model for short term forecasting of variance in about 90% of applications performed by the authors. Models 7, 8, and 9 can describe arbitrary nonlinearities with respect to variables of order 3–5 or higher. Fuzzy sets based approach to modeling supposes generating of a set of rules that

Table 2 Some linear and nonlinear models for describing process dynamics

№	Type of uncertainty	Reason for uncertainty
1	AR + polynomial of time	$y(k) = a_0 + \sum_{i=1}^{p} a_i y(k-i) + b_1 k + \cdots + b_m k^m + \varepsilon(k)$ $k = 0, 1, 2, \ldots$ is discrete time; $t = kT_s$; T_s is sampling time
2	Generalized bilinear model	$y(k) = a_0 + \sum_{i=1}^{p} a_i y(k-i) +$ $\sum_{j=1}^{q} b_j v(k-i) + + \sum_{i=1}^{m} \sum_{j=1}^{s} c_{ij} y(k-i) v(k-j) +$ $\varepsilon(k)$
3	Logistic regression	$\phi(x(k, z)) = \frac{1}{1+\exp(-x(k,z))}$ $x(k) = \alpha_0 + \alpha_1 z_1(k) + \ldots + \alpha_m z_m(k) + \varepsilon(k)$
4	Nonlinear extended econometric autoregression	$y_1(k) = a_0 + a_1 y_1(k-1) + b_{12} \exp(y_2(k)) + a_2 x_1 x_2 +$ $\varepsilon_1(k) y_2(k) =$ $c_0 + c_1 y_2(k-1) + b_{21} \exp(y_1(k)) + c_2 x_1 x_2 + \varepsilon_2(k)$
5	Generalized autoregression with conditional heteroscedasticity (GARCH)	$h(k) = \alpha_0 + \sum_{i=1}^{q} \alpha_i \varepsilon^2(k-i) + \sum_{i=1}^{p} \beta_i h(k-i)$
6	Exponential generalized autoregression with conditional heteroscedasticity (EGARCH)	$\log[h(k)] = \alpha_0 + \sum_{i=1}^{p} \alpha_i \frac{\lvert\varepsilon(k-i)\rvert}{\sqrt{h(k-i)}} +$ $\sum_{i=1}^{p} \beta_i \frac{\varepsilon(k-i)}{\sqrt{h(k-i)}} + + \sum_{i=1}^{q} \gamma_i \log[h(k-i)] + v(k)$
7	Nonparametric model with functional coefficients	$y(k) =$ $\sum_{i=1}^{p} \{\alpha_i + (\beta_i + \gamma_i y(k-d)) \cdot \exp(-\theta_i y^m(k-d))\} + \varepsilon(k)$
8	Radial basis function	$f_\theta(x(k)) = \sum_{i=1}^{M} \lambda_i \exp\left(-\frac{(x(k)-\mu_i)^2}{2\sigma_i^2}\right) + \varepsilon(k),$ $\theta = [\mu_i, \sigma_i, \lambda_i]^T$; $M = 2, 3, \ldots$
9	State-space representation	$\mathbf{x}(k) = \mathbf{F}[\mathbf{a}(k), \mathbf{x}(k-1)] + \mathbf{B}[\mathbf{b}(k), \mathbf{u}(k-d)] + \mathbf{w}(k)$
10	Neural networks	Selected (constructed) network structure
11	Fuzzy sets and neuro-fuzzy models	Combination of fuzzy variables and neural network model
12	Dynamic Bayesian networks	Probabilistic Bayesian network constructed with data and/or expert estimates
13	Multivariate distributions	Say, copula application for describing multivariate distribution
14	Immune systems	Immune algorithms and combined models

could describe with acceptable quality functioning of selected processes and formulate appropriate logical inference. Neural networks and fuzzy neural networks are suitable for modeling sophisticated nonlinear functions in conditions of availability of some unobservable variables. Dynamic Bayesian networks and multivariate distributions are statistical-and-probabilistic models that are suitable for describing complex multivariate processes (systems) with generating final result of their application in the form of conditional probabilities (probabilistic inference).

Architecture of IDSS. IDSS architecture is a generalized large-scale representation of basic system elements and connections between them. The system architecture gives a notion for the general purpose of system constructing and its basic functions (Fig. 4).

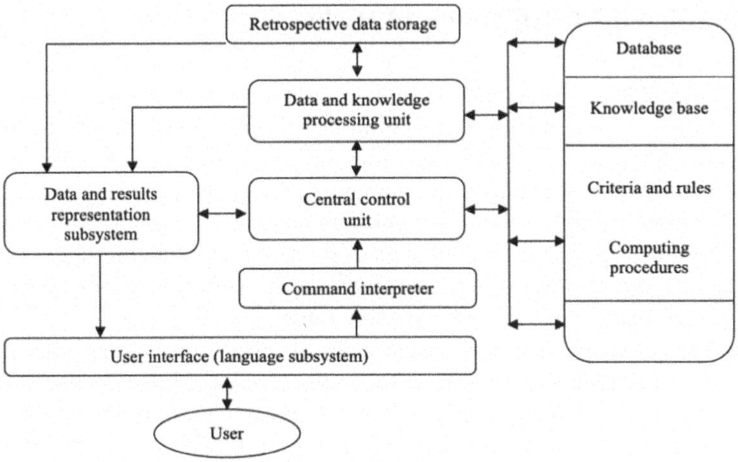

Fig. 4 IDSS architecture for estimation of financial risks

Functioning of the IDSS is controlled by user commands using the command interpreter which constitutes a part of the user interface. The user commands are directed to the central control unit that coordinates functioning of all system elements. Some of the specific commands and actions are as follows: expanding and modification of data and knowledge bases of the system; initiation and triggering of data and knowledge processing procedures; model structure and parameter estimation, risks and forecasts computing, generation of decision alternatives; visualization of intermediate and final results of computing; retrospective analysis of previous sessions of system application and decision making; comparing current results of data processing with the previous ones.

Generally the model constructing procedure in the frames of the IDSS proposed includes the following steps:

- collecting and performing preliminary data processing; at this stage the data is also analyzed on availability of the data uncertainties mentioned above and their elimination or minimization of their influence on the computational procedures;
- model structure and parameter estimation;
- application of the model constructed for risk and forecast estimation;
- if the result of risk estimation is satisfactory then we perform prediction of possible loss; if the result is not acceptable the model adaptation is performed and risk estimation is repeated;
- the possibility for risk minimization is considered.

4 Generalized Classification of Risks

Risks are available in every area of human activities, for example, in economy, finances, industry, construction, transportation systems, risks caused by military conflicts etc. We also meet risky situations in nature when earthquakes happen, hurricanes, flood, global changes of climate (Table 3). To get ready to risky situations it is necessary to study the possibilities for emerging the risky events (risk factors), construct appropriate mathematical models for predicting the events, and make decisions directed towards management of the situations. Some possibilities for coping with risky situations are shown in Table 3.

The processes associated with occurrence of risky events very often exhibit nonstationary and nonlinear behavior. Above we considered some possibilities for constructing models of such processes in the frames of the IDSS proposed. Some examples of such processes could be considered the following ones: variations of stock prices, currency exchange rate, especially for weak economies, market prices on energy resources etc. Most of these processes are heteroscedastic and require constructing special models both for the process itself and its variance. The variance forecast may be needed for computing, for example, market risk loss. Usually statistical data characterizing evolution of such processes in time is available for studying.

Possibilities for risk model adaptation. A substantial attention is paid today to analysis of the following financial risks: market risk, credit risk, operational and some types of actuarial risks. Preliminary studies performed showed that most of the process associated with risky events belong to the class of nonstationary and nonlinear (NNP) [13]. First of all to this class belong heteroscedastic processes (they are NNP by definition). That is why the IDSS proposed provides the possibility for constructing nonlinear models; the corresponding procedure illustrates Fig. 5; in fact this procedure is a part of general model constructing procedure given in some detail in Fig. 2.

To keep the quality (adequacy) of a model at high level in conditions of process nonstationarity it is necessary to develop and use adaptive estimation procedures. As initial parameters for analysis quality (adequacy) of a model can serve respective

Table 3 Generalized classification of risks

Origin of risk	Consequences of risk realization	Risk probability (frequency)	Possibility of coping with risks
– Finances/economy – Industry – Construction – Transportation systems – Medical risks – Military conflicts – Natural disasters	– Catastrophic loss – High loss (destruction) – Medium loss – Low loss	– High probability (very frequent) – Medium probability – Low probability	– Accept the risk – Taking measures to prevent and/or reduce risk – Transfer the risk to another side

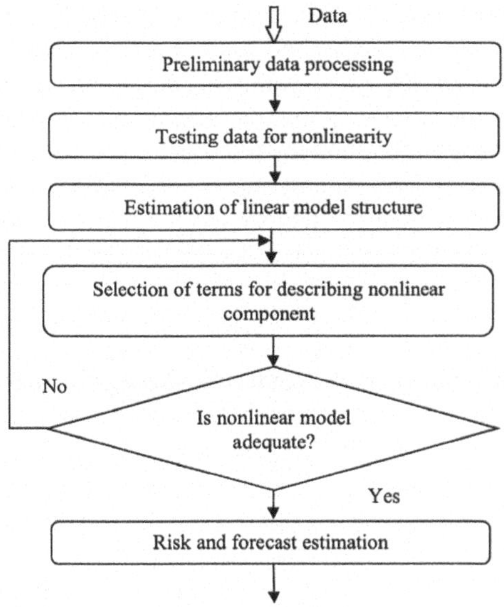

Fig. 5 Procedure illustrating sequence of actions necessary for constructing model of nonlinear process

statistical characteristics such as determination coefficient, Durbin-Watson statistic (DW), Akaike information criterion (AIC), mean absolute percentage error (MAPE is used for analyzing quality of forecasting) etc. To solve the model adaptation problem the following possibilities were implemented in the IDSS [13, 25, 26]:

- periodic analysis of data distribution type and its parameters so that to take into consideration this information when method of model parameter estimation is selected;
- automatized analysis of a partial autocorrelation function (PACF) for basic dependent variable with subsequent correction of a model structure performing adding and deleting lagged values;
- introducing into model possible independent variables and analysis of their influence on the model adequacy and forecasting quality; introducing leading indicators with actual appropriate lags;
- automatized selection of optimal weights in exponential smoothing procedures, search of similar trajectories etc.;
- automatized analysis of residuals for regression models aiming analysis of their information content with subsequent model structure correction;
- adaptive correction of state variables measurements using the method of hierarchic combining aiming increasing their quality.

The problem of parametric adaptation of a model to new data is solved thanks to repeated (recursive) parameter estimation with new data coming. For model parameter estimation different methods are applied what provides the possibility for creating extra candidate-models to be used for selecting the best model further on. The parameter estimation methods used in the IDSS are the following: ordinary least squares (LS), nonlinear LS (NLS), recursive least squares (RLS), maximum likelihood (ML), and Markov Chain Monte Carlo (MCMC). The last one provides the possibility for estimating parameters of linear and nonlinear in parameter models. Application of particular adaptation scheme depends on specific problem statement, quality requirements to models and risk estimates. Here we also should remember that each method of adaptive modeling, risk estimation and forecasting has its own specific features that should be taken into account in data processing procedures. As far as Bayesian networks are used in the IDSS its adaptation is also performed.

5 Bayesian Network Adaptation

One of the most powerful modern approaches to modeling and forecasting NNP is probabilistic-and-statistical Bayesian method that provides for constructing Bayesian networks (BN) and other probabilistic models [18–20]. The BN constructing procedure used in this study is based upon statistical analysis of data that characterize evolution of corresponding variables. The algorithm implementing adaptation scheme supposes that new data is coming in real time. Though there no much difference between the two possible modes of collecting necessary data. To explain the adaptation procedure introduce the following notation: $Z = \{X_1, \ldots, X_n\}$ is a set of BNnodes corresponding to the number of variables in the model; $E = \{(X_i, X_j)|X_i, X_j \in Z\}$ is a set off arcs in the network;X_i is a node of BN that corresponds to the observations of one variable; $n = |Z|$ is a number of BN nodes; r_i—is a number of values that could be accepted by the node, X_i; v_{ik} is k-th values of variable, X_i; Π_i is a set of predecessor nodes for the node, X_i; φ_i—is a set of possible initializations for Π_i; $q_i = |\varphi_i|$ is a number of possible initializations for Π_i; φ_{ij} is j-th initialization of predecessor node Π_i for the node X_i; B_S is BN structure; B_P represents probabilistic specification of BN, i.e. this is a part of model formal description representing probabilistic characteristics of BN; $\theta_{ijk} = p(X_i = v_{ik}|\varphi_{ij}, B_P)$ under condition that $\sum_k \theta_{ijk} = 1$; $f(\theta_{ij1}, \ldots, \theta_{ijr_i})$ is probability for the node X_i with initialization φ_{ij}; D_0 is initial observations database; S_0 is initial BN structure estimated after preliminary data processing containing in D_0; D_1—is a part of database that was not used to construct S_0; S_1 is BN structure that was estimated after adaptation of S_0 to the new data in D_1. The problem was stated in the following way: construct BN (graph) $G = Z, E$ with the structure, S_0, corresponding to the initial observation data base D_1. Thus, new structure of BN should correspond to D_1: $S_1 \Leftrightarrow D_1$. In this case statistical (or experimental) data may exhibit arbitrary distribution and the process described by the data may have nonstationary behavior: i.e. mathematical expectation, $M[X_i] \neq const$, and its variance $M\{X_i - M[X_i]\}^2 \neq const$.

Generally the adaptation procedure should obey the sequence given below.

(1) Constructing initial graph using available data.
(2) Correction of a model (graph) structure as follows:

   1. elimination of arcs that do not correspond to the new data available;

2. adding new arcs where necessary (according to local quality functional).

(3) Correction of the probabilistic part of a model, i.e. correction of conditional probability values in the conditional probability tables (CPT).

As far as at the initial stage of BN learning the probabilistic part of the model represent CPTs computed on the basis of frequency analysis of observations available, it would be more convenient to correct CPTs using the values of N_{ijk}. This way it will be possible to modify faster the information regarding probability distributions and the values of conditional probabilities themselves can be calculated by the Dirichlet formula in the following way [15]:

$$p(X_i = v_{ik}|\Pi_i = \varphi_{ij}) = \frac{N_{ijk} + 1}{N_{ij} + r_i} \tag{2}$$

While correcting BN structure the order of analyzing the nodes is determined by making use of the contribution of each node into general value of conditional probabilities [15, 17, 18]:

$$p(D_1|D_0, S_0) = \prod_{i=1}^{n} \frac{\prod_{s=1}^{R_i} \prod_{t=1}^{Q_i} \prod_{u=1}^{m_{its}} (N_{its} + u)}{\prod_{t=1}^{Q_i} \prod_{u=1}^{M_{it}} (N_{it} + r_i - 1 + u)}. \tag{3}$$

The essence of informational importance of the arcs is in the following: at the stage of arc, the analysis regarding its elimination is based on the value of the parameter, $K_{delete}(S_0)$, for current configuration of the predecessor nodes as well as the values of $K_{delete}(S_{-1}^m)$ for the configurations that represent results of elimination of the $M(1 \leq m \leq M)$ input arcs for the current node. If the condition $K_{delete}(S_{-1}^m) \leq K_{delete}(S_0)$ is true, then m-th arc remains in the current structure because its elimination results in lower value of local quality functional (for current node). Otherwise the arc is added to the list of arcs that will be tested further on the necessity of elimination. Then the list is sorted out according to growth of the value of, $K_{delete}(S_{-1}^m)$. The tactics of elimination and adding the arcs was used in the increment version of the adaptation algorithm shown below. As far as adaptation strategy is based upon analysis of the functional:

$$P(S_1|D_1, D_0, S_0) = \arg\max_{S} \frac{P(S|D_0)P(D_1|S, D_0)}{P(D_1|S_0, D_0)}, \tag{4}$$

the procedure of elimination and adding the arcs is performed as described below.

According to the idea of optimization the BN structure the tactics of eliminating the arcs should result in decreasing the first part of the nominator in expression for computing the value of, $P(S|D_0)$, because it reaches its maximum with $S = S_0$ when initial BN structure is formed. Thus, to get positive results for adaptation it is necessary to compensate the loss due to arc elimination by adding new arc. As far as K2 optimization algorithm requires as input an ordered sequence of nodes, the search of arcs for adding to the graph is performed in the way described. The arc estimation is performed by computing the value of local functional mentioned. Thus, the candidate for adding should determine the configuration of input arcs that shows maximum value of the local quality functional [17, 18]. Now consider some examples of IDSS application.

Preliminary data processing with optimal filter. Example—Application of extended and unscented Kalman filters to nonlinear data processing. As it was stressed above Kalman filters are suitable for fighting uncertainties in the form of two random processes: random state disturbances and measurement errors. Consider popular nonlinear econometric model of growth:

$$x_n = f(x_{n-1}) + u_n = 0.5x_{n-1} + \frac{25x_{n-1}}{1 + x_{n-1}^2} + 8\cos(1.2(n-1))$$
$$+ u_n y_n = h(x_n) + v_n = \frac{x_n^2}{20} + v_n \tag{5}$$

where $u_n\ N(0, \sigma_u^2)$, and $u_n\ N(0, \sigma_u^2)$. Using the model necessary data was generated shown in Fig. 6.

Result of application of unscented filter is shown in Fig. 7, where dotted line is filter output.

Result of application of extended Kalman filter is shown in Fig. 8, where dotted line is filter output.

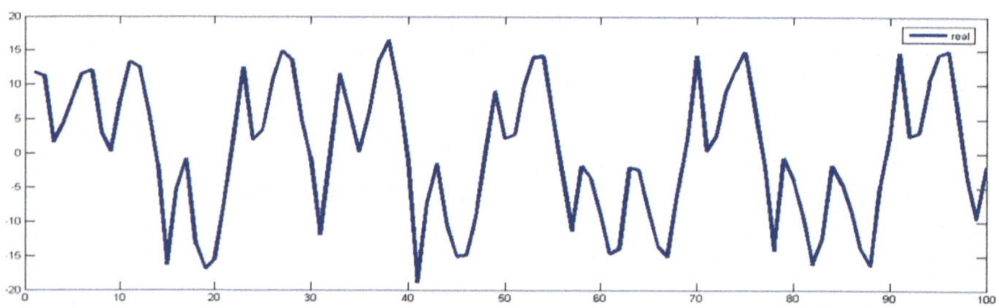

Fig. 6 Generated illustrative data for the state variable

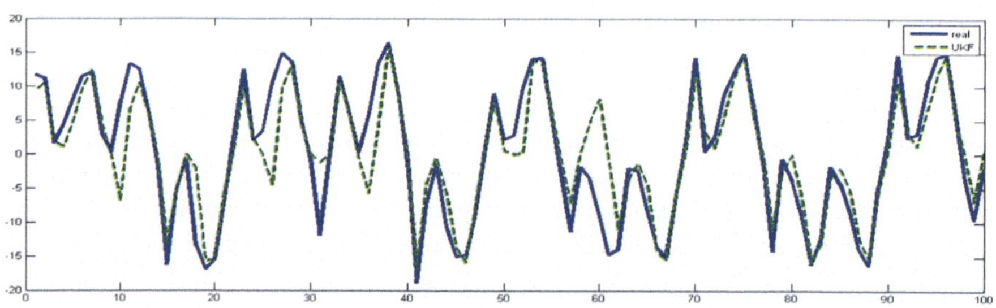

Fig. 7 Actual values of the state variable and unscented Kalman filter output

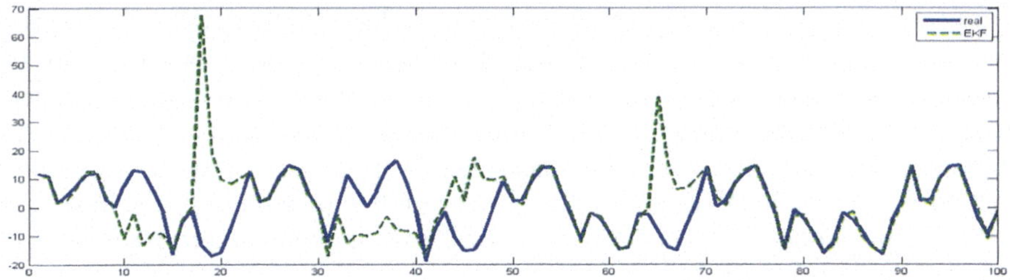

Fig. 8 Actual values of the state variable and extended Kalman filter output

Table 4 Mean squared errors of filtering for unscented Kalman filter

σ_v^2 | σ_u^2	0.20	0.40	0.60	0.80	1.00	1.20	1.40	1.60	1.80	2.0
0.20	26.95	23.81	40.63	55.89	71.01	78.17	72.94	89.42	97.23	116.17
0.40	22.18	33.86	47.43	44.22	49.71	53.97	46.79	81.86	73.87	86.72
0.60	26.55	39.15	32.82	52.41	57.49	54.15	60.87	85.22	54.72	83.37
0.80	32.13	25.14	48.40	54.05	52.37	52.64	50.15	56.61	58.36	51.20
1.00	31.54	43.44	38.81	53.11	45.10	60.41	58.57	47.03	73.23	63.92

Table 5 Mean squared errors of filtering for extended Kalman filter

σ_v^2 | σ_u^2	0.20	0.40	0.60	0.80	1.00	1.20	1.40	1.60	1.80	2.0
0.20	51.84	136.62	187.08	180.73	249.99	193.76	128.42	223.64	518.59	256.75
0.40	72.98	99.68	90.45	96.71	169.77	114.06	169.76	180.32	130.56	265.65
0.60	59.63	78.23	166.86	110.43	91.62	223.63	156.90	280.47	217.45	279.64
0.80	53.19	76.75	59.63	114.75	98.74	135.98	165.71	283.93	163.37	154.31
1.00	59.20	57.64	82.51	80.68	102.57	127.17	123.20	94.86	237.03	164.20

Mean squared errors of filtering are given in Tables 4 and 5 for different variances of state and measurement noise: σ_u^2, and σ_v^2 for both filters.

It can be easily seen that unscented filter provides much better results of filtering in this case because the function $f(x_n)$ exhibits substantial nonlinearity. The second order and higher members of Tailor expansion of the function have substantial weights which show that we cannot neglect the members. Their neglect results in much worse filtering quality as it was demonstrated by the example.

Market risk estimation. Example—The market, credit, operational and some actuarial ones are most widely spread risks to be used as example of analysis. Consider an example of market risk analysis using available statistical data that characterize stock prices for Microsoft, Apple and other known companies.

For a separate position that includes several financial instruments influenced by the same risk factor value-at-risk (VaR) for the time horizon of T days and confidence interval $(1 - \alpha)$ can be computed using the following expression [32]:

$$VaR = \lambda_{1-\alpha} V \sigma_k \sqrt{T}, \tag{6}$$

where $(1-\alpha)$ is the confidence interval for the value of VaR, and $\lambda_{(1-\alpha)}$ is respective quantile; V is current cost (volume) of financial position (product of current stock cost by the number of stock units); T is time horizon (usually it is estimated in days) for which the projection of VaR is estimated; σ_k is volatility (standard deviation) of a financial time series being analyzed. In the frames of the IDSS considered this parameter is modeled and forecasted with the model. For this purpose are often used the models given in Table 2—generalized autoregression with conditional heteroscedasticity (GARCH):

$$h(k) = \alpha_0 + \sum_{i=1}^{q} \alpha_i \varepsilon^2(k-i) + \sum_{i=1}^{p} \beta_i h(k-i), \tag{7}$$

and exponential GARCH (EGARCH):

$$\log[h(k)] = \alpha_0 + \sum_{i=1}^{p} \alpha_i \frac{|\varepsilon(k-i)|}{\sqrt{h(k-i)}} + \sum_{i=1}^{p} \beta_i \frac{\varepsilon(k-i)}{\sqrt{h(k-i)}}$$
$$+ \sum_{i=1}^{q} \gamma_i \log[h(k-i)] + v(k), \tag{8}$$

where $h(k) = \sigma_k^2$; $\varepsilon(k)$ is a random process determined by the residuals of low order (usually first or second order) autoregression for the main financial variable being studied. The VaR methodology has its advantages and disadvantages: advantage of VaR is in its simplicity of application and interpretation from financial point of view. This fact makes it an effective management instrument. Besides, VaR focuses on the distribution tail only, and thus it reacts on relatively rare loss only. The disadvantage of the method is that VaR does not provide information about maximum loss that can occur at some low frequency ignoring other points of the distribution tail. The practice of its use shows that there also are cases when VaR for a portfolio as a whole exceeds the sum of VaRs for each individual financial instrument. This fact is somewhat contradictory to commonly accepted views about the role of diversification regarding decreasing level of risk [33, 34].

Looking for the ways of avoiding the VaR drawbacks a new modification of VaR was proposed: conditional VaR (CVaR), which represents mathematical expectation of loss when it exceeds VaR. Thus CVaR contains information about part of the data distribution tail that is beyond the point that corresponds to VaR; formally CVaR is determined as follows [35]:

$$CVaR_\alpha = -\frac{1}{\alpha}\left(E\left[X 1_{\{X \leq x_\alpha\}}\right] + x_\alpha(\alpha - P[X \leq x_\alpha])\right)$$
$$x_\alpha = \inf\{x \in R : P(X \leq x) \geq \alpha\}, \tag{9}$$

$$A(x) = \begin{cases} 1, \; if x \in A, \\ 0, \; otherwise. \end{cases}$$

It can be seen that CVaR is always less than or equal to VaR; it means that CVaR is more conservative measure of financial risk. Dependence of CVaR on the tail part of a distribution provides the possibility for more correct analysis of financial risk at the moments of high external disturbances and shocks.

Besides, better reaction of CVaR to rare events, this measure of financial risk represents convex function with respect to the financial instruments, i.e. the sum of CVaRs for two financial instruments does not exceed CVaR for the combined portfolio of these instruments. This feature allows for the model to preserve likelihood in the context of risk diversification what means that CVaR more realistically reflects the diversification procedure than simpler VaR.

The data used characterize stock prices for Microsoft, Apple, Google, Dell, Cisco and HP companies within the period of time between 2005 and 2011. The data was divided into samples of 100 elements each. To perform market risk estimation the return of stocks was aggregated within each ten days and then compared to the recommended reserves computed with the VaR and CVaR methodologies. The number of cases when the aggregated return values exceeded computed reserves and the statistics characterizing exceeding surplus (breaks) are provided in tables below. The last row in tables gives mean frequency when VaR break occurred (i.e. when it exceeded the reserve); for the tables with statistic of non-realized financial resources the last row shows mean loss for columns (Table 6).

Table 6 characterizes typical features of CVaR: for all conditional variance models and loss distributions methodology CVaR provides better result of capital reserving with less frequency of occurring surplus capital than VaR. The models of EGARCH type exhibited better robustness with respect to substantial disturbances what can be seen in lesser number of samples with high number of breaks capital for VaR than for CVaR (the relation is about 2:3). The same computations were performed for all other companies mentioned above. Table 7 shows results of effectiveness of capital usage by Microsoft.

Table 7 shows that the use of EGARCH models for Microsoft data is generally better because it provides for higher effectiveness of capital usage. According to the expectations CVaR requires higher expenses; this is the payment for the lesser number of breaks.

The generalized results regarding model robustness with respect to the number of breaks is given in Table 8.

Table 8 shows that the most often breaks of VaR occur in the models that are based upon assumption of normal distribution for returns of financial instruments. The worst result we got for the EGARCH where probability of VaR break turned out to be 0.0607 what exceeds selected value of 0.05. The best results were achieved for EGARCH models with exponential distribution: 0.0089 for VaR, and 0.0041 for CVaR, respectively.

Table 6 Microsoft stocks: number of VaR breaks

MICROSOFT							
GARCH				EGARCH			
Normal		Exponential		Normal		Exponential	
VaR	CvaR	VaR	CvaR	VaR	CvaR	VaR	CvaR
0	0	0	0	4	2	0	0
27	19	30	28	2	0	0	0
0	0	0	0	2	1	0	0
5	1	1	0	3	1	0	0
0	0	0	0	8	0	0	0
0	0	0	0	0	0	0	0
6	1	0	0	7	2	0	0
14	12	11	8	4	2	0	0
0	0	0	0	4	1	0	0
0	0	0	0	6	2	0	0
1	0	0	0	3	3	0	0
0	0	0	0	1	0	0	0
0	0	0	0	9	5	0	0
0	0	0	0	1	0	0	0
18	17	25	22	4	3	2	1
0	0	0	0	1	0	0	0
7	1	0	0	13	13	8	3
0.05735	0.037	0.04926	0.04264	0.05294	0.02573	0.00735	0.00294

Table 7 Microsoft: excessive reserves

MICROSOFT							
GARCH				EGARCH			
Normal		Exponential		Normal		Exponential	
CvaR	VaR	CvaR	VaR	CvaR	VaR	CvaR	VaR
5797	7525	11,525	15,630	4703	5724	8691	11,881
4435	4079	17,098	22,986	5792	7341	11,083	14,920
11,536	13,861	19,248	24,773	8264	9611	13,030	16,547
11,089	12,906	17,788	23,177	10,584	12,421	17,344	22,626
5257	6909	10,736	14,662	4208	5384	8510	11,717
10,460	12,609	17,589	22,697	7485	8898	12,171	15,529
9177	11,512	17,347	23,436	7777	9192	13,903	18,880
5743	5938	8509	10,755	6707	7999	11,403	15,049
11,014	13,449	19,092	24,879	7484	8764	12,067	15,585
24,717	30,442	43,710	57,318	15,265	17,321	24,432	31,811
23,150	29,622	44,801	60,368	18,094	21,678	32,604	44,230
12,827	16,195	23,998	32,000	10,201	12,914	19,208	25,663
11,487	14,792	22,451	30,306	8823	9674	14,307	19,531
8194	10,097	14,508	19,032	6720	8204	11,745	15,376
7739	8044	22,102	30,436	6827	8032	10,941	14,125
6930	8231	11,243	14,333	4933	5736	7602	9515
11,926	14,324	21,397	28,660	12,339	11,296	10,915	13,076
10,675.1	12,972.6	20,184.8	26,791.0	8600.35	10,011.12	14,115.06	18,591.82

Table 8 Comparison of models regarding surplus capital given by VaR and CVaR

	GARCH				EGARCH			
	Normal		Exponential		Normal		Exponential	
	VaR	CvaR	VaR	CvaR	VaR	CvaR	VaR	CvaR
Microsoft	0.0574	0.0375	0.0493	0.0426	0.0529	0.0257	0.0074	0.0029
Google	0.0456	0.0191	0.0228	0.0184	0.064	0.0382	0.0081	0.0014
Dell	0.0324	0.0199	0.0243	0.0228	0.0493	0.0235	0.0051	0.0037
Cisco	0.0382	0.0191	0.011	0.011	0.0566	0.0368	0.0184	0.0118
HP	0.0713	0.0419	0.0162	0.0118	0.0809	0.0493	0.0059	0.0007
average	0.0489	0.0274	0.0247	0.0213	0.0607	0.0347	0.0089	0.0041

6 Discussion

The examples of the IDSS application showed that the Kalman filtering techniques (extended KF and unscented filter) may serve as appropriate instrumentation for fighting uncertainties of statistical (stochastic) types that are practically always available. Estimation of VaR on the basis of data distribution with long tails leads to substantially lesser number of breaks in VaR estimates. The exponential distribution used in the study demonstrated much lesser frequency of breaks of VaR estimates for both volatility models: GARCH and EGARCH. Mean values show improvement from 0.0489 to 0.0247for GARCH model, and from 0.0607 to 0.0089 for EGARCH in comparison to the use of normal distribution.

The parametric VaR models on the basis of EGARCH volatility model demonstrate lower effectiveness regarding protection against breaks in comparison to GARCH model and usage of normal distribution: 0.0489 and 0.0274 for GARCH against 0.0607 and 0.0347 for EGARCH. However, the results of EGARCH are improving substantially when exponential distribution is hired: 0.0247 and 0.0213 for GARCH against much lower values of about 0.0089 and 0.0041 for EGARCH.

The statistics that measures additional reserves for protection against breaks of VaR estimates shows that it is better to use EGARCH models because its application results in lower loss overestimation and leads to more effective use of capital: 13,231, 15,836, 24,456, 32,445 for GARCH against 10,221, 11,649, 16,160, 21,211 for EGARCH, respectively. The use of CVaR method resulted in much lower number of breaks for the measure of financial risk, though it led to the growth of extra capital reserves.

The model of stochastic volatility demonstrated higher effectiveness of extra reserve usage: 10,287.44 for stochastic volatility model against 11,028.59 for GARCH and 14,343.94 for EGARCH. However, on training sample the stochastic volatility model approached closer to the acceptable frequency of errors, 0.05, than VaR based upon GARCH and EGARCH: 0.042188 for stochastic volatility model against 0.026667 for both models GARCH and EGARCH.

7 Conclusions

The concept for adaptive risk estimation in the frames of specialized IDSS is proposed based on the principles of system analysis that is distinguished with combined usage of preliminary processing methods for data, adaptive statistical modeling, optimal state estimation and forecasting behavior of nonlinear nonstationary processes of various origin. The concept also supposes identification and taking into consideration possible uncertainties negatively influencing the computational procedures implemented in IDSS. The uncertainties are always available due to noisy measurements, influence of random external disturbances, short samples, missing measurements, low data information content etc. The concept of proposed implementation provides the following advantages for modeling nonlinear nonstationary processes, risk and forecast estimation: automatized search for adequate model decreases many times the search, what provides the possibility for increasing the probability for reaching better results. The search is optimized thanks to hiring several combined statistical criteria characterizing better the model quality; the IDSS integrates ideologically different methods of modeling, risk estimation and forecasting what creates conditions for further improvement of estimates thanks to the weighted combining of estimates computed by making use of different techniques. The models constructed with the IDSS are usually characterized by the higher degree of adequacy including the models of conditional variance.

The future research supposes enhancement of IDSS functional possibilities, study of other risk types and further refinement of risk estimates thanks to correction and modification of existing methods and development of new methods and models based upon application of the systemic methodology hired. The combined approach to risk estimates improvement thanks to new theoretical models together with improvement of computational algorithms seems to be promising direction for the future studies.

References

1. Holsapple, C., Winston, A.: Decision Support Systems. West Publishing Company (1996)
2. Bidyuk, I., Tymoshchuk, O., Kovalenko, A.: Decision Support System Development. National Technical University of Ukraine «Igor Sikorsky KPI» (2020)
3. Dovgij, S., Trofymchuk, O., Bidyuk, P.: DSS based on statistical and probabilistic procedures. Logos (2014)
4. Zgurovsky, M., Pankratova, N.: System Analysis: Problems, Methodology, Applications. Naukova Dumka (2005)
5. Van Gigch, P.: Applied General Systems Theory. Harper and Row Publishers (1978)
6. Turban, E., Aronson, J.: Decision Support Systems. Prentice Hall (2001)
7. Lukashin, P.: Adaptive Methods of Short-term Forecasting. Finances and Statistics Publishers (2003)
8. Bidyuk, P., Romanenko, V., Tymoshchuk, O.: Time Series Analysis. National Technical University of Ukraine «Igor Sikorsky KPI» (2012)
9. Tsay, R.: Analysis of Financial Time Series. Wiley (2010)
10. Altman, E., Avery, R., Eisenbeis, R., Sinkey, J.: Application of Classification Techniques in Business, Banking and Finance (1981)

11. Diebold, F.: Forecasting in Economics, Business, Finance and Beyond. University of Pennsylvania, 607p. (2015)
12. Hosmer, D., Lemeshow, S.: Applied Logistic Regression. Wiley (2000)
13. Terasvirta, T., Tjostheim, D., Granger, C.: Aspects of modeling nonlinear time series. In: Handbook of Econometrics, vol. 4, pp. 2919–2957. Elsevier Science, New York (1994)
14. Hansen, B.: Econometrics. University of Wisconsin, 427p. (2017)
15. Press, S.: Subjective and Objective Bayesian Statistics, 560p. Wiley, Hoboken (2013)
16. Rossi, P., Allenby, G., McCulloch, R.: Bayesian Statistics and Marketing, 348 p. Wiley, New Jersey (2005)
17. Cowell, R., Dawid, A., Lauritzen, S., Spiegelhalter, D.: Probabilistic Networks and Expert Systems, 323 p. Springer, New York (1999)
18. Ng, B.: Adaptive Dynamic Bayesian networks/2007 Joint Statistical Meetings, pp. 1–7. Salt Lake City (USA) (2007)
19. Zgurovsky, M., Bidyuk, P., Terentyev, O.: Method of constructing Bayesian networks based on scoring functions. Cybern. Syst. Anal. **44**, 219–224 (2008)
20. Zgurovsky, M., Bidyuk, P., Terentyev, O., Prosyankina-Zharova, T.: Bayesian Networks for Decision Support Systems. Lyra-K (2012)
21. Zgurovsky, M., Podladchikov, V.: Analytical Techniques of Kalman Filtering. Naukova Dumka Publishers (1995)
22. Trofymchuk, O., Bidyuk, P.: Probabilistic and statistical uncertainty processing using decision support systems. In: Visnyk of Lviv Polytechnic National University, pp. 237–248 (2015)
23. Kalman, R.: A new approach to linear filtering and prediction problems. ASME Trans. (1960)
24. Grewal, M., Andrews, A.: Kalman Filtering: Theory and Practice Using Matlab. Wiley & Sons Interscience (2008)
25. Maybeck, P.: Stochastic Models, Estimation, and Control. Academic Press (1979)
26. Hartikainen, J.: Optimal Filtering with Kalman Filters and Smoothers. Manual for Matlab toolbox EKF/UKF. Helsinki University of Technology, Department of Biomedical Engineering and Computational Science (2008)
27. Carlson, N.: Fast triangular formulation of the square root filter. AIAA J., pp. 1259–1265 (1973)
28. Linearand Nonlinear Filteringin Mathematical Finance an Overview: http://www.researchgate.net/publication/228384820_Linear_and_Nonlinear_Filtering_in_Mathematical_Finance_An_Overview/file/e0b49514c54ff0eee8.pdf, last accessed 2021/08/10
29. A Simplified Approach to Understanding the Kalman Filter Technique: http://www.thuemmel.co/FBMN-HP/download/ZRA/Kalman-Algo.pdf, last accessed 2021/08/13
30. Extended Kalman Filter Tutorial: http://users.ices.utexas.edu/~terejanu/files/tutorialEKF.pdf, last accessed 2021/07/20
31. Nong, Y.: The Handbook of Data Mining. Arizona State University Publishers (2003)
32. Kuznietsova, N., Bidyuk, P.: Financial Risk Analysis: Systemic Approach. Lyra-K (2020)
33. Miroshkin, O., Kovalov, S., Dmitrieva, O., Kushnarenko, V., Svyatnyy, V.: ParSimTech Research and Training Center. In: 2020 IEEE European Technology and Engineering Management (2020)
34. Bashkow, E., Dmitrieva, O., Huskova, N., Michalska, M., Amirgaliyev, Y.. Parallel multiple blocked methods of Bickart type. In: The International Society for Optical Engineering (2019)
35. Tymoshchuk, O., Huskova, V., Bidyuk, P.: A combined approach to modeling nonstationary heteroscedastic processes. In: Radioelectronics, Computer Science, Control (2019)

Chapter 9
An Approach to Reduction of the Number of Pair-Wise Alternative Comparisons During Individual and Group Decision-Making

Vitaliy Tsyganok, Oleh Andriichuk, Sergii Kadenko, Yaroslava Porplenko, and Oksana Vlasenko

Abstract Research on improvement of credibility of expert estimates and reduction of the number of pair-wise comparisons during decision-making support is extremely relevant, due to large time expenditures and high cost of experts' work. Analysis of results of theoretical research of human psycho-physiological limitations, that influence the credibility of expert estimates, indicates that the order of pair-wise comparisons, performed by experts, does influence the credibility of expert session results. We suggest the respective ways of improving the credibility of expert information during decision-making support. We also suggest a procedure for group expert session organization, which uses Combinatorial method of expert estimate aggregation. Based on information on preliminary ranking of alternatives, the procedure allows us to reduce the number of expert pair-wise comparisons, without compromising the credibility of expert session results. Suggested approaches provide the opportunity to improve existing decision-making support methods, and improve the algorithmic principles of pair-wise comparison-based decision support software development.

Keywords Decision support system · Expert · Pair comparisons · Weakly structured domain · Intangible factors

V. Tsyganok (✉) · O. Andriichuk · S. Kadenko · Y. Porplenko · O. Vlasenko
Institute for Information Recording of National Academy of Sciences of Ukraine, Kyiv, Ukraine
e-mail: tsyganok@ipri.kiev.ua

O. Andriichuk
e-mail: andriichuk@ipri.kiev.ua

Y. Porplenko
e-mail: daliss@ukr.net

V. Tsyganok · O. Andriichuk
National Technical University of Ukraine "Igor Sikorsky Kyiv Polytechnic Institute", Kyiv, Ukraine

V. Tsyganok · O. Andriichuk · O. Vlasenko
Taras Shevchenko National University of Kyiv, Kyiv, Ukraine

M. Zgurovsky and N. Pankratova (eds.), *System Analysis & Intelligent Computing*, Studies in Computational Intelligence 1022, https://doi.org/10.1007/978-3-030-94910-5_9

1 Introduction

Activity of any manager calls for decision-making on a daily basis. Decision-making is a special activity type that involves formulation of decision variants (alternatives), subsequent evaluation of their relative efficiency, and respective distribution of resources among these variants. Simpler decision types concern acceptance or rejection of an alternative, selection of the best alternative from a set, or ranking of alternatives.

In order to make complex decisions, we have to consider numerous (decades or hundreds of) interconnected factors that interact among themselves in a sophisticated way. In order to ensure high professional level of decisions, we need to aggregate the knowledge of multiple specialists (experts). However, due to psycho-physiological limitations, a human can only operate with 7–9 objects at a time [1]. Decision support systems (DSS) are one of the means, allowing us to overcome these limitations [2].

In order to make decisions in weakly-structured subject domains, managers, or decision-makers, are using DSS more and more frequently [2]. Such features of these subject domains, as uniqueness of management objects and functional environments, absence of benchmarks, and impact of human factor, bring us to the necessity to use information, obtained from experts (narrow-profile specialists in certain areas), alongside objective information [2]. At the same time, a considerable share of knowledge, used by organizations, comes from the specialists' minds [3, 4].

As we have already mentioned, human mind has certain psycho-physiological limitations [1]. As a result of these limitations, credibility of expert estimation might decrease. Particularly, expert estimates might not reflect the actual level of the expert's knowledge in the issue under consideration. Such discrepancies might substantially influence the adequacy of subject domain model, and, consequently, the quality of recommendations, produced by the DSS.

In our paper we are going to address some of human psycho-physiological features, that influence the credibility of expert estimates, and respective ways of improving this credibility in DSS.

2 Cognitive Biases, Associated with Expert Estimation Process

Improving the credibility of managerial decisions is an important and, at the same time, challenging task, because it primarily concerns the weakly structured subject domains. In order to ensure higher quality of decisions, decision-makers turn to DSS more and more often. Due to complexity and uniqueness of management objects, this process calls for involvement of experts [5]. Expert sessions, in their turn, are very resource-intensive, so all the costs should be justified and rewarded with high quality of obtained information. Credibility of expert estimates depends on multiple factors, which, if ignored, might have negative impact upon the adequacy of subject domain model, and worsen the quality of DSS-produced recommendations submitted to the decision-maker. Cognitive biases (CB), specific to expert estimation process, are one of the most significant factors, that influence the credibility of expert session results [6].

2.1 Sources of Cognitive Biases

It is assumed that expert judgments on specific issue under consideration are, generally, based on objectivity and unbiased attitudes. However, a series of studies in different domains [7] indicates that, in reality, experts are not immune to CB, even if they have the knowledge required to produce accurate judgments [6]. Studies [5, 7] allow us to single out eight CB sources:

(1) Data, with which the expert operates, might include certain fragments, that, quite possibly, trigger CB. For instance, voice data, including their intonation and content, might provoke certain emotions, which, in their turn, influence the decision-making process and its outcome.

(2) Reference materials might influence the expert's perception and interpretation of the data. As a result, the expert might be directed or guided towards a certain "desired" outcome, based on some manipulative input data or reference materials, and not on actual situation, inconsistent with this outcome.

(3) Context-specific information might provoke certain expectations within the expert. These expectations will influence data collection and interpretation strategies, as well as conclusions drawn, even if information is no longer relevant. Insignificant characteristic feature might be assigned larger weight, based on contextual information, implying its seeming importance. The opposite is also true. That is, CB emerge when the context of a certain problem leads to overestimation of certain aspects of analysis, their underestimation, or neglect: some estimate values are rejected as noise, abnormality, or strong deviations. This type of CB is unconscious, and cannot be controlled by the expert.

(4) Impact of similar past experience upon expert estimation process leads to distortions of the judgments. That is, expert estimates are heavily influenced by expectations, imposed by previous experience. As a result, estimation is conducted based on previous use-cases, rather than the current one.

(5) Organizational factors. CB might emerge, when the expert's decision can have impact (in terms of status, welfare etc.) upon himself or some organization/unit, which he sympathizes with or is liable to. Such CB might produce an effect of belonging (identifying with) or taking sides.

(6) Educative activities and workshops might also become the source of CB dissemination or strengthening.

(7) Personal factors. These include motivation, personal beliefs, proneness to risk (or risk evasion), tolerance/intolerance to ambiguity, urgency of decision, stress, fatigue, etc. Individual differences in perception of shapes, colors, and

other features of the same estimated object, might lead to CB. Technological solutions reduce these CB, but don't eliminate them. These solutions are calibrated and maintained by developers, but results of their operation are still interpreted by humans.

(8) Psychophysiology of human brain. Architectural features of human brain impose certain limitations upon volumes of information, which it can process, leading to CB witnessed during decision-making.

On Fig. 1 we can see eight sources of biases, which might "cognitively pollute" data sets, observations, testing strategies, and even experts' analyses and conclusions. These biases are organized into a taxonomy including three categories (top to bottom): sources, specific to particular use-case and analysis (category A); sources, specific to a particular individual, conducting the analysis (category B), and sources, specific to human nature (category C).

Detection of CB and neutralization of their impact is impossible without clear understanding of their nature. Erroneous perception of the nature of CB makes it impossible to neutralize them, so the first step towards minimization of CB impact is to acknowledge their existence and possible impact upon experts.

Misconceptions concerning CB in the context of expert estimation are as follows:

(1) CB are typical only for corrupted and unethical people, neglecting moral principles. In reality, CB are unrelated to personal ethical issues, integrity, or deliberate manipulations. They are inherent to all experts.

(2) CB result from experts' incompetence. Indeed, experts can be incompetent, but incompetence can be detected in advance. In our case, it is, usually, the organization of expert estimation process, that causes CB.

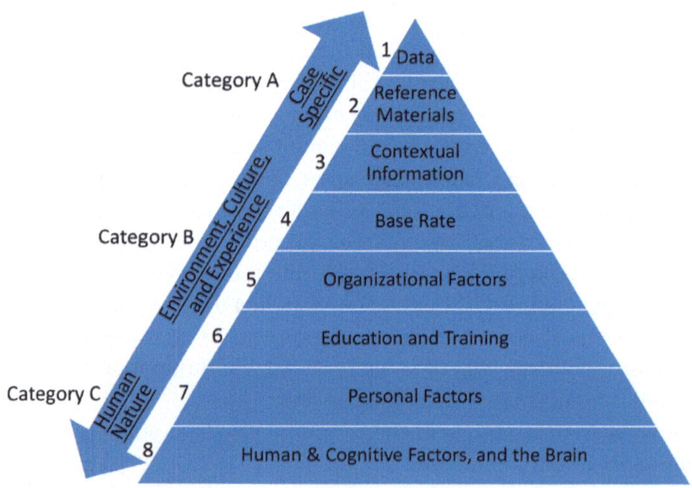

Fig. 1 Sources of cognitive biases

(3) Expert's immunity. General conviction, that an expert is objective and immune to CB a priory. In fact, no one is immune to CB, and experts are often even more prone to certain CB than others [7].
(4) Technologies eliminate CB. Well, they can reduce CB impact, but we should not forget that the system itself is built, programmed, and managed by humans, who also interpret the results of its work.
(5) Turning the blind eye on one's own CB. Experts more often notice CB in others rather than within themselves. As a result, many of them think that CB are not their flaw [8].
(6) Illusion of control. Even if an expert knows, which particular CB he is prone to, he might erroneously think that he can deliberately control them and neutralize their impact. In fact, by trying to reduce the impact of CB by his will, the expert only thinks of them more, and, thus, enhances the impact instead of reducing it [6].

Based on [7], the following recommendations for reducing the impact of CB can be formulated:

(1) Acknowledge the existence of biases and avoid the misconceptions concerning their nature.
(2) Take measures, making experts concentrate solely on the relevant data.
(3) Document all the procedures, in order to be able to detect errors if necessary.
(4) If possible, organize external control.
(5) Use masking methods, which prevent irrelevant information from getting within the scope of experts' attention.
(6) Use such methods as Linear Sequential Unmasking (LSU) to control the sequence, time, and linearity of information impact, in order to minimize "reversing" and biases, related to reference materials.
(7) Involve situational managers, who control, which information is accessed by a certain expert at a certain time.
(8) Use blind and double-blind checks when possible.
(9) Instead of having one "basic goal" or hypothesis, formulate a series of competing and alternative outcomes and hypotheses.
(10) Use differentiated approach, presenting several outcomes with different probabilities instead of one and the only outcome.

3 Quality of DSS Recommendations in Weakly-Structured Subject Domains

Based on the abovementioned human psychophysiological limitations and biases, as well as the specificity of weakly-structured subject domains, we can formulate a series of factors, that influence the quality of DSS recommendations, submitted to a decision-maker:

- Adequacy of subject domain model (no redundancy, ambiguity, contradictions);

- Credibility of decision support methods;
- Input data volume;
- Input data quality (completeness, detail, consistency, credibility, relevance).

The key factors in the list are adequacy of subject domain model and quality of input data. These factors are, in their turn, influenced by redundancy, ambiguity, and contradictions within the DSS knowledge base, as well as by completeness, credibility, and relevance of input data. In order to reduce the impact of negative factors, we propose to research and utilize certain psychological aspects of expert pair-wise comparisons.

3.1 Expert Pair-Wise Comparisons

For credibility's sake, expert estimation of alternatives is performed through pair-wise comparisons, obtained and processed using respective methods [9, 10]. Pair-wise comparisons are performed as part of an original expert estimation technology, that allows an expert to specify the degree of preference between alternatives from a given pair, and provides an opportunity to gradually improve the accuracy of preference value until the level of actual knowledge of the expert in the area under consideration is reached. Thus, we manage to avoid distortions of expert data, and, consequently, improve its credibility [11, 12].

Results of experimental research in the area of perception psychology include several principles [13, 14] that should be taken into account when designing DSS user interface for experts:

(1) equilibrium principle—visual images are closely connected to the feeling of balance: when an individual perceives visual images, he intuitively considers leftmost objects more significant than rightmost ones;
(2) simplicity principle—an individual has an inherent system of benchmarks he or she considers simple: when we consider colors, their hues can also have respective preferences (for example: red is easier for perception than blue; that is, red color surrounded by figures of different colors is perceived faster and better than blue).

During pair-wise comparisons, an expert indicates the degree of dominance of one alternative over another within a certain scale (multiplicative or additive). He indicates the scale grade that reflects his subjective image of absolute value of dominance of one alternative over the other.

Within the analytic hierarchy process [15], experts perform decomposition of an object's overall quality into presumably independent factors (criteria) that exert positive impact upon the main goal or criterion, and compare these factors (criteria) among themselves in pairs, using the fundamental scale. However, the order, in which pair-wise comparisons are performed, is not taken into consideration.

3.2 Psychological Aspects of Pair-Wise Comparisons

Let us consider psychological aspects of impact of alternative pairs' order in expert comparisons upon expert session results. Thanks to research [16, 17], we know that antonyms, such as "strong–weak", form bipolar pairs, where one adjective represents the positive extreme, and the other represents the negative. Such polarization reflects the fundamental system of both negative and positive feature coding, embedded into natural language. For instance, when evaluating the degree of some object's hardness in the context of hard-soft dichotomy, people select a measure of presence of a certain positive quality in the object. Similar elicitation of positive quality measure takes place when estimating values of length, weight, speed, and many other characteristics.

Besides estimation of length, similar experimental research on estimation of weight, brightness, loudness, duration, and other features was conducted [17]. This research disclosed the following patterns: connection between objective measures and subjective estimates is non-linear; respondents tend to overestimate the degree of the quality under consideration in the estimated object. And degree of this over-estimation is proportional to the average measure of the quality under consideration across all estimated objects.

Figure 2 shows the subjective estimation charts from the experiments of Stevens and Galanter [18]. Punctured line on the leftmost chart illustrates the experiment where the respondents estimated the lengths of 17 steel rods, evenly distributed between 4 and 111 cm, in an eleven-grade scale. Punctured line on the middle chart of Fig. 2 represents the experiment, where the respondents estimated the durations of 16 time intervals, evenly distributed within the range from 0.25 to 4 s, in a seven-grade scale.

The rightmost chart on Fig. 2 illustrates the experiment, where the respondents evaluated the squares of 9 rectangles in a five-grade scale. On chart I rectangle sizes were shifted towards minimum square (sorted from minimum to maximum), on chart II the rectangles were presented to experts in random order, while on chart III they were shifted towards maximum square (sorted from maximum to minimum).

Theoretical curves (diagonals) were built based on continuous model of subjective readiness for choice [19]. As we can see, respondents tend to subjectively overestimate the degrees of certain positive features in objects (punctured lines are convex).

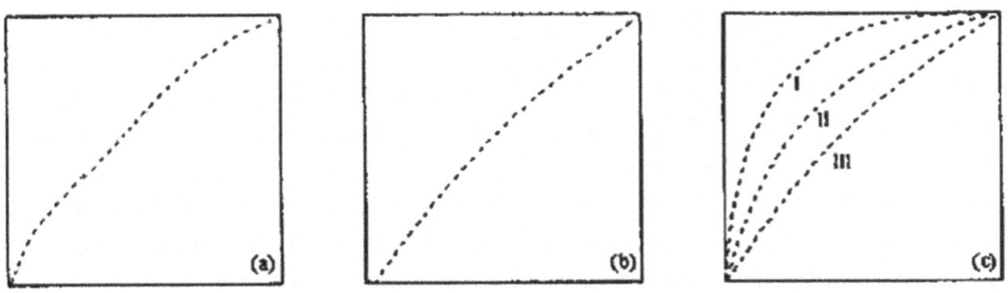

Fig. 2 Experimental research conducted by Stevens and Galanter

Increase of this subjective overestimation is inversely proportionate to the actual degree of this positive feature in the objects, previously presented for pair-wise comparisons (as true values were evenly distributed).

We can conclude, that if we present objects (alternative pairs) to an expert in a certain sequence (order), we can, actually, influence (increase or decrease) the credibility of estimation result.

4 A Method of Expert Pair-Wise Comparisons, Taking the Order of Alternatives into Consideration

The essence of the method is as follows. Let us assume, that a priori we know the ranking of n alternatives. Say, in the process of expert pair-wise comparisons we need to clarify the respective preference values in order to rate the alternatives according to their relative significance. Based on considerations, outlined in the previous section, in order to improve the credibility of expert estimates, we should present pairs of alternatives to the experts for comparison in a certain order.

This can be achieved if we draw a certain analogy between pair-wise comparisons and absolute estimation, which took place in the listed experiments [16–18]. Namely, during each pair-wise comparison (that is, in essence, a relative estimation of alternatives) the expert provides his estimate of a degree of preference of one alternative over the other. Thus, the key idea behind the method is that pair-wise comparisons should be performed in the order that ensures maximum difference of alternatives in pairs presented for comparisons. In this way we can get the most credible result, because expert judgments will be least distorted.

Let the alternatives be numbered according to their strict ranking (as follows):

$$a_1 > a_2 > \cdots > a_n,$$

where: a_i is the alternative number i, $i = \overline{1, n}$, n is the total number of alternatives.

In this case we suggest the following sequence of alternative pairs (for improving the accuracy of preference degrees during individual expert comparisons):

1st turn: (a_1, a_n);
2nd turn: (a_1, a_{n-1}) or (a_2, a_n);
3rd turn: (a_1, a_{n-2}) or (a_2, a_{n-1}) or (a_3, a_n);
…;
$n - 1$ turn: (a_1, a_2) or (a_2, a_3) or … or (a_{n-1}, a_n).

Within each turn, the respective alternative pairs will have equal priorities. That is why within each turn the order of alternative pairs can be arbitrary. Total number of “turns” is one point less than maximum alternative rank. So, in the first place, the expert should compare alternative pairs from the 1st turn, then—from the 2nd turn, the 3rd turn, etc. …, and, finally the turn # $(n - 1)$.

The advantage of the suggested method of expert pair-wise comparisons, taking the order of alternative pairs into consideration, is the opportunity to improve the credibility of estimates and, potentially, reduce the number of comparisons in the expert session. The disadvantage of the method is the need for preliminary ranking of alternatives.

5 Experimental Research of the Method

Participants of the listed psychological experiments estimated only the so-called “tangible” factors [15]. However, under expert data-based decision-making support, it is the “intangible factors” that are estimated [15]. These factors do not have known metric benchmarks or measurement units to compare alternatives with. That is why we conducted separate experimental research of how the sequence of pair-wise comparisons of alternatives influences the credibility of expert session results. Within our research the respondents had to deal with “intangible factors”. Instead of benchmarks we used subjective preferences of each respondent. Similar approach was used in [11, 12]. According to this approach, after completion of comparison sessions, the expert-respondent is shown the histograms with ratings of relative significance of alternatives. He should select the histogram, which represents his subjective preferences in the issue under consideration most adequately.

The experimental study of the impact of pair-wise comparison order upon credibility of expert session results included the following stages:

(1) definition of the problem, goal, or object, in which the expert-respondent feels sufficiently competent (i.e. has sustainable understanding of factors and connections between them) for further decomposition. So, in our research every expert considered his own individual problem;
(2) decomposition of the specified problem into 5–7 criteria (independent factors);
(3) ranking of the formulated criteria according to their importance;
(4) individual expert pair-wise comparisons of importance of criteria (see Fig. 3);

this phase (stage) was performed in three rounds, each featuring different sequences of expert pair-wise comparisons of alternatives: *A*, *B* and *C*, where sequence *A* was formed based on the suggested method (in order to study the effect of potential improvement of expert estimation credibility), sequence *B* was formed based on arbitrary order of expert pair-wise comparisons of alternatives (similarly to a control group or set), while sequence *C* was formed according to the order, inverse to the one, implied by the method (in order to study the effect of potential decrease of expert estimation credibility). A 2-week break was made before each subsequent round in order to prevent the experts from using their answers from previous rounds, and, thus, ensure independence of the estimates within each round. Finally, we should note that in each round the sequence of alternative pair presentation to the expert for comparison was chosen randomly (any of the sequences *A*, *B* or *C* could be

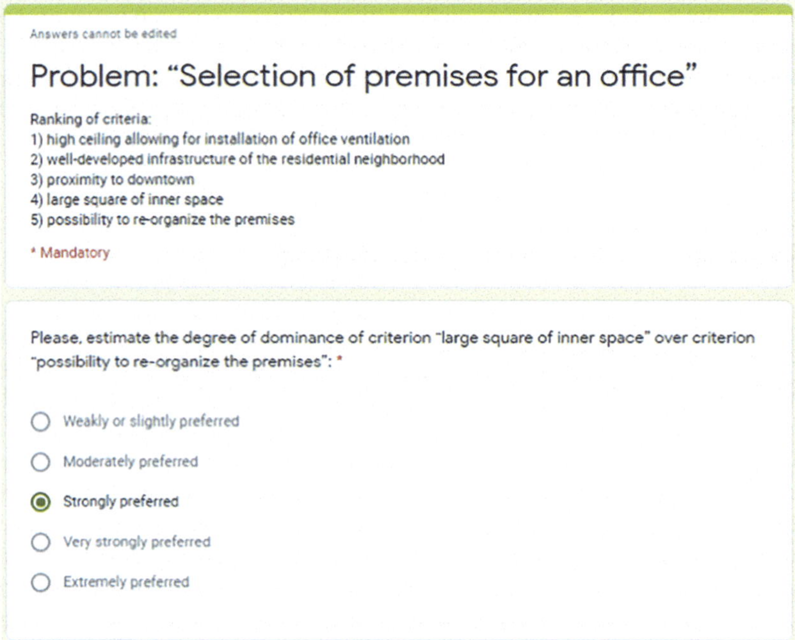
Answers cannot be edited

Problem: "Selection of premises for an office"

Ranking of criteria:
1) high ceiling allowing for installation of office ventilation
2) well-developed infrastructure of the residential neighborhood
3) proximity to downtown
4) large square of inner space
5) possibility to re-organize the premises

* Mandatory

Please, estimate the degree of dominance of criterion "large square of inner space" over criterion "possibility to re-organize the premises": *

- Weakly or slightly preferred
- Moderately preferred
- Strongly preferred
- Very strongly preferred
- Extremely preferred

Fig. 3 A screenshot of expert pair-wise comparison form

chosen in any round), but no sequence was presented to the expert for the second time.

(5) calculation of alternative ratings according to their relative significance for every sequence of expert pair-wise comparisons. For calculation, the eigenvector method was used [15]. Sequences of pair-wise comparisons, resulting in ordinal consistency violation for any series of comparisons (i.e. ranking from stage #3 does not correspond with the obtained rating) were rejected (filtered out);

(6) histogram choice: the expert chooses one of the three histograms of relative alternative significance (see Fig. 4). He is asked to pick a histogram, which most adequately represents his subjective preferences within the issue under consideration, and to explain his choice. Then, out of the remaining two histograms, he is asked to pick the histogram, reflecting his preferences better than the other one. This way the ranking of the sequences of alternative pairs was formed. These sequences were used to empirically confirm the practical value of the suggested method.

Let us stress a series of certain features and ways, ensuring the purity (unbiased nature) of the experiment:

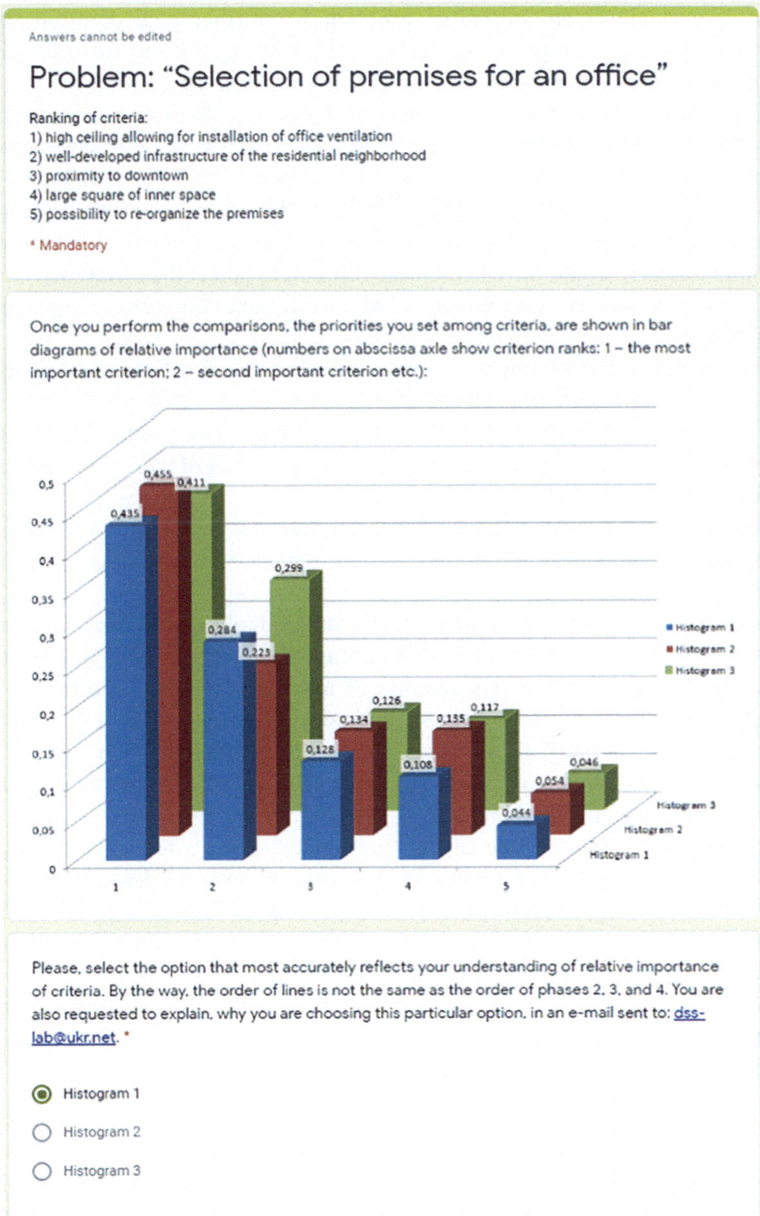

Answers cannot be edited

Problem: "Selection of premises for an office"

Ranking of criteria:
1) high ceiling allowing for installation of office ventilation
2) well-developed infrastructure of the residential neighborhood
3) proximity to downtown
4) large square of inner space
5) possibility to re-organize the premises

* Mandatory

Once you perform the comparisons, the priorities you set among criteria, are shown in bar diagrams of relative importance (numbers on abscissa axle show criterion ranks: 1 – the most important criterion; 2 – second important criterion etc.):

Please, select the option that most accurately reflects your understanding of relative importance of criteria. By the way, the order of lines is not the same as the order of phases 2, 3, and 4. You are also requested to explain, why you are choosing this particular option, in an e-mail sent to: dss-lab@ukr.net. *

- Histogram 1
- Histogram 2
- Histogram 3

Fig. 4 A screenshot of a form, where an expert selects a histogram of relative alternative importance rating, which most adequately represents his subjective preferences

- Every expert chose an individual problem for further consideration. It was chosen under the assumption that he had sufficient expertise to form sustainable understanding of preferences and connections between the factors;
- Only the problems, influenced by up to 5–7 factors (alternatives) were considered. On the one hand, this allowed us to reduce the workload placed upon an expert (5 alternatives—10 comparisons, 6 alternatives—15 comparisons, 7 alternatives—21 comparison), and, thus, minimize the errors resulting from loss of concentration on each round of stage #4. On the other hand, under smaller numbers of comparisons it would have been more difficult to detect the differences in relative alternative importance ratings, based on different comparison sequences on stage #4;
- A two weeks' intermission was made between rounds of stage #4 to ensure independence of results of each round;
- Within each round of stage #4 an expert could not revise previously input data (answers). This allowed us to eliminate conscious impact of pair-wise comparison values within one round;
- Within three rounds of stage #4 comparison sequences *A, B* and *C* were presented to an expert in random order; however, each sequence was presented to him only once;
- If, as a result of relative alternative significance calculation for each comparison sequence, using eigenvector method, we obtained at least one ordinal consistency violation, the whole respective ranking of alternative pairs was rejected. Such violations might be caused by errors made during initial problem definition, its decomposition, alternative ranking, answering the questions of each round of stage 4, as well as by changes of expert's priorities during stage 4, or reconsideration of the expert's competence level;
- When an expert chooses one of the three histograms, which most adequately reflects his subjective preferences in the issue under consideration, the order, in which the histograms are presented to the expert, is arbitrary, and does not coincide with the order of comparison sequences' presentation. This eliminates any deliberate impact upon the expert's choice of the histogram;
- On stage #6 the expert is required to explain his choice of relative alternative significance rating histogram. Thus, we verify whether the explanation does not contradict his ranking of pair-wise comparison sequences.

In order to ensure statistical credibility of the research, we calculated the necessary number of experiment instances. Evaluation of statistical credibility was conducted based on the central limit theorem. If we set the confidence probability value at $P_\beta = 0.9$ (i.e., the probability that the random variable value falls within confidence interval β), and confidence interval size for the given experimental study is $\beta = 0.1$, the minimum necessary number of experiment instances can be calculated based on the following inequality:

$$n \geq \frac{p \cdot (1-p)}{\beta^2} \left(F^{-1}(P_\beta)\right)^2, \quad (1)$$

where F^{-1} is the inverse Laplace function; p is the frequency of repetition of value of the random characteristic under consideration.

We select the value of p based on previously obtained experiment results as the "worst" probability/frequency (i.e. the one closest to 0.5). As a result of test (initial) experiment series, we collected 61 ranking of alternative pair sequences. After filtering (screening), the remaining set of test experiment series constituted 33 rankings of alternative pair sequences. The results of test experiment series are presented in Table 1.

Among the frequencies, defined based on the 1st line of the table $\{18/33 \approx 0.55;\ 9/33 \approx 0.27;\ 6/33 \approx 0.18\}$, the worst one according to the specified criterion is frequency $p = 0.55$, which we will input into the formula for calculation.

After inputting all the respective values into the formula, we get:

$$F^{-1}(0.9) \approx 1.65,$$

then:

$$\left(F^{-1}(0.9)\right)^2 \approx 2.72,$$

$$n \geq \frac{0.55 \cdot (1 - 0.55)}{(0.1)^2} 2.72 = 67.32,$$

and, finally, $n \geq 67.32$. It means, that in order to draw credible conclusions based on the experiment results, it is sufficient to perform at least 68 instances of the experiment.

Experts-respondents included Ph.D. students of the Institute for Information Recording of National Academy of Sciences of Ukraine, students of Taras Shevchenko National University of Kyiv, and students of the National Technical University of Ukraine "Igor Sikorsky Kyiv Polytechnic Institute". As of now, 120

Table 1 Test experiment series

Name of the sequence of expert pair-wise comparisons of alternatives	Number of respondents, that assigned the specified rank to the given sequence of expert pair-wise comparisons of alternatives		
	"1"	"2"	"3"
A	18	9	6
B	6	19	8
C	9	5	19

Table 2 Experiment results

Name of the sequence of expert pair-wise comparisons of alternatives	Number of respondents, that assigned the specified rank to the given sequence of expert pair-wise comparisons of alternatives		
	"1"	"2"	"3"
A	43	20	14
B	13	38	26
C	21	19	37

rankings of alternative pair sequences have been collected. After screening/filtering the set has been reduced to 77 rankings of alternative pair sequences.

Experiment results are presented in Table 2. As we can see from the table, the majority of respondents preferred the sequence of pair-wise comparisons (*A*), prescribed by the suggested method.

As we can see, based on Table 2, the sequence of pair-wise comparisons A was ranked "first" in 56% of cases, "second"—in 26% of cases, and "third"—in 18% of cases. This result allows us to confirm the adequacy of the method of expert pair-wise comparisons, taking the order of alternative pair presentation into account. It also indicates the value of the method in terms of practical application within DSS. It allows us to improve input data quality and adequacy of subject domain model. As a result, the overall quality of recommendations given to decision-maker based on expert data, is improved.

6 Usage of the Approach for Reduction of the Number of Expert Pair-Wise Comparisons During Estimation

One of the ways of expert estimation accuracy improvement is reduction of the number of pair-wise comparisons [20–22]. Similar research was conducted for estimation of "tangible factors" (length of segments, square of figures etc.). It indicated that when n alternatives were estimated, after the minimum number of pair-wise comparisons (necessary for building of at least one spanning tree) $(n - 1)$ was reached, the level of consistency started to gradually decrease, while the level of accuracy was, initially, growing, but then declined. At the same time, these studies did not take the order of pair-wise comparisons into consideration. Alongside the approach suggested in [20], we propose to use the described method of expert pair-wise comparisons, taking the order of alternative presentation into account, and, thus, further increase the level of accuracy, while reducing the minimum necessary number of pair-wise comparisons.

6.1 Reducing the Number of Pair-Wise Comparisons

As we know, pairwise comparisons are used to improve the credibility of expert session results. Generally, beside minimum amount of information required to calculate the priority vector, pair-wise comparison matrices contain significant numbers of additional comparisons, representing redundant information. Minimum number of comparisons, necessary for calculation of a relative priority vector, is $n - 1$ (for a matrix of dimensionality $n \times n$, formed by pair-wise comparisons of n alternatives. These $n - 1$ comparisons corresponding to matrix elements, should belong to any of spanning trees of the matrix's incidence graph. Otherwise, it is impossible to define the priority vector based on this pair-wise comparison set, because available information is insufficient for connecting all alternative weights with each other.

In practice, not all pair-wise comparisons can or should be obtained from the expert. An expert might refuse to compare some pair of alternatives, for instance, due to lack of knowledge on the alternatives, or to some conflict of interest, or biases associated with one of the alternatives in the pair. In such cases, during an expert session, we might have to deal with incomplete pair-wise comparison matrices.

Group expert sessions represent a special type of sessions, where groups of experts are involved by decision-makers in order to increase the credibility of expert estimation. Under group expert estimation it is necessary and sufficient to get at least one spanning tree for the unified set of pair-wise comparisons, provided by the whole expert group. Just like during individual comparisons, credibility of group expert session depends on redundancy of obtained expert information. In fact, credibility is ensured through redundancy of expert estimates. Both hiring of experts and long time spent on estimation make group expert sessions very costly.

So, in order to make expert sessions more efficient, one should find a compromise between accuracy (credibility) of obtained expert estimates and resource intensity of the session. The key resources which we are trying to save are time spent on estimation (session duration) and its cost. So, let us try to make an average expert session less time-consuming through reduction of the number of times the experts are addressed. In order to achieve this, we need to reduce the number of pair-wise comparisons of alternatives.

As we have shown in previous sections and in the experimental research, credibility of pair-wise comparisons depends on the difference between alternatives in a given pair, presented to an expert for estimation. Consequently, the order of pair-wise comparisons influences the credibility of the expert session and accuracy of alternative priority vector, calculated based on the pair-wise comparison matrix. That is why, we propose to start with the most credible pair-wise comparisons of alternatives. Moreover, in order to achieve the required credibility level, in the general case, we can limit the number of pair-wise comparisons (without enumerating all possible pairs), and, thus, significantly reduce the number of comparisons without loss of session credibility.

Let us now describe the procedure, ensuring credibility of an expert session under reduction of pair-wise comparison number.

6.2 Implementation of the Suggested Approach

The procedure, described below, is especially effective for group expert sessions. Application of the procedure is only possible if information on the order of alternatives in the ranking is available. So, preliminary ordinal estimation of alternatives is an essential step, which, in its turn also requires certain expenditures of resources (time and funds), which we are trying to reduce.

In order to save resources on group ordinal estimation, we propose to use the ranking of alternatives obtained by the first expert from the group, that estimates the alternatives. That is, in order to apply the approach, we have to order the set of alternatives, which are to be cardinally rated at the next phase (as their relative priority vector is calculated). So, this ordering can be obtained using one of alternative ranking methods. We should remind, that when experts estimate alternatives using scales with different numbers of grades [11, 12], every expert is initially requested to set the type of preference for each pair of alternatives (that is, ordinally compare them), and only then—specify the preference value, using more detailed cardinal scales, so, alternative ranking problem is resolved automatically.

As we mentioned above, experimental studies in the area of perception psychology allowed the researchers to formulate two hypotheses or principles based on gestalt approach [23]: equilibrium principle and simplicity principle.

According to equilibrium principle, if an expert has already specified ordinal preference on a pair of alternatives, then during further estimation the more significant alternative should be placed on the left, and the less significant one—on the right (as shown on Fig. 5).

Based on simplicity principle, when an expert specifies the degree of preference between alternatives, visual tips, representing verbal phrase or scale grade ("absolute dominance", "strong dominance", "weak dominance", other) should be displayed in red to facilitate better and faster perception (see Fig. 5).

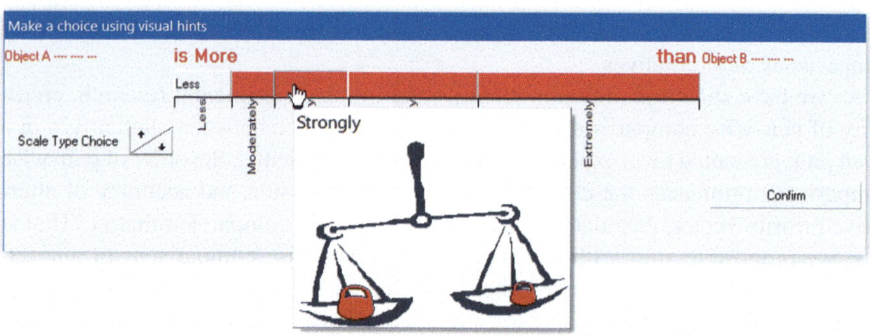

Fig. 5 Equilibrium and simplicity principles implemented within pair-wise comparison software interface

6.3 Idea and Formal Description of Group Estimation Procedure

We have: a ranking of alternatives *R*(*A*): $a_i \rightarrow a_j$, $i, j = (1\ldots n)$—non-strict order relation on the Cartesian square ($A \times A = A^2$) of a set of alternatives $A = \{a_i\}$.

We should: organize group estimation session in the form of incomplete pair-wise comparisons, performed by expert, in order to obtain relative alternative weights w_i.

The key requirements to expert estimation procedure in the context of this section are as follows:

(1) Desired level of credibility (adequacy of obtained alternative weights);
(2) Sufficient estimate consistency level;
(3) Opportunity to reduce the total number of performed pair-wise comparisons (save experts' time and session costs) without loss of credibility.

The idea, providing the basis of the procedure is that alternative pairs should be presented to each expert in the order, based on preliminary alternative ranking, that is, in the order of reduction of credibility of expert estimates (pair-wise comparisons), described in our experimental research. During group expert sessions, we suggest to calculate alternative weights using combinatorial method [24, 25], taking the relative competence of experts into consideration. The method is applicable to both complete and incomplete pair-wise comparison matrices. It also allows us to organize feedback with experts in order to achieve the level of consistency, sufficient for aggregation of expert data. Besides consideration of competence of the expert, who performed pair-wise comparisons, the weight of each comparison is calculated based on potential differences between alternatives, represented by their ranking. So, re-ordering of pair-wise comparisons during group estimation allows us to perform more credible comparisons in the first place, while less credible ones do not necessarily have to be performed, as they do not contribute significantly to the final expert session result. So, thanks to the suggested procedure, we can reduce the general number of pair-wise comparisons and the cost of group expert session.

These basic ideas provide the background for the following group estimation procedure:

- Based on alternative ranking *R*(*A*), we define non-strict ranking *R*(*P*) of alternative pairs from the set $P = \{(a_i, a_j)\}$, $i, j = (1\ldots n)$. Preference relation on set *P* is estimated during pair-wise comparisons. The process of ranking of *P* is described in detail in Sect. 4. Within each turn, the respective alternative pairs will have equal priorities, forming the unstrict order relationship.
- Each turn, where all alternative pairs within ranking *R*(*P*), have equal rank *r*, is assigned a weight $q_r(P)$.
- In the process of group expert session, alternative pairs are selected according to ranking *R*(*P*), so pairs with smaller rank *r* in this ranking (and with respective larger weights $q_r(P)$) are selected and presented to the experts in the first place.
- Group expert estimation procedure is meant to start when the ranking *R*(*A*) is built. The procedure is based on Combinatorial method of pair-wise comparison

aggregation. During aggregation, the method calculates the ratings of individual pair-wise comparison matrices (PCM) of experts, according to the following formula:

$$R_{kql} = c_k c_l s^{kq} s^l / \ln\left(\prod_{u.v} \max\left(\frac{a_{uv}^{kq}}{a_{uv}^l}; \frac{a_{uv}^l}{a_{uv}^{kq}}\right) + e - 1\right), \tag{2}$$

where k, l are the numbers of two experts ($k, l = 1 \ldots m$), whose matrices are compared; c_k, c_l are relative competence coefficients from the problem statement in Sect. 3 of this paper; k and l can be equal or different, i.e. ideally consistent PCM (ICPCM), based on the individual PCM of a given expert number k can be compared with this individual PCM and with individual PCMs built by other experts; q is the number of the respective ICPCM copy ($q = 1 \ldots mT_k$); s^{kq} is the average weight of scales, in which the pair-wise comparisons from the basic set (spanning tree) number q, reconstructed based on PCM of expert number k, were input (s^{kq} is calculated according to formula (3)); s^l is the average weight of scales, in which the elements of PCM of expert number l were built (s^l is calculated based on formula (4)).

$$s^{kq} = \left(\prod_{u=1}^{n-1} \log_2 N_u^{(kq)}\right)^{\frac{1}{n-1}} \tag{3}$$

So, weight of an individual PCM (in terms of the quantity of information), provided by expert number k, can be calculated as follows:

$$s^l = \left(\prod_{u,v=1;v>u}^{n} \log_2 N_{uv}^{(l)}\right)^{\frac{2}{n(n-1)}}, \tag{4}$$

where $N_{uv}^{(l)}$ is the number of grades in the scale, in which the element of the PCM $a_{uv}^{(l)}$ is provided.

The expression for aggregate priorities is as follows:

$$w_j^{aggregate} = \prod_{k,l=1}^{m}\left(\prod_{q^k=1}^{T_k}\left(w_j^{(kq_k l)}\right)^{\frac{R_{kq_k l}}{\sum_{u,p,v} R_{upv}}}\right); j = 1.n. \tag{5}$$

When aggregate priorities are calculated, we can either accept their values as the final result and the problem solution, or improve the level of estimates' agreement.

The procedure we suggest in our research implies the introduction of an additional multiplier—weight q_r—to the estimate of preference between alternatives for each element of individual expert PCM. The result of pair-wise comparisons is the vector of relative alternative priorities $W = \{w_i\}$, $i = (1 \ldots n)$, obtained using Combinatorial method. This method includes the procedure of feedback with the experts. During this feedback consistency of pair-wise comparisons is gradually improved until sufficient consistency level is reached by the whole set of obtained expert estimates.

Within the suggested procedure, the mechanism of group expert session completion after all generalized relative priorities of alternatives are calculated, remains the same as in Combinatorial aggregation method.

We should stress, that, thanks to assigning of greater priorities to more credible pair-wise comparisons, the resulting priority vector calculation process is still possible if we use smaller number of comparisons, without loss of credibility.

7 Further Research

In order to develop a new modification of pair-wise comparison method, taking the order of comparisons into consideration, based on the described approach, we need to perform additional experimental research. Besides that, termination of pair-wise comparison process when sufficient credibility level is reached, is a subject of a separate study.

In addition to the abovementioned usage of the suggested method for expert pair-wise comparison number reduction during estimation, which requires respective experimental corroboration, we also intend to dedicate further research on improvement of credibility of expert estimation through pair-wise comparisons to studies of impact of previous expert session experience, particularly, to fine-tuning of expert estimates based on their previously obtained values.

8 Conclusions

We have substantiated the relevance of research on improvement of expert estimation credibility in DSS. We have presented the results of theoretical studies of human psychophysiological features, influencing the credibility of expert estimates (namely: the equilibrium principle, the simplicity principle, and previous estimation experience). We suggested the respective ways of improving this credibility of information within DSS (particularly, through taking of the listed peculiarities into consideration when developing expert interface of DSS, as well as when defining the order, in which alternative pairs are presented to the expert for comparisons). The conducted experimental research confirmed the adequacy and practical value of the suggested method of expert pair-wise comparisons, taking the order of alternatives into account. The suggested approach and group expert session procedure, based on Combinatorial method of expert estimate aggregation, and taking the initial ranking of alternatives into account, allow us to reduce the number of expert pair-wise comparisons without compromising the credibility of estimation results.

References

1. Miller, G.A.: The magical number seven, plus or minus two: some limits on our capacity for processing information. The Psychol. Rev. **63**(2), 81–97 (1956)
2. Lee, D.T.: Expert decision-support systems for decision-making. J. Inf. Technol. **3**(2), 85–94 (1988)
3. Shabrina, V., Silvianita, A.: Factors analysis on knowledge sharing at Telkom Economic and Business School (TEBS) Telkom University Bandung. Procedia. Soc. Behav. Sci. **169**, 198–206 (2015)
4. Jaziri-Bouagina D.: Handbook of Research on Tacit Knowledge Management for Organizational Success (Advances in Knowledge Acquisition, Transfer, and Management), 1st edn. IGI Global, Hershey, PA (2017)
5. Andriichuk O., Tsyganok V., Kadenko S., Porplenko Y.: Experimental Research of Impact of Order of Pairwise Alternative Comparisons upon Credibility of Expert Session Results. In: 2nd IEEE International Conference on System Analysis & Intelligent Computing (SAIC), pp. 1–5. IEEE, Kyiv, Ukraine (2020)
6. Arnott, D.: Cognitive biases and decision support systems development: a design science approach. Inf. Syst. J. **16**(1), 55–78 (2006)
7. Itiel, E.D.: Cognitive and human factors in expert decision making: six fallacies and the eight sources of bias. Anal. Chem. **92**(12), 7998–8004 (2020)
8. Holmgren, M., Kabanshi, A., Marsh, J.E., Sörqvist, P.: When A+B < A: cognitive bias in experts' judgment of environmental impact. Front. Psychol. **9**, 1–6 (2018)
9. David, H.A.: The Method of Paired Comparisons. Oxford University Press, New York (1988)
10. Totsenko, V.G., Tsyganok, V.V.: Method of paired comparisons using feedback with expert. J. Autom. Inf. Sci. **31**(7–9), 86–96 (1999)
11. Tsyganok, V.V., Kadenko, S.V., Andriichuk, O.V.: Using different pair-wise comparison scales for developing industrial strategies. Int. J. Manag. Decis. Mak. **14**(3), 224–250 (2015)
12. Tsyganok, V.V., Kadenko, S.V., Andriichuk, O.V.: Usage of scales with different number of grades for pair comparisons in decision support systems. Int. J. Anal. Hierarchy Process **8**(1), 112–130 (2016)
13. Dumper K., Jenkins W., Lacombe A., Lovett M., Perimutter M.: Introductory Psychology. Pressbooks, Washington State University (2014)
14. Goldstein, S., Naglieri, J.A.: Encyclopedia of Child Behavior and Development. Springer, US (2011)
15. Saaty, T.L.: Principia Mathematica Decernendi: Mathematical Principles of Decision Making. RWS Publications, Pittsburgh, Pennsylvania (2010)
16. Osgood, C.E., Susi, G.J., Tannenbaum, P.H.: The Measurement of Meaning. University of Illinois Press, Urbana (1957)
17. Kelly, G.A.: The Psychology of Personal Construct. Norton, New York (1955)
18. Stevens, S.S., Galanter, E.H.: Ratio Scales and Category Scales for a Dozen Perceptual Continua. J. Exp. Psychol. **54**(6), 377–411 (1957)
19. Lefebvre, V.A.: Algebra of Conscience. Springer, Netherlands (2001)
20. Wedley, W.C.: Fewer comparisions—efficiency via sufficient redundancy. In: Proceedings of the 10th International Symposium on the Analytic Hierarchy/Network Process. Multi-criteria Decision Making, pp. 1–15, Pitsburg, Pensilvania, USA (2009)
21. Harker, P.T.: Shortening the comparison process in the AHP. Math. Modell. **8**, 139–141 (1987)
22. Whitaker, R.: Validation examples of the analytic hierarchy process and analytic network process. Math. Comput. Model. **46**, 840–859 (2007)
23. Pospelov D.A.: Metaphor, image and symbol in the cognition of the world. Novosti iskusstvennogo intellekta 1, 91–114 (1998) (in Russian)

24. Tsyganok V., Kadenko S., Andriichuk O., Roik P.: Combinatorial method for aggregation of incomplete group judgments. In.: Proceedings of 2018 IEEE 1st International Conference on System Analysis & Intelligent Computing (SAIC), pp. 25–30. IEEE, Kyiv, Ukraine (2018).
25. Kadenko, S., Tsyganok, V., Szádoczki, Z., Bozóki, S.: An update on combinatorial method for aggregation of expert judgments in AHP. Production **31**, 1–17 (2021)

Part II
Computational Intelligence and Intelligent Computing Technologies

Chapter 10
Enhancing the Relevance of Information Retrieval in Internet Media and Social Networks in Scenario Planning Tasks

Michael Zgurovsky, Andrii Boldak, Dmytro Lande, Kostiantyn Yefremov, Ivan Pyshnograiev, Artem Soboliev, and Oleh Dmytrenko

Abstract The problem of enhancing the relevance of information retrieval in the Internet media and social networks is solved in the paper, given the compliance of search results with the user's information needs. The mathematical tools technique and algorithms for determining the proximity of message semantics are proposed on basis of the model of message semantics representation in the form of Directed Weighted Network of Terms. A new method of information retrieval of text messages is developed, which provides for the formulation of user's information needs in the form of text in the natural language. The microservice architecture of the system is offered, in which the specified methods are implemented. The developed system forms part of the analytical and expert environment of the World Data Center for Geoinformatics and Sustainable Development and is used to solve problems of data mining from Internet media and social networks.

Keywords Internet-media · Social networks · Information retrieval · Text mining · DWNT · Microservices

---

M. Zgurovsky · A. Boldak · D. Lande · K. Yefremov (✉) · I. Pyshnograiev · A. Soboliev
National Technical University of Ukraine "Igor Sikorsky Kyiv Polytechnic Institute", Kyiv 03056, Ukraine
e-mail: k.yefremov@wdc.org.ua

M. Zgurovsky
e-mail: mzz@kpi.ua

M. Zgurovsky · A. Boldak · D. Lande · K. Yefremov · I. Pyshnograiev · A. Soboliev · O. Dmytrenko
World Data Center for Geoinformatics and Sustainable Development, Kyiv 03056, Ukraine

D. Lande · O. Dmytrenko
Institute for Information Recording of National Academy of Sciences of Ukraine, Kyiv 03113, Ukraine

M. Zgurovsky and N. Pankratova (eds.), *System Analysis & Intelligent Computing*, Studies in Computational Intelligence 1022, https://doi.org/10.1007/978-3-030-94910-5_10

1 Introduction

Modern IT technologies allow each individual to consume information, and, moreover, to create and disseminate it, thus expressing his/her own attitude to various phenomena, which are important from his/her point of view. The global coverage of all spheres of human activity by the information and communication environment allows us to consider the media content of this environment as an informational presentation of social phenomena, socially significant events, social thought, etc. Fast and adequate perception of social phenomena and events using powerful information retrieval engines in Internet media and social networks allows us to raise scenarios for further course of events and phenomena, perform their full-scale scenario modeling in order to make balanced and reasoned decisions.

One of the most effective modern tools for studying the media content is the so-called Open-Source Intelligence (OSINT), which includes the search, collection and analysis of requested information obtained from public sources [1]. The effectiveness of OSINT in analytical work is determined primarily by the properties of technical means that ensure the quick delivery of up-to-date information, the ability to manage bulk information, the ability to process and analyze data, their easy further use and reuse, etc. [2, 3].

On the other hand, the social information and communication environment can be a platform for negative manipulative effects aimed at increasing psycho-emotional and socio-psychological stresses in the society, the distortion of ethical and moral norms, the moral and political disorientation of people, etc., which in turn, can result in negative transformations of the moral-political and socio-psychological climate in the society, escalation of social conflicts and confrontations.

Development of the methodological basis for revealing, storage, processing and analysis of information received from media sources is an important task aimed at overcoming the so-called "information gap" [4]. In order to solve it, the approach to quantitative assessment of the social transformation effectiveness was proposed in [5, 6], based on the assessment of the proximity of the vector of public expectations, $\vec{S}$, and the vector of government action, $\vec{G}$. This model takes into account the synergy of the society, which results from interaction of native citizens and can be both positive, when government actions meet public expectations, and negative, when mistakes and failures of the government lead to loss of people's hope for positive changes (apathy) and, therefore, generate new failures.

There is a link to a web application in [5, 6], where this approach is implemented in order to analyze public attitude to government actions related to the escalation of quarantine measures caused by the COVID-19 pandemic spread in Ukraine. A test set of data collected over 200 days from the most popular 11 social media and 100 Ukrainian news websites was used (the sample volume was over 2 million messages).

The PRO ET CONTRA web application is presented in [7], it is a part of the integrated online platform "Advanced Analytics", which is used to solve problems related to the analysis of sentiments in information messages. Analysis of information retrieval performed with the use of this application revealed a gap between search results and user's information needs [8–10].

One of the reasons for the mismatch of query results and user's information needs is the imperfection of the query language, which limits the ability to formulate the semantics of the user's information needs.

Therefore, the task of this paper is to enhance the relevance of information retrieval in online media and social networks. This problem is solved by using a semantic model for user's information needs, which is developed on the basis of analytical processing of their textual description in natural language, and assessing the closeness of such a model to semantic models of messages received from Internet media and social networks. In order to implement such search capabilities, the task consists in creating a highly productive, fault-tolerant, distributed information system.

2 Reduction of Inconsistency Between User's Information Needs and the Results of User's Inquiry in Search Engine

The possibilities of expressing the semantics of the user's information needs in traditional query languages used in search engines are limited by the contextual independence of such languages and their formal syntax. In order to overcome this shortcoming, it is necessary to proceed to more powerful forms of expression of semantics, which are semantic networks.

A semantic model for a text in natural language known as Directed Weighted Network of Terms, DWNT, or a Network of Terms, is proposed in [11]. The nodes of this network are the key terms (words and phrases) of the text, and the edges of the network define the semantic-syntactic links between these terms. The construction of the DWNT is based on a predefined set of terms and estimates of their importance [12], it is realized in three stages: building a non-directional network using the algorithm for constructing a horizontal visibility graph for time series (Horizontal Visibility Graph algorithm—HVG) [13, 14], determination directions of links [15] and determination of link weights [12].

In order to develop a method of information retrieval using the DWNT, it is necessary to define operations on DWNT, algorithms for DWNT manipulation, and the method itself.

2.1 Operations on DWNT

Let's introduce the operations on DWNT, which are needed in order to compare (reveal compatibility of) semantics specified by them.

The DWNT unary reduction operation $N \subseteq V \times V \times W$, here N is the DWNT, V is the set of terms, W is the set of weights, with the threshold t, is the mapping:

$$R(N,t) = N_R,$$

$$\forall v_i, v_j \in V : w_R(v_i, v_j) = w(v_i, v_j) - t,$$

$$V_R = V - V_-,$$

$$V_- = \{v_- : w_R(v_-, v_j) \le 0 \wedge w_R(v_j, v_-) \le 0, v_-, v_j \in V\},$$

here N_R is the resultant DWNT, $w(v_i, v_j)$ and $w_R(v_i, v_j)$ is the value of the link weight for the terms v_i, v_j in the original and final DWNT.

The DWNT N unary link indication operation is:

$$L(N) = N_L,$$
$$\forall v_i, v_j \in V : w_L(v_i, v_j) = \begin{cases} 1 & w(v_i, v_j) > 0, \\ 0 & w(v_i, v_j) \le 0. \end{cases}$$

The DWNT N_1 and N_2 binary union operation:

$$N_\cup = N_1 \cup N_2,$$
$$V_\cup = V_1 \cup V_2,$$
$$\forall v_i, v_j \in V_\cup : w_\cup(v_i, v_j) = w_1(v_i, v_j) + w_2(v_i, v_j).$$

The DWNT N_1 and N_2 binary intersection operation:

$$N_\cap = N_1 \cap N_2,$$
$$V_\cap = V_1 \cap V_2,$$
$$\forall v_i, v_j \in V_\cap : w_\cap(v_i, v_j) = w_1(v_i, v_j) \cdot w_2(v_i, v_j).$$

The DWNT N_1 to DWNT N_2 binary complementary operation:

$$N_2/N_1 = N_/,$$

$$V_/ = V_2 - V_1,$$

$$\forall v_i, v_j \in V_/ : w_/(v_i, v_j) = w_2(v_i, v_j) \cdot w_1(v_i, v_j).$$

As the DWNT N matrix norm we use the Frobenius norm:

$$N = \sqrt{\sum_{i=1}^{|V|}\sum_{j=1}^{|V|} w\left(v_i, v_j\right)^2}.$$

2.2 DWNT Manipulation Algorithms

Assume that the algorithm for DWNT iterative structure for messages $M = \langle m_k \rangle, k = \overline{1, |M|}$, that sequentially come to the system, is based on the use of the union operation.

Suppose there is a sequence of text messages M, which is mapped into a sequence of elements $\langle N_k \rangle$, corresponding to each m_k DWNT. Then DWNT N_M for the M sequence can be written as follows:

$$N_M = \bigcup_{k=1}^{|M|} N_k.$$

The algorithm for evaluating the compatibility of semantics for DWNT N_Q and DWNT N_S set of messages (for example, those coming from one source of information, or those that are combined by a certain time interval or a certain topic) is based on estimating the weighted matrix norm of inclusion of N_Q in N_S:

$$d(S, Q) = \frac{L(R\left(N_Q, t_Q\right)/\left(L(R(N_S, t_S) \cap L(R\left(N_Q, t_Q\right)\right)\right)}{L(R\left(N_Q, t_Q\right)}. \quad (1)$$

The correspondence of the set of messages S to the information needs specified by the query Q, is calculated as follows:

$$u(S, Q) = 1 - d(S, Q). \quad (2)$$

2.3 Information Message Search Method Using DWNT

Text data can be mapped in the DWNT network not only for messages, but also for presentation of user's information needs in the text form. That is, some intent document (or set of documents) can be considered as an information intent, that meets the information needs of the user. If the user does not have such a document, he/she can model this document by writing the desired information in natural language.

If the DWNT N_Q network is built for the intent document, then, using formulas (1) or (2), we can estimate conformity of messages and (or) their sets to the user's information intentions which are presented in the intent document.

The proposed method has two significant advantages compared with traditional search models:

- Firstly, the search is not based on an artificially generated query, but on the text-based document provided by the user, which is almost impossible to apply in other search models;
- Secondly, all the resulting documents meet the user's information need, which is unattainable in other models that provide a high level of relevance of the result to the query.

These advantages are achieved by possible reducing the completeness of the search (for example, if the synonyms will not be taken into account when constructing the DWNT networks), and by increasing the indexing time needed to design the DWNT network for all database documents as well.

3 Design of the Data Collection and Mining System for the Data Obtained from Internet Media and Social Networks

Data collection and mining systems for the data obtained from Internet media and social networks must ensure the maintenance and improvement of their operationability in view of permanent changes in the information and communication environment analyzed by such systems and changes in user requirements. Such systems are not only systems of collective use, they are systems of collective development.

Therefore, the need for reuse of software components, their independent development, their independent placement in the system and support by individual development teams, determine the choice of microservice architecture for the implementation of such systems.

3.1 Defining Data-Driven Workflows

In such a system, there are two types of independent data-driven workflows:

1. Intrasystem periodic workflows related to data extraction from various information sources, with their further processing and storage of results in the data warehouse;
2. Workflows related to the processing of queries that are performed through the facade of the system, and based on the processing of a sample of data taken from the data warehouse.

In order to implement the workflows of the first group, it is important to ensure the possibility to expand the functionality of the system. For the workflows of the second group, it is important to ensure a high level of availability of system resources.

We apply functional decomposition and scrutinize the workflows of the first group as pipelines:

$$F = \{P_i\}, i = \overline{1, m}, P_i = \langle A_{i,1}, A_{i,2}, \ldots, A_{i,n_i} \rangle,$$

here F is the system functionality as a set of pipelines P_i (m is the number of pipelines, i is the pipeline index), each of which is determined by a sequence of operations $A_{i,j} (j = \overline{1, n_i})$.

Operations $A_{i,j}$ are software components that implement the mapping:

$$A_{i,j} \subseteq X_{i,j} \times Y_{i,j},$$

here $X_{i,j}$, $Y_{i,j}$ is the specification of the initial data and results of the operation $A_{i,j}$, respectively.

The condition for combining operations A_1, A_2 in the sequence $\langle A_1, A_2 \rangle$ is the compatibility of specifications:

$$Y_1 \subseteq X_2. \tag{3}$$

It should be noted that $\{A_i\} \cap \{A_j\} \neq \emptyset$, here $\{A_i\}$, $\{A_j\}$ are the sets of operations that form part of the pipelines P_i, P_j, respectively. Moreover, the set $A = \bigcup_{j=1}^{m} \{A_j\}$ is not definitively known at the system design stage.

The system deploys stateless microservices that implement operations $A_{i,j}$, whose interaction is realized through messages. Hence, it becomes possible to use a serial-parallel scheme of operations in the framework of implementation of the choreography of distributed transactions associated with data-driven workflows.

The base node of the system is a pipeline of microservices, which controls the life cycle of microservice instances that run in individual processes of the operating environment. The pipeline of microservices also provides an external API for configuring and managing microservice instances. The workflow controller uses this API to configure pipelines based on the availability of microservices stored in the local storage.

Microservices, as well as pipelines, use a self-registration template in the service registry. Asynchronous registration messages and health control messages are processed by the microservice status monitor.

Therefore, the transactions related to the configuration of data processing pipelines are carried out due to the orchestration of local transactions of microservice pipelines, which is performed by the workflow controller in the microservice registry.

The microservice registry provides external RESTfull APIs for monitoring the system readiness and for managing workflows.

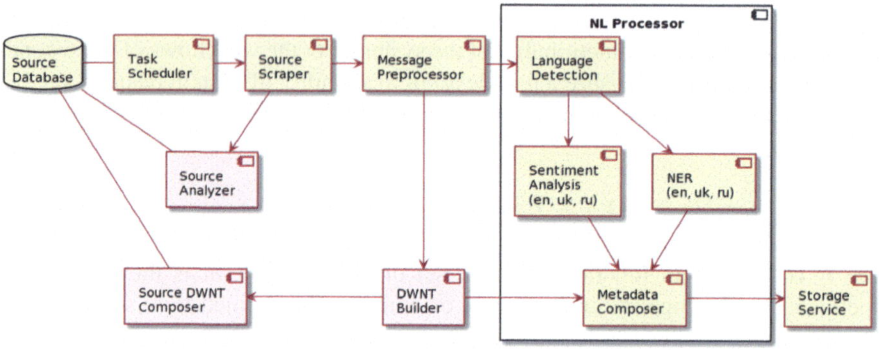

Fig. 1 Basic and auxiliary workflows of data processing

The ability to configure workflows as a sequence of operations causes the verification of the compatibility of the latter according to formula (3), and, as a consequence, the need for a formal description of the microservice types. In this case, we use a specification similar to OpenAPI [16], which includes sections of the instance configuration, setting up incoming and outgoing queues of messages.

3.2 Implementation of Workflows

The basic and auxiliary workflows of data processing are developed in our system. The main workflow (components with yellow background in Fig. 1) periodically fetches data from media sources according to the task queue, which is formed by the scheduler based on data from the store of media sources. The scraper, which is based on the use of parsers of common data formats and headless browsers to implement complex protocols of interaction with information sources, performs message sampling. At the stage of data pre-processing, messages are normalized: they are transformed into a structured JSON document, enriched with source metadata and metadata of the sampling process.

The next stage is the parallel processing of natural language texts (supported by NLP and Sentiment Analysis for English, Ukrainian and Russian languages). This step ends by synchronizing parallel operations and combining their results, then the processing results are passed to the save service.

In the proposed system with microservice architecture, it is quite easy to integrate auxiliary workflows (components with red background in Fig. 1). In particular, the analysis of links contained in messages is used in order to identify new sources and assess the stability of their news traffic [17], which is used to assess the accuracy of information published by the source. Also, microservices for building DWNT messages and DWNT compositions for messages from a single source have been deployed. Hence, it was possible to implement the method of information retrieval of messages proposed in the previous section using the DWNT semantic model.

4 Example of Information Retrieval System Application

The proposed system of collection, control and analytical processing of data obtained from information Internet sources forms part of an integrated analytical and expert environment of the World Data Center for Geoinformatics and Sustainable Development [18], which is used to solve problems related to drafting strategic management decisions in the field of economy, public relations and national security.

In particular, this analytical and expert environment has been used in order to assess public sentiment over the past 6 months regarding the imposition of quarantine restrictions due to the COVID-19 spread. In order to identify events related to these public sentiment, we used the PRO ET CONTRA web application [7], which selected messages from two retrieval engines: the traditional full-text retrieval engine with query language support and the experimental system in which the proposed semantic search method is based on the use of the DWNT.

On the query "anti ~ lockdown protest covid" made in the first retrieval engine, 24,153 messages were received. The respective news traffic is shown in Fig. 2a.

Two documents were selected [19, 20] for the experimental system, which reflect the user's information needs. The DWNT network was built for them (Fig. 3); it was used to search for messages. As a result, 15,698 messages were received. The respective news traffic is shown in Fig. 2b.

The data on news traffic, shown in Fig. 2, is comparable to the data on public sentiment published in the Global Protest Tracker [21]. It is obvious that public sentiment in regard to the COVID-19 outbreak has a potential impact on governance and policy at the local, national and international levels.

Global Protest Tracker uses the results of expert analysis of reports from news sources, such as Al Jazeera, the Atlantic, Balkan Insight, BBC, Bloomberg, CNN, DW News, the Economist, Euronews, Financial Times, Foreign Affairs, Foreign Policy, France24, the Guardian, the Nation, NBC News, New York Times, NPR, Reuters, Radio Free Europe/Radio Liberty, Vox, Wall Street Journal, Washington Post and World Politics Review.

During the last 6 months, Global Protest Tracker has recorded two powerful surges in public sentiment:

- In France, in July 2021 there were protests (with more than 150,000 participants) for 1 day, motivated by concerns about the adoption of excessive restrictions by the government and the introduction of vaccination passports;
- In July 2021, long-lasting protests (with more than 4000 participants) took place in Australia against the imposition of quarantine restrictions in many regions of the country.

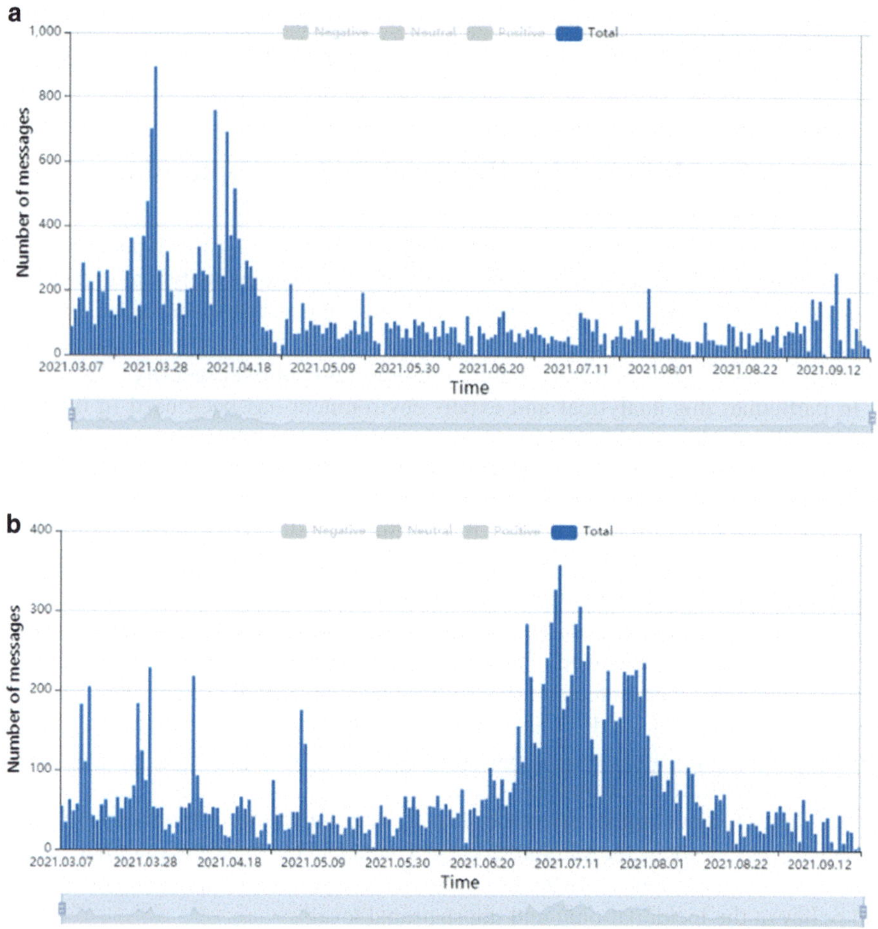

Fig. 2 **a** News traffic selected by query. **b** News traffic selected by DWNT of intent document

We see in Fig. 2a that the news traffic selected by query, does not increase in the specified period of public sentiment activation (July 2021), however, the news traffic selected by the proposed method and shown in Fig. 2b, demonstrates the increase during the same period.

Further analysis of the texts of messages confirmed their compliance with the topic of public sentiment regarding quarantine restrictions related to the introduction of vaccination passports. Also, this sample of messages contained reports of protests in Malaysia in August 2021, related to public dissatisfaction with ineffective actions of the authorities, which led to an increase in coronavirus cases.

Let's evaluate the usefulness of the sample with regard to the user's information needs using weighted entropy:

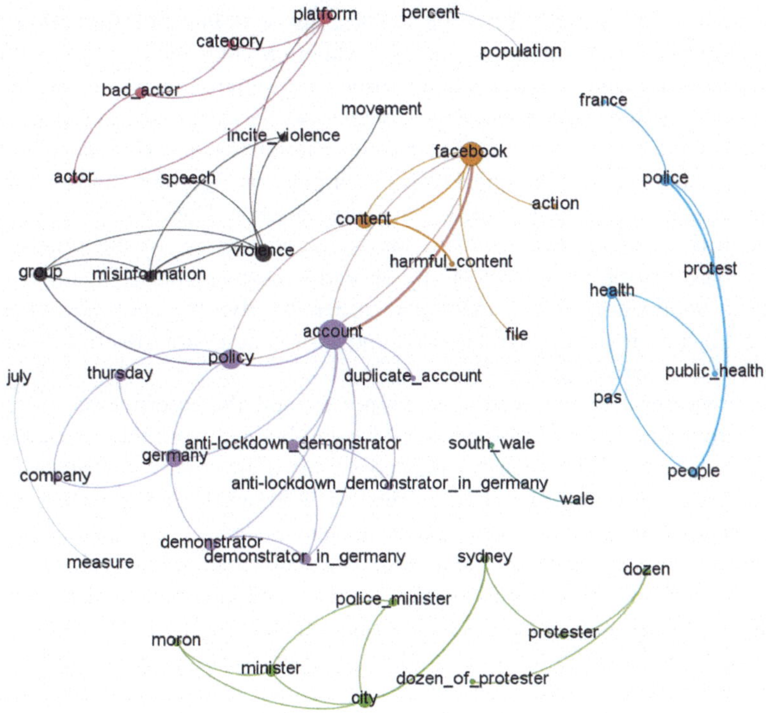

Fig. 3 DWNT of the intent document

$$U = \frac{-p \cdot \log(p) - (1-p) \cdot \log(1-p)}{\log(2)},$$

here U is the weighted entropy (usefulness) of the sample of messages in the range [0,1] (value 1 corresponds to the case when all messages meet the information needs); p is the proportion of sample messages that meet the information needs.

If we assume that the messages recorded during July–August 2021, meet the user's information needs (there are 3861 of such messages for the sample based on the query, and 8300 for the sample based on intent document), then we have for the sample based on the query $U = 0.634$, and for the sample based on intent document $U = 0.998$. That is, the selectivity of search results increases when using the proposed method of information retrieval, which is estimated by weighted entropy, by more than 57%.

5 Conclusions

1. The approach to the implementation of the procedure of information retrieval of text messages is proposed in the paper, which is based on the use of user's

information query in the form of a text description of his/her information needs in natural language (intent document), which provides the better matching of search results with semantics of the user's information needs compared with full-text search engines (which use context-independent query languages).
2. Based on the semantic model of text representation in natural language, DWNT, a mathematical tool technique, DWNT manipulation algorithms and an information retrieval method developed, which are based on determining the semantic similarity of texts. According to the weighted entropy of the relevance of messages distribution the selectivity of search results increased by almost 57% due to application of this method in order to solve the applied problem of revealing public sentiment related to imposition of quarantine restrictions during the last 6 months.
3. The microservice architecture is proposed, and the information system of collecting and analytical processing of data obtained from the Internet media resources and social networks with a possibility of information retrieval, is realized. This architecture is based on the use of intent document, which creates the preconditions for its use in experimental research of new methods of data collection, storage and analytical processing, information retrieval, etc.
4. The developed system forms part of an integrated analytical and expert environment based on the concept of Information and Analytical Situation Center of the World Data Center for Geoinformatics and Sustainable Development.
5. Future research should focus on improving the efficiency of the information retrieval method using DWNT, ways to implement operations on DWNT, methods of indexing the data specified by semantic graph models.

Acknowledgements This research was partially supported by the Ministry of Education and Science of Ukraine. We thank our colleagues from the ISC WDS World Data Center for Geoinformatics and Sustainable Development, National Technical University of Ukraine "Igor Sikorsky Kyiv Polytechnic Institute", who provided insight and expertise that greatly assisted the research.

References

1. ATP 2-22.9: Army Techniques Publication No. 2-22.9 (FMI 2-22.9). Headquarters Department of the Army Washington, DC (2012)
2. Lande, D., Shnurko-Tabakova, E.: OSINT as a part of cyber defense system. Theor. Appl. Cybersecur. **1**, 103–108 (2019). https://doi.org/10.20535/tacs.2664-29132019.1.169091
3. Dodonov, O.G., Lande, D.V., Nesterenko, O.V., Berezin, B.O.: Approach to forecasting the effectiveness of public administration using OSINT technologies. In: Proceedings of the XIX International Scientific and Practical Conference Information Technologies and Security (ITS 2019), pp. 230–233 (2019)
4. Zgurovsky, M.Z., Zaychenko, Y.P.: Big Data: onceptual Analysis and Applications. Springer, Cham (2020). https://doi.org/10.1007/978-3-030-14298-8
5. Zgurovsky, M., Lande, D., Boldak, A., Yefremov, K., Perestyuk, M.: Linguistic analysis of internet media and social network data in the problems of social transformation assessment. Cybern. Syst. Anal. **57**, 228–237 (2021)

6. Zgurovsky, M., Boldak, A., Lande, D., Yefremov, K., Perestyuk, M.: Predictive online analysis of social transformations based on the assessment of dissimilarities between government actions and society's expectations. In: 2020 IEEE 2nd International Conference on System Analysis & Intelligent Computing (SAIC). IEEE (2020). https://doi.org/10.1109/SAIC51296.2020.9239186
7. PRO ET CONTRA v.2.0 Internet Media Analytics. http://open-wdc-dev.herokuapp.com/design/proEtContra%20v.2.0/. Last accessed 22 Nov 2021
8. Broder, A.: A taxonomy of web search. ACM SIGIR Forum. **36**, 3–10 (2002). https://doi.org/10.1145/792550.792552
9. Donato, D., Donmez, P., Noronha, S.: Toward a deeper understanding of user intent and query expressiveness. In: ACM SIGIR, Query Representation and Understanding Workshop (2011)
10. Jansen, B., Booth, D., Spink, A.: Determining the informational, navigational and transactional intent of Web queries. Inf. Proc. Manage. **44**, 1251–1266 (2008)
11. Lande, D.V., Dmytrenko, O.O. Creating the directed weighted network of terms based on analysis of text corpora. IEEE 2nd International Conference on System Analysis & Intelligent Computing (SAIC) (5–9 Oct 2020, Kyiv) (2020). https://doi.org/10.1109/SAIC51296.2020.9239182
12. Lande, D., Dmytrenko, O.: Using Part-of-Speech tagging for building networks of terms in legal sphere. In: Proceedings of the 5th International Conference on Computational Linguistics and Intelligent Systems (COLINS 2021). Volume I: Main Conference Kharkiv, Ukraine, 22–23 Apr 2021. CEUR Workshop Proceedings (ceur-ws.org), vol. 2870. pp. 87–97 (2021)
13. Luque, B., Lacasa, L., Ballesteros, F., Luque, J.: Horizontal visibility graphs: Exact results for random time series. Phys. Rev. E **80** (2009)
14. Gutin, G., Mansour, T., Severini, S.: A characterization of horizontal visibility graphs and combinatoris on words. Physica A **390**, 2421–2428 (2011)
15. Lande, D.V., Dmytrenko, O.O., Radziievska, O.H.: determining the directions of links in undirected networks of terms, In: CEUR Workshop Proceedings (ceur-ws.org). Vol. 2577 urn:nbn:de:0074-2318-4. Selected Papers of the XIX International Scientific and Practical Conference Information Technologies and Security (ITS 2019), vol. 2577, pp. 132–145 (2019). ISSN 1613-0073
16. OpenAPI Specification. https://swagger.io/specification/. Last accessed 22 Nov 2021
17. Soboliev, A.M.: Identification of information sources on global Internet that disseminate inaccurate information. Data Record. Storage Proc. **21**, 56–68 (2019). https://doi.org/10.35681/1560-9189.2019.21.3.183717
18. World Data Center for Geoinformatics and Sustainable Development. http://wdc.org.ua/. Last accessed 22 Nov 2021
19. Covid restrictions over Delta variant trigger protests in Europe, Australia. https://www.hindustantimes.com/world-news/covid-restrictions-over-delta-variant-trigger-protests-in-europe-australia-101627152365258.html. Last accessed 22 Nov 2021
20. Facebook bans German accounts under new 'social harm' policy. https://kdvr.com/news/technology/facebook-bans-german-accounts-under-new-social-harm-policy/. Last accessed 22 Nov 2021
21. Global Protest Tracker. https://carnegieendowment.org/publications/interactive/protest-tracker. Last accessed 22 Nov 2021

Chapter 11
Breathmonitor: AI Sleep Apnea Mobile Detector

Anatolii Petrenko

Abstract This proposal describes the on-line AI system for diagnosis and monitoring sleep apnea at home, based on the processing of human respiratory signals from an accelerometer and pressure transducer composition by using Deep machine learning and alternative louder analytics.

Keywords Monitoring · Respiratory diseases · Deep learning · Accelerometer · Sleep apnea · Convolutional neural network

1 Introduction

The need for diagnosis and observation of different respiratory diseases or their exacerbations has contributed to the development of different direct and indirect methods of respiration measuring: the use of a spirometer in which the patient should breathe; determination of oxygen concentration in blood, which is impossible in real time; analysis of the tracheal sounds, which does not really fit the patient's calm state, since the sound changes very little. At the same time, the recent development of IoT technologies and Deep learning using a convolutional neural network (CNN) have created an opportunity to combine them and build a system of continuous breathing monitoring. Such a system can be used to detect or predict and prevent exacerbations of dangerous conditions in various daily human activities (such as walking, sleeping, and other physical activity).

In particular, sleep apnea is a common disease that affects both children and adults. It is characterized by periods of cessation of breathing (apnea) and periods of decline in respiration (hypopnea). Both types of events have similar pathophysiology and are generally considered equal in their impact on patients. The most common form of sleep apnea, called obstructive sleep apnea, is due to partial or complete collapse of the upper respiratory tract. Obstructive apnea is caused by mechanical stresses

A. Petrenko (✉)
System Design Department, National Technical University of Ukraine "Igor Sikorsky Kyiv Polytechnic Institute", Kyiv, Ukraine
e-mail: petrenko@cad.kiev.ua

M. Zgurovsky and N. Pankratova (eds.), *System Analysis & Intelligent Computing*,
Studies in Computational Intelligence 1022, https://doi.org/10.1007/978-3-030-94910-5_11

on the throat, central sleep apnea is the inability of the brain to send a signal to the diaphragm.

There are several methods of quantifying the severity of respiratory distress, such as measuring the amount of sleep apnea and hypopnea per hour (i.e., sleep apnea index AHI), the severity of oxygen starvation during sleep (oximetry, SpO_2), or the degree of daytime sleepiness [1]). When AHI $\geq$ 5% then 24% of men and 9% of women aged 30–60 years are suspected in sleep apnea. Diagnosis of sleep apnea can use many signals from different sensors (polysomnography, polysomnography) when during night examination in the clinic data on respiratory flow, respiratory motion, SpO_2, posture, electroencephalography (EEG), electromyography (EMG), electrooculography (EOG), and electrocardiography (ECG) are measured. Since such a procedure is very expensive (average device cost $2625) and is not possible at the home, only one or two signals (ECG, SpO_2 oximetry, sound snoring spectrum, etc.) are used to perform "portable" sleep apnea diagnosis in the home setting. Of course, home diagnosis is inferior to laboratory accuracy in diagnosing the disease (80–84% vs. 90–94%) [2].

There is now a great demand for real-time monitoring of the state of the patient's respiratory system in the home environment. For patients, such systems allow home-based measurement of the disease, with the physician, relatives (and/or ambulance) alerted automatically if the patient's vital signs are close to a dangerous limit. For the doctor it is possible to remotely monitor the patient" condition, promptly change the plan of his treatment, maintain contact with the patient, as well as the opportunity to consult with colleagues and specialists in the mode of television sessions with confidential transfer of patient data. In May 2018, an on-line system based on the use of a single-channel ECG signal, which can operate for 46 h with an accuracy of 88%, was proposed [3]. It identifies the RR intervals and RS amplitudes of the ECG time series of signals, their filtering to eliminate motion and muscle noise, and uses the SVM (Support Vector Machines) classifier. In August 2018, another on-line system was announced, also based on ECG signals, using their statistical properties and classifying apnea every minute for 18 h [4]. In order to eliminate the need for electrical contact with the human body necessary to receive ECG signals, a startup was launched in Canada in September 2018, which involves the use of an infrared video camera and infrared sources to capture chest movement from the video and to determine breathing rate and heart rate [5].

2 Proposed Solution

The rapid development of IoT technologies and Deep learning using a convolutional neural network (CNN) make it possible to offer an alternative approach to creating systems for continuous breathing monitoring when combined. Such systems can be used to detect or predict and prevent exacerbations of dangerous conditions in various daily human activities (such as walking, sleeping, and other physical-altering activities). This approach is based on the receipt of breathing signals by fixing the

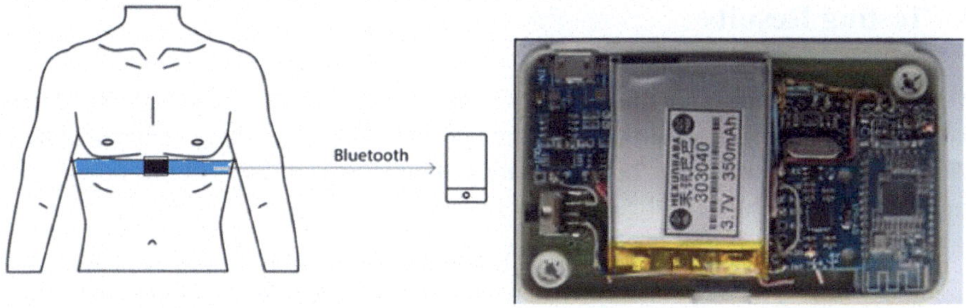

Fig. 1 Respiratory sensor

movement of the chest without the need for electrical contact of the sensor with the human body (Fig. 1). To do this, it is proposed to use a full-body smart sensor, built, say, based on an accelerometer that is placed, in a band or in a pocket of clothing that fits the patient's body, and sends data to the client's smartphone three times per second throughout the year [6]. It essentially replaces the SleepStripTM Portable Diagnostic Device [7], which is one of the main devices for controlling a patient's breathing, whether during a laboratory nightly polysomnography session or during a home sleep study. This device is actually placed on the face of the patient and contains two airflow sensors (oral and nasal thermistors), located directly under the nose and above the upper lip of the patient for breathing, and it must be worn for at least five hours of sleep.

An additional feature of the proposed solution is the use (other than an accelerometer) of pressure sensors or flexible resistive sensors to confidently monitor the breathing condition of a person who is accustomed to sleeping on their stomach. If patients are children, then you can use a belt for, say, Tactilus type "fabric cloth" diapers, which is actually a matrix clock sensor on the surface.

Innovative in our offer is the analytics of the system of continuous monitoring of the breath, built on the basis of modern technologies of IoT and Deep learning with the help of a convolutional neural network (CNN). In this case, the decisions are consistent with the technology of mobile medicine (mHealth) using cellular sensors and smartphones to process their signals as computing edge-nodes, but CNN opens important additional opportunities to improve the quality of signal processing of sensors in the presence of interfering signals (noise) from other sources and instrumental errors of the device. The input signals are pre-normalized with respect to the axis of rotation to reduce the effect of noise on the results, since the accelerometer measures the gravitational acceleration of G and the linear acceleration of A.

A method of converting one-dimensional (1d) accelerometer signals into two-dimensional (2d) graphical images processed by CNN with multiple processing layers is proposed, whereby the accuracy of determining the breathing pattern in different situations for different physical states of patients is increased compared to the case when two-dimensional conversion accelerometer signals are not used.

3 Testing Results

As a result of our study, not only was it possible to apply CNN to highlight breathing patterns (normal; fast shallow breathing; deep labored breathing; Cheyn-stokes) using accelerometer signals, but also demonstrated the usefulness of pre-processing the signal. In particular, in the course of research:

- A test group of 18 users (9 people each gender) was studied, eight of them 25-year-olds, four 40-year-olds, and four more 55-year-olds. All users included in the test group had an average level of physical activity. Measurements were made using a smartphone with a built-in appendix and a sensor attached to the chest, for the regimes of quiet breathing, breathing after squeezing, breathing while running, deep breathing, rapid breathing, delayed breathing. Each action was recorded for 5 min (for re-cording after squeezing, the data was recorded for 30 s, followed by a series of 10 pushes). When recording accelerometer data with a frequency of 30 Hz (30 values per second) for 5 min, 9000 values are recorded for each activity and for each person. Thus, the complete data set includes 432,000 values sufficient to test the proposed methods and hypotheses.
- After the recording session, the data was stored on a mobile phone and then marked off by adding an activity column. The resulting dataset was divided into test and training parts. As the measurements were made with generalized data describing a human activity, separately individual data collected from one participant was stored to further validate the model. Because different people have different lung volumes, chest movement, and other physiological parameters, this means that if the technique is true for a person not included in the test group, then it can be applied to other people.
- For the experiments, a sliding window strategy for signal processing was used, which was to split the time series signal into short fragments. When the data is taken at a frequency of 30 observations per second, the size of the sliding window is equal to 90 samples, which corresponds to 3 s of observations, and the step size of the window, equal to 32 samples, is 1.06 s. Larger or smaller step sizes can be used either to increase accuracy or to reduce the required computing power or power consumption, which is a critical parameter.
- CNN was trained using iterative optimization using a backpropagation algorithm. The most widely used optimization methods are stochastic gradient descent (SGD) and the YellowFin optimizer. The experiment used the YellowFin optimizer since it has the lowest learning error and dynamic learning speed, and SGD typically accumulates error after 300 epochs.
- Using CNN, the accuracy of breathing patterns of about 89% is realized. Most misclassifications are related to recognizing the difference between running and breathing quickly, as well as between push-ups and deep breathing. This can be explained by the fact that delayed and calm breathing signals are less likely to be accompanied by other actions related to chest movement and therefore are more unique. At the same time, when fast breathing and running (as well as pressing and deep breathing), the movements of the chest are more identical and, in addition,

the removed signal is contaminated with a huge amount of side physical activity, which makes the respiratory movement of the chest to be more difficult recognized.

- The apnea signals in the studies were simulated by holding patients' breath for a different number of seconds, and because they have a specific shape, CNN learned and recognized them better than the breathing patterns discussed above (accuracy is above 94%). Now, with the aim of processing real apnea signals, we have transferred the system to proceeding of biomedical signals which are freely available from the PhysioNet website [8].

4 Implementation

There are three consumer groups who are dissatisfied with present sleep apnea diagnostic device situation:

- *Patients* (want to take measurements at home and get early diagnosis by recognizing their breathing patterns before symptoms of the disease appear).
- *Private and family doctors* (they need the possibility of remote continuous monitoring of the disease and the patient's recovery, as well as the diagnostic process with the possibility of remote communication).
- *Medical clinics* (for example, to provide complex treatment and complex services by integrating external data into electronic medical records).

The use of a software-defined network (SDN) with task allocation capabilities, which is proposed in this project, considers the amount of data that is then transmitted to the cloud for processing. The proposed structure can use different routing and data distribution algorithms, which allows you to create a flexible system that is most relevant to the specific case of use. The introduction of high-speed mobile networks, portable sensors, and actuators, energy-efficient microprocessors, and communication protocols has led to the widespread adoption of the Internet of Things (IoT), an adaptive network that connects real-world things (including environmental sensors, and health data) human beings, and executive mechanisms, actively interact with him) between themselves and the Internet. Lightweight and portable, most of these devices are resource-intensive in terms of computing and network power, which by definition means the need for an intermediate computing layer between them and the cloud.

As discussed in academia, this intermediate-level can use 3 additional approaches.

Mist Computing, which is performed at the edge of the network by the most intelligent sensors and starters. Only pre-processed data is sent over the network, and IoT devices do not depend on an Internet connection.

- Fog computing. The layer with computing, network gateway and storage capabilities covers the network from the point of data creation to the place of their storage, which allows decentralized computation of the collected data. Any device that supports the necessary network technologies, storage capabilities, and network capabilities can be used as a fog node.

- Edge computing defines any computing and network resource between a data source and a data center (cloud or local) as a computing node. Usually, edge calculations are used as a general term for all 3 levels.

The transfer of part of the data processing from the cloud to the edge level creates many resource constraints, such as computing power constraints, lack of dynamic horizontal scalability, and power consumption constraints, but brings the following benefits:

- *Location-awareness.* Edge computing systems know the context in which the calculations are performed
- *Reduction of delay.* The classic cloud computing approach with aggregation of data on a smart hub, packet or streaming to the cloud for processing and retrieval results synchronously or asynchronously introduces two pitfalls critical to real-time applications: network delay and possible network failure. Moving data processing to the edge can help solve these problems and has already found use in a variety of systems, including:

Security. Any data transfer is subject to human attack in the middle, and data protection requires energy-intensive encryption algorithms.

- Elimination of bandwidth restrictions. Some data, especially storage media, require a high-speed communication channel. By processing it at the edge level, we eliminate the need for this expensive transmission. For example, smart calls that unlock doors with face identification technology can process the video stream locally instead of sending it to the cloud.
- Reduction of energy consumption. Data transfer is much more expensive in terms of energy consumption than basic processing, so unloading cleaning, aggregation, and extraction operations to the fog and mist layers can increase the battery life of the IoT device.
- Cost reduction. Edge calculations help to use the maximum available resources, which leads to increased economic efficiency.

Modern cases of use and architecture of edge calculations are investigated in [9] with the new architecture of edge networks on the basis of graphs and their estimation. It reduces and even eliminates bottlenecks on web gateways, reducing the amount of data transmitted through them, because they are already pre-processed by foggy nodes. Accordingly, the outgoing traffic decreases.

5 Conclusions

Thus, for the on-line continuous diagnosis and monitoring of sleep apnea at home, a new innovative approach is proposed, based on the processing of human respiratory signals. But in contrast to similar applications [10, 11] instead of the nasal and oral airflow sensors, oximetry sensors (SpO_2), heart rate, or electrocardiogram, it uses an accelerometer and pressure sensor composition to fix the chest movement. Instead of traditional statistical and violet-conversion methods for processing respiratory signals, it has implemented alternative louder analytics through the application of Deep machine learning methods, which essentially allows the system to be tuned to diagnose and monitor other respiratory diseases. As a result, the product became cheaper, more reliable, and comfortable than the existing ones.

In our preliminary estimation, the market is attractive for attracting investors [12, 13]. According to estimates released as early as 2012 by the UN, the global population with respiratory problems has reached nearly 809.4 million and is expected to increase more than double by 2050. Statistics also show that about 100 million of them are suspected of having sleep apnea, of which more than 80% remain undiagnosed (in the US alone, between 18 and 50 million people). The prevalence of apnea in children is 0.7–3%, with the peak incidence of pre-school children. Sleep apnea is present in 5–7% of people over 30, and in those over 65, the incidence of the disease can reach 60%.

There is no direct analog to the proposed BreathMonitor system, given the proposed analytics for signal processing and pattern recognition. In October 2018 our startup proposal "Identifying human breathing patterns using deep convolutional neural networks (CNN)" won the SIKORSKY CHALLENGE 2018 Innovation Projects Startup Contest in the Best Technology Startup nomination.

Additionally, a multimodular approach to diagnostics is investigated now using signals from several different sensors to tune a neural network.

Thus, we have a strong idea, a good team, and we are looking for investments to scale the product needed today by hundreds of thousands of people at high risk due to the coronavirus pandemic.

Acknowledgements Postgraduate students Igor Pysmennyi and Roman Kyslyi and student Oleg Boloban took an active part in this research.

References

1. Almazaydeh, L., Elleithy, K., Faezipour, M.: A panoramic study of obstructive sleep apnea detection technologies. https://www.academia.edu/23577845/A_Panoramic_Study_of_Obstructive_Sleep_Apnea_Detection_Technologies. Last accessed 23 Dec 2021
2. Flemons, W.W., MD; Littner, M.R., Rowley, J.A., Gay, P.: Home diagnosis of sleep apnea: a systematic review of the literature. CHEST **124**(4), 1573–1579 (2003)
3. Surrel, G., Aminifar, A., Rincton, F., Murali, S., Atienza, D.: Online obstructive sleep apnea detection on medical wearable sensors. IEEE Trans. Biomed. Circ. Syst. **99**, 1–12 (2018)
4. Alsalamah, M., Amin, S., Palade, V.: Clinical practice for diagnostic causes for obstructive sleep apnea using artificial intelligent neural networks. In: International Conference for Emerging Technologies in Computing Proceedings, London, UK, August 23–24 (2018)
5. Non-contact sleep monitoring and sleep apnea detection. In: Intelligent Assistive Technology and Systems Lab, Toronto Universitry (2018). http://www.iatsl.org/projects/sleep_apnea.html. Last accessed 23 Sep 2021

6. Petrenko, A., Kyslyi, R., Pysmennyi, I.: Detection of human respiration patterns using deep convolution neural networks. Eastern-Eur. J. Enterp. Technol. **4/9**(94), 5–17 (2018)
7. Shochat, T., Hadas, N., Kerkhofs, M., et al.: The SleepStripTM: an apnoea screener for the early detection of sleep apnoea syndrome. Eur. Respir. J. **19**, 121–126 (2002)
8. https://physionet.org/about/database/. Last accessed 21 Sep 2021
9. Pysmennyi, A., Petrenko, A., Kyslyi, R.: Graph-based fog computing network model. Appl. Comput. Sci. **16**(4), 5–20 (2020)
10. Várady, S., Micsik, T., Benedek, S., Benyó, Z.: A novel method for the detection of apnea and hypopnea events in respiration signals. IEEE Trans. Biomed. Eng. **49**(9), 936–942 (2002)
11. Almazaydeh, L., Elleithy, K., Faezipour, M., Abushakra, A.: Apnea detection based on respiratory signal classification. Procedia Comput. Sci. 310–316 (2013)
12. Market parameters. https://databridgemarketresearch.com/reports/global-sleep-apnea-devices-market/. Last accessed 21 Sep 2021
13. https://www.businesswire.com/news/home/20210811005652/en/Global-Sleep-Apnea-Devices-Market-2020-to-2027---Key-Drivers-and-Challenges---ResearchAndMarkets.com. Last accessed 23 Sep 2021

Chapter 12
Structure Optimization and Investigations of the Hybrid GMDH-Neo-fuzzy Neural Networks in Forecasting Problems

Yevgeniy Bodyanskiy, Yuriy Zaychenko, Olena Boiko, Galib Hamidov, and Anna Zelikman

Abstract The hybrid evolving GMDH- neo-fuzzy system was suggested and investigated. The application of GMDH based on self-organization principle enables to build optimal structure of neo-fuzzy system and train weights of neural network in one procedure. The suggested approach allows to prevent the drawbacks of deep learning such as vanishing or exploding of gradient. As a node of neo-fuzzy system neo-fuzzy neuron with small number of tunable parameters is suggested. This enables to cut training time and accelerate convergence of training. The experimental studies of hybrid neo-fuzzy network were carried out in the task of forecasting of industrial output index, share prices and NASDAQ index. The forecasting efficiency of the suggested hybrid neo-fuzzy system in macro-economy and financial sphere was estimated, its structure was optimized and sensitivity to variation of tuning parameters was investigated.

Keywords Hybrid GMDH-neo fuzzy network · Deep learning · Forecasting

1 Introduction

Last decades deep learning networks cause the great interest for scientists in the field of neural networks (NN), they find a lot of applications for solution of various problems in the field of computational intelligence, e.g., forecasting, pattern recognition, medical images processing and express diagnostics. Many training algorithms were developed and investigated, the drawbacks of deep learning were discovered, and possible ways of their mitigation were suggested.

Y. Bodyanskiy · O. Boiko
Kharkiv National University of Radio Electronics, Kharkiv, Ukraine

Y. Zaychenko (✉) · A. Zelikman
Institute for Applied System Analysis, Igor Sikorsky Kyiv Polytechnic Institute, Kyiv, Ukraine
e-mail: zaychenkoyuri@ukr.net

G. Hamidov
Information Technologies Department Azershig, Baku, Azerbaijan

M. Zgurovsky and N. Pankratova (eds.), *System Analysis & Intelligent Computing*,
Studies in Computational Intelligence 1022, https://doi.org/10.1007/978-3-030-94910-5_12

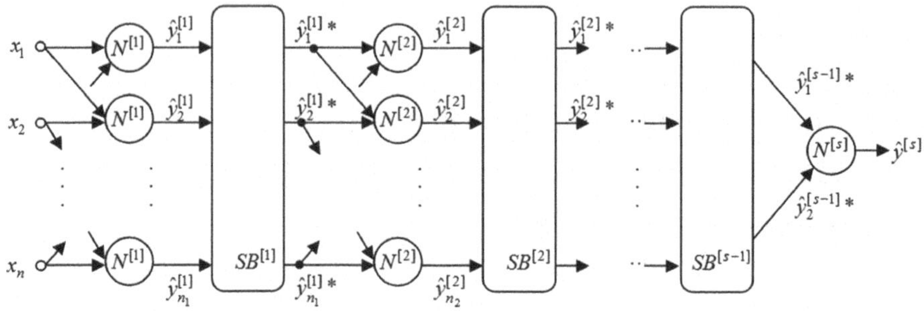

Fig. 1 Evolving GMDH-network

But the conventional deep learning algorithms adjust only weights of connection in NN and don't change the network structure. At the same time the efficiency of training can be substantially increased if not only the weights values are adapted during training process but the structure of neural network itself as well.

For this aim method of self-organization—so-called Group Method of Data Handling (GMDH) [1–3] may be successfully applied. The main advantages of GMDH are following: its ability to construct the model structure automatically in the process of algorithm run; work with short sample size; high speed of training due to small number of inputs.

Due to these properties, the application of GMDH for deep NN structure optimization seems very attractive.

In the previous year GMDH-neural networks with active neurons [4–6], R neurons [7], Q neurons [8], N-adalines [9], GMDH-wavelet-neuro-fuzzy systems [11, 12] and GMDH-neo-fuzzy systems [13] were elaborated.

Last two years new class of hybrid neural networks using this approach were suggested and investigated—hybrid deep GMDH—neuro-fuzzy networks [14] which confirmed high efficiency in problem of forecasting in economy.

But the drawback of GMDH- neuro-fuzzy networks lies in that it's necessary to adapt not only output weights but membership function parameters as well in these networks. The efficiency of training can be substantially increased if to train only rule weights. For this the application of neo-fuzzy neurons as nodes of hybrid networks seems very promising.

The goal of this paper is to develop new hybrid deep learning neo-fuzzy network and investigate its efficiency in the problem of forecasting in economy and financial sphere.

2 The Description of the Evolving Hybrid GMDH-Neo-Fuzzy Network

The evolving hybrid GMDH-network architecture is presented in Fig. 1. To the system's input layer a $(n \times 1)$-dimensional vector of input signals is fed. After that this signal is transferred to the first hidden layer. This layer contains $n_1 = c_n^2$ nodes, and each of these neurons has only two inputs.

At the outputs $N^{[1]}$ of the first hidden layer the output signals are formed. Then these signals are fed to the selection block of the first hidden layer.

It selects among the output signals $\hat{y}_l^{[1]}$ n_1* ($n_1* = F$ is so called freedom of choice) most precise signals by some chosen criterion (mostly by the mean squared error $\sigma^2_{y_l^{[1]}}$).Among these n_1* best outputs of the first hidden layer $\hat{y}_l^{[1]}*$ n_2 pairwise combinations $\hat{y}_l^{[1]}*$, $\hat{y}_p^{[1]}*$ are formed. These signals are fed to the second hidden layer, that is formed by neurons $N^{[2]}$. After training these neurons output signals of this layer $\hat{y}_l^{[2]}$ are transferred to the selection block $SB^{[2]}$ which choses F best neurons by accuracy (e.g. by the value of $\sigma^2_{y_l^{[2]}}$) if the best signal of the second layer is better than the best signal of the first hidden layer $\hat{y}_1^{[1]}*$. Other hidden layers forms signals similarly. The system evolution process continues until the best signal of the selection block $SB^{[s+1]}$ appears to be worse than the best signal of the previous s th layer, that is $\sigma^2_{y_l^{[s+1]}} > \sigma^2_{y_l^{[s]}}$. Then we return to the previous layer and choose its best node neuron $N^{[s]}$ with output signal $\hat{y}^{[s]}$.And moving from this neuron (node) along its connections backwards and sequentially passing all previous layers we finally get the structure of the GMDH-neo-fuzzy network.

It should be noted that in such a way not only the optimal structure of the network may be constructed but also well-trained network due to the GMDH algorithm. Besides, since the training is performed sequentially layer by layer the problems of high dimensionality as well as vanishing or exploding gradient are avoided. This is very important for deep learning networks.

3 Neo-fuzzy Neuron with Small Number of Adjusted Parameters as a Node of Hybrid GMDH-System

Let's introduce into consideration the architecture of the node that is presented in Fig. 2 and is suggested as a neuron of the proposed GMDH-system. As a node of this structure a neo-fuzzy neuron (NFN) by Takeshi Yamakawa and co-authors in [15] is used. The authors of the NFN admit among its most important advantages: the high rate of learning, computational simplicity, the possibility of finding the global minimum of the learning criterion in real time. The neo-fuzzy neuron is a nonlinear multi-input single-output system shown in Fig. 2. The main difference of this node from the general neo-fuzzy neuron structure is that each node uses only two inputs.

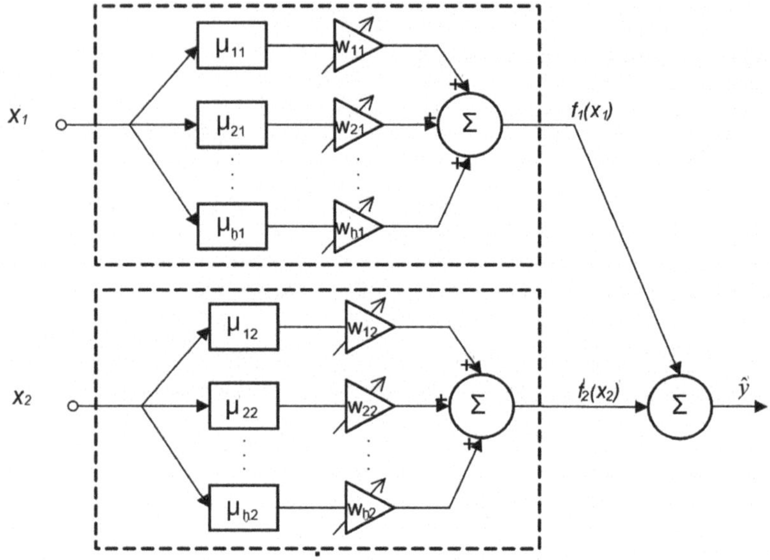

Fig. 2 The neo-fuzzy neuron

It realizes the following mapping:

$$\hat{y} = \sum_{i=1}^{2} f_i(x_i). \tag{1}$$

where x_i is the input i ($i = 1, 2, \ldots, n$), $\hat{y}$ is a system output. Structural blocks of neo-fuzzy neuron are nonlinear synapses NS_i which perform transformation of input signal in the form.

$$f_i(x_i) = \sum_{j=1}^{h} w_{ji}\mu_{ji}(x_i). \tag{2}$$

and realize fuzzy inference: *if* x_i *is* x_{ji} *then the output is* w_{ji},where x_{ji} is a fuzzy set which membership function is μ_{ji}, w_{ji} is a synaptic weight in consequent [15].

3.1 The Neo-fuzzy Neuron Learning Algorithm

The learning criterion (goal function) is the standard local quadratic error function:

$$E(k) = \frac{1}{2}(y(k) - \hat{y}(k))^2 = \frac{1}{2}e(k)^2 = \frac{1}{2}\left(y(k) - \sum_{i=1}^{2}\sum_{j=1}^{h} w_{ji}\mu_{ji}(x_i(k))\right)^2. \quad (3)$$

It is minimized via the conventional stochastic gradient descent algorithm.

For the purpose of increasing training speed Kaczmarz-Widrow-Hoff optimal one-step algorithm [11, 12] can be used which possesses both smoothing and filtering properties.

In case we have priori defined data set the training process can be performed in a batch mode for one epoch using conventional least squares method.

$$w^{[1]}(N) = \left(\sum_{k=1}^{N} \mu^{[1]}(k)\mu^{[1]T}(k)\right)^{+} \sum_{k=1}^{N} \mu^{[1]}(k)y(k) = P^{[1]}(N)\sum_{k=1}^{N} \mu^{[1]}(k)y(k). \quad (4)$$

where $(\bullet)^+$ means pseudo inverse of Moore–Penrose, $y(k)$ denotes external reference signal (real value).

If training observations are fed sequentially in on-line mode, the recurrent form of the LSM can be used in the form

$$\begin{cases} w_l^{ij}(k) = w_l^{ij}(k-1) + \dfrac{\mathrm{P}^{ij}(k-1)\left(y(k) - \left(w_l^{ij}(k-1)\right)^{\mathrm{T}}\varphi^{ij}(x(k))\right)\varphi^{ij}(x(k))}{1 + \left(\varphi^{ij}(x(k))\right)^{\mathrm{T}}\mathrm{P}^{ij}(k-1)\varphi^{ij}(x(k))}, \\ \mathrm{P}^{ij}(k) = \mathrm{P}^{ij}(k-1) - \dfrac{\mathrm{P}^{ij}(k-1)\varphi^{ij}(x(k))\left(\varphi^{ij}(x(k))\right)^{\mathrm{T}}\mathrm{P}^{ij}(k-1)}{1 + \left(\varphi^{ij}(x(k))\right)^{\mathrm{T}}\mathrm{P}^{ij}(k-1)\varphi^{ij}(x(k))}. \end{cases} \quad (5)$$

4 Results

For experimental investigations three forecasting problems were considered which are described below.

4.1 Problem Of Industrial Prices Forecast

For this problem Producers Price Index (PPI) in Ukraine in the period 2017–2018 years was taken as output data. As input variables were taken: wholesale price index (WPI), money aggregate M2, money aggregate M0, import, export.

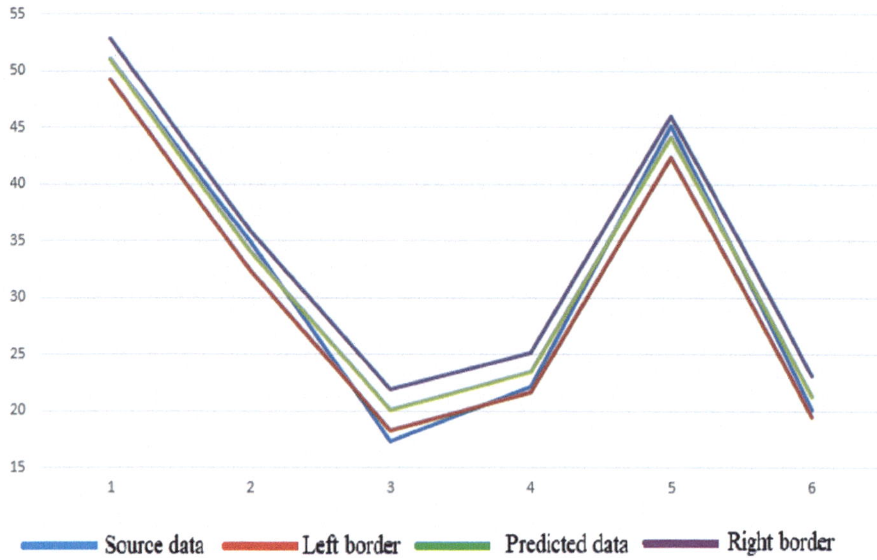

Fig. 3 MSE = 1.02, MAPE = 4.33%

Investigation of Sensitivity to the Ratio Training/Test Sample In the first experiment the sensitivity of the ratio training /test sample to forecasting accuracy was investigated. The forecasting results with ratio 50/50 are presented in Fig. 3.

The forecasting results for the ratio 60/40 is presented in Fig. 4.

The results of the experiment are presented in the Table 1.

The corresponding results are shown in Fig. 5 and Fig. 6.

From these results it follows that the best accuracy was obtained with the ratio training/test sample 60/40.

Investigation of Sensitivity to Inputs Number This experiment was performed with ratio of training sample 50% and 60%, the number of inputs n was taken 3, 4, 5.

For ratio training/test 50/50 the corresponding results are presented in Fig. 7 for n = 5.

The total results by criterion MSE for different ratio are presented in Table 2 and are shown in Fig. 8.

The accuracy results by criterion MAPE are shown in Table 3.

After experiments it was established that the best accuracy was obtained for 3 inputs for training/test ratios 50/50 and 60/40 by both criteria MSE and MAPE.

Other experiments for forecasting Producers Price Index (PPI) in Ukraine in the period 2013–2020 years (quarterly) took as input variables: industrial production index, dollar rate, external investment, money aggregate M2.

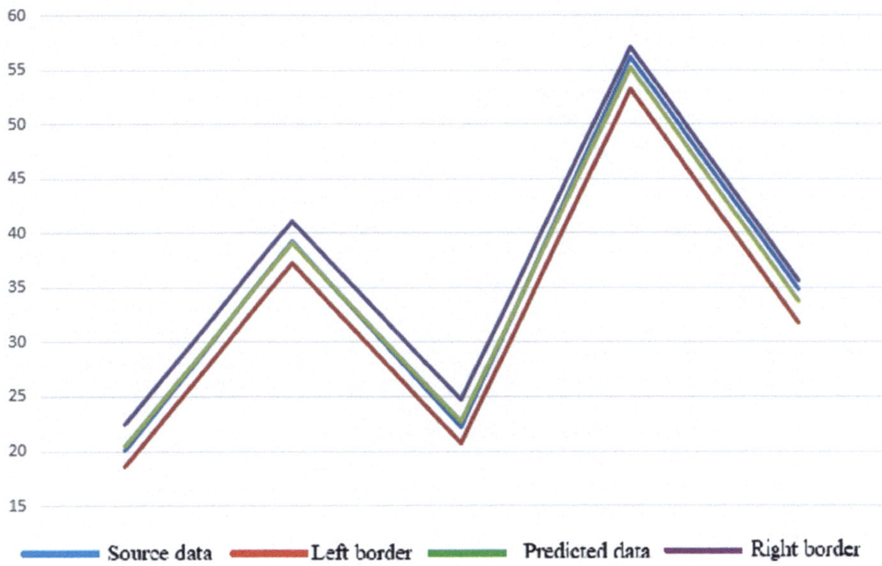

Fig. 4 MSE = 0.55, MAPE = 1.053%

Table 1 MSE and MAPE for different ratios training/test sample for industrial prices forecasting

Criteria/ ratio training/ test	50/50	60/40	70/30	80/20
MSE	1.3237	0.6237	0.791	0.6352
MAPE	1.542%	0.7212%	1.67%	0.7342%

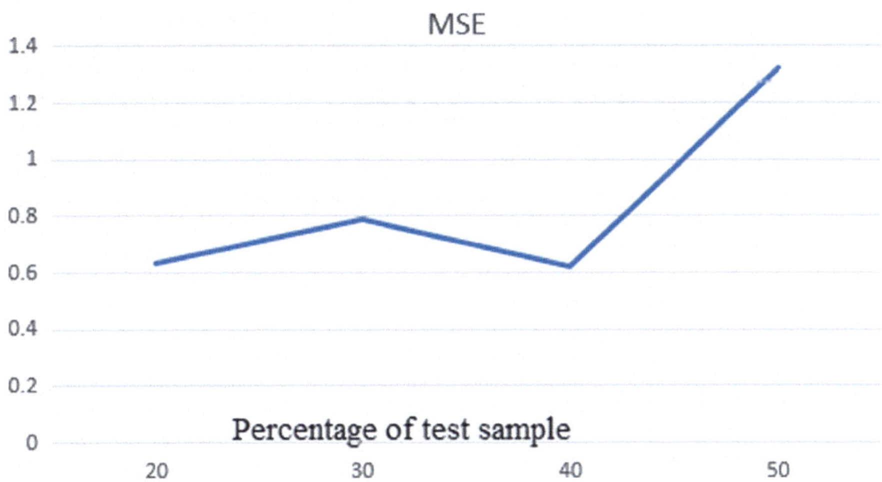

Fig. 5 MSE for different traning/test ratio

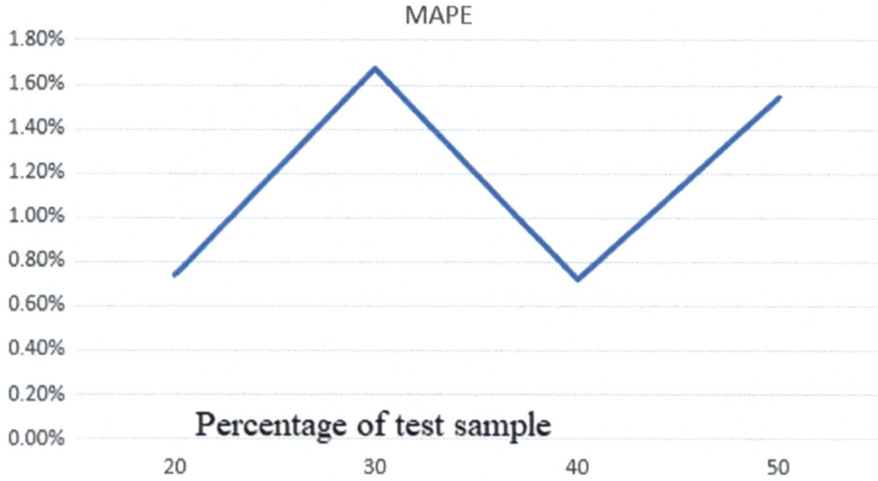

Fig. 6 MAPE for different training/test ratio

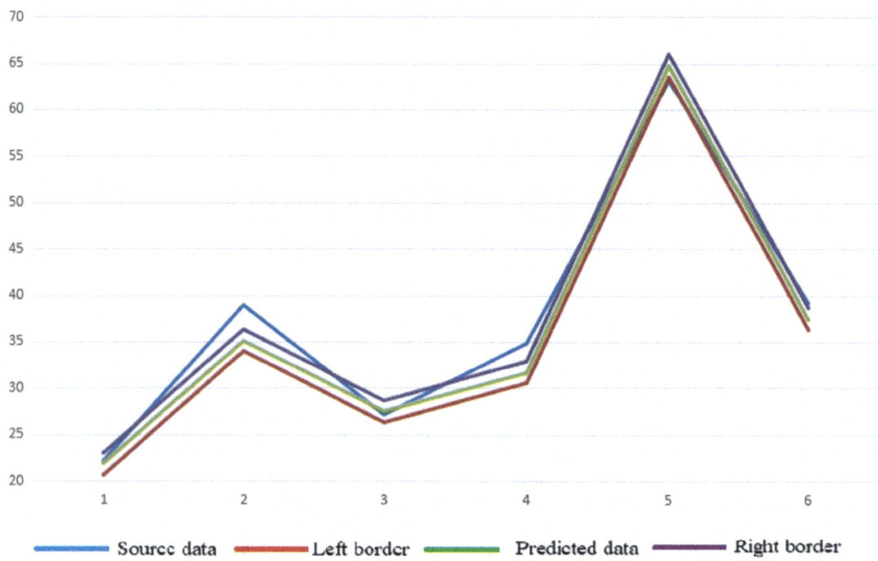

Fig. 7 MSE = 4.15 MAPE = 3.75%

Table 2 Criterion MSE for different number of inputs for industrial prices forecasting

Inputs number/ training sample, %	50%	60%
2	3.33703	1.28
3	1.06	0.55
4	4.7	1.486
5	4.15	4.247

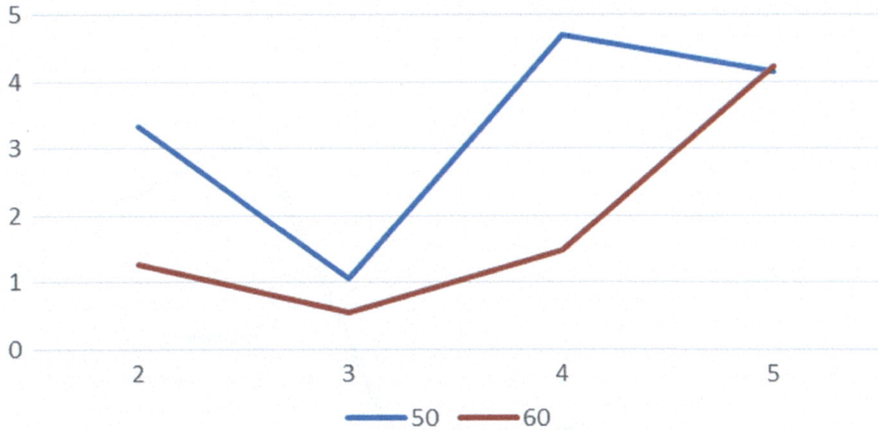

Fig. 8 MSE versus number of inputs for ratio training/test 50/50 and 60/40

Table 3 Criterion MAPE for different number of inputs for industrial prices forecasting

Inputs number/ training sample, %	50%	60%
2	4.765	3.176
3	2.57	1.053
4	3.24	3.761
5	3.75	4.006

Investigation of Sensitivity to the Ratio Training/Test Sample In the first experiment the sensitivity of the ratio training /test sample to forecasting accuracy was investigated. Input number 4 was used for the experiment, the final results of the experiment are presented in Table 4.

The corresponding results are shown in Figs. 9 and 10.

From these results it follows that the best accuracy was obtained with the ratio training/test sample 70/30.

Investigation of Sensitivity to Inputs Number This experiment was performed with ratio of training sample 60% and 80%, the number of inputs n was taken 3, 4, 5, 6.

The total results by criterion MSE for different ratio are presented in Table 5 and are shown in Fig. 11.

Table 4 MSE and MAPE for different ratios training/test sample

Criteria/ ratio training/ test	50/50	60/40	70/30	80/20
MSE	3.16297	3.81893	0.72267	1.34762
MAPE	1.16%	1.41%	0.63%	1.00%

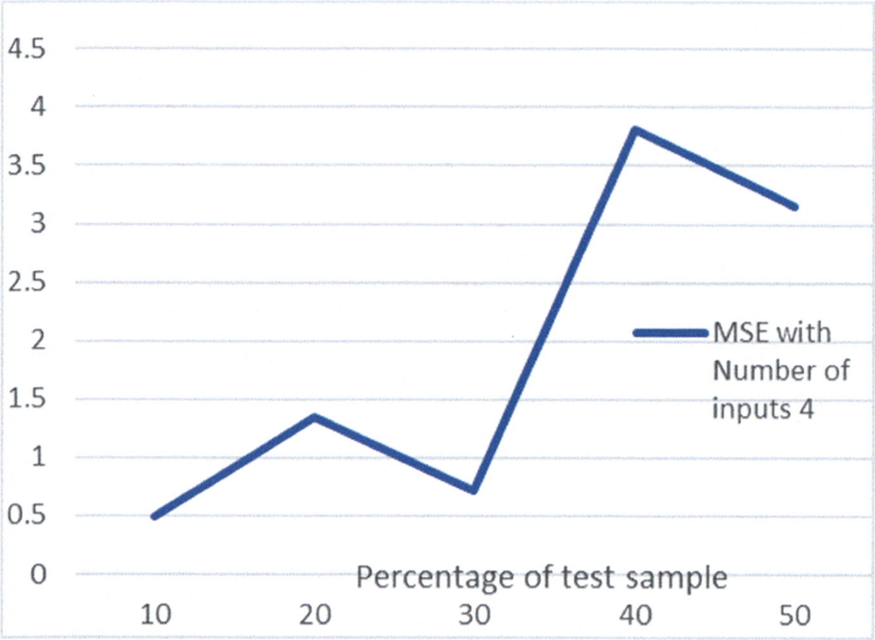

Fig. 9 MSE for different training/test ratio

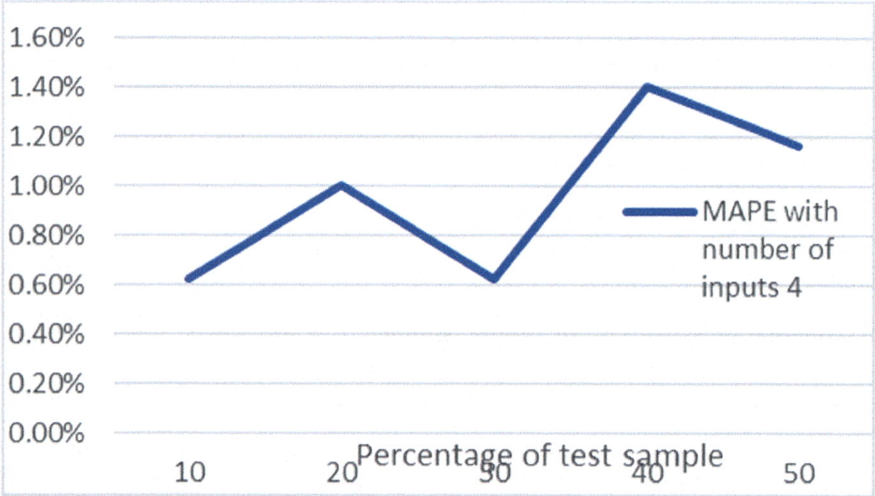

Fig. 10 MAPE for different training/test ratio

Table 5 Criterion MSE for different number of inputs

Inputs number/ training sample, %	60%	80%
3	4.6346925	3.167755
4	3.8189265	1.347618
5	1.9264174	1.121255
6	1.6645216	1.103059

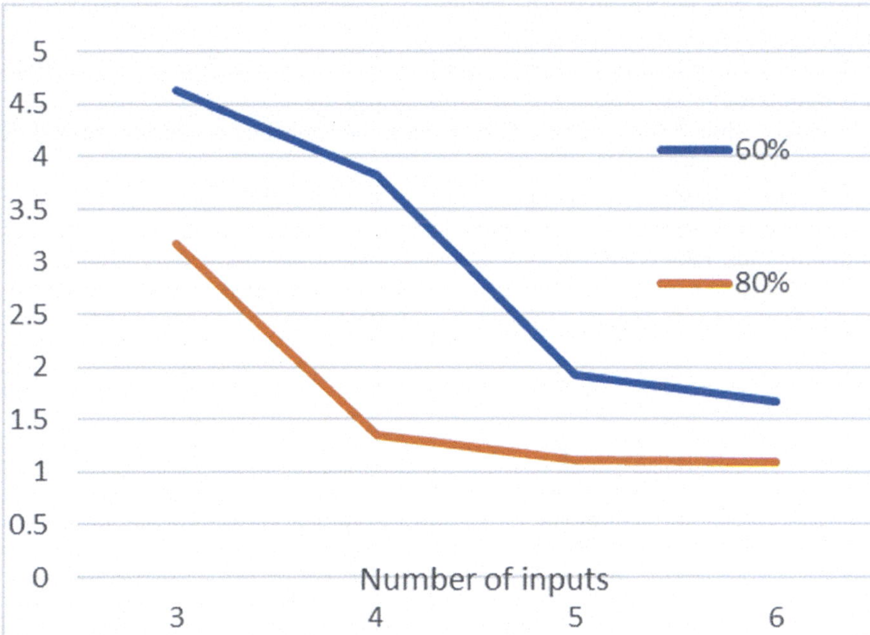

Fig. 11 MSE versus number of inputs for ratio training/test 60/40 and 80/20

The accuracy results by criterion MAPE are shown in Table 6.

After experiments it was established that the best accuracy was obtained for 6 inputs for both training/test ratios 60/40 and 80/20 by both MSE and MAPE criteria.

Table 6 Criterion MAPE for different number of inputs

Inputs number/ training sample, %	60%	80%
3	1.98%	1.31%
4	1.41%	1.00%
5	0.90%	1.01%
6	0.84%	0.94%

4.2 Problem Share Prices Forecasting

In this problem Google shares close prices since August till December 2019 were taken as output variable. As input variables were taken: Open (prices), Low (minimal daily price), High (the highest daily price) and Volume -number price changes between observations.

Number of inputs n = 3. In the first experiment the accuracy sensitivity to ratio training/ test was investigated. The results are presented in Fig. 12 for ratio 50/50 and the total results are shown in Table 7.

From presented results it follows that the best ratio training/test sample is 60/40 by both criteria.

In order estimate forecasting efficiency of hybrid GMDH-neo-fuzzy network it was compared with GMDH method.

The corresponding results are presented in Tables 8 and 9.

From these experiments it was found the best forecasting accuracy was obtained for training/ test ratio 60/40.

The comparison results with classical GMDH has confirmed that hybrid GMDH_ neo-fuzzy network has better accuracy than GMDH.

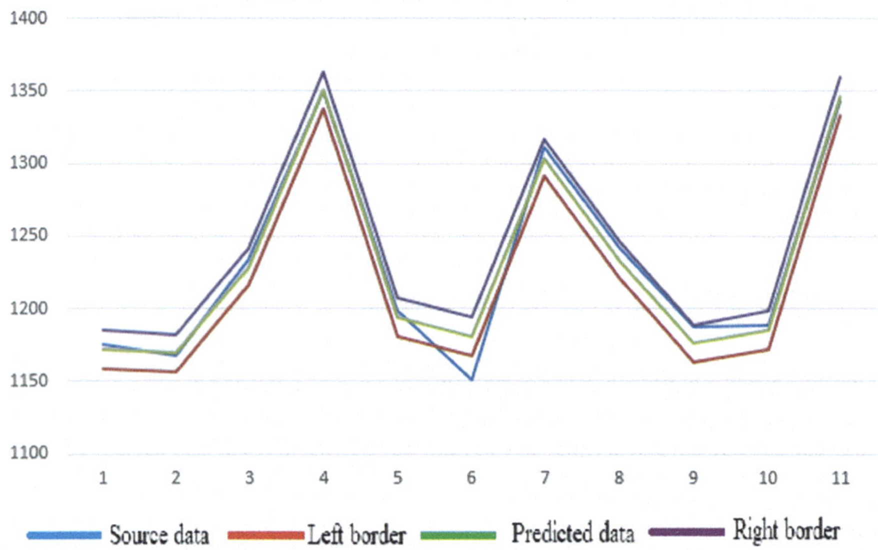

Fig. 12 MSE = 70.31, MAPE = 0.56%

Table 7 Forecasting accuracy for different ratio training/test for share prices forecasting

Criteria/ ratio training/ test	50/50	60/40	70/30	80/20
MSE	70.31	45.67	60.51	47.51
MAPE	0.56%	0.46%	0.5%	0.49%

Table 8 MSE for GMDH-neo-fuzzy network and GMDH for different ratios training/test

Ratio training/test	MSE hybrid neo-fuzzy network	MSE GMDH
50	70.31	72.34
60	45.67	51.46
70	60.51	62.2
80	47.51	52.79

Table 9 Comparative accuracy MAPE for hybrid GMDH-neo-fuzzy network and GMDH

Criteria/ ratio training/ test	MAPE hybrid neo-fuzzy network	MAPE GMDH
50	0.56	0.64
60	0.46	0.8
70	0.5	0.89
80	0.49	1.18

In the next experiment the accuracy sensitivity to number of inputs was investigated The number of inputs n varied from 2 to 10. The results are presented in Fig. 13 (for n = 2), and Table 10.

After experiments it was established the best forecasting accuracy was obtained for inputs number n = 4, ratio 60/40.

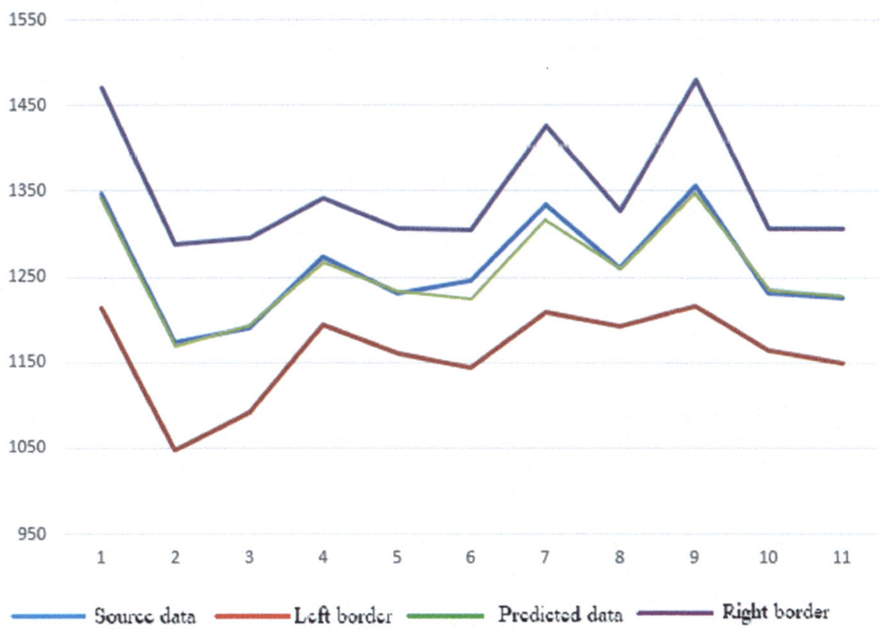

Fig. 13 MSE = 58.62, MAPE = 0.5% for number of inputs n = 2

Table 10 Criteria MSE and MAPE versus number of inputs for share prices forecasting

Criteria/number of inputs	2	3	4	5
MSE	58.62	61.713	42.52	45.39
MAPE	0.5%	0.481%	0.38%	0.41%

In the next experiments the accuracy sensitivity to the type of membership functions (MF) was explored. During experiments the following MF were investigated: triangular, Gaussian and bell-wise. The results are presented in Table 11. The number of inputs was n = 4 under ratio 60/40.

As it follows, the best accuracy was obtained for Gaussian MF.

Investigation of error versus number of cascades In the next experiment it was investigated the GMDH property to construct optimal structure of hybrid neo-fuzzy network. In this experiment the following parameters were taken:ratio training/test 60/40 and number of inputs n = 4 and Gaussian MF.

The process of structure generation and change of error at test sample which obtained in the run of GMDH algorithm of structure synthesis of hybrid neo-fuzzy network are shown in Fig. 14.

Table 11 MSE and MAPE for different membership functions

Membership function	Triangular	Gaussian	Bell-wise
MSE	48.13	35.21	61.03
MAPE	0.417%	0.38%	0.504%

	0	1	2	3
(0, 1)	68.3614331983088			
(0, 2)	75.1526466511595			
(0, 3)	75.1526466511437			
(1, 2)	84.9465302237544			
(1, 3)	99.0762919906586			
(2, 3)	137.4251018604620			
((0, 1), (0, 3))		50.8990395656736		
((0, 1), (0, 2))		50.8990395657113		
((0, 1), (1, 2))		59.3327382908490		
((0, 3), (0, 2))		75.3479176470599		
((0, 3), (1, 2))		58.5786426183977		
((0, 2), (1, 2))		58.5786426183945		
(((0, 1), (0, 3)), ((0, 1), (0, 2)))			50.8990395657047	
(((0, 1), (0, 3)), ((0, 2), (1, 2)))			50.8990395657001	
(((0, 1), (0, 3)), ((0, 3), (1, 2)))			50.8990395656716	
(((0, 1), (0, 2)), ((0, 2), (1, 2)))			50.8990395657365	
(((0, 1), (0, 2)), ((0, 3), (1, 2)))			50.8990395656752	
(((0, 2), (1, 2)), ((0, 3), (1, 2)))			58.5786426183992	
((((0, 1), (0, 3)), ((0, 3), (1, 2))), (((0, 1), (0, 2)), ((0, 3), (1, 2))))				50.8990395656870
((((0, 1), (0, 3)), ((0, 3), (1, 2))), (((0, 1), (0, 3)), ((0, 2), (1, 2))))				50.8990395656936
((((0, 1), (0, 3)), ((0, 3), (1, 2))), (((0, 1), (0, 3)), ((0, 1), (0, 2))))				50.8990395657056
((((0, 1), (0, 2)), ((0, 3), (1, 2))), (((0, 1), (0, 3)), ((0, 2), (1, 2))))				50.8990395656936
((((0, 1), (0, 2)), ((0, 3), (1, 2))), (((0, 1), (0, 3)), ((0, 1), (0, 2))))				50.8990395657056
((((0, 1), (0, 3)), ((0, 2), (1, 2))), (((0, 1), (0, 3)), ((0, 1), (0, 2))))				50.8990395656936

Fig. 14 MSE value at each cascade of hybrid neo-fuzzy network

Analysing the presented results the process of structure construction can be observed after which the optimal structure of deep hybrid neo-fuzzy network was built with minimal MSE at the test sample (marked with yellow colour). The optimal structure has 3 layers for this problem.

4.3 Forecasting Index NASDAQ

In the next experiments efficiency of hybrid neo-fuzzy network in forecasting index NASDAQ was explored. The data was taken in the period from 13.11.17 till 29.11.19. The whole sample consisted of 510 points. The initial data was taken from site www.finanz.ru. As an output variable the closing price of the index next day was taken.

In the first experiment the accuracy dependence on number of inputs was investigated. In Table 12 the forecasting results are presented under different inputs number with 8 membership functions per variable (parameter h) and ratio training/test = 70/30.

In the next experiment the investigation of error dependence on number of MF per variable (parameter h) was performed. Number of inputs was n = 5, training/test ratio was 70/30. The results are presented in Fig. 15.

Table 12 Forecasting MAPE versus different inputs number

Inputs number	2	3	4	5	6	7	8	9	10
MAPE	5.24	4.71	4.33	3.91	4.22	4.72	5.24	5.53	5.85

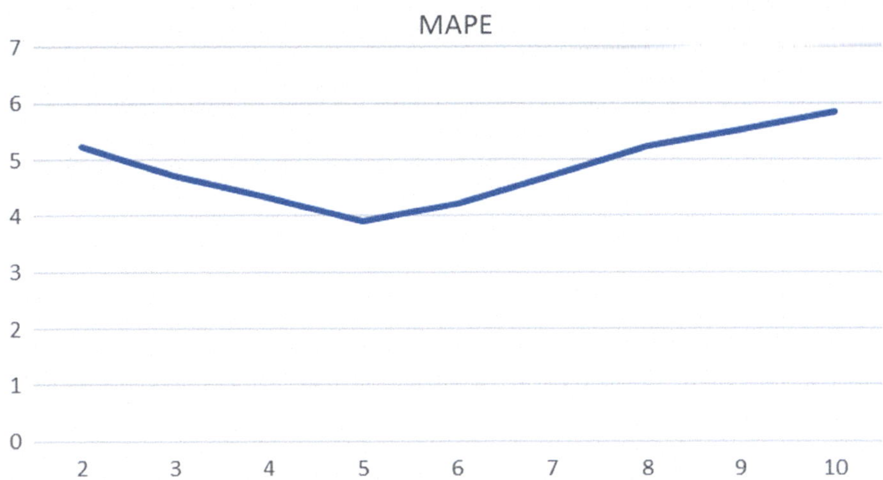

Fig. 15 MAPE versus number of membership functions h per variable

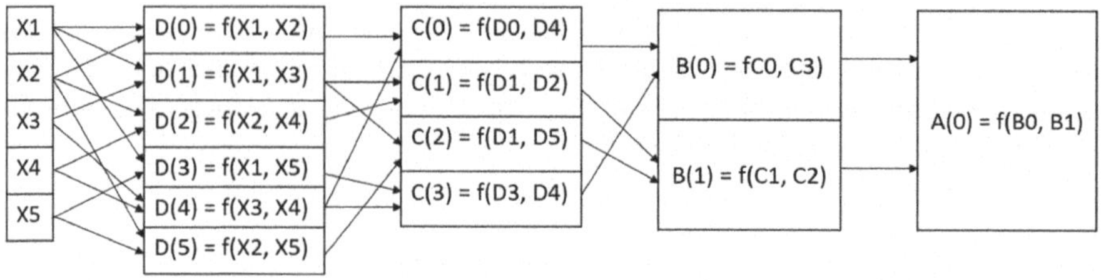

Fig. 16 The optimal structure of hybrid neo-fuzzy network

Analyzing these results, one may conclude that with MF number rise MAPE first falls, then attains minimum and after then begins to rise. That fully matches to self-organization principle of GMDH method [6]. The best value was obtained with the following parameters values: number of inputs n = 5, h = 5, number of layers 4 and MAPE value is 3.91.

The optimal structure generated by GMDH was such: 6 neurons at the first layer, 4 neurons at the second, 2 neurons at the third layer and one neuron at the last (4) layer. All 5 inputs were used in the structure. Freedom choice was taken F = 5. The generated optimal network structure is shown in Fig. 16.

For forecasting efficiency estimation of the hybrid network, it was compared with a cascade neo-fuzzy network [15] and GMDH at the same data. In the cascade neo-fuzzy network, the following parameters values were used: number of inputs n = 9, number of rules 9, cascades number is 3. The comparative forecasting results are presented in the Table 13, training sample—70%.

Analyzing these results one can easily conclude the suggested hybrid neo-fuzzy network has the best accuracy, the second one is GMDH method, and the worst is the cascade neo-fuzzy network.

Other experiments that were held for NASDAQ forecasting problem were related to sensitivity to training/test ratio investigation. In this problem NASDAQ index values for 3 months were taken as output variable. As input variables Google stocks, Microsoft stocks, Intel stocks and pre-history data were taken.

Number of inputs n = 4. The results for forecasting based only on stocks inputs are shown in the Table 14.

From presented results it follows that the best ratio training/test sample is 60/40 by both criteria.

Table 13 MAPE values for different forecasting methods for NASDAQ index forecasting

Inputs number/method	Hybrid GMDH- neo-fuzzy network	GMDH	Cascade neo-fuzzy neural network
4 inputs	4.31	4.19	6.04
5 inputs	3.91	4.11	6.09
6 inputs	4.36	5.53	8.01
7 inputs	4.77	6.26	8.68

Table 14 Forecasting accuracy for different ratio training/test for NASDAQ index forecasting

Criteria/ ratio training/test	50/50	60/40	70/30	80/20
MSE	40,803.7	33,481.1	39,036.3	38,215.7
MAPE	2.27%	1.95%	2.35%	2.25%

Table 15 Forecasting accuracy for case with one previous value

Criteria/ ratio training/test	50/50	60/40	70/30	80/20
MSE	19,075.8	17,738.4	16,283.4	12,959.2
MAPE	1.37%	1.22%	1.31%	1.07%

Table 16 Forecasting accuracy for case with two previous values

Criteria/ ratio training/test	50/50	60/40	70/30	80/20
MSE	17,342.5	12,732.6	13,449.9	23,928.1
MAPE	1.38%	0.97%	1.19%	1.54%

In order to estimate impact of adding pre-history next experiments were held:

1. To the existing input variables one previous value was added (results are shown in Table 15);
2. To the existing input variables two previous values were added (results are shown in Table 16);
3. Three history values only were used as input values (results are shown in Table 17).

The comparison of forecasting results for different sets of input variables is presented on Figs. 17 and 18.

After experiments it was established that the best forecasting accuracy was obtained in case of two previous values added with ratio 60/40. For the case with only three previous values as well as for the case without pre-history, best accuracy was obtained in case of 60/40 ratio. And for case with only one previous value added best accuracy was obtained in case of 80/20 training/test ratio.

Table 17 Forecasting accuracy for case with only three previous values

Criteria/ ratio training/test	50/50	60/40	70/30	80/20
MSE	16,324.7	9766.43	13,336.4	20,989.6
MAPE	1.34%	1.01%	1.31%	1.46%

Fig. 17 Comparison of MSE for NASDAQ index forecasting results

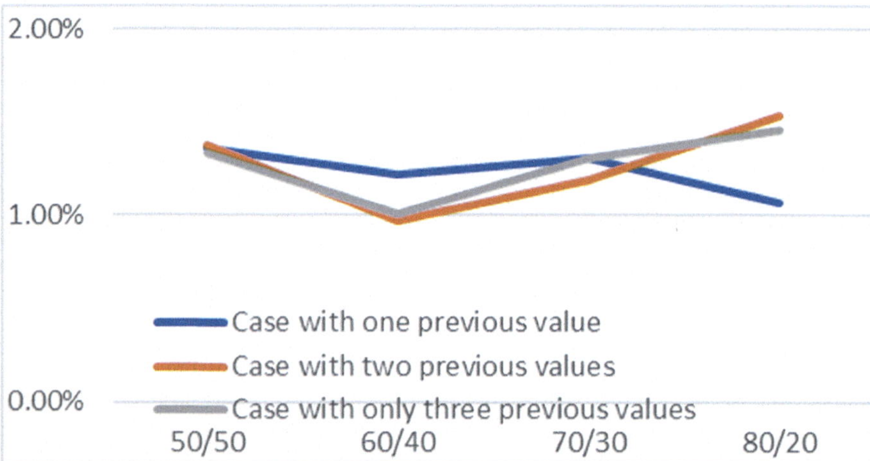

Fig. 18 Comparison of MAPE for NASDAQ index forecasting results

5 Conclusion

In this paper new class of deep learning networks –hybrid GMDH-neo-fuzzy network was suggested and investigated in forecasting Google share prices, NASDAQ index and producers price index. This class of hybrid networks differs from previous ones by that as a node of hybrid network is used neo-fuzzy neuron with two inputs.

During experiments the optimal parameters of hybrid neo-fuzzy network were found: number of inputs, training/ test sample ratio and type of membership functions.

The best results were obtained when the ratio training /test sample is 60/40, number of inputs n = 4 by both criteria MSE and MAPE and the best MF type.

The optimal structure of hybrid neo-fuzzy network was generated using GMDH method.

The comparison experiments of the hybrid neo-fuzzy network with alternative methods GMDH and neo-fuzzy cascade network were performed which had shown the suggested hybrid network has higher forecasting accuracy.

In a whole it was established the developed GMDH-neo-fuzzy network appears to be very perspective for forecasting in the financial sphere and macroeconomy, in particular for share prices and index forecasting at stock exchanges.

Besides it has the following advantages over conventional deep learning networks: high speed of training, possibility to find optimal structure and absence of vanishing and explosion of gradient.

Additionally it has advantages over earlier developed GMDH-neuro-fuzzy network [14]—more simple and efficient training procedures.

References

1. Ivakhnenko, A.G.: Self-learning systems of recognition and automatic control. Kiev (1969)
2. Ivakhnenko, A.G., Stepashko, V.S.: Disturbance tolerance of modeling. Kiev (1985)
3. Zaychenko, Yu.: The fuzzy group method of data handling and its application for economical processes forecasting. Sci. Inq. **7**(1), 83–96 (2006)
4. Ivakhnenko, A.G., Ivakhnenko, G.A., Mueller, J.A.: Self-organization of the neural networks with active neurons. Pattern Recogn. Image Anal. **4**(2), 177–188 (1994)
5. Ivakhnenko, A.G., Wuensch, D., Ivakhnenko, G.A.: Inductive sorting-out GMDH algorithms with polynomial complexity for active neurons of neural networks. Neural Netw. **2**, 1169–1173 (1999)
6. Ivakhnenko, G.A.: Self-organisation of neuronet with active neurons for effects of nuclear test explosions forecastings. Syst. Anal. Mode. Simul. SAMS) **20**, 107–116 (1995)
7. Ohtani, T.: Automatic variable selection in RBF network and its application to neurofuzzy GMDH. In: Proceedings of Fourth International Conference on Knowledge-Based Intelligent Engineering Systems and Allied Technologies, vol. 2, pp. 840–843 (2000)
8. Bodyanskiy, Ye., Vynokurova, O., Pliss, I.: Hybrid GMDH-neural network of computational intelligence. In: Proceedings 3rd International Workshop on Inductive Modeling, Krynica, Poland, pp. 100–107 (2009)
9. Pham, D.T., Liu, X.: Neural Networks for Identification, Prediction and Control. Springer, London (1995)
10. Bodyanskiy, Ye., Vynokurova, O.A., Dolotov, A.I.: Self-learning cascade spiking neural network for fuzzy clustering based on group method of data handling. J. Automa. Inf. Sci. **45**(3), 23–33 (2013)
11. Bodyanskiy, Ye., Vynokurova, O., Dolotov, A. and Kharchenko, O.: Wavelet-neuro-fuzzy network structure optimization using GMDH for the solving forecasting tasks. In: Proceedings of 4th International Conference on Inductive Modelling ICIM 2013, Kyiv, pp. 61–67 (2013)
12. Bodyanskiy, Ye., Vynokurova, O., Teslenko, N.: Cascade GMDH-wavelet-neuro-fuzzy network. In: Proceedings of 4th International Workshop on Inductive Modeling «IWIM 2011», Kyiv, Ukraine, pp. 22–30 (2011)

13. Bodyanskiy, Ye., Zaychenko, Yu., Pavlikovskaya, E., Samarina, M., Viktorov, Ye.: The neo-fuzzy neural network structure optimization using the GMDH for the solving forecasting and classification problems. In: Proceedings of International Workshop on Inductive Modeling, Krynica, Poland, pp. 77–89 (2009)
14. Bodyanskiy, Ye., Boiko, O., Zaychenko, Yu., Hamidov, G.: Evolving Hybrid GMDH-Neuro-Fuzzy Network and Its Applications. In: Proceedings of the International Conference SAIC 2018, Kiev, pp. 134–139 (2018)
15. Yamakawa, T., Uchino, E., Miki, T., Kusanagi, H.: A neo-fuzzy neuron and its applications to system identification and prediction of the system behavior. In: Proceedings of 2nd International Conference on Fuzzy Logic and Neural Networks «LIZUKA-92», pp. 477–483 (1992)

Chapter 13
The Method of Deformed Stars as a Population Algorithm for Global Optimization

Vitaliy Snytyuk, Maryna Antonevych, Anna Didyk, and Nataliia Tmienova

Abstract In this paper, a new method of deformed stars for global optimization based on the ideas and principles of the evolutionary paradigm was proposed. The two-dimensional case was developed and then extended for n-dimensional case. This method is based on the assumption of rational use of potential solutions groups, which allows increasing the rate of convergence and the accuracy of result. Populations of potential solutions are used to optimize the multivariable function, as well as their transformation, the operations of deformation, rotation and compression. The obtained results of experiments allow us to conclude that the proposed method is applicable to solving problems of finding optimal (suboptimal) values, including non-differentiated functions. The advantages of the developed method in comparison of genetic algorithms, evolutionary strategies and differential evolution as the most typical evolutionary algorithms were shown. The experiments were conducted using several well-known functions for global optimization (Ackley's function, Rosenbrock's saddle, Rastrigin's function).

Keywords Global optimization · Method of deformed stars · Polyextreme function · Evolution · Solutions

1 Introduction

The problem of continuous and discrete optimization of functional dependencies is well-known as the methods of its solution. To start with, two approaches are used: the first is the classical integro-differential approach, the technologies of the second approach are called stochastic optimization. In the first case, the function whose optimal value is sought is subject to tight constraints, and the solution of the problem using the second approach does not guarantee finding a global optimum and requires significant computing resources.

New problems that have emerged in recent years, as well as the fact that functional dependencies can be set not only analytically, but also tabularly or algorithmically,

---

V. Snytyuk (✉) · M. Antonevych · A. Didyk · N. Tmienova
Taras Shevchenko National University of Kyiv, Kyiv 04116, Ukraine

M. Zgurovsky and N. Pankratova (eds.), *System Analysis & Intelligent Computing*,
Studies in Computational Intelligence 1022, https://doi.org/10.1007/978-3-030-94910-5_13

and they can be undifferentiated and multi-extremal, triggered the emergence and relevance of the use of new methods and algorithms that attributed to the evolutionary paradigm of simulation. It is known that the main evolutionary paradigms appeared in the second half of the twentieth century. These methods have been improved, they were used to creating hybrid technologies for solving optimization problems of practice. The development of new methods, based on the principles of natural evolution and recommended for use in solving certain classes of problems was continued. The examples of the methods are Differential Evolution [1], Harmony Search [2], Cuckoo Search [3], Symbiotic Organisms Search [4], Genetic Algorithms [5], Firefly Algorithm [6], Water Cycle Algorithm [7], Cooperative Algorithms [8], Tabu Search [9, 10], Grey Wolf optimization [11], Ant Colony Optimization [12] etc.

Important features of such methods are imitation of the technology inspired by natural evolution, and their adaptation to environmental conditions.

Below we will consider the method of deformed stars for two-dimensional case and for n-dimensional case for problem of global optimization.

2 Method of Deformed Stars for Multi-extremal Optimization. Two-Dimensional Case

2.1 Problem Formulation for One-, Two-Dimensional Cases

Consider the problem (one-dimensional case) [13]:

$$f(x) \to \max, \ x \in [a, b], \ \ a, b \in R \tag{1}$$

In the two-dimensional case, we write the problem as:

$$f(x_1, x_2) \to \min \tag{2}$$

$$x_1 \in [p_1, p_2], \ x_2 \in [q_1, q_2], \ p_1, p_2, q_1, q_2 \in R.$$

About characteristics of the functions $f(x)$, $f(x_1, x_2)$ nothing is known.

The function can be given tabularly, and then we can get its model in the form of a neural network, or algorithmically.

Since the problem of multifactor optimization is considered and, given that the number of vertices in a "star" can be arbitrary, we limit ourselves to three- and four- and five-vertex stars.

2.2 The Method of Triangular Deformed Stars (MODS-3)

We will solve the problem (2) by the method of deformed stars. As with other population methods, we will work with sets of potential solutions. Let the number of the current iteration: $t = 0$.

We generate an initial population of potential solutions $P_t = \{(x_1^1, x_2^1), (x_1^2, x_2^2), \ldots, (x_1^n, x_2^n)\}$, uniformly distributed in a rectangle $\Omega = [p_1, p_2]\times[q_1, q_2]$, $|P_t| = n$. For all $\left(x_1^j, x_2^j\right) \in P_t$ we find $f_j = f\left(x_1^j, x_2^j\right)$, that form the population $F_t = \{f_j\}_{j=1}^n$.

We choose the value $i, j, k = random\{1, 2, \ldots, n\}, i \neq j \neq k$. The points (x_1^i, x_2^i), $\left(x_1^j, x_2^j\right)$ and (x_1^k, x_2^k) form a triangle (do not lie on a straight line). Among these three points we find one in which the value of the function f is minimal, because there is a high probability that the optimal value will be in direction of the best point. At the same time, it is necessary to explore in more detail its neighborhood. We find the coordinates (x_1^0, x_2^0) of the triangle "center":

$$x_1^0 = \left(x_1^i + x_1^j + x_1^k\right)/3,\ x_2^0 = \left(x_2^i + x_2^j + x_2^k\right)/3. \tag{3}$$

Perform the compression of triangle and transfer to the distance a in direction of the best point. Without limiting the generality, we consider (x_1^i, x_2^i) to be such a point. To do this, we calculate the distance R from (x_1^0, x_2^0) to (x_1^i, x_2^i). Set the parameter r—the point offset distance, $r = coef \cdot R$, where $coef \in (0, 1)$. Then the ratio of segments is as follows: $a = r/R$. We shift the best vertex:

$$x_1^{i'} = (1 + a) \cdot x_1^i - a \cdot x_1^0, \quad x_2^{i'} = (1 + a) \cdot x_2^i - a \cdot x_2^0. \tag{4}$$

Then we shift other vertices:

$$\begin{aligned} x_1^{j'} &= \left(x_1^i + a \cdot x_1^j\right)/(1 + a), x_2^{j'} = \left(x_2^i + a \cdot x_2^j\right)/(1 + a), \\ x_1^{k'} &= \left(x_1^i + a \cdot x_1^k\right)/(1 + a), x_2^{k'} = \left(x_2^i + a \cdot x_2^k\right)/(1 + a). \end{aligned} \tag{5}$$

In Fig. 1 ABC is the parent triangle, point B is the best. A new triangle formed by potential solutions-offsprings is $A_1B_1C_1$. If at least one of points lies inside the rectangle Ω, then this point is recorded to the population P_z.

If at least one point is outside the rectangle, then we need to move this point to the edge of rectangular area and then add the modified point to the population P_z. For all $(x_1^{i'}, x_2^{i'})$, $\left(x_1^{j'}, x_2^{j'}\right)$, $(x_1^{k'}, x_2^{k'})$ we find $f(x_1^{i'}, x_2^{i'})$, $f\left(x_1^{j'}, x_2^{j'}\right)$ and $f(x_1^{k'}, x_2^{k'})$, that form the population $F_z = \{f_k^z\}_{k=1}^{3 \cdot k}$.

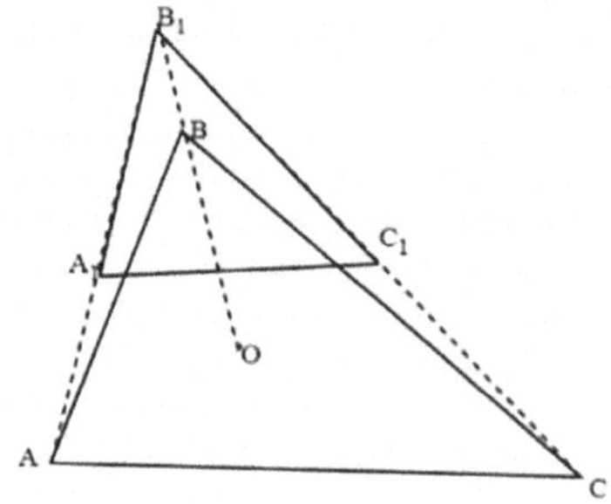

Fig. 1 Transform above the triangle

Then we generate elements of the new population P_s. There are points (x_1^i, x_2^i), $\left(x_1^j, x_2^j\right)$ and (x_1^k, x_2^k), that form a triangle. We rotate the triangle by an angle β around the best vertex. In this case, the vertex (x_1^i, x_2^i) remains unchanged, while the others are changed by the formulas (6 and 7).

$$\begin{aligned} x_1^{j'} &= x_1^j + \left(x_1^j - x_1^i\right) \cdot \cos\beta - \left(x_2^j - x_2^i\right) \cdot \sin\beta, \\ x_2^{j'} &= x_2^j + \left(x_1^j - x_1^i\right) \cdot \sin\beta - \left(x_2^j - x_2^i\right) \cdot \cos\beta, \end{aligned} \tag{6}$$

$$\begin{aligned} x_1^{k'} &= x_1^k + (x_1^k - x_1^i) \cdot \cos\beta - (x_2^k - x_2^i) \cdot \sin\beta, \\ x_2^{k'} &= x_2^k + (x_1^k - x_1^i) \cdot \sin\beta - (x_2^k - x_2^i) \cdot \cos\beta. \end{aligned} \tag{7}$$

The points $\left(x_1^{j'}, x_2^{j'}\right)$ and $(x_1^{k'}, x_2^{k'})$ are written to the population P_s provided that they belong to the rectangle Ω. Otherwise, we modify them [14] and record in a population P_s. For all $\left(x_1^{j'}, x_2^{j'}\right)$ and $(x_1^{k'}, x_2^{k'})$ we find $f\left(x_1^{j'}, x_2^{j'}\right)$, $f\left(x_1^{k'}, x_2^{k'}\right)$, that form the population $F_s = \{f_k^s\}_{k=1}^{2\cdot k/3}$. Generate another new population P_w. There are points (x_1^i, x_2^i), $\left(x_1^j, x_2^j\right)$ and (x_1^k, x_2^k), that form a triangle. Let's turn the vertices of the triangle by an angle α around the "center" of the triangle.

Turn the vertex (x_1^i, x_2^i). We obtain a new vertex with coordinates:

$$\begin{aligned} x_1^{i'} &= (x_1^i - x_1^0) \cdot \cos\alpha - (x_2^i - x_2^0) \cdot \sin\alpha + x_1^0, \\ x_2^{i'} &= (x_1^i - x_1^0) \cdot \sin\alpha + (x_2^i - x_2^0) \cdot \cos\alpha + x_2^0. \end{aligned} \tag{8}$$

Other points are shifted similarly. The points $(x_1^{i'}, x_2^{i'})$, $\left(x_1^{j'}, x_2^{j'}\right)$ and $(x_1^{k'}, x_2^{k'})$ are written to the population P_w provided that they belong to the rectangle Ω. Otherwise, act according to [14] and record them in the population P_w. For all $(x_1^{i'}, x_2^{i'})$, $\left(x_1^{j'}, x_2^{j'}\right)$, $(x_1^{k'}, x_2^{k'})$ we find $f(x_1^{i'}, x_2^{i'})$, $f\left(x_1^{j'}, x_2^{j'}\right)$ and $f\left(x_1^{k'}, x_2^{k'}\right)$, that form the population $F_w = \{f_k^w\}_{k=1}^{3\cdot k}$.

To aggregate the elements of the populations we obtain new sets $P_t \cup P_z \cup P_s \cup P_w = P$ and $F_t \cup F_z \cup F_s \cup F_w = F$. Elements of population F are sorted by ascending. Change the number of the current iteration $t = t + 1$.

Create a new population P_t from elements of the population P that correspond to the first n elements of population F.

If the stop criterion is not fulfilled, the iterative process continues. If the stop criterion is satisfied, then the value of the potential solution, which corresponds to the minimum value of the function f, will be the solution of the problem (2).

The stop conditions of the algorithm are similar to the condition for the one-dimensional method.

The next version of MODS we are going to consider is using four points for creating a "star".

2.3 The Method of Quadrangular Deformed Stars (MODS-4)

Consider the method of deformed stars based on the representation of parental solutions as a quadrangle. Let $t = 0$ (iteration number). Generate the initial population of potential solutions $P_t = \{(x_1^1, x_2^1), (x_1^2, x_2^2), \ldots, (x_1^n, x_2^n)\}$, uniformly distributed in the interval $\Omega = [p_1, p_2] \times [q_1, q_2]$, $|P_t| = n$.

For all $\left(x_1^j, x_2^j\right) \in P_t$ we find $f_j = f\left(x_1^j, x_2^j\right)$, that form the populations. Form rectangles as follows (method 1, method 2 optional).

Method 1 Generate 4 random unique points u, w, m, h, that $u, w, m, h = random\{1, \ldots n\}$, $u \neq w \neq m \neq h$. As result, we obtain quadrangles like in Fig. 2.

Method 2 Form rectangles in which points are written in the correct order (Fig. 3). Select two points from the whole population $u, w = random\{1, \ldots n\}$, $u \neq w$.

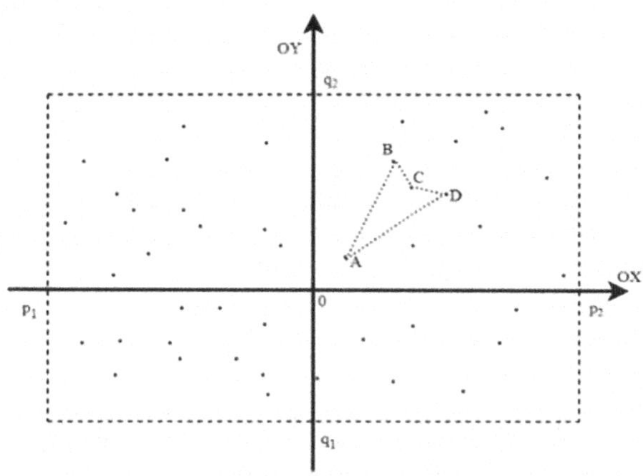

Fig. 2 Generation of four points

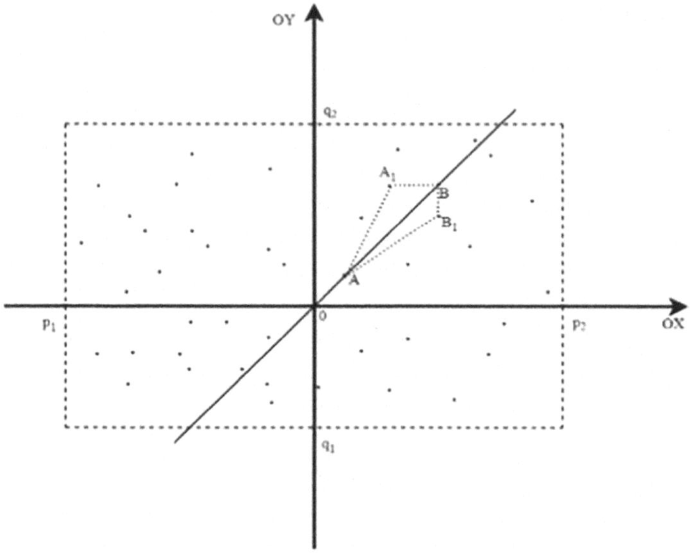

Fig. 3 Another option to create a quadrangle

Draw a line between these two points, call them A and B, which will be the diagonal of the future quadrangle. The following two points should be selected such that one will lie above, and another—lower the line AB, in Fig. 3 it will be A_1 and B_1. If it turns out that there is no point above or below the line AB, then one of the points A or B must be regenerated. As a result, we obtain k quadrangles $R_i = \{(x_1^u, x_2^u), (x_1^w, x_2^w), (x_1^m, x_2^m), (x_1^h, x_2^h),\}$, $i = \overline{1, k}$. Denote the set of all quadrangles as $R = \{R_1, R_2, R_3, ..., R_k\}$.

Next, generate elements of the new population P_s from the elements R. They are obtained after compressing and moving each figure R_i, $i = \overline{1, k}$, to a distance a towards the best point (x_1^b, x_2^b), $b \in \{u, w, m, h\}$, $f(x_1^b, x_2^b) = f_b$, $f_b = \min\{f_u, f_w, f_m, f_h\}$.

To perform these actions, you need to find the "center" of the quadrangle (x_1^0, x_2^0). There are two ways to determine this point.

Method 1 In the first case, the "center" coincides with the center of mass of the quadrangle. It lies at the intersection of the two straight lines, which we obtain by applying the distribution property of the mass centers. Divide the quadrangle into two triangles by the diagonal. The center of mass of the quadrangle lies on a straight line connecting the centers of gravity of these triangles. This line is the first of two searched lines. Another line will be obtained in the same way, breaking the quadrangle into two triangles, but with another diagonal (Fig. 4). In this embodiment, it is very important to write the vertices in the correct order.

Method 2 Another option is to apply the idea that was used in the three-dimensional case. It is much easier to use:

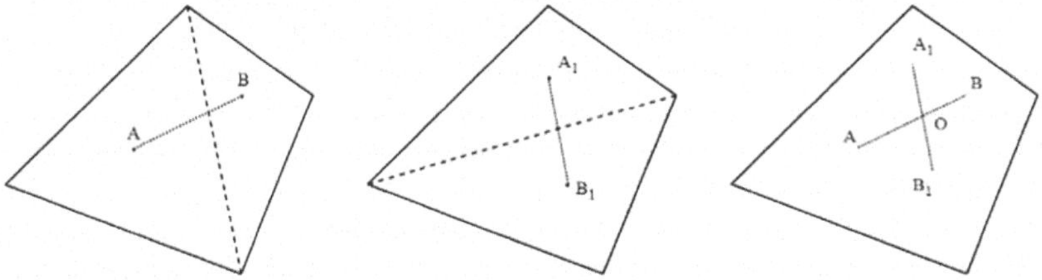

Fig. 4 The center of mass of the quadrangle

$$x_1^0 = \left(x_1^u + x_1^w + x_1^m + x_1^h\right)/4,$$
$$x_2^0 = \left(x_2^u + x_2^w + x_2^m + x_2^h\right)/4. \tag{9}$$

Calculate the distance d between the best point $\left(x_1^b, x_2^b\right)$ and the center $\left(x_1^0, x_2^0\right)$. The parameter $a = coef \cdot d$, where $coef$ is the coefficient on the interval [0, 1]. The coordinates of the shifted best point will be found using the formula of dividing the segment in the given ratio. The new point B^* will lie on the continuation of the line passing through the points $\left(x_1^b, x_2^b\right)$ and $\left(x_1^0, x_2^0\right)$.

Denote the points as follows: point $O = \left(x_1^0, x_2^0\right)$, point $B = \left(x_1^b, x_2^b\right)$, point $B^* = \left(x_1^*, x_2^*\right)$. Then $|OB| = d$, $|BB^*| = a$, $B^* \in OB$.

Then, the segment OB^* is divided as $\frac{OB}{BB^*} = \frac{d}{a} = \lambda$.

So the coordinates of the point $B^*\left(x_1^*, x_2^*\right)$ will be determined as follows:

$$x_1^* = (1 + \lambda) \cdot x_1^b - \lambda \cdot x_1^0,$$
$$y_1^* = (1 + \lambda) \cdot y_1^b - \lambda \cdot y_1^0. \tag{10}$$

If the point $B^*\left(x_1^*, x_2^*\right)$ beyond rectangle Ω, then we need to change it so that it is midway between the point B and the boundary of the rectangle Ω, perform appropriate actions for each of its coordinates. The other three points, call them A^*, C^*, D^*, will lie on straight lines AB^*, CB^*, DB^* at a distance, which will divide these segments in a given ratio λ from A, C, D respectively, where A, C, D—the other three points of the quadrangle R_i, $i = \overline{1, k}$, which were not the best. The coordinates A^*, C^*, D^* we can find using the formula of dividing the segment in a given ratio, the coordinates of the ends of the segment of which are known.

Obviously, if the point $B^*\left(x_1^*, x_2^*\right)$ is in the area Ω, then the points A^*, C^*, D^* also will not go beyond it. For all sets of new points $\{A^*, B^*, C^*, D^*\}$ we find $f(A^*)$, $f(B^*)$, $f(C^*)$, $f(D^*)$, that form the population $F_s = \left\{f_j^s\right\}$, $j = \overline{1, 4k}$.

Form elements of a new population P_{p_1}. For each quadrangle R_i, $i = \overline{1, k}$, rotate around the best vertex $B\left(x_1^b, x_2^b\right)$ by an angle β, $\beta \in [0°; 360°]$. In this case, the vertex $B\left(x_1^b, x_2^b\right)$ will remain unchanged, so it will not be written to the new population, and all other points will have new coordinates. For example, let's take a point $A\left(x_1^A, x_2^A\right)$ and do the following transformation:

$$\begin{aligned} x_1^{A*} &= x_1^A + \left(x_1^A - x_1^b\right) \cdot \cos\beta - \left(x_2^A - x_2^b\right) \cdot \sin\beta, \\ x_2^{A*} &= x_2^A + \left(x_1^A - x_1^b\right) \cdot \sin\beta + \left(x_2^A - x_2^b\right) \cdot \cos\beta. \end{aligned} \tag{11}$$

Similar steps for all other vertices. If a new point belongs to the rectangle Ω, it is written to the new population P_{p_1}.

If the point is out of the area Ω, then the transformations described earlier in the two-dimensional case must be performed to return the point to the rectangle Ω, and then record it to the population P_{p_1}. So we get the population $F_{p_1} = \left\{f_j^{p_1}\right\}, j = \overline{1,\ 3k}$.

Form the elements of a new population P_{p_2}. To do this, for each quadrangle, rotate around the "center" $O\left(x_1^0, x_2^0\right)$ by some angle γ, $\quad \gamma \in [0°;\ 360°]$. Get the population $F_{p_2} = \left\{f_j^{p_2}\right\}, j = \overline{1, 4k}$.

To aggregate the elements of the populations, we obtain new sets $P_t \cup P_s \cup P_{p_1} \cup P_{p_2} = P,\ F_t \cup F_s \cup F_{p_1} \cup F_{p_2} = F$.

Elements of set P are sorted by ascending of population F. Change the number of the current iteration.

As result, we have a new population P_t that corresponds to the first n elements of population P.

If the stop condition is not fulfilled, the iterative process continues. If the stop condition is satisfied, then the value of the potential solution, which corresponds to the minimum value of the function f, will be the solution of the problem (2).

Similar is the implementation technology of the method of deformed stars: five-dimensional case (MODS-5). The results of the experiments are described in Sect. 4.1 of this article.

In the next section we are going to consider method of deformed stars for n-dimensional case.

3 Method of Deformed Stars for Global Optimization, n-Dimensional Case

3.1 Formulation of the Problem

Let us consider the problem of finding the minimum value of the multivariable function [15, 16]:

$$\begin{aligned} &f(x_1, x_2, \ldots, x_n) \to \min, \\ &\overline{x} = (x_1, x_2, \ldots, x_n), \\ &\overline{x} \in D \subset R^n, \end{aligned} \tag{12}$$

where n is the dimension of the search space, D is some area.

The problem (12) is equivalent to the search problem:

$$Arg \min f(\overline{x}),$$
$$x \in D. \tag{13}$$

Let some unknown point $x^* = \left(x_1^*, x_2^*, \ldots, x_n^*\right) \in D$ be its solution.

Note that the function f can be undifferentiated, polyextreme, as well as given tabularly or algorithmically. In such cases, the methods of classical optimization are ineffective.

We propose to use the idea that underlies all evolutionary algorithms, namely the population approach, which consists of solving problems (12), and (13) by evolution the population of potential solutions.

In contrast to the evolutionary strategy [17], where one parental solution generates a set of offsprings solutions from which the best ones are chosen, and the genetic algorithm [18], where a pair of parental solutions generates offsprings solutions, we propose not to be limited to one or two solutions. The reason for this is the fact that simultaneously taken into account and analyzed the number of potential solutions contains more information about the study area and will allow the generation of offsprings solutions more efficiently.

Another hypothesis is that there is a motion direction of the potential solution of the problem (12), and (13), which will allow with high probability and quickly finding the optimal solution of the problem, and there is an area in neighborhood of potential solution to be investigated in more detail because it has solutions that are better than the parent's potential solution.

The main constructive elements of the solution search for the problem (12), and (13) are compression, rotation and some other transformations of figures formed by the connection of points, which are the potential solutions and which we will call deformed stars.

So, let's consider the main idea of MODS for n-dimensional case.

3.2 The Idea of MODS in n-Dimensional Case

Let the initial population of potential solutions P_0 consist of vectors $(\overline{x}^1, \overline{x}^2, \ldots, \overline{x}^m)^T$. Usually, $m \in \{20, 21, \ldots, 50\}$, but this is not mandatory and determined by the size of the problem, the search area, the requirement for data representativeness and computational time constraints.

Suppose that our stars are triangles in n-dimensional space. Obviously, the maximum number of possible triangles is C_n^3, it is 1140 in the case of 20 points, and it is 19,600 for 50 points. This number of triangles is significant for calculations, so we will be limited, for example, to m triangles that will be generated randomly using a uniform distribution and without any other restrictions. Note that all triangles will lie in the area D.

We denote the set of triangles by T and $T = \{T_1, T_2, \ldots, T_m\}$. Each triangle is defined by three vertices, that is $T_i = \left(\overline{x}^{1i}, \overline{x}^{2i}, \overline{x}^{3i}\right)$, $i = \overline{1, m}$. Put in accordance with each vertex of the triangle the value of the function f, namely $T_i \to F_i$, where $F_i = \left(f^{1i}, f^{2i}, f^{3i}\right)$, $i = \overline{1, m}$.

Let us solve the search problems $\forall i \in \{1, \ldots, m\}$ $f^i_{min} = \min\limits_k f^{ki}$ and $\overline{x}^i_{min} = Arg\, \underset{k}{min}\, f^{ki}$.

For each triangle we will find its centroid:

$$\overline{c_i} = \left(\frac{1}{3}\sum_{j=1}^{3} x_1^{ji}, \frac{1}{3}\sum_{j=1}^{3} x_2^{ji}, \ldots, \frac{1}{3}\sum_{j=1}^{3} x_m^{ji}\right), \ i = \overline{1, m}, \tag{14}$$

or

$$\overline{c_i} = \left(\overline{x}_1^i, \overline{x}_2^i, \ldots, \overline{x}_m^i\right). \tag{15}$$

Draw a line through the points $\overline{x}^i_{min}$ and $\overline{c_i}$. We assume that the solution of problem (12), and (13) will most likely be in the direction of the vector $\left(\overline{c_i}, \overline{x}^i_{min}\right)$. Thus, the point $\overline{x}^i_{min}$ of the triangle T_i is mapped to a new point $\overline{y}^{1i}$, the coordinates of which are calculated by the formula

$$y_k^{1i} = 2x^i_{\min k} - c_{ik}, \ k = \overline{1, m}, \ i = \overline{1, m}. \tag{16}$$

Coordinates of the second point $\overline{y}^{2i}$ and third point $\overline{y}^{3i}$ are $y_k^{2i} = \frac{1}{2}\left(x_k^{2i} + y_k^{1i}\right)$ and $y_k^{3i} = \frac{1}{2}\left(x_k^{3i} + y_k^{1i}\right)$ respectively.

In the general case:

$$y_k^{1i} = \frac{1}{k-1}\left(kx^i_{\min k} - c_{ik}\right), \tag{17}$$

$$y_k^{2i} = \frac{1}{k}\left((k-1)x_k^{2i} + y_k^{1i}\right), \tag{18}$$

$$y_k^{3i} = \frac{1}{k}\left((k-1)x_k^{3i} + y_k^{1i}\right), \tag{19}$$

where k is a method parameter.

The points $\overline{y}^{1i}, \overline{y}^{2i}, \overline{y}^{3i}$ form a new triangle R_i. A larger value of k will correspond to a triangle of a smaller area.

Thus, the new points are in the neighborhood of the previous best solution among the points of the triangle. In order not to lose the already found point—the optimal solution, as well as to explore the surrounding area, we will rotate the initial triangle T_i around the point where the value of the function is the best, in this case around $\overline{x}^{1i}$.

How can a rotation in n-dimensional space be made? It is known [19] and was previously implemented [20] that in two-dimensional space the coordinates (x', y') as a result of rotation of a point (x, y) by an angle are as follows:

$$x' = x\cos\alpha - y\sin\alpha,\\ y' = x\sin\alpha + y\cos\alpha. \tag{20}$$

In three-dimensional space, the rotation matrices around the axes x, y and z are as follows:

$$M_x(\alpha) = \begin{pmatrix} 1 & 0 & 0 \\ 0 & \cos\alpha & -\sin\alpha \\ 0 & \sin\alpha & \cos\alpha \end{pmatrix},\quad M_y(\alpha) = \begin{pmatrix} \cos\alpha & 0 & \sin\alpha \\ 0 & 1 & 0 \\ -\sin\alpha & 0 & \cos\alpha \end{pmatrix},$$

$$M_z(\alpha) = \begin{pmatrix} \cos\alpha & -\sin\alpha & 0 \\ \sin\alpha & \cos\alpha & 0 \\ 0 & 0 & 1 \end{pmatrix}. \tag{21}$$

Any rotation can be achieved by the composition of the three rotations described above.

In n-dimensional space there is a rotation only in a certain plane. The rotation matrix in the plane $x_k x_l$ in n-dimensional space is as follows:

$$M_{k,l}(\alpha) = \begin{pmatrix} 1 & 0 & \overset{k}{0} & \cdots & 0 & \overset{l}{0} & 0 & 0 \\ 0 & 1 & 0 & \ldots & 0 & 0 & 0 & 0 \\ 0 & 0 & \cos\alpha & \ldots & 0 & -\sin\alpha & 0 & 0 \\ 0 & 0 & 0 & \ldots & 0 & 0 & 0 & 0 \\ \vdots & \vdots & \vdots & \ldots & \vdots & \vdots & \vdots & \vdots \\ 0 & 0 & \sin\alpha & \ldots & 0 & \cos\alpha & 0 & 0 \\ 0 & 0 & 0 & \ldots & 0 & 0 & 1 & 0 \\ 0 & 0 & 0 & \ldots & 0 & 0 & 0 & 1 \end{pmatrix}. \tag{22}$$

Thus, the coordinates of the new point will coincide with the old coordinates except for the k-th and l-th coordinates, which will be as follows:

$$x'_k = x_k\cos\alpha - x_l\sin\alpha,\\ x'_l = x_k\sin\alpha + x_l\cos\alpha,\\ x'_i = x_i,\ i \neq k, l. \tag{23}$$

So, the point $\overline{x}^{1i}$ will belong to the new triangle Q_i, $\overline{t}^{1i} = \overline{x}^{1i}$. The coordinates of the second point $\overline{t}^{2i}$ and third point $\overline{t}^{3i}$ are calculated as follows:

$$t_k^{vi} = x_k^{vi}\cos\alpha - x_l^{vi}\sin\alpha,$$

$$t_l^{vi} = x_k^{vi} \sin\alpha + x_l^{vi} \cos\alpha,$$
$$t_j^{vi} = x_j^{vi},\ j \neq k, l,\ v \in \{2, 3\}. \tag{24}$$

Another way to explore the neighborhood of the best vertex of a triangle is to compress the vectors $(\overline{x}^{2i}, \overline{x}^{1i})$ and $(\overline{x}^{3i}, \overline{x}^{1i})$ in the direction of the point $\overline{x}^{1i}$ at which the function becomes the minimal value.

Then we will get a new triangle $U_i = (\overline{z}^{1i}, \overline{z}^{2i}, \overline{z}^{3i})$, where $\overline{z}^{1i} = \overline{x}^{1i}$, $z_k^{2i} = \frac{k \cdot x_k^{1i} + x_k^{2i}}{1+k}$, $z_k^{3i} = \frac{k \cdot x_k^{1i} + x_k^{3i}}{1+k}$, k is a compression coefficient. A larger value of k corresponds to a better approximation of the points $\overline{z}^{2i}$ and $\overline{z}^{3i}$ to $\overline{x}^{1i}$.

When converting a triangle T_i to a triangle R_i, as well as to triangle Q_i, it can happen that some points go beyond the area D, which in the general case is a rectangular hyperparallelepiped, $D^0 = \times_{i=1}^m [a_i, b_i]$.

Suppose that there is a potential solution $\overline{x}^l = (x_1^l, x_2^l, \ldots, x_m^l)$ and $\exists j : x_j^l < a_j$ or $x_j^l > b_j$. Then in the first case $x_j^l = b_j - (a_j - x_j^l) = x_j^l + (b_j - a_j)$, and in the other case $x_j^l = x_j^l + (a_j - b_j)$.

The disadvantage of this approach is the need for numerous checks, but at the same time, the diversity of the population of solutions will be ensured.

The algorithm of MODS for n-dimensional case will be as follows.

3.3 Algorithm of the Method of Deformed Stars

Step 1. Perform the initialization of the algorithm parameters. $t = 0$.

Step 2. Generate m potential solutions in the area D (population P_t).

Step 3. Form m triangles (stars) $T_i,\ i = \overline{1, m}$.

Step 4. For each T_i find the vertex in which the function f acquires the minimum value and consider it as the first vertex.

Step 5. Find the centroid of each triangle.

Step 6. For each triangle T_i we find a modified triangle $R_i, i = \overline{1, m}$.

Step 7. For each triangle T_i we find the triangle Q_i obtained by rotating T_i around the first vertex.

Step 8. For each triangle T_i we find the triangle U_i obtained as a result of the operation of compressing the vertices to the first vertex.

Step 9. $t = t + 1$. We form a population P_t that will include all potential solutions from P_{t-1}, as well as the vector of solutions $(\overline{y}^{1i}, \overline{y}^{2i}, \overline{y}^{3i}, \overline{t}^{2i}, \overline{t}^{3i}, \overline{z}^{2i}, \overline{z}^{3i})\ \forall i \in \{1, \ldots, m\}$. So $|P_t| = 8m$.

Step 10. Check whether all elements P_t belong to D. If this is not the case, then perform the necessary transformations.

Step 11. For all potential solutions from P_t, find the values of the function f and sort the potential solutions in ascending order of these values.

Step 12. We leave in P_t only the best m solutions and check the fulfillment of the stop criteria.

Step 13. If the stop condition is not executed, then go to step 3, else we have the end of algorithm.

Note that the stop criteria can be:

I. achieving a given number of iterations;
II. the maximum distance between solutions in one population is less than the specified distance;
III. the absolute value of the deviation of minimum values of the function in neighboring populations is less than the specified value;
IV. others.

When implementing the method of deformed stars, a much smaller number of steps are performed in the wrong direction, in contrast to the genetic algorithm and evolutionary strategy as classical representatives of the evolutionary paradigm.

The accuracy of the obtained solutions is, on average, higher than the accuracy of competing algorithms due to a deeper study of the area D.

The results of the experiments are described in Sect. 4.2 of this article.

4 The Experimental Results

4.1 The Two-Dimensional Case. MODS-3, MODS-4, MODS-5

To determine the efficiency of the deformed stars method, a comparative analysis of its results was performed with the results of classical evolutionary methods: genetic algorithm (GA), evolutionary strategy (ES), differential evolution (DE).

The experiments were performed for different stop conditions of the algorithm: by the number of iterations (100 and 300), by the closeness of the differences between the mean and maximum values of fitness functions for the neighboring populations.

The average execution time of each of the algorithms at 100 iterations (averaged over 100 program launches) was investigated, and the number of resulting values that reached the specified accuracy and those that did not reach was calculated.

All algorithms were tested for ten functions: Ackley, Beale, Goldstein-Price, Booth, Levi N.13, Three-hump camel, Matyas, McCormick, Schaffer N. 2 and Bukin N.6 (Fig. 5).

Convergence plots for one function Levi N.13 are constructed (Figs. 6, 7 and 8). The average number of iterations is determined by which each algorithm finds the solution with acceptable accuracy.

According to the data (Fig. 5), at 100 iterations results MODS-3, 4-MODS, MODS-5 is more accurate, and other algorithms in many cases have not found a solution with acceptable accuracy.

Test function	Global minimum	100 iterations, accuracy $=10^{-6}$										
		Genetic algorithm	Accuracy achieved	Evolution strategy	Accuracy achieved	Differential evolution	Accuracy achieved	MODS-3	Accuracy achieved	MODS-4	Accuracy achieved	MODS-5	Accuracy achieved
	$f(x_1,x_2)$	$f(x_1,x_2)$	N^a	$f(x_1,x_2)$	N	$f(x_1,x_2)$	N	$f(x_1,x_2)$	N	$f(x_1,x_2)$	N	$f(x_1,x_2)$	N
Ackley	0	0,0002	-	0,0	72	0,0	81	0,0	46	0,0	19	0,0	26
Beale	0	0,0128	-	0,0	16	0,0015	-	0,0	56	0,0	15	0,0	22
Goldstein–Price	3	3,0018	-	3,0002	-	3,1193	-	3,0	27	3,0	14	3,0	19
Booth	0	0.0	14	0.0	15	0.0101	-	0.0	32	0.0	12	0.0	20
Levi N.13	0	0,0	15	0,0	20	0,0000	-	0,0	65	0,0	15	0,0	25
Three-hump camel	0	0,000	-	0,0	9	0.0	80	0,0	22	0,0	10	0,0	13
Matyas	0	0.0009	-	0,0	12	0.0011	-	0.0	34	0,0	9	0,0	19
McCormick	-1.9133	-1.9132	-	0.0	-	-4.897	-	-1.9132	-	-1.9132	-	-1.9132	-
Schaffer N. 2	0	0.0	79	0.0	22	0,0001	-	0.0	46	0.0	16	0,0	17
Bukin N.6	0	0.0014	-	0.0069	-	0.8597	-	0.0017	-	0.0025	-	0,0042	-
F_{ser}, accuracy $=10^{-6}$													
Beale	0	0.0	14	0.3976	-	0.0018	-	0.0	53	0.0	16	0.0	21
Goldstein–Price	3	3.0	-	5.1224	-	3.0258	-	3.0	39	3.0	12	3.0	17
Booth	0	0.0	16	0.8913	-	0.0079	-	0.0	38	0.0	10	0,0	17
Levi N.13	0	0,0	-	0.5064	-	0,0000	-	0.0	66	0.0	17	0,0	31
Three-hump camel	0	0.0	15	0.7692	-	0.0019	-	0.0	28	0.0	10	0,0	12
Matyas	0	0,0	11	0.1628	-	0,0134	-	0,0	39	0,0	9	0,0	16
McCormick	-1.9133	-1.913	-	-1.8901	-	-1.911	-	-1.9132	-	-1.9132	-	-1.9132	-
Schaffer N. 2	0	0.0	18	0.4018	-	0,0081	-	0.0	53	0.0	17	0,0	21
Bukin N.6	0	0,2952	-	5.27	-	3.0309	-	0.0078	-	0,0003	-	0,0093	-

Fig. 5 Experiments results

Using the second condition, namely the limited absolute difference between the mean values of the fitness function of two neighboring populations, there is a similar trend. MODS-3 showed the best average run time (in seconds) per 100 runs.

Experiments show that, in general, MODS-4, MODS-5 methods have about 50 iterations enough to find a sufficiently accurate solution, which cannot be said about GA, ES or DE (Figs. 6, 7 and 8).

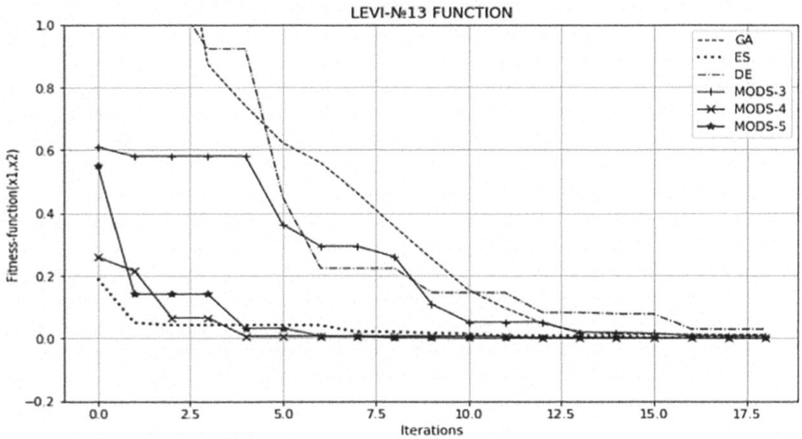

Fig. 6 Stop condition by 100 iterations

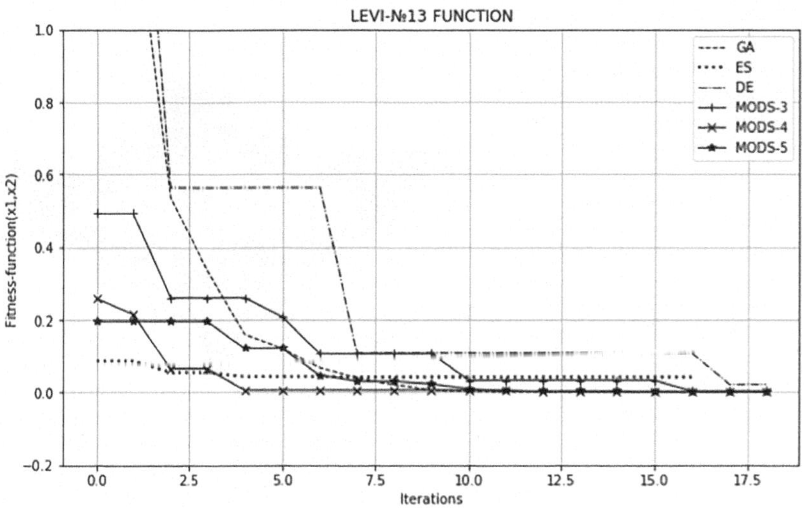

Fig. 7 Stop condition by the closeness of the differences between the mean values of fitness functions for the neighboring populations

4.2 The n-Dimensional Case

Ackley's function, Rosenbrock's saddle, and Rastrigin's function were chosen for the experiments. Ackley's function has the form (Fig. 9):

$$f(x) = 20 + e - 20e^{-0,2\sqrt{\frac{1}{n}\sum_{i=1}^{n} x_i^2}} - e^{\frac{1}{n}\sqrt{\sum_{i=1}^{n}\cos(2\pi x_i)}}, x \in (-30; 30), \ \min \to x_i = 0.$$

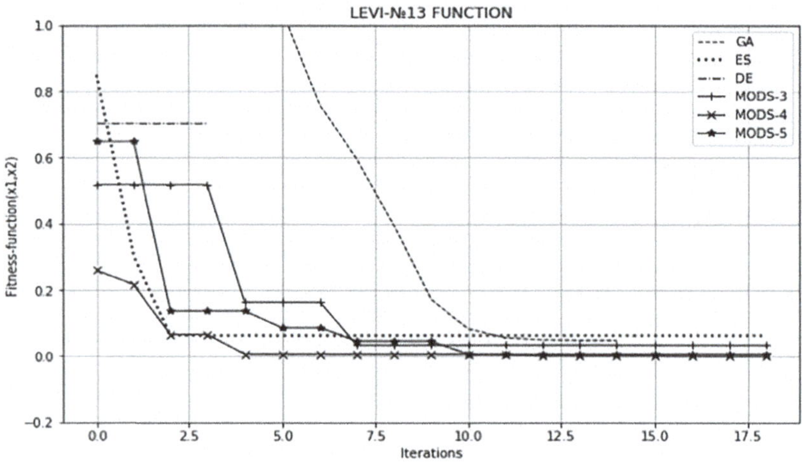

Fig. 8 Stop condition by the closeness of the differences between the maximum values of fitness functions for the neighboring populations

Fig. 9 Ackley's function

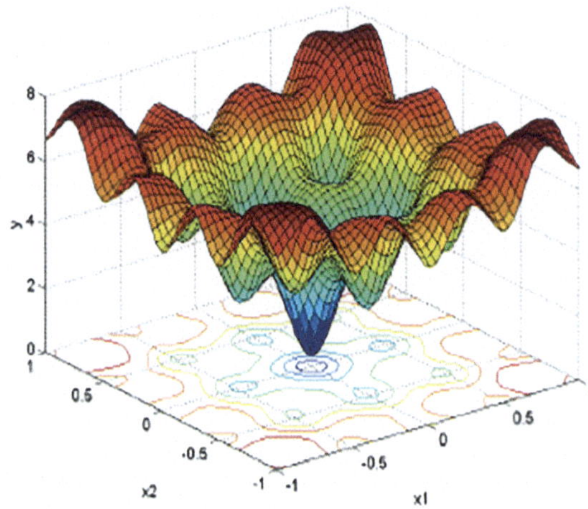

Rosenbrock's saddle (DeJong2) has the form (Fig. 10):

$f(x) = \sum_{i=1}^{n-1} \left(100(x_{i+1} - x_i^2)^2 + (x_i - 1)^2\right)$, $x \in (-2,048;\ 2,048)$, $\min \rightarrow x_i =$ 0.

Rastrigin's function has the form (Fig. 11):

$$f(x) = 10n + \sum_{i=1}^{n} \left(x_i^2 - 10\cos(2\pi x_i)\right), \quad x \in (-5, 12; 5, 12).$$

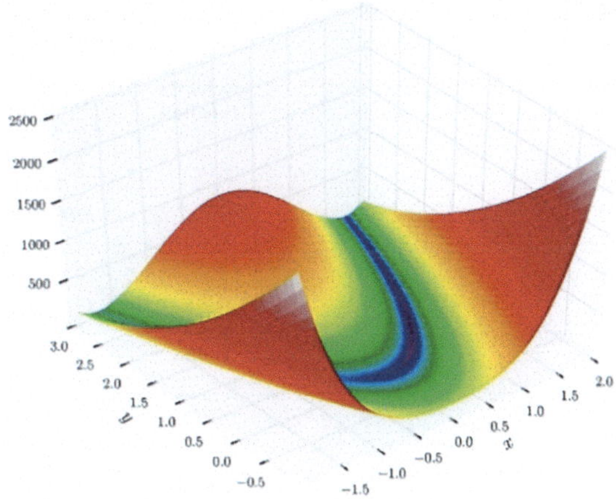

Fig. 10 Rosenbrock's saddle (DeJong2)

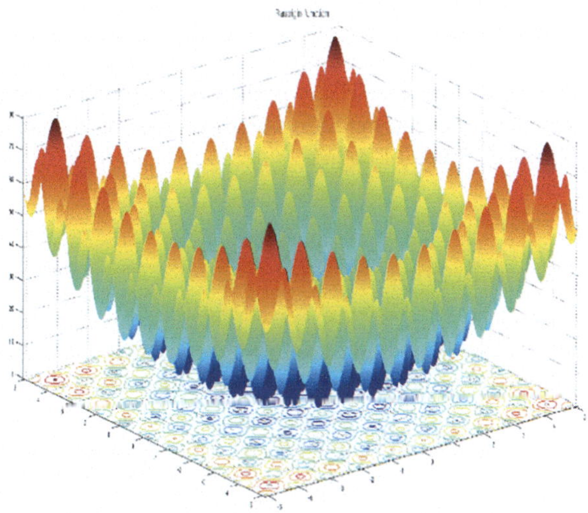

Fig. 11 Rastrigin's function

The modeling results are summarized in Fig. 12 (GA—Genetic Algorithm, ES—Evolutionary Strategy, MODS—Method of Deformed Stars).

The criteria used for modeling are described above. Analysis of the table data shows that MDS, except for one case, demonstrated the benefits of use. All three functions have a minimum zero value and they are polyextreme. The genetic algorithm, as well as the evolutionary strategy, demonstrate approximately the same accuracy. It should be noted that the execution of each of the algorithms was limited by a given number of iterations or the expected deviation of the function values. The method of deformed stars has demonstrated the best computational characteristics

Fig. 12 Optimal values of function

	Method	Criteria 1	Criteria 2	Criteria 3
Ackley's function	*GA*	0,027	0,101	0,034
	ES	0,00017	0,0214	0,0273
	MDS	0,0001	0,0034	0,0173
Rosen-brock's suddle	*GA*	0,0021	0,0145	0,027
	ES	0,00937	0,1374	0,0374
	MDS	0,00011	0,0102	0,0024
Rastrigin's function	*GA*	0,0093	0,0014	0,0372
	ES	0,016	0,024	0,012
	MDS	0,0003	0,0034	0,0134

in almost every case. The experiments were performed for the function of ten variables. Note that the modeling was performed on a computer with a Core I7 9700 K processor, 32 GB of RAM.

5 Conclusions

The developed method of deformed stars (MODS) can be used to optimize complex functional dependencies. The obtained results showed its advantages and effectiveness, in particular, that the desired result can be obtained faster than with using other evolutionary methods. It was considered only some variants of the method. In particular, in one-dimensional case we can consider also pairs, triple points and investigate the effectiveness of the method. In the two-dimensional case, we can do the same by moving the star towards the vertex that has the best value of target function. Also, we considered MODS generalized to the n-dimensional case. It is suggested to use groups of potential solutions (stars) to find the optimal solution. These groups are modified, rotated and compressed, which allows avoiding falling into the local optimums and more deeply exploring their neighborhoods. MODS demonstrates convincing results in its effectiveness. Its main idea is a deeper investigation of the area of argument changes, taking into account the interplay of potential solutions, the focus of the search and the ease of implementation. The method is parametric and allows a considerable amount of tuning, which can significantly improve both the speed of convergence and the accuracy of solving optimization problems. The advantage of this method is fewer steps in the wrong directions and deep scanning of promising areas. The results of experiments showed the effectiveness of the method in finding global optimums of polyextreme functions.

References

1. Storn, R., Price, K.: Differential evolution—a simple and efficient heuristic for global optimization over continuous. J. Global Optim. **11**, 341–359 (1997)
2. Geen, Z.W., Kim, J.H., Loganathan, G.V.: A new heuristic optimization algorithm: Harmony search. SIMULATION **76**(2), 60–68 (2001)
3. Yang, X.-S., Deb, S.: Cuckoo search via Levy flights. In: Proceedings of World Congress on Nature and Biological Inspired Computing, pp. 201–214. IEEE Publications, India, USA (2009)
4. Cheng, M.-Y., Prayogo, D.: Symbiotic organisms search: a new metaheuristic optimization algorithm. Comput. Struct. **139**, 98–112 (2014)
5. Vasconcelos, J.A., Ramirez, R.H.C., Takahashi, R.R.: Saldanha improvements in genetic algorithms. IEEE Trans. Mag. **37**(5) (2001)
6. Yang, X.-S.: Firefly algorithms for multimodal optimization. In: Stochastic Algorithms: Foundations and Applications, SAGA, Lecture Notes in Computer Sciences, vol. 5792, pp. 169–178 (2009)
7. Eskandar, H., Sadollah, A., Bahreininejad, A., Hamdi, M.: Water cycle algorithm—a novel metaheuristic optimization method for solving constrained engineering optimization problems. Comput. Struct. **110–111**, 151–166 (2012)
8. Wolpert, D.H., Macready, W.G.: No free lunch theorems for optimization. IEEE Trans. Evol. Comput. **1**(1) (1997)
9. Glover, F.: Heuristics for integer programming using surrogate constraints. Decis. Sci. **8**(1), 156–166 (1977)
10. Glover, F.: Tabu search—Part I. ORSA J. Comput. **1**, 190–206 (1989)
11. Guha, D., Roy, P.K., Banerjee, S.: Load frequency control of large scale power system using quasi-oppositional grey wolf optimization algorithm. Eng. Sci. Technol. Int. J. **19**(4) (2015)
12. Dorigo, M., Stützle, T.: Ant Colony Optimization. A Bradford Book, 1st edn. First Printing (2004)
13. Snytyuk, V.: Method of deformed stars for multi-extremal optimization. One- and two-dimensional cases. In: International Conferences Mathematical Modeling and Simulation of Systems. MODS 2019, Advances in Intelligent Systems and Computing, vol. 1019. Springer, Cham (2019)
14. Bull, L., Holland, O., Blackmore, S.: On meme-gene coevolution. Artif. Life **6**, 227–235 (2000)
15. Antonevych, M., Didyk, A., Snytyuk, V.: Choice of better parameters for method of deformed stars in n-dimensional case, Kyiv, IT & I, 17–20 (2020)
16. Tmienova, N., Snytyuk, V.: Method of deformed stars for global optimization. In: 2020 IEEE 2nd International Conference on System Analysis & Intelligent Computing (SAIC), pp. 259–262 (2020)
17. Beyer, H.-G., Schwefel, H.-P.: Evolution strategies: a comprehensive introduction. J. Nat. Comput. **1**(1), 3–52 (2002)
18. Srinivas, M., Patnaik, L.M.: Genetic algorithms: a survey. Computer **27**(6), 17–26 (1994)
19. Hestenes, D.: New Foundations for Classical Mechanics. Kluwer Academic Publishers, Dordrecht (1999)
20. Antonevych, M., Didyk, A., Snytyuk, V.: Optimization of functions of two variables by deformed stars method. In: 2019 IEEE International Conference on Advanced Trends in Information Theory (ATIT), Kyiv, Ukraine, pp. 475–480 (2019)

Chapter 14
Guaranteed Estimation of Solutions to First Order Compatible Linear Systems of Periodic Ordinary Differential Equations with Unknown Right-Hand Sides

Oleksandr Nakonechnyi and Yuri Podlipenko

Abstract We elaborate the technique of guaranteed estimation of solutions to first-order linear systems of ordinary differential equations with periodic boundary conditions and unknown right-hand sides that belong to some special bounded set and satisfy to the corresponding compatibility condition. We obtain optimal, in certain sense, estimates of solutions to above-mentioned problems from indirect noisy observations of these solutions on a finite system of intervals and points.

Keywords Guaranteed estimates · Noisy observations · Periodic boundary conditions

1 Introduction

The problems of state estimation of dynamical systems from observations with unknown deterministic errors of phase coordinates were first posed in [1]. In the class of linear transformations of the observed functions, guaranteed estimates of the phase coordinates were obtained in it under special constraints on the observation errors. In [2], a general theory of guaranteed estimates of solutions of Cauchy problems for ordinary differential equations under uncertainty was constructed. These results were further developed in [3–6].

The formulation of problems of estimating the parameters of differential equations with periodic boundary conditions under uncertainty is new and research in this direction has not been carried out previously.

The aim of the present paper is to extend methods of guaranteed estimation for values of functionals from solutions to linear periodic ordinary differential equations with unknown right-hand sides from indirect noisy observations of these solutions under the assumption that above solutions are not unique and exist only if right-hand sides of equations satisfy the corresponding compatibility condition. Notice that this work is a continuation of our earlier studies set forth in [7–10], where we elaborate

O. Nakonechnyi · Y. Podlipenko (✉)
Taras Shevchenko National University of Kyiv, Kyiv, Ukraine

M. Zgurovsky and N. Pankratova (eds.), *System Analysis & Intelligent Computing*,
Studies in Computational Intelligence 1022, https://doi.org/10.1007/978-3-030-94910-5_14

the guaranteed estimation method for the case of uniquely solvable linear periodic ordinary differential equations. For some other types of equations, the guaranteed estimation problems were investigated in [11–13].

For solving the estimation problems we use observations that are linear transformations of unknown solution on a finite system of intervals and points perturbed by additive random noises. We also suppose that correlation functions of random noises in observations, as well as right-hand sides of equations, are unknown and belong to some ellipsoids in the corresponding function spaces.

Our approach to estimation problem makes it possible to obtain optimal estimates both for the unknown solutions and for linear functionals from them, that is, estimates looked for in the class of linear estimates with respect to observations for which the maximal mean square error taken over all the realizations of perturbations from above-mentioned sets takes its minimal value. Such estimates are called the guaranteed or minimax estimates.

It is proved that guaranteed estimates and estimation errors are expressed explicitly via the solutions of special boundary value problems for systems of linear ordinary differential equations with periodic boundary conditions for which the unique solvability is established.

To do this, we reduce the guaranteed estimation problem to certain problem of optimal control of adjoint equation with the quadratic performance criteria under the restrictions on control. Solving this optimal control problem, we obtain the above mentioned boundary value problems that generate the minimax estimates.

2 Preliminaries and Auxiliary Results

Let vector-function $x(t) \in \mathbb{C}^n$ be a solution of the following problem

$$\frac{dx(t)}{dt} = Ax(t) + B(t)f(t), \quad t \in (0, T), \tag{1}$$

$$x(0) = x(T), \tag{2}$$

where $A(t) = \left[a_{ij}(t)\right]$ and $B(t) = \left[b_{ij}(t)\right]$ are $n \times n$ and $n \times r$ -matrices, correspondingly, with elements $a_{ij}(t)$ and $b_{ij}(t)$ which are square integrable and piecewise continuous complex-valued functions on $(0, T)$, and $[0, T]$, respectively, $f(t)$ is a complex-valued vector-function such that $f \in (L^2(0, T))^r$. By a solution of this problem, we mean a function[1] $x(t) \in W_2^1(0, T)$ that satisfies Eq. (1) almost everywhere on $(0, T)$ and the boundary conditions (2).

[1] $W_2^1(0, T)$ is a space of functions absolutely continuous on $[0, T]$ for which the derivative that exists almost everywhere on $(0, T)$ belongs to space $L^2(0, T)$.

Denote by $X(t)$ a matrix-valued function $X(t) = [x_1(t), \dots, x_n(t)]$ whose columns are linearly independent solutions $x_1(t), \dots, x_n(t)$ of the homogeneous system

$$\frac{dx(t)}{dt} = A(t)x(t) \tag{3}$$

such that $X(0) = E_n$, where E_n is the unit $n \times n$-matrix. In this case $X(t)$ is said to be a normalized fundamental matrix of Eq. (3).

Further we will assume that the following condition is valid

$$\det[E_n - X(T)] = 0.$$

Then the problem

$$\frac{dx(t)}{dt} = A(t)x(t), \quad t \in (0, T), \quad x(0) = x(T) \tag{4}$$

has d linearly independent solutions $\varphi_i(t)$, $i = 1, \dots, d$,[2] where d is the deficiency index of matrix $X(T) - E_n$. We also will assume that the following compatibility condition

$$\int_0^T (f(t), B^*(t)\psi_i(t))_r dt = 0, \quad i = 1, \dots, d, \tag{5}$$

holds, where $\psi_i(t)$, $i = 1, \dots, d$ are linearly independent solutions of problem

$$-\frac{dz(t)}{dt} = A^* z(t), \quad t \in (0, T), \quad z(0) = z(T), \tag{6}$$

adjoint to (4) (as is known [14], under the assumption made on matrix $[X(T) - E_n]$, problem (6) has also d linearly independent solutions $\psi_1, \dots, \psi_d$). Here and in what follows, we will denote by Λ^* the matrix complex conjugate and transpose of a matrix Λ and by $(\cdot, \cdot)_n$ the inner product in $\mathbb{C}^n$.

Then problem (1), and (2) is solvable for $f \in (L^2(0, T))^r$ satisfying (5) (see [14]). If $x(t)$ is a solution to (1), and (2) then

$$x(t) + \sum_{i=1}^{d} \gamma_i \varphi_i(t),$$

[2] These solutions are determined by $\varphi_i(t) = X(t)c_i$, $i = 1, \dots, d$, where $c_1, \dots, c_d$, are linearly independent solutions of the equation, $[X(T) - E_n]c = 0$.

where $\gamma_i \in \mathbb{C}$, $i = 1, \ldots, d$, are arbitrary constants, is also a solution.

We also assume that matrix $B(t)$ is such that vectors $B^*(t)\psi_i(t)$, $i = 1, \ldots, d$, are linearly independent.

Further, the following assertion will be frequently used. If vector-functions $f(t) \in \mathbb{C}^n$ and $g(t) \in \mathbb{C}^n$ are absolutely continuous on the closed interval $[t_1, t_2]$, then the following integration by parts formula is valid

$$(f(t_2), g(t_2))_n - (f(t_1), g(t_1))_n = \int_{t_1}^{t_2} [(f(t), \frac{dg(t)}{dt})_n + \left(g(t), \frac{df(t)}{dt}\right)_n \Big] dt. \quad (7)$$

Lemma 1 *Suppose Q is a bounded positive*[3] *Hermitian (self-adjoint) operator in a complex (real) Hilbert space H with bounded inverse Q^{-1}. Then, the generalized Cauchy–Bunyakovsky inequality*

$$|(f, g)_H| \leq (Q^{-1}f, f)_H^{1/2}(Qg, g)_H^{1/2} \quad (f, g \in H) \quad (8)$$

is valid. The equality sign in (8) is attained at the element

$$g = \frac{Q^{-1}f}{(Q^{-1}f, f)_H^{1/2}}.$$

For a proof we refer to [15].

3 Problem Statement

In this section, assuming that vector-function f in the right-hand side of (1) is unknown, we formulate the problem of minimax estimation of unknown solution from its indirect noisy observations on a finite system of points and intervals.

Let t_i, $i = 1, \ldots, N$, $(0 < t_1 < \cdots < t_N < T)$ be a given system of points on the closed interval $[0, T]$ with $t_0 = 0$ and $t_{N+1} = T$ and let Ω_j, $j = 1, \ldots, M$, be a given system of subintervals of $[0, T]$. The problem is to estimate the expression

$$l(x) = \int_0^T (x(t), l_0(t))_n dt, \quad (9)$$

from observations of the form

$$y_j^{(1)}(t) = H_j^{(1)}(t)x(t) + \xi_j^{(1)}(t), \quad t \in \Omega_j, \quad j = 1, \ldots, M, \quad (10)$$

[3] That is $(Qf, f)_H > 0$ when $f \neq 0$.

$$y_i^{(2)} = H_i^{(2)} x(t_i) + \xi_i^{(2)}, \quad i = 1, \ldots, N, \tag{11}$$

in the class of estimates

$$\widehat{l(x)} = \sum_{j=1}^{M} \int_{\Omega_j} (y_j^{(1)}(t), u_j^{(1)}(t))_l dt + \sum_{i=1}^{N} (y_i^{(2)}, u_i^{(2)})_m + c, \tag{12}$$

linear with respect to observations (10) and (11); here $x(t)$ is the state of a system described by problem (1), and (2), $l_0 \in (L^2(0,T))^n$, $H_j(t)$ are $l \times n$ matrices with the elements that are complex-valued piecewise continuous functions on $\overline{\Omega}_j$ and complex numbers, correspondingly, $u_j^{(1)}(t)$ are vector-functions belonging to $(L^2(\Omega_j))^l$, $u_i^{(2)} \in \mathbb{C}^m$, $c \in \mathbb{C}$. We also assume that at least one of conditions (i) or (ii) below is satisfied:

(i) there exists at least one integer j_0, $1 \leq j_0 \leq M$, such that the operator generated by matrix $H_{j_0}^{(1)}(t)$ is injective on $span\{\varphi_1, \ldots, \varphi_d\}$;
(ii) there exists at least one integer i_0, $1 \leq i_0 \leq N$, such that the operator generated by matrix $H_{i_0}^{(2)}$ is injective on $span\{\varphi_1, \ldots, \varphi_d\}$.

For this, it suffices to assume that one of the conditions (a) or (b) is fulfilled:

(a) $l \geq d$ and there exists at least one integer j_0, $1 \leq j_0 \leq M$, such that

$$rank H_{j_0}^{(1)}(t) \geq d \quad \forall t \in \overline{\Omega}_{j_0};$$

(b) $m \geq d$ and there exists at least one integer i_0, $1 \leq i_0 \leq N$, such that

$$rank H_{i_0}^{(2)} \geq d.$$

We suppose that unknown vector-function $f \in G_1$, where

$$G_1 = .\{f \in (L^2(0,T))^r : \int_0^T (f(t), B^*(t)\psi_i(t))_r dt = 0, \quad i = 1, \ldots, d,$$

$$\int_0^T (Q(t)(f(t) - f_0(t)), f(t) - f_0(t))_r dt \leq 1.\}, \tag{13}$$

and $\xi := \left(\xi_1^{(1)}(\cdot), \ldots, \xi_M^{(1)}(\cdot), \xi_1^{(2)}, \ldots, \xi_N^{(2)}\right) \in G_2$, where $\xi_j^{(1)}(t)$ and $\xi_i^{(2)}$ are estimation errors in (10) and (11), respectively, that are realizations of random vector-functions $\xi_j^{(1)}(t) = \xi_j^{(1)}(\omega, t) \in \mathbb{C}^l$ and random vectors $\xi_i^{(2)} = \xi_i^{(2)}(\omega) \in \mathbb{C}^m$ and G_2

denotes the set of random elements ξ, whose components are uncorrelated,[4] have zero means, $\mathbb{E}\xi_j^{(1)}(\cdot)=0$, and $\mathbb{E}\xi_i^{(2)}=0$, with finite second moments $\mathbb{E}\|\xi_j^{(1)}(\cdot)\|^2_{(L^2(\Omega_j))^l}$ and $\mathbb{E}|\xi_i^{(2)}|^2$, and unknown correlation matrices $R_j^{(1)}(t,s)=\mathbb{E}\xi_j^{(1)}(t)(\xi_j^{(1)})^*(s)$ and $R_i^{(2)}=\mathbb{E}\xi_i^{(2)}(\xi_i^{(2)})^*$ satisfying the conditions correspondingly.[5] Here in (13), $f_0\in(L^2(0,T))^r$ is a prescribed vector such that $\int_0^T(f_0(t),B^*(t)\psi_i(t))_n dt=0$, $D_j^{(1)}(t)$, $j=1,\ldots,M$, and $Q(t)$ are Hermitian positive definite $l\times l$ and $r\times r$ -matrices with complex-valued piecewise continuous elements on $\overline{\Omega}_j$ and $[0,T]$, respectively, $D_i^{(2)}$, $i=1,\ldots,N$, are Hermitian positive matrices with constant elements for which there exist their inverse matrices $(D_j^{(1)})^{-1}(t)$, $Q^{-1}(t)$, and $(D_i^{(2)})^{-1}$.

$$\sum_{j=1}^{M}\int_{\Omega_j}Tr\Big[D_j^{(1)}(t)R_j^{(1)}(t,t)\Big]dt\le 1 \text{ and } \sum_{i=1}^{N}Tr\Big[D_i^{(2)}R_i^{(2)}\Big]\le 1, \tag{14}$$

Set $u:=\left(u_1^{(1)}(\cdot),\ldots,u_M^{(1)}(\cdot),u_1^{(2)},\ldots,u_N^{(2)}\right)\in(L^2(\Omega_1))^l\times\cdots\times(L^2(\Omega_M))^l\times\mathbb{C}^{N\times m}=:H$.

Norm and inner product in space H are defined by

$$\|u\|_H=\{\sum_{j=1}^{M}\|u_j^{(1)}(\cdot)\|^2_{(L^2(\Omega_j))^l}+\sum_{i=1}^{N}\|u_i^{(2)}\|_m^2\}^{1/2}$$

and

$$(u,v)_H=\sum_{j=1}^{M}(u_j^{(1)}(\cdot),v_j^{(1)}(\cdot))_{(L^2(\Omega_j))^l}+\sum_{i=1}^{N}(u_i^{(2)},v_i^{(2)})_m\quad\forall u,v\in H,$$

respectively.

Definition 1 The estimate

$$\widehat{l(x)}=\sum_{j=1}^{M}\int_{\Omega_j}(y_j^{(1)}(t),\hat{u}_j^{(1)}(t))_l dt+\sum_{i=1}^{N}(y_i^{(2)},\hat{u}_i^{(2)})_m+\hat{c},$$

in which elements $\hat{u}_j^{(1)}(\cdot)$, $\hat{u}_i^{(2)}$, and a number $\hat{c}$ are determined from the condition will be called the minimax estimate of expression (9).

---

[4] That is, it is assumed that $\mathbb{E}(\xi_i^{(2)},v)_m(\xi_j^{(1)}(\cdot),v(\cdot))_{(L^2(\Omega_j))^l}=0\ \forall v\in\mathbb{R}^m$, $v(\cdot)\in(L^2(\Omega_j))^l$, $i=1,\ldots N$, $j=1,\ldots M$.

[5] $TrD:=\sum_{i=1}^{l}d_{ii}$ denotes the trace of the matrix $D=\{d_{ij}\}_{i,j=1}^{l}$.

$$\inf_{u\in H, c\in\mathbb{C}} \sigma(u,c) = \sigma\left(\hat{u},\hat{c}\right), \text{ where } \sigma(u,c) = \sup_{f\in G_1, \xi\in G_2} \mathbb{E}|l(x) - \widehat{l(x)}|^2,$$

The quantity

$$\sigma := \{\sigma\left(\hat{u},\hat{c}\right)\}^{1/2}$$

will be called the error of the minimax estimation of $l(x)$.

4 Main Results

In this section we deduce systems of linear ordinary differential equations that generate minimax estimates.

For any fixed $u = \left(u_1^{(1)}(\cdot), \ldots, u_M^{(1)}(\cdot), u_1^{(2)}, \ldots, u_N^{(2)}\right) \in U$, introduce the vector-function $z(t; u)$ as a unique solution to the problem[6]

$$-\frac{dz(t;u)}{dt} = A^*(t)z(t;u) + l_0(t) - \sum_{j=1}^{M} \chi_{\Omega_j}(t)(H_j^{(1)})^*(t)u_j^{(1)}(t)$$

$$\text{for a.e. } t \in (0,T), \tag{15}$$

$$\Delta z(\cdot; u)|_{t=t_i} = z(t_i+0; u) - z(t_i; u) = (H_i^{(2)})^* u_i^{(2)}, \quad i = 1, \ldots, N, \tag{16}$$

$$z(T; u) = z(0; u), \quad \int_0^T (\tilde{Q}(t)z(t;u), \psi_i(t))_n dt = 0, \quad i = 1, \ldots d, \tag{17}$$

where $\tilde{Q}(t) = B(t)Q^{-1}(t)B^*(t)$, $\chi_\Omega(t)$ is a characteristic function of the set Ω,

$$U = \{u \in H : \int_0^T (\varphi_k(t), l_0(t))_n dt - \sum_{j=1}^{M} \int_{\Omega_j} (\varphi_k(t), (H_j^{(1)})^*(t)u_j^{(1)}(t))_n dt$$

$$- \sum_{i=1}^{N} (\varphi_k(t_i), (H_i^{(2)})^* u_i^{(2)})_n = 0 \quad \forall k = 1, \ldots d\}.$$

---

[6] Here and in what follows we assume that if a function is piecewise continuous then it is continuous from the left.

Lemma 2 *Finding the minimax estimate of functional $l(x)$ is equivalent to the problem of optimal control of the system (15)–(17) with the cost function*

$$I(u) := \inf_{c\in\mathbb{C}} \sigma(u, c) = \int_0^T (\tilde{Q}(t)z(t; u), z(t; u))_n dt$$

$$+\sum_{j=1}^{M}\int_{\Omega_j}((D_j^{(1)})^{-1}(t)u_j^{(1)}(t), u_j^{(1)}(t))_l dt + \sum_{i=1}^{N}((D_i^{(2)})^{-1}u_i^{(2)}, u_i^{(2)})_m \to \inf_{u\in U}, \quad (18)$$

where for any fixed $u \in U$, the infimum over c is attained at

$$c = \int_0^T (f_0(t), B^*(t)z(t; u))_r dt. \quad (19)$$

Proof Show first that set U is nonempty. Denote by $C : (W_2^1(0, T))^n \to H$ a linear operator defined by in space H, where $\gamma_i = \int_0^T (l_0(t), \varphi_i(t))_n dt$, $i = 1, \dots, d$. From (a)–(b) it follows that $C\varphi_i$, $i = 1, \dots, d$, are linearly independent in H.

$$C = \left(H_1^{(1)}(\cdot)\varphi(\cdot)\Big|_{\Omega_1}, \dots, H_M^{(1)}(\cdot)\varphi(\cdot)\Big|_{\Omega_M}, H_1^{(2)}\varphi(t_1), \dots, H_N^{(2)}\varphi(t_N)\right)$$
$$\forall\varphi \in \left(W_2^1(0, T)\right)^n.$$

It is easy to see that U is the intersection of d hyperplanes

$$(C\varphi_i, u)_H = \gamma_i \quad (20)$$

Denote by $span\{C\varphi_1, \dots, C\varphi_d\}$ a subspace in H spanned over vectors $C\varphi_1, \dots, C\varphi_d$ and prove that there is one and only one element $u_0 \in span\{C\varphi_1, \dots, C\varphi_d\}$ that belongs to set U. To this end, representing u_0 as $u_0 = \sum_{j=1}^d \beta_j C\varphi_j$, where $\beta_j \in \mathbb{C}$, and substituting this into (20), we see that u_0 belongs to U if and only if the linear equation system

$$\sum_{j=1}^{d} \overline{\beta}_j (C\varphi_i, C\varphi_j)_H = \gamma_i, i = 1, \dots, d, \quad (21)$$

with respect to unknowns β_j is solvable. Since $C\varphi_j$, $j = 1, \dots, d$, are linearly independent then $det[(C\varphi_i, C\varphi_j)_H]_{i,j=1}^d \neq 0$ and the system (20) has unique solution $\overline{\beta}_1^0, \dots, \overline{\beta}_d^0$. Consequently, the element $u_0 = \sum_{j=1}^d \beta_j^0 C\varphi_j$, belongs, as well as $u_0 + u^\perp$ for any $u^\perp \in H \ominus span\{C\varphi_1, \dots, C\varphi_d\}$, to set U. Thus $U \neq \emptyset$.

Next, let us show that for every fixed $u \in U$, function $z(t; u)$ can be uniquely determined from equalities (15)–(17). Indeed, the condition $u \in U$ coincides, according to [10, Theorem 4.2], with the solvability condition for problem (15) and (16). Let $z_0(t; u) \in (W_2^1(0, T))^n$ be a solution to this problem, e.g. a solution that satisfies the condition

$$(z_0(0; u), \psi_k(0))_n = 0, \quad k = 1, \dots, d;$$

then the function

$$z(t; u) := z_0(t; u) + \sum_{i=1}^{d} c_i \psi_i(t), \tag{22}$$

also satisfies (15) and (16) for any $c_i \in \mathbb{C}$, $i = 1, \dots, d$. Let us prove that coefficients c_i, $i = 1, \dots, d$, can be chosen so that this function would also satisfy (17). Substituting expression (22) for $z(u)$ into (17), we obtain a system of d linear algebraic equations with d unknowns $c_1, \dots, c_d$:

$$\sum_{i=1}^{d} \alpha_{ij} c_i = \beta_j(u), \quad j = 1, \dots, d, \tag{23}$$

where

$$\alpha_{ij} = (\tilde{Q}\psi_i, \psi_j)_{(L^2(0,T))^n}, \tag{24}$$

$$\beta_j = -(\tilde{Q} z_0(\cdot; u), \psi_j)_{(L^2(0,T))^n}.$$

Notice that matrix $[\alpha_{ij}]_{i,j=1}^d$ of system (23) is non-singular. Indeed, since Q^{-1} is a Hermitian matrix satisfying the inequality

$$\int_0^T (Q^{-1}(t)v(t), v(t))_r dt \geq \beta \|v\|^2_{(L^2(0,T))^r} \text{ for all } v(\cdot) \in (L^2(0, T))^r, \tag{25}$$

we may introduce in the space $(L^2(0, T))^r$ the inner product given by $\langle v, w\rangle = \int_0^T (Q^{-1}(t)v(t), w(t))_r dt$ for all $v(\cdot), w(\cdot) \in (L^2(0, T))^r$.
Therefore, we have

$$\alpha_{ij} = (\tilde{Q}\psi_i, \psi_j)_{(L^2(0,T))^n} = (Q^{-1}B^*\psi_i, B^*\psi_j)_{(L^2(0,T))^r} = B^*\psi_i, B^*\psi_j.$$

Hence, $[\alpha_{ij}]_{i,j=1}^d$ is the Gram matrix of the set of linearly independent vectors $B^*\psi_1, \dots, B^*\psi_d$ in the space $(L^2(0, T))^r$ with respect to the inner product (25).

Thus, $\det[\alpha_{ij}] \neq 0$ and system (23) has unique solution, $c_1, \dots, c_d$. Consequently, problem (15)–(17) is uniquely solvable. Indeed, we have shown that there exists a solution to problem (15)–(17); let us prove that this solution is unique. Assume that there are two solutions to this problem, $z_1(t; u)$ and $z_2(t; u)$. Then

$$-\frac{dz_1(t;u)}{dt} = A^*(t)z_1(t;u) + l_0(t) - \sum_{j=1}^{M} \chi_{\Omega_j}(t)(H_j^{(1)})^*(t)u_j^{(1)}(t)$$

for a.e. $t \in (0, T)$, (26)

$$\Delta z_1(\cdot; u)|_{t=t_i} = (H_i^{(2)})^* u_i^{(2)}, \quad i = 1, \dots, N, \quad z_1(T; u) = z_1(0; u), \tag{27}$$

$$\int_0^T (\tilde{Q}(t)z_1(t;u), \psi_i(t))_n dt = 0, \quad i = 1, \dots d, \tag{28}$$

and

$$-\frac{dz_2(t;u)}{dt} = A^*(t)z_2(t;u) + l_0(t) - \sum_{j=1}^{M} \chi_{\Omega_j}(t)(H_j^{(1)})^*(t)u_j^{(1)}(t) for\ a.e. t \in (0, T), \tag{29}$$

$$\Delta z_2(\cdot; u)|_{t=t_i} = (H_i^{(2)})^* u_i^{(2)}, \quad i = 1, \dots, N, \quad z_2(T; u) = z_2(0; u), \tag{30}$$

$$\int_0^T (\tilde{Q}(t)z_2(t;u), \psi_i(t))_n dt = 0, \quad i = 1, \dots d, \tag{31}$$

Subtracting (29)–(31) from equalities (26)–(28) and setting $\overline{z}(t; u) = z_1(t; u) - z_2(t; u)$, we obtain

$$-\frac{d\overline{z}(t;u)}{dt} = A^*(t)\overline{z}(t;u) \text{ for a.e. } t \in (0, T),$$
$$z(T; u) = z(0; u),$$

$$\int_0^T (\tilde{Q}(t)\overline{z}(t;u), \psi_i(t))_n dt = 0, \quad i = 1, \dots d, \tag{32}$$

Since $\overline{z}(t; u)$ solves homogeneous problem (15)–(17), this function has the form

$$\bar{z}(t;u) = \sum_{i=1}^{d} c_i \psi_i(t). \tag{33}$$

Substituting (33) into (32), we obtain

$$\sum_{i=1}^{d} c_i \int_0^T (\tilde{Q}\psi_i(t), \psi_j(t))_n dt = 0, \quad j = 1, \ldots, d,$$

or, in line with (24),

$$\sum_{i=1}^{d} c_i \alpha_{ij} = 0, \quad j = 1, \ldots, d.$$

We see that coefficients c_i satisfy a linear homogeneous algebraic equation system with nonsingular matrix $[\alpha_{ij}]_{i,j=1}^{d}$; therefore, this system has only the trivial solution $c_i = 0, i = 1, \ldots, d$. This proves the unique solvability of problem (15)–(17).

Further, we compute the quantity $\inf\limits_{c \in \mathbb{C}} \sigma(u, c)$ using (15)–(17). Note that $\sigma(u, c)$ will be finite if and only if $u \in U$. In fact, since any solution $x(t)$ of (1), and (2) can be written as $x(t) = x_0(t) + \varphi(t)$, where $\varphi \in N(A) := span\{\varphi_1, \ldots, \varphi_d\}$ and $x_0(t)$ is the unique solution to this problem satisfying the condition

$$(x_0(0), \varphi_k(0))_n = 0, \quad k = 1, \ldots, d,$$

we have

$$\sigma(u, c)$$
$$= \sup_{f \in G_1, \xi \in G_2} \mathbb{E}\left|l(x) - \widehat{l(x)}\right|^2 = \sup_{f \in G_1, \xi \in G_2} \sup_{\varphi \in N(A)} \mathbb{E}\left|l(x_0 + \varphi) - \widehat{l(x_0 + \varphi)}\right|^2. \tag{34}$$

Taking into account (9)–(12), and the fact that for arbitrary $u \in H$, we have

$$|l(x) - \widehat{l(x)}| = |\int_0^T (x(t), l_0(t))_n dt$$
$$- \sum_{i=1}^{N} (y_i^{(2)}, u_i^{(2)})_m - \sum_{j=1}^{M} \int_{\Omega_j} (y_j^{(1)}(t), u_j^{(1)}(t))_l dt - c|$$
$$= |\int_0^T (x(t), l_0(t) - \sum_{j=1}^{M} \chi_{\Omega_j}(t)(H_j^{(1)})^*(t) u_j^{(1)}(t))_n dt$$

$$-\sum_{j=1}^{M}\int_{\Omega_j}(\xi_j^{(1)}(t),u_j^{(1)}(t))_l dt$$

$$-\sum_{i=1}^{N}(x(t_i),(H_i^{(2)})^*u_i^{(2)})_n-\sum_{i=1}^{N}(\xi_i^{(2)},u_i^{(2)})_m-c|$$

$$=|\int_0^T(x_0(t),l_0(t)-\sum_{j=1}^{M}\chi_{\Omega_j}(t)(H_j^{(1)})^*(t)u_j^{(1)}(t))_n dt$$

$$-\sum_{i=1}^{N}(x_0(t_i),(H_i^{(2)})^*u_i^{(2)})_n$$

$$-\sum_{j=1}^{M}\int_{\Omega_j}(\xi_j^{(1)}(t),u_j^{(1)}(t))_l dt-\sum_{i=1}^{N}(\xi_i^{(2)},u_i^{(2)})_m-c$$

$$+\int_0^T(\varphi(t),l_0(t)-\sum_{j=1}^{M}\chi_{\Omega_j}(t)(H_j^{(1)})^*(t)u_j^{(1)}(t))_n dt$$

$$-\sum_{i=1}^{N}(\varphi(t_i),(H_i^{(2)})^*u_i^{(2)})_n|. \tag{35}$$

Since function $\varphi(t)$ may be an arbitrary element of space $N(A)$, then the right-hand side of (35) will be finite if and only if $u\in U$ and

$$\sup_{\varphi\in N(A)}\mathbb{E}\Big|l(x_0+\varphi)-\widehat{l(x_0+\varphi)}\Big|^2$$

$$=\mathbb{E}|\int_0^T(x_0(t),l_0(t)-\sum_{j=1}^{M}\chi_{\Omega_j}(t)(H_j^{(1)})^*(t)u_j^{(1)}(t))_n dt$$

$$-\sum_{j=1}^{M}\int_{\Omega_j}(\xi_j^{(1)}(t),u_j^{(1)}(t))_l dt-\sum_{i=1}^{N}(x_0(t_i),(H_i^{(2)})^*u_i^{(2)})_n$$

$$-\sum_{i=1}^{N}(\xi_i^{(2)},u_i^{(2)})_m-c|^2.$$

Hence

$$\sigma(u,c)=\sup_{f\in G_1,\xi\in G_2}\mathbb{E}|l(x)-\widehat{l(x)}|^2$$

$$= \sup_{f\in G_1,\xi\in G_2} \mathbb{E}|\int_0^T (x_0(t), l_0(t) - \sum_{j=1}^M \chi_{\Omega_j}(t)(H_j^{(1)})^*(t)u_j^{(1)}(t))_n dt$$

$$-\sum_{j=1}^M \int_{\Omega_j} (\xi_j^{(1)}(t), u_j^{(1)}(t))_l dt - \sum_{i=1}^N (x_0(t_i), (H_i^{(2)})^* u_i^{(2)})_n$$

$$-\sum_{i=1}^N (\xi_i^{(2)}, u_i^{(2)})_m - c|^2 \tag{36}$$

and

$$\inf_{u\in H, c\in\mathbb{C}} \sigma(u,c) = \inf_{u\in U, c\in\mathbb{C}} \sigma(u,c).$$

Therefore, further we will assume that $u \in U$.

Next, for each $i = 1, \ldots, N+1$, denote by $z_i(t;u)$ the restriction of function $z(t;u)$ to a subinterval (t_{i-1}, t_i) of the interval $(0, T)$ and extend it from this subinterval to the ends t_{i-1} and t_i by continuity. Then, due to (15) and (16),

$$-\frac{dz_i(t;u)}{dt} = A^*(t)z_i(t;u) + l_0(t) - \sum_{j=1}^M \chi_{\Omega_j}(t)(H_j^{(1)})^*(t)u_j^{(1)}(t)$$

$$\text{for a.e.} t \in (t_{i-1}, t_i), \quad i = 1, \ldots, N+1, \tag{37}$$

$$z_{i+1}(t_i;u) = z_i(t_i;u) + (H_i^{(2)})^* u_i^{(2)}, \quad i = 1, \ldots, N, \quad z_{N+1}(T;u)$$
$$= z_1(0;u). \tag{38}$$

Transform the expression under the absolute value sign in (36). From the integration by parts formula and (37), and (38), we obtain

$$\int_0^T (x_0(t), l_0(t) - \sum_{j=1}^M \chi_{\Omega_j}(t)(H_j^{(1)})^*(t)u_j^{(1)}(t))_n dt$$

$$-\sum_{i=1}^N (x_0(t_i), (H_i^{(2)})^* u_i^{(2)})_n$$

$$-\sum_{i=1}^N (\xi_i^{(2)}, u_i^{(2)})_m - \sum_{j=1}^M \int_{\Omega_j} (\xi_j^{(1)}(t), u_j^{(1)}(t))_l dt - c$$

$$= \sum_{i=1}^{N+1} \int_{t_{i-1}}^{t_i} (x_0(t), -\frac{dz_i(t;u)}{dt} - A^*(t)z_i(t;u))_n dt$$

$$- \sum_{i=1}^{N} (x_0(t_i), (H_i^{(2)})^* u_i^{(2)})_n$$

$$- \sum_{i=1}^{N} (\xi_i^{(2)}, u_i^{(2)})_m - \sum_{j=1}^{M} \int_{\Omega_j} (\xi_j^{(1)}(t), u_j^{(1)}(t))_l dt - c$$

$$= \sum_{i=1}^{N+1} ((x_0(t_{i-1}), z_i(t_{i-1};u))_n - (x_0(t_i), z_i(t_i;u))_n)$$

$$+ \sum_{i=1}^{N+1} \int_{t_{i-1}}^{t_i} (\frac{dx_0(t)}{dt} - A(t)x_0(t), z_i(t;u))_n dt$$

$$- \sum_{i=1}^{N} (x_0(t_i), z_{i+1}(t_i;u) - z_i(t_i;u))_n$$

$$- \sum_{i=1}^{N} (\xi_i^{(2)}, u_i^{(2)})_m - \sum_{j=1}^{M} \int_{\Omega_j} (\xi_j^{(1)}(t), u_j^{(1)}(t))_l dt - c$$

$$= (x(t_0), z_1(t_0;u))_n - (z_1(x(t_1), t_1;u))_n$$

$$+ \sum_{i=2}^{N} (x_0(t_{i-1}), z_i(t_{i-1};u))_n - (x_0(t_i), z_i(t_i;u))_n)$$

$$+(x_0(t_N), z_{N+1}(t_N))_n - (x_0(t_{N+1}), z_{N+1}(t_{N+1}))_n$$

$$+ \sum_{i=1}^{N+1} \int_{t_{i-1}}^{t_i} (B(t)f(t), z_i(t;u))_n dt$$

$$- \sum_{i=1}^{N} (x_0(t_i), z_{i+1}(t_i;u) - z_i(t_i;u))_n$$

$$- \sum_{i=1}^{N} (\xi_i^{(2)}, u_i^{(2)})_m - \sum_{j=1}^{M} \int_{\Omega_j} (\xi_j^{(1)}(t), u_j^{(1)}(t))_l dt - c.$$

Taking into account that

$$\sum_{i=2}^{N}(x_0(t_{i-1}), z_i(t_{i-1}; u))_n + (x_0(t_N), z_{N+1}(t_N))_n$$

$$= \sum_{i'=1}^{N-1}(x_0(t_{i'}), z_{i'+1}(t_{i'}; u))_n + (x_0(t_N), z_{N+1}(t_N))_n$$

$$= \sum_{i=1}^{N}(x_0(t_i), z_{i+1}(t_i; u))_n,$$

from latter equalities and (36), we have

$$l(x) - \widehat{l(x)} = \int_0^T (B(t)f(t), z(t; u))_n dt$$

$$-\sum_{i=1}^{N}(\xi_i^{(2)}, u_i^{(2)})_m - \sum_{j=1}^{M}\int_{\Omega_j}(\xi_j^{(1)}(t), u_j^{(1)}(t))_l dt - c. \tag{39}$$

The latter equality yields

$$\mathbb{E}\Big[l(x) - \widehat{l(x)}\Big] = \int_0^T (B(t)f(t), z(t; u))_n dt - c. \tag{40}$$

Taking into consideration the known relationship

$$\mathbb{D}\eta = \mathbb{E}|\eta|^2 - |\mathbb{E}\eta|^2 \tag{41}$$

that couples the dispersion $\mathbb{D}\eta = \mathbb{E}|\eta - \mathbb{E}\eta|^2$ of random variable η with its expectation $\mathbb{E}\eta$, in which η is determined by right-hand side of (39) and noncorrelatedness of $\xi_i = (\xi_1^{(i)}, \dots, \xi_m^{(i)})^T$ and $\xi_j(\cdot) = \xi_1^{(j)}(\cdot), \dots, \xi_l^{(j)}(\cdot))^T$, from the equalities (39) and (40) we find

$$\mathbb{E}\Big|l(x) - \widehat{l(x)}|^2 = \Big|\int_0^T (B(t)f(t), z(t; u))_n dt - c|^2$$

$$+\mathbb{E}|\sum_{i=1}^{N}(\xi_i^{(2)}, u_i^{(2)})_m + \sum_{j=1}^{M}\int_{\Omega_j}(\xi_j^{(1)}(t), u_j^{(1)}(t))_l dt|^2$$

$$= \left| \int_0^T (B(t)(f(t) - f_0(t)), z(t;u))_n dt + \int_0^T (B(t)f_0(t), z(t;u))_n dt - c \right|^2$$

$$+ \mathbb{E}\left|\sum_{i=1}^{N} (\xi_i^{(2)}, u_i^{(2)})_m\right|^2 + \mathbb{E}\left|\sum_{j=1}^{M} \int_{\Omega_j} (\xi_j^{(1)}(t), u_j^{(1)}(t))_l dt\right|^2.$$

Thus,

$$\inf_{c\in\mathbb{C}} \sigma(u,c) = \inf_{c\in\mathbb{C}} \sup_{f\in G_1, \xi\in G_2} \mathbb{E}|l(x) - \widehat{l(x)}|^2 =$$

$$= \inf_{c\in\mathbb{C}} \sup_{f\in G_1, \xi\in G_2} \left| \int_0^T (B(t)(f(t) - f_0(t)), z(t;u))_n dt \right.$$

$$\left. + \int_0^T (B(t)f_0(t), z(t;u))_n dt - c \right|^2$$

$$+ \sup_{\xi\in G_2} \left(\mathbb{E}\left|\sum_{i=1}^{N} (\xi_i^{(2)}, u_i^{(2)})_m\right|^2 + \mathbb{E}\left|\sum_{j=1}^{M} \int_{\Omega_j} (\xi_j^{(1)}(t), u_j^{(1)}(t))_l dt\right|^2\right) \tag{42}$$

Set

$$y := \int_0^T (B(t)(f(t) - f_0(t)), z(t;u))_n dt$$

$$= \int_0^T (f(t) - f_0(t), B^*(t)z(t;u))_r dt,$$

$$d = c - \int_0^T (B(t)f_0(t), z(t;u))_n dt.$$

Then Lemma 1 and (13) imply

$$|y| \le \left| \int_0^T (\tilde{Q}(t)z(t;u), (t)z(t;u))_n dt \right|^{\frac{1}{2}}$$

$$\int_0^T (Q(t)(f(t) - f_0(t)), f(t) - f_0(t))_r dt \Big|^{\frac{1}{2}}$$

$$\leq \left| \int_0^T (\tilde{Q}(t)z(t;u), z(t;u))_n dt \right|^{1/2} =: l. \tag{43}$$

The direct substitution shows that last inequality is transformed to an equality at

$$f^{(0)} := f_0 \pm \frac{Q^{-1}(\cdot)B^*z(\cdot;u)}{(Q^{-1}(\cdot)B^*z(\cdot;u), B^*z(\cdot;u))^{1/2}_{(L^2(0,T))^r}}.$$

Vector-fuction $f^{(0)} \in G_1$, because, obviously, the second condition in (13) is fulfilled; in addition,

$$(f^{(0)}, B^*\psi_k)_{(L^2(0,T))^n}$$

$$= (Q^{-1}(\cdot)B^*z(\cdot;u), B^*z(\cdot;u))_{(L^2(0,T))^r})^{-1/2} \times \left(Q^{-1}B^*z(\cdot;u), B^*\psi_k)_{(L^2(0,T))^n}\right)$$

$$+(f_0, B^*\psi_k)_{(L^2(0,T))^n} = 0 \quad \forall \psi_k, \quad k = 1, \ldots, d,$$

which yields, by virtue of (17), the validity of the first condition (13). Therefore, taking into account the equality

$$\inf_{d\in\mathbb{C}} \sup_{|y|\leq l} |y-d|^2 = l^2,$$

we find

$$\inf_{c\in\mathbb{C}} \sup_{f\in G_1} \left| \int_0^T (B(t)(f(t)-f_0(t)), z(t;u))_n dt \right.$$

$$\left. + \int_0^T (B(t)f_0(t), z(t;u))_n dt - c \right|^2$$

$$= l^2 = \int_0^T (\tilde{Q}(t)z(t;u), z(t;u))_n dt, \tag{44}$$

where the infimum over c is attained at

$$c = \int_0^T (f_0(t), B^*(t)z(t;u))_r dt. \tag{45}$$

Calculate the last term on the right-hand side of (42). Applying Lemma 1, we have

$$\mathbb{E}\left|\sum_{i=1}^{N}(\xi_i^{(2)}, u_i^{(2)})_m\right|^2 \le \mathbb{E}\left[\sum_{i=1}^{N}(D_i^{-1}u_i^{(2)}, u_i^{(2)})_m \cdot \sum_{i=1}^{N}(D_i\xi_i, \xi_i)_m\right]$$
$$= \sum_{i=1}^{N}(D_i^{-1}u_i^{(2)}, u_i^{(2)})_m \cdot \mathbb{E}\left[\sum_{i=1}^{N}(D_i\xi_i, \xi_i)_m\right]. \tag{46}$$

Transform the last factor on the right-hand side of (46):

Let $d_{jk}^{(i)}$ and $r_{jk}^{(i)}$ $(j, k = 1, \ldots, m)$ be the elements of matrices $D_i^{(2)}$ and $R_i^{(2)}$, respectively. Then $r_{jk}^{(i)} = \mathbb{E}\xi_{ij}^{(2)}\overline{\xi}_{ik}^{(2)}$, where $\xi_{ij}^{(2)}$ and $\xi_{ik}^{(2)}$ are j-th and k-th coordinates of vector $\xi_i^{(2)}$, respectively. Therefore,

$$\mathbb{E}\left[\sum_{i=1}^{N}(D_i^{(2)}\xi_i, \xi_i)_m\right] = \sum_{i=1}^{N}\mathbb{E}\left(\sum_{j=1}^{m}\sum_{k=1}^{m}d_{jk}^{(i)}\xi_{ik}^{(2)}\overline{\xi}_{ij}^{(2)}\right) = \sum_{i=1}^{N}\sum_{j=1}^{m}\sum_{k=1}^{m}d_{jk}^{(i)}\mathbb{E}\xi_{ik}^{(2)}\overline{\xi}_{ij}^{(2)}$$
$$= \sum_{i=1}^{N}\sum_{j=1}^{m}\sum_{k=1}^{m}d_{jk}^{(i)}r_{kj}^{(i)} = \sum_{i=1}^{N}Tr\left[D_i^{(2)}R_i^{(2)}\right].$$

Analogously,

$$\mathbb{E}|\sum_{j=1}^{M}\int_{\Omega_j}(\xi_j(t), u_j^{(1)}(t))_l dt|^2$$

$$\le \sum_{j=1}^{M}\int_{\Omega_j}((D_j^{(1)})^{-1}(t)u_j^{(1)}(t), u_j^{(1)}(t))_l dt \cdot \mathbb{E}\left[\sum_{j=1}^{M}\int_{\Omega_j}(D_j^{(1)}(t)\xi_j^{(1)}(t), \xi_j^{(1)}(t))_l dt\right]$$

and

$$\mathbb{E}\left[\sum_{j=1}^{M}\int_{\Omega_j}(D_j^{(1)}(t)\xi_j^{(1)}(t), \xi_j^{(1)}(t))_l dt\right] = \sum_{j=1}^{M}\int_{\Omega_j}Tr\left[D_j^{(1)}(t)R_j^{(1)}(t,t)\right]dt.$$

Taking into account (14), we deduce from here

$$\mathbb{E}\left|\sum_{i=1}^{N}(u_i^{(2)}, \xi_i^{(2)})_m\right|^2 + \mathbb{E}|\sum_{j=1}^{M}\int_{\Omega_j}(\xi_j^{(1)}(t), u_j^{(1)}(t))_l dt|^2$$

$$\leq \sum_{i=1}^{N}((D_i^{(2)})^{-1}u_i^{(2)}, u_i^{(2)})_m + \sum_{j=1}^{M}\int_{\Omega_j}((D_j^{(1)})^{-1}(t)u_j^{(1)}(t), u_j^{(1)}(t))_l dt.$$

It is not difficult to check that here, the equality sign is attained at the element

$$\xi = \left(\xi_1^{(1)}(\cdot), \ldots, \xi_M^{(1)}(\cdot), \xi^{(2)}, \ldots, \xi_N^{(2)}\right) \in G_2$$

with

$$\xi_j^{(1)}(t) = \frac{\eta_2(D_j^{(1)})^{-1}(t)u_j^{(1)}(t)}{\left[\sum_{j=1}^{M}\int_{\Omega_j}\left((D_j^{(1)})^{-1}(t)u_j^{(1)}(t), u_j^{(1)}(t)\right)_l dt\right]^{1/2}}, \quad j = 1, \ldots, M,$$

$$\xi_i^{(2)} = \frac{\eta_1(D_i^{(2)})^{-1}u_i}{\left[\sum_{i=1}^{N}\left((D_i^{(2)})^{-1}u_i, u_i\right)_m\right]^{1/2}}, \quad i = 1, \ldots, N,$$

where η_1 and η_2 are uncorrelated random variables such that $\mathbb{E}\eta_i = 0$ and $\mathbb{E}|\eta_i|^2 = 1$, $i = 1, 2$. Hence,

$$\sup_{\xi \in G_2}\left(\mathbb{E}\left|\sum_{i=1}^{N}(\xi_i^{(2)}, u_i^{(2)})_m\right|^2 + \mathbb{E}\left|\sum_{j=1}^{M}\int_{\Omega_j}(\xi_j^{(1)}(t), u_j^{(1)}(t))_l dt\right|^2\right)$$

$$= \sum_{i=1}^{N}((D_i^{(2)})^{-1}u_i^{(2)}, u_i^{(2)})_m + \sum_{j=1}^{M}\int_{\Omega_j}((D_j^{(1)})^{-1}(t)u_j^{(1)}(t), u_j^{(1)}(t))_l dt. \tag{47}$$

The statement of the lemma follows now from (42), (44), (45) and (47). The proof is complete. □.

In the proof of Theorem 1 stated below, it will be shown that solving the optimal control problem (15)–(18) is reduced to solving some system of impulsive differential equations with periodic boundary conditions.

Theorem 1 *The minimax estimate $\widehat{\widehat{l(x)}}$ of expression $l(x)$ has the form*

$$\widehat{\widehat{l(x)}} = \sum_{j=1}^{M}\int_{\Omega_j}(y_j^{(1)}(t), \hat{u}_j^{(1)}(t))_l dt + \sum_{i=1}^{N}(y_i^{(2)}, \hat{u}_i^{(2)})_m + \hat{c} = l(\hat{x}), \tag{48}$$

where

$$\hat{u}_j^{(1)}(t) = D_j^{(1)}(t)H_j^{(1)}(t)p(t), \quad j = 1, \ldots, M,$$

$$\hat{u}_i = D_i^{(2)} H_i^{(2)} p(t_i), \quad i = 1, \ldots, N, \quad \hat{c} = \int_0^T (f_0(t), B^*(t)\hat{z}(t))_r dt, \tag{49}$$

and vector-functions $p(t)$, $\hat{z}(t)$, and $\hat{x}(t)$ are determined from the solution of the systems of equations

$$-\frac{d\hat{z}(t)}{dt} = A^*(t)\hat{z}(t) + l_0(t)$$
$$-\sum_{j=1}^{M} \chi_{\Omega_j}(t)(H_j^{(1)})^*(t)D_j^{(1)}(t)H_j^{(1)}(t)p(t)$$
$$for\ a.e.\ \ t \in (0, T), \tag{50}$$

$$\Delta\hat{z}\big|_{t=t_i} = (H_i^{(2)})^* D_i^{(2)} H_i^{(2)} p(t_i), \quad i = 1, \ldots, N, \quad \hat{z}(T) = \hat{z}(0), \tag{51}$$

$$\int_0^T (\tilde{Q}(t)\hat{z}(t), \psi_k(t))_n dt = 0, \quad k = 1, \ldots d, \tag{52}$$

$$\frac{dp(t)}{dt} = A(t)p(t) + \tilde{Q}(t)\hat{z}(t) \quad for\ a.e.\ \ t \in (0, T), \quad p(0) = p(T) \tag{53}$$

$$\int_0^T (\varphi_k(t), l_0(t))_n dt - \sum_{j=1}^{M} \int_{\Omega_j} (\varphi_k(t), (H_j^{(1)})^*(t)D_j^{(1)}(t)H_j^{(1)}(t)p(t))_n dt$$
$$-\sum_{i=1}^{N} (\varphi_k(t_i), (H_i^{(2)})^* D_i^{(2)} H_i^{(2)} p(t_i))_n = 0, \quad k = 1, \ldots d, \tag{54}$$

and

$$-\frac{d\hat{p}(t)}{dt} = A^*(t)\hat{p}(t) - \sum_{j=1}^{M} \chi_{\Omega_j}(t)(H_j^{(1)})^*(t)D_j^{(1)}(t)\Big[H_j^{(1)}(t)\hat{x}(t) - y_j^{(1)}(t)\Big]$$
$$for\ a.e.\ \ t \in (0, T), \tag{55}$$

$$\Delta\hat{p}\big|_{t=t_i} = (H_i^{(2)})^* D_i^{(2)}\Big[H_i^{(2)}\hat{x}(t_i) - y_i^{(2)}\Big], \quad i = 1, \ldots, N, \quad \hat{p}(T) = \hat{p}(0), \tag{56}$$

$$\int_0^T (\tilde{Q}(t)\hat{p}(t), \psi_k(t))_n dt = 0, \quad k = 1, \dots d, \tag{57}$$

$$\frac{d\hat{x}(t)}{dt} = A(t)\hat{x}(t) + \tilde{Q}(t)\hat{p}(t) + B(t)f_0(t) \quad for\ a.e.\ \ t \in (0, T), \tag{58}$$

$$\hat{x}(0) = \hat{x}(T), \tag{59}$$

$$\begin{aligned}&\sum_{j=1}^{M}\int_{\Omega_j} (\varphi_k(t), (H_j^{(1)})^*(t)D_j^{(1)}(t)\Big[H_j^{(1)}(t)\hat{x}(t) - y_j^{(1)}(t)\Big])_n dt\\ &+\sum_{i=1}^{N}(\varphi_k(t_i), (H_i^{(2)})^* D_i^{(2)}\Big[H_i^{(2)}\hat{x}(t_i) - y_i^{(2)}\Big])_n\\ &= 0, \quad k = 1, \dots d,\end{aligned} \tag{60}$$

respectively. Problems (50)–(54) and (55)–(60) are uniquely solvable. Equations (55)–(60) are fulfilled with probability 1.

The minimax estimation error σ is determined by the formula

$$\sigma = [l(p)]^{1/2}. \tag{61}$$

Proof Show first that the solution $z(t; u)$ of problem (15)–(17) continuously depends on $u \in U$, i.e. that for any sequence $\{u_k\}$ convergent to u in set U there holds the relation

$$\lim_{u_k \to u} z(t; u_k) = z(t; u) \quad in \quad (L^2(0, T))^n. \tag{62}$$

To this end, put

$$v_k = u - u_k, \quad \tilde{z}(t; v_k) = z(t; u) - z(t; u_k). \tag{63}$$

Then $v_k \in V$, where

$$\begin{aligned}V = \{v \in H : &\sum_{j=1}^{M}\int_{\Omega_j} (\varphi_k(t), (H_j^{(1)})^*(t)v_j^{(1)}(t))_n dt\\ &+\sum_{i=1}^{N}(\varphi_k(t_i), (H_i^{(2)})^* v_i^{(2)})_n = 0 \ \ \forall k = 1, \dots d\}.\end{aligned}$$

it is easy to see that $\tilde{z}(t; v_k)$ solves the problem

$$-\frac{d\tilde{z}(t;v_k)}{dt}=A^*(t)\tilde{z}(t;v_k)$$
$$-\sum_{j=1}^{M}\chi_{\Omega_j}(t)(H_j^{(1)})^*(t)v_{j,k}^{(1)}(t)dt,\quad t\in(0,T),\quad t\neq t_i, \tag{64}$$

$$\Delta\tilde{z}(\cdot;v_k)|_{t=t_i}=(H_i^{(2)})^*v_{i,k}^{(2)},\quad i=1,\dots,N,\quad \tilde{z}(T;v_k)=\tilde{z}(0;v_k), \tag{65}$$

$$\int_0^T(\tilde{Q}\tilde{z}(t;v_k),\psi_i(t))_n dt=0,\quad i=1,\dots d. \tag{66}$$

Proceeding similarly to the beginning of the proof of Lemma 2, we conclude that the unique solution $\tilde{z}(t;v_k)$ to BVP (64)–(66) can be represented in the form

$$\tilde{z}(t;v_k)=\tilde{z}_0(t;v_k)+\sum_{i=1}^{d}c_i(v_k)\psi_i(t), \tag{67}$$

where $\tilde{z}_0(t;v_k)$ is the solution to problem (64)–(66) satisfying the condition

$$(\tilde{z}_0(T;v_k),\psi_i(T))_n=0,\quad i=1,\dots,d,$$

and the coefficients $c_i(v_k)\in\mathbb{C}$ are uniquely determined from the linear algebraic equation system

$$\sum_{i=1}^{d}\alpha_{ij}c_i(v_k)=\beta_j(v_k),\quad j=1,\dots,d$$

by the formulas

$$c_i(v_k)=\frac{d_i(v_k)}{D}, \tag{68}$$

where $D=\det[\alpha_{ij}]_{i,j=1}^{d}$,

$$d_i(v_k)=\det\begin{bmatrix}\alpha_{1,1} & \cdots & \alpha_{i-1,1} & \beta_1(v_k) & \alpha_{i+1,1} & \cdots & \alpha_{d,1}\\ \alpha_{1,2} & \cdots & \alpha_{i-1,2} & \beta_2(v_k) & \alpha_{i+1,2} & \cdots & \alpha_{d,2}\\ \cdots & \cdots & \cdots & \cdots & \cdots & \cdots & \cdots\\ \alpha_{1,d} & \cdots & \alpha_{i-1,d} & \beta_d(v_k) & \alpha_{i+1,d} & \cdots & \alpha_{d,d}\end{bmatrix},$$

elements $\alpha_{i,j}$, $i,j=1,\dots,d$, of non-singular matrix $\left[\alpha_{i,j}\right]$ are determined from (24), and

$$\beta_j(v_k) = -\int_0^T (\tilde{Q}(t)\tilde{z}_0(t; v_k), \psi_j(t))_n dt, \; j = 1, \ldots, d.$$

It follows from [17, Theorem 4.21] that $\tilde{z}_0(t; v_k)$ is expressed via the generalized Green's function $G(t, s)$ of Eq. (6) as

$$\tilde{z}_0(t; v_k) = -\sum_{j=1}^{M} G(t, s)(H_j^{(1)})^*(s)v_{j,k}^{(1)}(s)ds + \sum_{i=1}^{M} G(t, t_i)(H_i^{(2)})^* v_{i,k}^{(2)}, \tag{69}$$

where

$$G(t, s) = \begin{cases} W(t, 0)\widetilde{B}^{-1}W(T, s) + W(t, s) & \text{if } \; 0 < s < t \le T, \\ W(t, 0)\widetilde{B}^{-1}W(T, s) & \text{if } \; 0 < t \le s \le T, \end{cases}$$

$W(t, s) = Y(t)Y^{-1}(s)$ is the Cauchy matrix for Eq. (6), the matrix

$$\tilde{B} = \left[E_n - M - \Phi(0)\Psi^*(0)\right]$$

is non-singular, and $M = W(T, 0)$, $\Phi = [\varphi_1, \ldots, \varphi_d]$ and $\Psi = [\psi_1, \ldots, \psi_d]$, and by $Y(t)$ we denote a matrix-valued function $Y(t) = [y_1(t), \ldots, y_n(t)]$ whose columns are linearly independent solutions $y_1(t), \ldots, y_n(t)$ of the homogeneous system (6) such that $Y(0) = E_n$. Notice that $Y(t) = [X^*(t)]^{-1}$.

The representation (69) implies the inequality

$$\|\tilde{z}_0(t; v_k)\|_{(L^2(0,T))^n} \le a\|v_k\|_H, \tag{70}$$

where a is a constant independent of v_k.

Expanding determinant $d_i(v_k)$ entering the right-hand side of (68) in elements of the i th column, we have

$$c_i(v_k) = \gamma_1^{(i)} b_1(v_k) + \gamma_2^{(i)} b_2(v_k) + \cdots + \gamma_d^{(i)} b_d(v_k), \; i = 1, \ldots, d,$$

where

$$\gamma_j^{(i)} = \frac{A_{ji}}{D}, \quad i, j = 1 \ldots, d,$$

are constants independent of v_k, A_{ji} is the algebraic complement of the element of the i th column that enters the j th row of determinant $d_i(v_k)$ which is independent of v_k. Applying the generalized Cauchy–Bunyakovsky inequality to the representation for $c_i(v_k)$, we obtain

$$|c_i(v_k)| \leq \sum_{j=1}^{d} \left|\gamma_j^{(i)}\right| \left|b_j(v_k)\right| = \sum_{j=1}^{d} \left|\gamma_j^{(i)}\right| \left| \int_0^T (\tilde{Q}(t)\tilde{z}_0(t; v_k), \psi_j(t))_n dt \right|$$

$$\leq \sum_{j=1}^{d} \left|\gamma_j^{(i)}\right| \left| \int_0^T (\tilde{Q}(t)\tilde{z}_0(t; v_k), \tilde{z}_0(t; v_k))_n dt \right|^{1/2} \times \left| \int_0^T (\tilde{Q}(t)\psi_j(t), \psi_j(t))_n dt \right|^{1/2}$$

$$= A_i \left| \int_0^T (\tilde{Q}(t)\tilde{z}_0(t; v_k), \tilde{z}_0(t; v_k))_n dt. \right|^{1/2}, \tag{71}$$

where

$$A_i = \sum_{j=1}^{d} \left|\gamma_j^{(i)}\right| \left| \int_0^T (\tilde{Q}(t)\psi_j(t), \psi_j(t))_n dt \right|^{1/2}, \quad i = 1, \ldots, d,$$

are constants independent of v_k. Next, taking into consideration the estimate

$$\left| \int_0^T (\tilde{Q}(t)\tilde{z}_0(t; v_k), \tilde{z}_0(t; v_k))_n dt \right|^{1/2}$$

$$\leq (\|\tilde{Q}(t)\tilde{z}_0(\cdot; v_k)\|_{(L^2(0,T))^n} \|\tilde{z}_0(\cdot; v_k)\|_{(L^2(0,T))^n})^{1/2}$$

$$\leq \max_{0 \leq t \leq T} \|\tilde{Q}(t)\|^{1/2} \|\tilde{z}_0(\cdot; v_k)\|_{(L^2(0,T))^n} \leq a \max_{0 \leq t \leq T} \|\tilde{Q}(t)\|^{1/2} \|v_k\|_H,$$

which is obtained with the help of the Cauchy–Bunyakovsky inequality and (70), we see that the following inequalities

$$|c_i(v_k)| \leq C_i \|v_k\|_H, i = 1, \ldots, d, \tag{72}$$

hold, where $C_i = A_i a \max_{0 \leq t \leq T} \left\{ \|\tilde{Q}(t)\|^{1/2} \right\}$ are constants independent of v_k.

Using inequalities (72), (70) and representation (67) of solution $\tilde{z}(t; v)$ to BVP (64)–(66), we will prove that this solution satisfies the inequality

$$\int_0^T \|\tilde{z}(t; v_k)\|_n^2 dt \leq K \|v_k\|_H^2, \tag{73}$$

where K is a constant independent of v_k. Taking into notice the inequality $\|a + b\|^2 \leq 2(\|a\|^2 + \|b\|^2)$ which is valid for any elements a and b from a normed space, we have

$$\int_0^T \|\tilde{z}(t; v_k)\|_n^2 dt = \int_0^T \|\tilde{z}_0(t; v_k) + \sum_{i=1}^d c_i(v_k)\psi_i(t)\|_n^2 dt$$

$$\leq 2\left(\int_0^T \|\tilde{z}_0(t; v_k)\|^2 dt + \sum_{i=1}^d |c_i(v_k)|^2 \int_0^T \|\psi_i(t)\|_n^2 dt\right)$$

$$\leq 2a^2\|v_k\|_H^2 + 2\sum_{i=1}^d C_i^2\|v_k\|_H^2 \int_0^T \|\psi_i(t)\|_n^2 dt = K\|v_k\|_H^2,$$

where

$$K = 2a^2 + 2\sum_{i=1}^d C_i^2 \int_0^T \|\psi_i(t)\|_n^2 dt.$$

Passing to the limit in (73) as $v_k \to 0$ and taking into account (63), we receive (62).

This implies the continuity of functional $I(u)$ on set U and, hence, its lower (upper) semicontinuity on this set. Since $I(u)$ is also a strictly convex functional on the closed convex set U satisfying the condition

$$I(u) \geq \int_0^T (\tilde{Q}(t)z(t; u), z(t; u))_n dt + \sum_{j=1}^M \int_{\Omega_j} ((D_j^{(1)})^{-1}(t)u_j^{(1)}(t), u_j^{(1)}(t))_l dt$$

$$+\sum_{i=1}^N ((D_i^{(2)})^{-1}u_i^{(2)}, u_i^{(2)})_m \geq c\|u\|_H^2 \quad \forall u \in U, \quad c = const,$$

then by Remark 1.1 to Theorem 1.1 from [16] we see that there exists the unique element $\hat{u} \in U$ such that

$$I(\hat{u}) = \inf_{u \in U} I(u).$$

Therefore, and taking into account that $\{U\} = \hat{u} + \{V\}$, for any fixed $v \in V$ and $\tau \in \mathbb{R}$ the functions $s_1(\tau) := I(\hat{u} + \tau v)$ and $s_2(\tau) := I(\hat{u} + i\tau v)$ reach their minimums at a unique point $\tau = 0$, so that

$$\frac{d}{d\tau} I(\hat{u} + \tau v).|_{\tau=0} = 0 \text{ and } \frac{d}{d\tau} I(\hat{u} + i\tau v).|_{\tau=0} = 0, \tag{74}$$

where $i = \sqrt{-1}$. Since $z(t; \hat{u} + \tau v) = z(t; \hat{u}) + \tau\tilde{z}(t; v)$ and $z(t; \hat{u} + i\tau v) = z(t; \hat{u}) + i\tau\tilde{z}(t; v)$, where $\tilde{z}(t; v)$ is a unique solution to problem (15), and (16) at $l_0 = 0$ and $u = v$, from (18) and (74), we obtain

$$0 = Re\{\sum_{i=1}^{N+1}\int_{t_{i-1}}^{t_i}(\tilde{Q}(t)z_i(t; \hat{u}), \tilde{z}_i(t; v))_n dt$$

$$+\sum_{i=1}^{N}((D_i^{(2)})^{-1}\hat{u}_i^{(2)}, v_i^{(2)})_m + \sum_{j=1}^{M}\int_{\Omega_j}((D_j^{(1)})^{-1}(t)\hat{u}_j^{(1)}(t), v_j^{(1)}(t))_l dt\},$$

and

$$0 = Im\{\sum_{i=1}^{N+1}\int_{t_{i-1}}^{t_i}(\tilde{Q}(t)z_i(t; \hat{u}), \tilde{z}_i(t; v))_n dt$$

$$+\sum_{i=1}^{N}((D_i^{(2)})^{-1}\hat{u}_i^{(2)}, v_i^{(2)})_m + \sum_{j=1}^{M}\int_{\Omega_j}((D_j^{(1)})^{-1}(t)\hat{u}_j^{(1)}(t), v_j^{(1)}(t))_l dt\},$$

where $\tilde{z}_i(t; v)$ have the same sense as $z_i(t; u)$, $i = 1, \ldots, N + 1$. Whence,

$$0 = \sum_{i=1}^{N+1}\int_{t_{i-1}}^{t_i}(\tilde{Q}(t)z_i(t; \hat{u}), \tilde{z}_i(t; v))_n dt$$

$$+\sum_{i=1}^{N}((D_i^{(2)})^{-1}\hat{u}_i^{(2)}, v_i^{(2)})_m + \sum_{j=1}^{M}\int_{\Omega_j}((D_j^{(1)})^{-1}(t)\hat{u}_j^{(1)}(t), v_j^{(1)}(t))_l dt. \quad (75)$$

Let $p(t)$ be a solution of the problem[7]

$$\frac{dp(t)}{dt} = A(t)p(t) + \tilde{Q}(t)z(t; \hat{u}) \quad for\ a.e.\ \ t \in (0, T), \quad (76)$$

$$p(0) = p(T). \quad (77)$$

Then we have

[7] Relationship (17) at $u = \hat{u}$ coincides with the solvability condition for this problem by virtue of (5).

$$\sum_{i=1}^{N+1}\int_{t_{i-1}}^{t_i}(\tilde{Q}(t)z_i(t;\hat{u}),\tilde{z}_i(t;v))_n dt=\sum_{i=1}^{N+1}\int_{t_{i-1}}^{t_i}(\frac{dp(t)}{dt}-A(t)p(t),\tilde{z}_i(t;v))_n dt$$

$$=\sum_{i=1}^{N+1}((p(t_i),\tilde{z}_i(t_i;v))_n-(p(t_{i-1}),\tilde{z}_i(t_{i-1};v))_n)$$

$$+\sum_{i=1}^{N+1}\int_{t_{i-1}}^{t_i}(p(t),-\frac{d\tilde{z}_i(t;v)}{dt}-A^*(t)\tilde{z}_i(t;v))_n dt$$

$$=\sum_{i=1}^{N}(p(t_i),\tilde{z}_i(t_i;v))_n+(p(t_{N+1}),\tilde{z}_{N+1}(t_{N+1};v))_n-(p(t_0),\tilde{z}_1(t_0;v))_n$$

$$-\sum_{i=2}^{N+1}(p(t_{i-1}),\tilde{z}_i(t_{i-1};v))_n=\sum_{i=1}^{N}(p(t_i),\tilde{z}_i(t_i;v))_n-\sum_{i=1}^{N}(p(t_i),\tilde{z}_{i+1}(t_i;v))_n.$$

$$=-\sum_{j=1}^{M}\int_{\Omega_j}(p(t),(H_j^{(1)})^*(t)v_j^{(1)}(t))_n dt+\sum_{i=1}^{N}(p(t_i),\tilde{z}_i(t_i;v)-\tilde{z}_{i+1}(t_i;v))_n$$

$$=-\sum_{j=1}^{M}\int_{\Omega_j}(p(t),(H_j^{(1)})^*(t)v_j^{(1)}(t))_n dt-\sum_{i=1}^{N}(p(t_i),(H_i^{(2)})^*v_i^{(2)})_n. \tag{78}$$

Let $Q_0 : H \to H$ be a linear operator defined by the formula

$$Q_0u=\left(D_1^{(1)}(\,)u_1^{(1)}(\cdot),\dots,D_M^{(1)}(\cdot)u_M^{(1)}(\cdot),D_1^{(?)}u_1^{(2)},\dots,D_N^{(2)}u_N^{(2)}\right)\forall u\in H.$$

Then the inverse operator Q_0^{-1} of Q_0 is given by

$$Q_0^{-1}u=((D_1^{(1)})^{-1}(\cdot)u_1^{(1)}(\cdot),\dots,(D_M^{(1)})^{-1}(\cdot)u_M^{(1)}(\cdot),$$

$$(D_1^{(2)})^{-1}u_1^{(2)},\dots,\left(D_N^{(2)})^{-1}u_N^{(2)}\right)\forall u\in H.$$

Equalities (78) and (75) yield that for any $v \in V$,

$$(Q_0^{-1}\hat{u}-Cp,v)_H=0. \tag{79}$$

Let us show that in the set of solutions to problem (76), and (77) there is only one, $p(t)$, for which

$$Q_0^{-1}\hat{u}-Cp\in V. \tag{80}$$

Indeed, condition (80) means that for any $1 \le i \le d$ the equalities

$$(C\varphi_i(t), Q_0^{-1}\hat{u} - Cp)_H = 0 \tag{81}$$

hold.

Since general solution $p(t)$ to BVP (76), and (77) has the form

$$p(t) = p_0(t) + \sum_{j=1}^{d} a_j \varphi_j(t),$$

where $p_0(t)$ is the solution to this problem satisfying the condition

$$(p_0(0), \psi_i(0))_n = 0, \quad i = 1, \dots, d,$$

and $a_j \in \mathbb{C}$ $(j = 1, \dots, d)$ are arbitrary numbers, we conclude that in line with (81), function $p(t)$ satisfies condition (80) if $(a_1, \dots, a_d)^T$ is a solution to the uniquely solvable linear algebraic equation system

$$\sum_{i=1}^{d} a_i (C\varphi_i, C\varphi_j)_H = (C\varphi_j, Q_0^{-1}\hat{u} - Cp_0)_H, \quad j = 1, \dots, d,$$

where matrix $[(C\varphi_i, C\varphi_j)_H]_{i,j=1}^{d}$ has a non-zero determinant because it is the Gram matrix of the system of linearly independent elements $C\varphi_1, \dots, C\varphi_d$. It is easy to see that the unique solvability of this system yields the existence of the unique function $p(t)$ that satisfies condition (80) and Eqs. (76) and (77).

Setting in (79) $v = Q_0^{-1}\hat{u} - Cp$ we have $Q_0^{-1}\hat{u} - Cp = 0$, so that
x $\hat{u} = Q_0 Cp$

$$= (D_1^{(1)}(t)H_1^{(1)}(t)p(t), \dots, D_j^{(1)}(t)H_j^{(1)}(t)p(t), \dots, D_M^{(1)}(t)H_M^{(1)}(t)p(t),$$

$$D_1^{(2)}H_1^{(2)}p(t_1), \dots, D_i^{(2)}H_i^{(2)}p(t_i), \dots, D_N^{(2)}H_N^{(2)}p(t_N)).$$

The latter equality means that

$$\hat{u}_j^{(1)}(t) = D_j^{(1)}(t)H_j^{(1)}(t)p(t), \quad j = 1, \dots, M.$$

$$\hat{u}_i^{(2)} = D_i^{(2)}H_i^{(2)}p(t_i), \quad i = 1, \dots, N. \tag{82}$$

Since $\hat{u} \in U$, vector-function $p(t)$ satisfies Eq. (54).

Setting $u = \hat{u}$ in (45) and (15)–(17) and denoting $\hat{z}(t) = z(t; \hat{u})$, we see that $\hat{z}(t)$ and $p(t)$ satisfy system (50)–(54); the unique solvability of this system follows from the fact that functional $I(u)$ has one minimum point $\hat{u}$.

Now let us establish that $\sigma^2 = \sigma(\hat{u}, \hat{c}) = l(p)$. Substituting expression (82)–(18), we obtain

$$\sigma(\hat{u}, \hat{c}) = I(\hat{u}) = \sum_{i=1}^{N+1} \int_{t_{i-1}}^{t_i} (\tilde{Q}(t)\hat{z}_i(t), \hat{z}_i(t))_n dt$$

$$+ \sum_{i=1}^{N} (H_i^{(2)} p(t_i), D_i^{(2)} H_i^{(2)} p(t_i))_m$$

$$+ \sum_{j=1}^{M} \int_{\Omega_j} (H_j^{(1)}(t) p(t), D_j^{(1)}(t) H_j^{(1)}(t) p(t))_l dt \tag{83}$$

However,

$$\sum_{i=1}^{N+1} \int_{t_{i-1}}^{t_i} (\tilde{Q}(t)\hat{z}_i(t), \hat{z}_i(t))_n dt = \sum_{i=1}^{N+1} \int_{t_{i-1}}^{t_i} (\frac{dp(t)}{dt} - A(t)p(t), \hat{z}_i(t))_n dt$$

$$= \sum_{i=1}^{N+1} ((p(t_i), \hat{z}_i(t_i))_n - (p(t_{i-1}), \hat{z}_i(t_{i-1}))_n)$$

$$+ \sum_{i=1}^{N+1} \int_{t_{i-1}}^{t_i} (p(t), -\frac{d\hat{z}_i(t)}{dt} - A^*(t)\hat{z}_i(t))_n dt$$

$$= \sum_{i=1}^{N+1} \int_{t_{i-1}}^{t_i} (p(t), l_0(t) - \sum_{i=1}^{N} (p(t_i), (H_i^{(2)})^* D_i^{(2)} H_i^{(2)} p(t_i))_n$$

$$- \int_0^T \sum_{j=1}^{M} \chi_{\Omega_j}(t)(p(t), (H_j^{(1)})^*(t) D_j^{(1)}(t) H_j^{(1)}(t) p(t))_n dt$$

$$= l(p) - \sum_{i=1}^{N} (H_i^{(2)} p(t_i), D_i^{(2)} H_i^{(2)} p(t_i))_m$$

$$- \sum_{j=1}^{M} \int_{\Omega_j} (H_j^{(1)}(t) p(t), D_j^{(1)}(t) H_j^{(1)}(t) p(t))_l dt \tag{84}$$

The representation (61) follows from (83) and (84).
Prove that

$$\widehat{\widehat{l(x)}} = l(\hat{x}). \tag{85}$$

We should note, first of all that unique solvability of problem (55)–(60) at realizations $y_j^{(1)}(t)$, $j = 1, \ldots, M$, and $y_i^{(2)}$, $i = 1, \ldots, N$, that belong with probability 1 to the space $(L^2(\Omega_j))^l$ and $\mathbb{C}^m$, respectively, can be proved similarly as to the problem (50)–(54).

Denote by $\hat{p}(t)$ restrictions of $\hat{p}(t)$ to (t_{i-1}, t_i), $i = 1, \ldots, N+1$, extended by continuity to their ends. Using (1), (49), (55), and (56), we have

$$\widehat{\widehat{l(x)}} = \sum_{i=1}^{N} (y_i^{(2)}, \hat{u}_i^{(2)})_m + \sum_{j=1}^{M} \int_{\Omega_j} (y_j^{(1)}(t), \hat{u}_j^{(1)}(t))_l dt + \hat{c}$$

$$= \sum_{i=1}^{N} (y_i^{(2)}, D_i^{(2)} H_i^{(2)} p(t_i))_m + \sum_{j=1}^{M} \int_{\Omega_j} (y_j^{(1)}(t), D_j^{(1)}(t) H_j^{(1)}(t) p(t))_l dt + \hat{c}$$

$$= \sum_{i=1}^{N} ((H_i^{(2)})^* D_i^{(2)} y_i^{(2)}, p(t_i))_n$$

$$+ \int_0^T \sum_{j=1}^{M} \chi_{\Omega_j}(t) ((H_j^{(1)})^*(t) D_j^{(1)}(t) y_j^{(1)}(t), p(t))_l dt + \hat{c}$$

$$= \sum_{i=1}^{N} (\hat{p}_i(t_i) - \hat{p}_i(t_{i+1}) + (H_i^{(2)})^* D_i^{(2)} H_i^{(2)} \hat{x}(t_i), p(t_i))_n$$

$$+ \int_0^T (f_0(t), B^*(t)\hat{z}(t))_n dt + \int_0^T (-\frac{d\hat{p}(t)}{dt} - A^*(t)\hat{p}(t)$$

$$+ \sum_{j=1}^{M} \chi_{\Omega_j}(t) (H_j^{(1)})^*(t) D_j^{(1)}(t) H_j^{(1)}(t) \hat{x}(t), p(t))_n dt \tag{86}$$

From (50)–(54) and (55)–(60), we obtain

$$\int_0^T (-\frac{d\hat{p}(t)}{dt} - A^*(t)\hat{p}(t), p(t))_l dt = \sum_{i=1}^{N+1} \int_{t_{i-1}}^{t_i} (-\frac{d\hat{p}_i(t)}{dt} - A^*(t)\hat{p}_i(t), p(t))_n dt$$

$$= \sum_{i=1}^{N+1} ((\hat{p}_i(t_{i-1}), p(t_{i-1}))_n - (\hat{p}_i(t_i), p(t_i))_n)$$

$$+\sum_{i=1}^{N+1}\int_{t_{i-1}}^{t_i}(\hat{p}_i(t), \frac{dp(t)}{dt} - A(t)p(t))_n dt = (\hat{p}_1(t_0), p(t_0))_n + \sum_{i=2}^{N+1}(\hat{p}_i(t_{i-1}), p(t_{i-1}))_n$$

$$-\sum_{i=1}^{N}(\hat{p}_i(t_i), p(t_i))_n - (\hat{p}_{N+1}(t_{N+1}), p(t_{N+1}))_n + \sum_{i=1}^{N+1}\int_{t_{i-1}}^{t_i}(\hat{p}_i(t), \tilde{Q}(t)\hat{z}_i(t))_n dt$$

$$= \sum_{i=1}^{N}(\hat{p}_{i+1}(t_i), p(t_i))_n - \sum_{i=1}^{N}(\hat{p}_i(t_i), p(t_i))_n + \sum_{i=1}^{N+1}\int_{t_{i-1}}^{t_i}(\tilde{Q}(t)\hat{p}_i(t), \hat{z}_i(t))_n dt$$

$$= \sum_{i=1}^{N}(\hat{p}_{i+1}(t_i) - \hat{p}_i(t_i), p(t_i))_n$$

$$+\sum_{i=1}^{N+1}\int_{t_{i-1}}^{t_i}(\frac{d\hat{x}(t)}{dt} - A(t)\hat{x}(t), \hat{z}_i(t))_n dt - \sum_{i=1}^{N+1}\int_{t_{i-1}}^{t_i}(B(t)f_0(t), \hat{z}_i(t))_n dt. \qquad (87)$$

However,

$$\sum_{i=1}^{N+1}\int_{t_{i-1}}^{t_i}(\frac{d\hat{x}(t)}{dt} - A(t)\hat{x}(t), \hat{z}_i(t))_n dt = \sum_{i=1}^{N+1}((\hat{x}(t_i), \hat{z}_i(t_i))_n - \big(\hat{x}(t_{i-1}), \hat{z}_i(t_{i-1}))_n\big)$$

$$+\sum_{i=1}^{N+1}\int_{t_{i-1}}^{t_i}(\hat{x}(t), -\frac{d\hat{z}_i(t)}{dt} - A^*(t)\hat{z}_i(t))_n dt$$

$$= \sum_{i=1}^{N}(\hat{x}(t_i), \hat{z}_i(t_i))_n - \sum_{i=2}^{N+1}(\hat{x}(t_{i-1}), \hat{z}_i(t_{i-1}))_n$$

$$+\int_0^T(\hat{x}(t), l_0(t) - \sum_{j=1}^{M}\chi_{\Omega_j}(t)(H_j^{(1)})^*(t)D_j^{(1)}(t)H_j^{(1)}(t)p(t))_n dt$$

$$= l\big(\hat{x}\big) + \sum_{i=1}^{N}(\hat{x}(t_i), \hat{z}_i(t_i) - \hat{z}_{i+1}(t_i))_n$$

$$-\sum_{j=1}^{M}\int_{\Omega_j}(\hat{x}(t), (H_j^{(1)})^*(t)D_j^{(1)}(t)H_j^{(1)}(t)p(t))_n dt$$

$$= l(\hat{x}) - \sum_{i=1}^{N} (\hat{x}(t_i), \left(H_i^{(2)})^* D_i^{(2)} H_i^{(2)} p(t_i)\right))_n$$

$$- \sum_{j=1}^{M} \int_{\Omega_j} (\hat{x}(t), (H_j^{(1)})^*(t) D_j^{(1)}(t) H_j^{(1)}(t) p(t))_n dt. \tag{88}$$

The representation (85) follows from (86)–(88). □.

Remark 1 In the representation $\widehat{\widehat{l(x)}} = l(\hat{x})$ of the minimax mean square estimate of $l(x)$, vector-function $\hat{x}(t)$ does not depend on a specific form of functional l (see Eqs. (55)–(60)). Therefore, $\hat{x}(t)$ can be taken as a good estimate of solution $x(t)$ to problem (1), and (2) which is observed.

Remark 2 The above results can be generalized to the case when $l(x)$ has the form

$$l(x) = \int_0^T (x(t), l_0(t))_n dt + \sum_{r=1}^{q} (x(\tau_r), a_r)_n,$$

where $a_r \in \mathbb{C}$, $r = 1, \dots, q$ are prescribed complex numbers, τ_r, $r = 1, \dots, q$, $(0 < \tau_1 < \dots < \tau_q < T)$ are a given system of points on the interval $(0, T)$, $l_0 \in (L^2(0, T))^n$ is a given vector-function.

We conclude our paper by considering an important special case when the problem (4) has only the trivial solution. In this case problem (1), and (2) is uniquely solvable for any $f \in (L^2(0, T))^r$, the following condition

$$\det[E_n - X(T)] \neq 0$$

holds, conditions (i) and (ii) are missing, and the set G_1 takes the form

$$G_1 = \{f \in (L^2(0, T))^r : \int_0^T (Q(t)(f(t) - f_0(t)), f(t) - f_0(t))_r dt \leq 1\}.$$

Then the above analysis leads to the conclusion that in the considered case, the following assertion is valid.

Theorem 2 *The minimax estimate* $\widehat{\widehat{l(x)}}$ *of expression* $l(x)$ *has the form*

$$\widehat{\widehat{l(x)}} = \sum_{j=1}^{M} \int_{\Omega_j} (y_j^{(1)}(t), \hat{u}_j^{(1)}(t))_l dt + \sum_{i=1}^{N} (y_i^{(2)}, \hat{u}_i^{(2)})_m + \hat{c} = l(\hat{x}), \tag{89}$$

where

$$\hat{u}_j^{(1)}(t) = D_j^{(1)}(t)H_j^{(1)}(t)p(t), \quad j = 1, \ldots, M,$$

$$\hat{u}_i^{(2)} = D_i^{(2)}H_i^{(2)}p(t_i), \quad i = 1, \ldots, N, \quad \hat{c} = \int_0^T (f_0(t), B^*(t)\hat{z}(t))_r dt, \tag{90}$$

and vector-functions $p(t)$, $\hat{z}(t)$, and $\hat{x}(t)$ are determined from the solution of the systems of equations

$$-\frac{d\hat{z}(t)}{dt} = A^*(t)\hat{z}(t) + l_0(t) - \sum_{j=1}^{M} \chi_{\Omega_j}(t)(H_j^{(1)})^*(t)D_j^{(1)}(t)H_j^{(1)}(t)p(t)$$

for a.e. $t \in (0, T)$, (91)

$$\Delta\hat{z}\big|_{t=t_i} = (H_i^{(2)})^* D_i^{(2)} H_i^{(2)} p(t_i), \quad i = 1, \ldots, N, \quad \hat{z}(T) = \hat{z}(0), \tag{92}$$

$$\frac{dp(t)}{dt} = A(t)p(t) + \tilde{Q}(t)\hat{z}(t) \quad for\ a.e. \quad t \in (0, T), \quad p(0) = p(T) \tag{93}$$

and

$$-\frac{d\hat{p}(t)}{dt} = A^*(t)\hat{p}(t) - \sum_{j=1}^{M} \chi_{\Omega_j}(t)(H_j^{(1)})^*(t)D_j^{(1)}(t)\Big[H_j^{(1)}(t)\hat{x}(t) - y_j^{(1)}(t)\Big]$$

for a.e.$t \in (0, T)$, (94)

$$\Delta\hat{p}\big|_{t=t_i} = (H_i^{(2)})^* D_i^{(2)}\Big[H_i^{(2)}\hat{x}(t_i) - y_i^{(2)}\Big], \quad i = 1, \ldots, N, \quad \hat{p}(T) = \hat{p}(0), \tag{95}$$

$$\frac{d\hat{x}(t)}{dt} = A(t)\hat{x}(t) + \tilde{Q}(t)\hat{p}(t) + B(t)f_0(t) \text{ for a.e. } t \in (0, T), \tag{96}$$

$$\hat{x}(0) = \hat{x}(T), \tag{97}$$

respectively. Problems (91)–(93) and (94)–(97) are uniquely solvable. Equations (94)–(97) are fulfilled with probability 1.

The minimax estimation error σ is determined by the formula

$$\sigma = [l(p)]^{1/2}.$$

Show that in the case when observations are pointwise, that is, $H_j^{(1)}(t) = 0$ and $\xi_j^{(1)}(t) = 0$, $j = 1, \dots, M$, in (10), the determination of functions $p(t)$ and $\hat{z}(t)$ from (91) to (93) can be reduced to solving some system of linear algebraic equations. To do this, let us represent function $\hat{z}(t)$ in the form

$$\hat{z}(t) = z(t; \hat{u}) = \overline{z}^{(0)}(t) + \sum_{i=1}^{N} \overline{z}^{(i)}(t; \hat{u}), \tag{98}$$

where functions $\overline{z}^{(0)}(t)$ and $\overline{z}^{(i)}(t; u)$, $i = 1, \dots, N$, solve the problems

$$-\frac{d\overline{z}^{(0)}(t)}{dt} = A^*(t)\overline{z}^{(0)}(t) + l_0(t), \quad t \in (0, T), \quad \overline{z}^{(0)}(T) = \overline{z}^{(0)}(0),$$

and

$$-\frac{d\overline{z}^{(i)}(t; \hat{u})}{dt} = A^*(t)\overline{z}^{(i)}(t; \hat{u}), \quad t \in (0, T), \quad t \neq t_i,$$

$$\overline{z}^{(i)}(t_i + 0; \hat{u}) - \overline{z}^{(i)}(t_i; \hat{u}) = (H_i^{(2)})^* \hat{u}_i^{(2)}, \quad \overline{z}^{(i)}(T; \hat{u}) = \overline{z}^{(i)}(0; \hat{u}),$$

respectively. It is easy to see that

$$\overline{z}^{(0)}(t) = Z(t)(E_n - Z(T))^{-1} \int_0^T Z(T)Z^{-1}(s)l(s)ds + \int_0^t Z(t)Z^{-1}(s)l(s), \tag{99}$$

$$\overline{z}^{(i)}(t; \hat{u}) = Z(t)M_i(t)Z^{-1}(t_i)H_i^* \hat{u}_i, \tag{100}$$

where $Z(t) = [X^*(t)]^{-1}$, $M_i(t) = (E_n - Z(T))^{-1}Z(T) + \chi_{(t_i,T)}(t)E_n$.

In fact, (99) is the direct corollary of the Cauchy formula. For proof of (100), we note that if $0 < t < t_i$ then

$$\overline{z}^{(i)}(t; \hat{u}) = Z(t)\overline{z}^{(i)}(0; \hat{u}) \tag{101}$$

and $\overline{z}^{(i)}(t_i - 0; \hat{u}) = Z(t_i)\overline{z}^{(i)}(0; \hat{u})$. If $t_i < t < T$ then

$$\overline{z}^{(i)}(t; \hat{u}) = Z(t)Z^{-1}(t_i)\overline{z}^{(i)}(t_i + 0; \hat{u}) = Z(t)Z^{-1}(t_i)(\overline{z}^{(i)}(t_i - 0; \hat{u}) + \left((H_i^{(2)})^* \hat{u}_i^{(2)}\right)$$

$$= Z(t)Z^{-1}(t_i)Z(t_i)\overline{z}^{(i)}(0; \hat{u}) + Z(t)Z^{-1}(t_i)(H_i^{(2)})^* \hat{u}_i^{(2)}$$

$$= Z(t)\overline{z}^{(i)}(0; \hat{u}) + Z(t)Z^{-1}(t_i)(H_i^{(2)})^* \hat{u}_i^{(2)}. \tag{102}$$

Setting in (102) $t = T$, we find

$$\bar{z}^{(i)}(T; \hat{u}) = Z(T)\bar{z}^{(i)}(0; \hat{u}) + Z(T)Z^{-1}(t_i)(H_i^{(2)})^* \hat{u}_i^{(2)}$$

Due to the equality $\bar{z}^{(i)}(T; \hat{u}) = \bar{z}^{(i)}(0; \hat{u})$, it follows from here that

$$\bar{z}^{(i)}(0; \hat{u}) = Z(T)\bar{z}^{(i)}(0; \hat{u}) + Z(T)Z^{-1}(t_i)(H_i^{(2)})^* \hat{u}_i^{(2)}$$

and

$$\bar{z}^{(i)}(0; \hat{u}) = (E_n - Z(T))^{-1} Z(T)Z^{-1}(t_i)(H_i^{(2)})^* \hat{u}_i^{(2)}.$$

Substituting this expression into (101) and (102), we obtain

$$\bar{z}^{(i)}(t; \hat{u}) = Z(t)(E_n - Z(T))^{-1} Z(T)Z^{-1}(t_i)(H_i^{(2)})^* \hat{u}_i^{(2)}, \quad \text{if } 0 < t < t_i \tag{103}$$

and

$$\bar{z}^{(i)}(t; \hat{u}) = Z(t)[(E_n - Z(T))^{-1} Z(T) + E_n]Z^{-1}(t_i)(H_i^{(2)})^* \hat{u}_i^{(2)}, \quad \text{if } t_i < t < T. \tag{104}$$

Combining (103) and (104), we get (100).

Further, using the Cauchy formula, (98), (99), and (100), we obtain from Eq. (93) that

$$p(t) = X(t)(E_n - X(T))^{-1} \int_0^T X(T)X^{-1}(s)\tilde{Q}(s)z(s)ds$$

$$+ \int_0^t X(t)X^{-1}(s)l(s)\tilde{Q}(s)z(s)ds \tag{105}$$

$$= X(t)(E_n - X(T))^{-1} \int_0^T X(T)X^{-1}(s)\tilde{Q}(s)\bar{z}^{(0)}(s)ds$$

$$+ \int_0^t X(t)X^{-1}(s)l(s)\tilde{Q}(s)\bar{z}^{(0)}(s)ds$$

$$+X(t)(E_n - X(T))^{-1}X(T) \sum_{k=1}^{N} \int_0^T X^{-1}(s)\tilde{Q}(s)Z(s)M_k(s)dsZ^{-1}(t_k)(H_k^{(2)})^* \hat{u}_k^{(2)}$$

$$+X(t)\sum_{k=1}^{N}\int_0^t X^{-1}(s)\tilde{Q}(s)Z(s)M_k(s)ds Z^{-1}(t_k)(H_k^{(2)})^*\hat{u}_k^{(2)}$$

$$= X(t)C_0(t) + X(t)\sum_{k=1}^{N}[(E_n - X(T))^{-1}X(T)C_k(T) + C_k(t)]Z^{-1}(t_k)(H_k^{(2)})^*\hat{u}_k^{(2)}, \tag{106}$$

where

$$C_0(t) = (E_n - X(T))^{-1}\int_0^T X(T)X^{-1}(s)\tilde{Q}(s)\overline{z}^{(0)}(s)ds + \int_0^t X^{-1}(s)l(s)\tilde{Q}(s)\overline{z}^{(0)}(s)ds,$$

$$C_k(t) = \int_0^t X^{-1}(s)\tilde{Q}(s)Z(s)M_k(s)ds.$$

Setting in (106) $t = t_i$ and $\hat{u}_k^{(2)} = D_k^{(2)}H_k^{(2)}p(t_k)$, $i = 1, \dots, N$, $k = 1, \dots, N$, we arrive at the following system of linear algebraic equations for determination of unknown quantities $p(t_i)$:

$$p(t_i) = X(t_i)C_0(t_i)$$

$$+X(t_i)\sum_{k=1}^{N}[(E_n - X(T))^{-1}X(T)C_k(T) + C_k(t_i)]Z^{-1}(t_k)(H^{(2)})_k^*D_k^{(2)}H_k^{(2)}p(t_k)$$

or

$$p(t_i) + \sum_{k=1}^{N}\alpha_{ik}p(t_k) = b_i, \quad i = 1, \dots, N, \tag{107}$$

where

$$\alpha_{ik} = -X(t_i)\sum_{k=1}^{N}[(E_n - X(T))^{-1}X(T)C_k(T) + C_k(t_i)]Z^{-1}(t_k)(H^{(2)})_k^*D_k^{(2)}H_k^{(2)},$$

$$b_i = X(t_i)C_0(t_i).$$

Finding $p(t_i)$ from (107), we determine $\hat{u}_i$, $\overline{z}^{(i)}(t;\hat{u})$, $i = 1, \dots, N$, $\overline{z}^{(0)}(t)$, $\hat{z}(t)$, $p(t)$, and c according to (90), (100), (99), (98), and (105), respectively.

In a similar way we can deduce a system of linear algebraic equations via whose solution the functions $\hat{x}(t)$ and $\hat{p}(t)$ satisfying (94)–(97) are expressed.

References

1. Krasovskii, N.N.: Theory of Motion Control. Nauka, Moscow (1968)
2. Kurzhanskii, A.B.: Control and Observation under Uncertainties. Nauka, Moscow (1977)
3. Chernousko, F.L.: State Estimation for Dynamic Systems. CRC Press, Boca Raton (1994)
4. Kurzhanski, A.B., Valyi, I.: Ellipsoidal Calculus for Estimation and Control. Birkhauser (1997)
5. Kurzhanski, A.B., Varaiya P.: Dynamics and Control of Trajectory Tubes. Theory and Computation. Birkhauser (2014)
6. Kurzhanski, A.B., Daryin, A.N.: Dynamic Programming for Impulse Feedback and Fast Controls. Springer (2020)
7. Nakonechniy, O.G., Podlipenko, Yu.K.: The minimax approach to the estimation of solutions to first order linear systems of ordinary differential periodic equations with inexact data. arXiv: 1810.07228V1, 1–19 (2018)
8. Nakonechnyi, O., Podlipenko, Y.: Guaranteed estimation of solutions of first order linear systems of ordinary differential periodic equations with inexact data from their indirect noisy observations. In: Proceedings of 2018 IEEE First International Conference on System Analysis and Intelligent Computing (SAIC). IEEE (2018)
9. Nakonecnyi, O., Podlipenko, Y.: Optimal estimation of unknown right-hand sides of first order linear systems of periodic ordinary differential equations from indirect noisy observations of their solutions. In: Proceedings of 2020 IEEE 2nd International Conference on System Analysis Intelligent Computing, pp. 44–47 (2018)
10. Nakonechnyi, O., Podlipenko, Y.: Guaranteed a posteriori estimation of unknown right-hand sides of linear periodic systems of ODEs, Appl. Anal., Published online: 29 Apr 2021. https://doi.org/10.1080/00036811.2021.1919647
11. Nakonechnyi, O., Podlipenko, Y., Shestopalov, Y.: Guaranteed a posteriori estimation of uncertain data in exterior Neumann problems for Helmholtz equation from inexact indirect observations of their solutions. Inverse Prob. Sci. Eng. **29**(4), 525–535 (2021)
12. Podlipenko, Y., Shestopalov, Y.: Mixed variational approach to finding guaranteed estimates for solutions and right-hand sides of the second-order linear Elliptic equations under incomplete data. Minimax Theory Its Appl. **1**(2), 197–244 (2016)
13. Nakonechnyi, O., Podlipenko, Y., Shestopalov, Y.: The minimax estimation method for a class of inverse Helmholtz transmission problems. Minimax Theory Its Appl. **4**(2), 305–328 (2019)
14. Yakubovich, V.A., Starzhinskii, V.M.: Linear Differential Equations with Periodic Coefficients, vol. 1. Wiley (1975)
15. Hutson, V., Pym, J., Cloud, M.: Applications of Functional Analysis and Operator Theory, 2nd edn. Elsevier, Amsterdam (2005)
16. Lions, J.L.: Optimal Control of Systems Described by Partial Differential Equations. Springer, Berlin Heidelberg New York (1971)
17. Bainov, D., Simeonov P.: Impulsive Differential Equations Periodic Solutions and Applications. Wiley (1993)

Chapter 15
Application of the Theory of Optimal Set Partitioning for Constructing Fuzzy Voronoi Diagrams

Elena Kiseleva, Olga Prytomanova, Liudmyla Hart, and Oleg Blyuss

Abstract The paper demonstrates the possibility of constructing fuzzy Voronoi diagrams based on a unified approach. This approach implies formulating a continuous problem of optimal partitioning of a set from n-dimensional Euclidean space into subsets with a quality criterion that provides the appropriate form of the Voronoi diagram. The theory of optimal set partitioning is a universal mathematical apparatus for constructing Voronoi diagrams, which is based on the following general idea. The original problems of optimal set partitioning, which are mathematically formulated as infinite-dimensional optimization problems, are reduced through the Lagrange functional to auxiliary finite-dimensional nonsmooth maximization problems or to nonsmooth maximin problems, for the numerical solution of which modern efficient optimization methods are used. A feature of this approach is the fact that the solution of the original infinite-dimensional optimization problems can be obtained analytically in an explicit form, and the analytical expression can include parameters sought as the optimal solution of the above auxiliary finite-dimensional optimization problems with nonsmooth objective functions. The paper proposes an algorithm for constructing one of the variants of fuzzy Voronoi diagrams, when the set of points forming a Voronoi cell may be fuzzy. The algorithm is developed based on the synthesis of methods from the theory of optimal set partitioning and the theory of fuzzy sets. The proposed algorithm is implemented in software; its work is demonstrated by examples of constructing standard and additively weighted diagrams with fuzzy Voronoi cells.

Keywords Fuzzy Voronoi diagram · Generating points · Optimal set partitioning · Nondifferentiable optimization · Shor's r-algorithm

E. Kiseleva (✉) · O. Prytomanova · L. Hart
Oles Honchar Dnipro National University, Gagarin Avenue, 72, 49010 Dnipro, Ukraine
e-mail: kiseleva47@ukr.net

O. Prytomanova
e-mail: olgmp@ua.fm

O. Blyuss
University of Hertfordshire, Mosquito Way, Hatfield AL10 9EU, UK

M. Zgurovsky and N. Pankratova (eds.), *System Analysis & Intelligent Computing*, Studies in Computational Intelligence 1022, https://doi.org/10.1007/978-3-030-94910-5_15

1 Introduction

The theory of optimal set partitioning (OSP) today is a powerful tool for solving many practically important problems, which are reduced in mathematical formulation to continuous problems of optimal partitioning of a set from Euclidean space. These problems can be linear or nonlinear, static or dynamic, deterministic or stochastic, under certainty conditions or fuzzy. The solution of a number of model problems from these classes often leads to mathematical objects, which are called Voronoi diagrams or Dirichlet tessellations [1].

The Voronoi diagram is broadly described as follows. If a finite set of N points (*generating points*) is specified on a plane, then for each element of this set, the so-called *Voronoi cell* can be defined as a set of plane points for which this element is the closest among all points of the set. The boundaries of the Voronoi cells form a partition of the plane, which is called the Voronoi diagram. The mathematical structure under consideration is named after Georgy Voronoi [2]; the cells are also known as Thiessen polytopes [3] or two-dimensional Dirichlet honeycombs [4].

Voronoi diagrams in two and three dimensions are used in various fields of applied sciences: crystallography, physics, astronomy, chemistry, microbiology, computer graphics, in solving problems of artificial intelligence, pattern recognition, ophthalmology. In condensed matter physics, such mosaics are also known as Wigner–Seitz blocks [5]. For general lattices in Lie groups, cells are simply called fundamental regions [6, 7]. In the case of general metric spaces, cells are often called metric fundamental polygons. Other equivalent names for this mathematical structure (or its specific important applications) are Voronoi polytope, Voronoi polygons, domains of influence, Voronoi decomposition, Voronoi mosaics, Dirichlet tessellations. It should be noted that cells are understood as selected parts of a continual system, which are periodically located in the volume of the system, and each particle or vacancy occupies a modified Wigner–Seitz cell surrounded by a Voronoi polytope [8].

There are other areas of application of Dirichlet-Voronoi partitions. Informative reviews of applications of Voronoi diagrams in various fields are given in Refs. [9, 10]. A broad overview of recent technical results and applications, with several hundred references to the computational geometry literature, can be found in Refs. [11, 12].

Voronoi diagrams are fairly well-studied objects, and for them many different algorithms have been obtained that work with the optimal asymptotics $O(N \log(N))$; some of these algorithms even run $O(N)$ on average. However, all of these algorithms are quite complex. As for the Voronoi diagram for the case of a space of arbitrary dimension, its construction is associated with significant algorithmic problems. Indeed, it is known [9] that for a given number N of generating points, the number of elements required to describe the Voronoi diagram grows exponentially depending on the dimension of the space.

The mathematical apparatus for constructing Voronoi diagrams, which has a number of advantages over the known approaches described in scientific literature, is the optimal set partitioning theory developed by E.M. Kiseleva [1, 13]. In Refs. [14, 15], based on the methods of the OSP theory, algorithms were proposed for constructing both the standard Voronoi diagram with clear parameters and its various generalizations. It was shown in Refs. [14, 15] that for the appropriate formulation of the continuous linear problem of optimal set partitioning, the solution of this problem leads to one or another variant of the Voronoi diagram of a given number of generating points. Moreover, the algorithms to solve continuous linear OSP problems do not depend on the dimension of Euclidean space E_n containing a bounded set, which should be partitioned into subsets, and do not depend on the geometry of this set. The complexity of the implementation of algorithms for constructing Voronoi diagrams based on the methods of the OSP theory does not increase with an increase in the number N of generating points, and the speed of auxiliary iterative procedures for nondifferentiable optimization allows solving problems of large dimensions. Such algorithms are applicable to constructing Voronoi diagrams of a given number of generating points not only with their fixed location, but also with an optimal placement of these points in a bounded set of space E_n. The approach proposed in Refs. [14, 15] has a high degree of versatility, since it allows to easily construct not only already known Voronoi diagrams, but also new ones. In particular, models and methods for solving continuous OSP problems can be generalized to cases of fuzzy assignment of the initial parameters of the problem or fuzzy partition of the set, due to which the resulting Voronoi diagrams can also be fuzzy (fuzzy Voronoi diagrams).

The work [16] describes an algorithm for constructing a multiplicatively weighted Voronoi diagram in the presence of fuzzy parameters with an optimal placement of a finite number N of generating points in a bounded set Ω from n-dimensional Euclidean space E_n ($n \geq 2$). The algorithm was developed based on the synthesis of methods for solving OSP problems [17], neuro-fuzzy technologies [18] and modifications of Shor's r-algorithm to solve nonsmooth optimization problems [19, 20].

In this paper, we consider one of the types of fuzzy Voronoi diagrams, in which the set of points forming the Voronoi cells are fuzzy sets. To "remove" uncertainty in such problems (i.e., to formalize uncertain information), the apparatus of the theory of fuzzy sets is used, which is based on the concept of a fuzzy set introduced by L.A. Zade [21], as well as the apparatus of fuzzy logic.

2 Materials and Methods

2.1 Definition and Basic Properties of the Voronoi Diagram

Let us give a formal definition of the Voronoi diagram. Let $M \subset E_n$ be a finite set (ordered collection) of points, the elements of which will be denoted $\tau_1, \tau_2, \ldots, \tau_N$. The Voronoi cell (Voronoi polytope) associated with the point $\tau_i \in M (i = 1, \ldots, N)$ is the set $Vor(\tau_i) \subset E_n$ in Euclidean space such that:

$$Vor(\tau_i) = \left\{x \in E_n : r(x, \tau_i) \leq r\left(x, \tau_j\right),\ j = 1, \ldots, N(j \neq i)\right\}, i = 1, \ldots, N, \tag{1}$$

where $r \equiv r(x, y)$ is a given distance function for all $x, y \in E_n$, and this is usually the standard Euclidean metric:

$$r(x, \tau_i) = \|x - \tau_i\|_n = \sqrt{\sum_{j=1}^{n}\left(x^{(j)} - \tau_i^{(j)}\right)^2}; \tag{2}$$

$x = \left(x^{(1)}, \ldots, x^{(n)}\right)$, $\tau_i = \left(\tau_i^{(1)}, \ldots, \tau_i^{(n)}\right)$, $i = 1, \ldots, N$. Sometimes cases of other metrics are also considered, for example, Manhattan or Chebyshev ones. Next, we consider the case of the Euclidean metric.

Obviously, the Voronoi cell associated with the point $\tau_i (i = 1, \ldots, N)$ is the intersection of closed half-spaces Π_j, $j = 1, \ldots, N(j \neq i)$, where the half-space Π_j contains the point τ_i and is bounded by the connected perpendicular to the n-dimensional segment $\left[\tau_i, \tau_j\right] = \left\{x \in E_n : \tau_i^{(\ell)} \leq x^{(\ell)} \leq \tau_j^{(\ell)}, \ell = 1, \ldots, \mathrm{n}\right\}$. Therefore, the set $Vor(\tau_i)$ is a convex polytope (possibly unbounded). Voronoi cells form a normal filling of Euclidean space, i.e. fill E_n completely and without overlapping and adjoin each other along the entire edges. The partition of E_n into Voronoi cells associated with a given collection of points is unambiguous.

A Voronoi diagram of a finite set M of points in a plane is a partition of the plane into disjoint convex polygons obtained by constructing a Voronoi cell (Voronoi polygon) for each point $\tau_i (i = 1, \ldots, N)$ from the set M. The vertices (nodes) of the Voronoi polygons are called *the vertices of the Voronoi diagram*, and their edges are called *the edges of the Voronoi diagram*.

Here are some important properties of the Voronoi diagram [9].

Statement 1. Let a set M be such that no four points of M lie on one circle periphery and no three points of M lie on one straight line. Then exactly three edges are joined at each vertex of the Voronoi diagram of the set M. In other words, each vertex V of the Voronoi diagram is the center of a circle $C(V)$ on periphery of which exactly three points of the set M lie.

Statement 2. Let the conditions of Statement 1 be satisfied. Then the interior of the circle $C(V)$ does not contain points of the set M.

Statement 3. The Voronoi diagram is a planar graph, so it has $O(N)$ vertices and edges, where $N = |M|$.

Statement 4. A Voronoi cell $Vor(\tau_i)$ is infinite if and only if the point $\tau_i (i = 1, \ldots, N)$ lies on the boundary of the convex hull of the set M.

Points belonging, according to the definition, to several Voronoi cells at once, are usually assigned to several cells at once (in the case of the Euclidean metric, the set of such points has a measure of zero).

Thus, the Voronoi diagram of a finite set M of points in a plane is a partition of the plane such that each region of this partition forms a set of points closer to one of the elements of the set M than to any other element of this set. The basis for generalizing the definition of a Voronoi diagram can be such its elements [9, 22]: a set of points, a plane, one element, as well as formulas by which the proximity of two points is determined. Different applications of the Voronoi diagram cause different its variations. Next, we consider some Voronoi diagram generalizations most often encountered in practice.

2.2 *Some Variations of the Voronoi Diagram*

Standard (classical) Voronoi diagram. A Voronoi diagram of a finite set M is the following set:

$$Vor(M) = \bigcup_{\tau_i \in M} Vor(\tau_i), \tag{3}$$

wherein

$$mes\big(Vor(\tau_i) \cap Vor(\tau_j)\big) = 0, \quad i, j = 1, ..., N(i \neq j),$$

where $Vor(\tau_i)(i = 1, \ldots, N)$ is determined by relation (1); $mes(\cdot)$ is the Lebesgue measure.

Furthest point Voronoi diagram. In search of a data structure that provides an efficient solution to the problem of finding k-nearest neighbors, one may want to determine regions containing points closer to some given subset of k elements of a given set M than to any other subset with the same number of elements. If we put $k = |M| - 1$, then we get the furthest point Voronoi diagram. The development of effective algorithms for constructing such diagrams, as well as questions of their practical application, are studied, for example, in the works [8, 9, 23].

Additively weighted Voronoi diagram. One of the visual ways to obtain a Voronoi diagram is to grow crystals. If the crystals start growing at the same time and at the same rate, then a certain number of growing circles are obtained, which, when they meet, form straight lines. Ultimately, the entire plane is filled, with each crystal precisely occupying the Voronoi region of its origin. In fact, crystals start growing at different times. Even if they grow at the same rate, but the growth begins at different times, then the boundaries between them are already hyperbolic segments. Such situation corresponds to the so-called additively weighted Voronoi diagram [23]. This diagram is defined as the ordinary (standard) Voronoi diagram, but each point in it has its own weight a_i $(i = 1, \ldots, N)$, which is added to the distance function when the latter is measured:

$$AWVor(\tau_i) = \{x \in E_n : r(x, \tau_i) - a_i \le r\big(x, \tau_j\big) - a_j, \ \ \forall j \neq i\}, i = 1, \ldots, N.$$

One of the most interesting facts of using the additively weighted Voronoi diagram is its use in animation [24].

Multiplicatively weighted Voronoi diagram. The case when all crystals start growing simultaneously, but at different rates, corresponds to the so-called multiplicatively weighted Voronoi diagram [1, 24]. Let us give its definition.

Let $\tau_1, \ldots, \tau_N$ be points in $E_n (n \geq 1)$, which are assigned positive numbers (weights) $w_1, \ldots, w_N > 0$. The weighted (multiplicatively weighted) Voronoi region (Voronoi cell) associated with the point τ_i $(i = 1, \ldots, N)$ is the set of points in Euclidean space, the weighted distance from which to the point τ_i does not exceed the weighted distance to any other given point:

$$MWVor(\tau_i) = \left\{ x \in E_n : \frac{r(x, \tau_i)}{w_i} \leq \frac{r(x, \tau_j)}{w_j}, \ \forall j \neq i, \ j = 1, \ldots, N \right\},$$
$$i = 1, \ldots, N.$$

Regions $MWVor(\tau_i), i = 1, \ldots, N$, form a weighted (multiplicatively weighted) Voronoi diagram of points $\tau_1, \ldots, \tau_N$.

Power Voronoi diagram (also, *Laguerre diagram* or *Dirichlet cell complex*). There is another generalization of the Voronoi diagram. Let M be a set of N points in space E_n and let a real number w_i be associated with each point τ_i of the set M $(i = 1, \ldots, N)$. For each point $\tau_i \in M$, we define the set

$$GVor(\tau_i) = \left\{ x \in E_n : r^2(x, \tau_i) - w_i \leq r^2(x, \tau_j) - w_j, \ \forall j \neq i, \ j = 1, \ldots, N \right\},$$
$$i = 1, \ldots, N$$

called the generalized Voronoi region associated with the point τ_i (with respect to the set M). The generalized Voronoi diagram (power diagram) of the set M is the following set:

$$GVor(M) = \bigcup_{\tau_i \in M} GVor(\tau_i).$$

Voronoi diagrams with constraints on the powers of generating points. In Ref. [25], Voronoi diagrams with constraints on the powers of generating points are studied. In this case, the power of a point τ_i $(i = 1, \ldots, N)$ of a set $M \subset \Omega$ $(\Omega \subset E_n)$ is the weighted area of the Voronoi cell $Vor(\tau_i)$, determined by the following formula:

$$|Vor(\tau_i)| = \int_{x \in Vor(\tau_i)} \rho(x) dx,$$

where $\rho(x) > 0$, $x \in \Omega$, is a given density function. The Voronoi diagram with constraints on the powers $|Vor(\tau_i)|$ of the generating points $\tau_i, i = 1, \ldots, N$ is known as a Voronoi diagram on the set $\Omega \subset E_n$, in which each point $\tau_i \in M$ has its own power $b_i > 0$ $(i = 1, \ldots, N)$ or, equivalently, a diagram that satisfies the following condition:

$$\int\limits_{x \in Vor(\tau_i)} \rho(x)dx = b_i,\ i = 1, \ldots, N,$$

wherein $b_1, \ldots, b_N$ are given positive numbers such that

$$S = \int\limits_{x \in \Omega} \rho(x)dx = \sum_{i=1}^{N} b_i,\ 0 < b_i \leq S,\ i = 1, \ldots, N.$$

Note that diagrams of this kind were considered in the scientific literature long before the appearance of the work [25]. Thus, in Ref. [26], continuous problems of optimal set partitioning with constraints on the powers of subsets are presented, the mathematical formulations of which include, up to notation, the above conditions on the weighted areas of Voronoi cells. Optimal partitions in such problems are the so-called Voronoi diagrams with constraints on the powers of generating points.

Dynamic Voronoi diagrams. Another generalization of Voronoi diagrams is dynamic Voronoi diagrams. For example, in Ref. [27] a problem of controlling a small aircraft in the presence of winds is interpreted as a problem of constructing a dynamic Voronoi diagram (ZVD-problem). In this problem, the nodes (generating points) of the diagram are not fixed, but are moving objects that must be achieved in a minimum time. This problem is known as the Zermelo navigation problem. Dynamic Voronoi diagrams are a powerful tool in the design of modern geographic information systems. On the use of this toolkit in solving pattern recognition problems, technology, computer science, see, for example, in Refs. [23, 24].

Voronoi diagram of a finite number of generating points optimally placed in a bounded set. In cases when the generating points $\tau_i, i = 1, \ldots, N$ must be placed (selected) in a bounded set $\Omega \subset E_n$, another variant of the Voronoi diagram can be introduced, namely, a Voronoi diagram of a finite number of points optimally placed in a bounded set [14, 15]. A Voronoi diagram of a finite number of generating points $\tau_i, i = 1, \ldots, N$ optimally placed in a bounded set Ω, we call a partition of the set Ω into N subsets such that the total weighted distance from the points of the set Ω to the corresponding generating points is the smallest, i.e. the functional

$$F(\{\tau_1, \ldots, \tau_N\}) = \sum_{i=1}^{N} \int\limits_{Vor(\tau_i)} (r(x, \tau_i)/w_i + a_i)\rho(x)dx$$

takes on a minimum value. By specifying the parameters $a_1, \ldots, a_N, w_1, \ldots, w_N$ and the functions $r(x, \tau_i), \rho(x)$ in the expression for the functional $F(\{\tau_1, \ldots, \tau_N\})$, one can obtain various variants of the Voronoi diagram with the optimal placement of the generating points (additively weighted diagram, multiplicatively weighted diagram, etc.).

Fuzzy (approximate) Voronoi diagrams. Voronoi diagram for N points in n-dimensional space requires $O\left(N^{[n/2]}\right)$ storage space. Therefore, it is often not possible to store Voronoi diagrams for $n > 2$. As an alternative, one can use approximate (fuzzy) Voronoi diagrams [28], where Voronoi cells have fuzzy boundaries, each of which can be approximated. Fuzzy Voronoi diagrams also appear in the case when any point $\tau_i (i = 1, \ldots, N)$ of the set M has fuzzy coordinates or when the weights of the distance functions defining Voronoi cells are given fuzzy. Thus, two main types of fuzzy Voronoi diagrams can be distinguished: Voronoi diagrams with fuzzy parameters and Voronoi diagrams, in which the sets of points forming Voronoi cells are fuzzy sets.

Now we present a unified approach to the construction of various versions of the Voronoi diagram, based on the theoretical apparatus of continuous problems of optimal set partitioning.

2.3 Continuous Problems of Optimal Set Partitioning. Voronoi Diagrams as a Result of Solving Continuous OSP Problems

Let us formulate a continuous linear problem of optimal partitioning of a set from n-dimensional Euclidean space E_n into subsets with constraints and previously unknown coordinates of some specific for each subset points called "centers" of subsets [29].

Let Ω be a bounded, Lebesgue measurable set of n-dimensional Euclidean space. A set of Lebesgue measurable subsets $\Omega_1, \ldots, \Omega_N$ of the set $\Omega \subset E_n$ is called a possible partition of the set Ω into its disjoint subsets $\Omega_1, \ldots, \Omega_N$ if $\bigcup_{i=1}^{N} \Omega_i = \Omega$, $mes(\Omega_i \cap \Omega_j) = 0, \forall i, j = 1, \ldots, N (i \neq j)$, where $mes(\cdot)$ means the Lebesgue measure. We denote the class of all possible partitions of the set Ω into disjoint subsets $\Omega_1, \ldots, \Omega_N$ by Σ_Ω^N, i.e.

$$\Sigma_\Omega^N = \left\{ (\Omega_1, \ldots, \Omega_N) : \bigcup_{i=1}^{N} \Omega_i = \Omega, mes(\Omega_i \cap \Omega_j) = 0,\ i, j = 1, \ldots, N\ (i \neq j) \right\}.$$

Let us introduce the functional

$$F(\{\Omega_1, \ldots, \Omega_N\}, \{\tau_1, \ldots, \tau_N\}) = \sum_{i=1}^{N} \int_{\Omega_i} (c(x, \tau_i) + a_i)\rho(x)dx,$$

$$\{\Omega_1, \ldots, \Omega_N\} \in \Sigma_\Omega^N, \quad \{\tau_1, \ldots, \tau_N\} \in \Omega^N,$$

where $x = \left(x^{(1)}, \ldots, x^{(n)}\right) \in \Omega$, $\tau_i = \left(\tau_i^{(1)}, \ldots, \tau_i^{(n)}\right) \in \Omega_i$, $i = 1, \ldots, N$; $c(x, \tau_i)$ is a given real function bounded on $\Omega \times \Omega$ and measurable by x for any fixed $\tau_i (i = 1, \ldots, N)$; $\rho(x)$ is a given non-negative function bounded and measurable on

Ω; $a_1, \ldots, a_N$ are given numeric parameters. Here and below, integrals are understood in the Lebesgue sense. We assume that the measure of the set of boundary points of the subsets Ω_i, $i = 1, .., N$ is equal to zero.

A continuous linear problem of optimal partitioning of a set Ω from E_n into its disjoint subsets $\Omega_1, \ldots, \Omega_N$ (some of which may be empty) *under constraints in the form of equalities and inequalities with finding the coordinates of the centers* $\tau_1, \ldots, \tau_N$ *of these subsets, respectively*, is the following problem.

Problem 1 [17]. Find $\min\limits_{\{\Omega_1,\ldots,\Omega_N\},\{\tau_1,\ldots,\tau_N\}} F(\{\Omega_1, \ldots, \Omega_N\}, \{\tau_1, \ldots, \tau_N\})$,

$$\{\Omega_1, \ldots, \Omega_N\} \in \Sigma_\Omega^N, \quad \{\tau_1, \ldots, \tau_N\} \in \Omega^N,$$

under the conditions

$$\int\limits_{\Omega_i} \rho(x)dx = b_i, \; i = 1, \ldots, p; \int\limits_{\Omega_i} \rho(x)dx \le b_i, \; i = p + 1, ..., N,$$

where the coordinates $\tau_i^{(1)}, \ldots, \tau_i^{(n)}$ of the center $\tau_i \in \Omega_i$ $(i = 1, \ldots, N)$ are previously unknown; $b_1, \ldots, b_N$ are given non-negative numbers such that

$$S = \int\limits_{\Omega} \rho(x)dx \le \sum_{i=1}^{N} b_i, \; 0 \le b_i \le S, \; i = 1, \ldots, N.$$

A partitioning $\left\{\Omega_1^*, \ldots, \Omega_N^*\right\} \in \Sigma_\Omega^N$, which minimizes the functional $F(\{\Omega_1, \ldots, \Omega_N\}, \{\tau_1, \ldots, \tau_N\})$ under the above conditions, is called *an optimal partitioning* of the set $\Omega \subset E_n$ into N subsets in Problem 1. Wherein a collection $\left[\tau_1^*, \ldots, \tau_N^*\right] \subset \Omega^N$ of the centers $\tau_i^* \subset \Omega_i^*$, $i = 1, \ldots, N$, is called *a set of optimal centers* of the subsets Ω_i^*, $i = 1, \ldots, N$, in Problem 1.

Below, we present three particular statements of Problem 1, in which all conditions on the functions included in the statements of corresponding problems remain the same as in Problem 1.

Problem 1.1 *Continuous problem of optimal partitioning of a set* Ω from E_n into its disjoint subsets $\Omega_1, \ldots, \Omega_N$ (some of which may be empty) *without constraints with given coordinates of the centers* $\tau_1, \ldots, \tau_N$ *of the subsets* $\Omega_1, \ldots, \Omega_N$, *respectively*. Find

$$\min_{\{\Omega_1,\ldots,\Omega_N\}\in\Sigma_\Omega^N} F(\{\Omega_1, \ldots, \Omega_N\}),$$

where

$$F(\{\Omega_1, \ldots, \Omega_N\}) = \sum_{i=1}^{N} \int\limits_{\Omega_i} (c(x, \tau_i) + a_i)\rho(x)dx,$$

$x = \left(x^{(1)}, \ldots, x^{(n)}\right) \in \Omega$, $\tau_i = \left(\tau_i^{(1)}, \ldots, \tau_i^{(n)}\right) \in \Omega_i$, $i = 1, \ldots, N$; the coordinates $\tau_i^{(1)}, \ldots, \tau_i^{(n)}$ of the center $\tau_i \in \Omega_i$ $i = 1, \ldots, N$ are fixed; $a_1, \ldots, a_N$ are given numeric parameters.

Problem 1.2 *Continuous problem of optimal partitioning of a set* Ω from E_n into its disjoint subsets $\Omega_1, \ldots, \Omega_N$ (some of which may be empty) *without constraints with the placement of the centers* $\tau_1, \ldots, \tau_N$ *of the subsets* $\Omega_1, \ldots, \Omega_N$, *respectively.* Find

$$\min_{\{\Omega_1,\ldots,\Omega_N\},\{\tau_1,\ldots,\tau_N\}} F(\{\Omega_1, \ldots, \Omega_N\}, \{\tau_1, \ldots, \tau_N\}),$$
$$\{\Omega_1, \ldots, \Omega_N\} \in \Sigma_\Omega^N, \quad \{\tau_1, \ldots, \tau_N\} \in \Omega^N,$$

where

$$F(\{\Omega_1, \ldots, \Omega_N\}, \{\tau_1, \ldots, \tau_N\}) = \sum_{i=1}^{N} \int_{\Omega_i} (c(x, \tau_i) + a_i)\rho(x)dx,$$

$x = \left(x^{(1)}, \ldots, x^{(n)}\right) \in \Omega$, $\tau_i = \left(\tau_i^{(1)}, \ldots, \tau_i^{(n)}\right) \in \Omega_i$, $i = 1, \ldots, N$; the coordinates $\tau_i^{(1)}, \ldots, \tau_i^{(n)}$ of the center $\tau_i \in \Omega_i$ $i = 1, \ldots, N$ are previously unknown (to be determined); $a_1, \ldots, a_N$ are given numeric parameters.

Problem 1.3 *Continuous problem of optimal partitioning of a set* Ω from E_n into its disjoint subsets $\Omega_1, \ldots, \Omega_N$ (some of which may be empty) *under constraints in the form of equalities and inequalities with given coordinates of the centers* $\tau_1, \ldots, \tau_N$ *of the subsets* $\Omega_1, \ldots, \Omega_N$, *respectively.* Find

$$\min_{\{\Omega_1,\ldots,\Omega_N\}\in\Sigma_\Omega^N} F(\{\Omega_1, \ldots, \Omega_N\})$$

under the conditions

$$\int_{\Omega_i} \rho(x)dx = b_i, \ i = 1, \ldots, p; \int_{\Omega_i} \rho(x)dx \le b_i, \ i = p + 1, ..., N,$$

where $F(\{\Omega_1, \ldots, \Omega_N\}) = \sum_{i=1}^{N} \int_{\Omega_i} (c(x, \tau_i) + a_i)\rho(x)dx$, $x = \left(x^{(1)}, \ldots, x^{(n)}\right) \in \Omega$, $\tau_i = \left(\tau_i^{(1)}, \ldots, \tau_i^{(n)}\right) \in \Omega_i$, $i = 1, \ldots, N$; the coordinates $\tau_i^{(1)}, \ldots, \tau_i^{(n)}$ of the center $\tau_i \in \Omega_i$ $(i = 1, \ldots, N)$ are fixed; $a_1, \ldots, a_N$, $b_1, \ldots, b_N$ are given non-negative numbers such that

$$S = \int_{\Omega} \rho(x)dx \le \sum_{i=1}^{N} b_i, \ 0 \le b_i \le S, \ i = 1, \ldots, N.$$

The formulated Problem 1 and its special cases 1.1–1.3 are strictly mathematically formalized. However, the real situations, for which Voronoi diagrams and the corresponding models of optimal set partitioning are created, are often characterized by some degree of uncertainty. In these cases, the quality of decisions made in optimization models of set partitioning directly depends on the completeness of taking into account all the uncertain factors that are significant for the consequences of the decisions made.

In decision-making processes, there are a number of situations that have some degree of uncertainty and require such a mathematical apparatus for their description, which would a priori include the possibility of the appearance of uncertainty and would allow this uncertainty to be taken into account using a suitable method.

The class of problems of optimal set partitioning, which is associated with the need to take into account uncertainty factors that are not of a probabilistic-statistical nature, is considered in the papers [17, 29–31]. These are OSP problems, in which some parameters included in the model description are fuzzy, imprecise, underdetermined, or there is an unreliable mathematical description of some dependencies in the model (for example, the demand functions and the cost functions for transporting a unit of production in infinite-dimensional transport problems). In addition, these are OSP problems, in which the criteria and (or) systems of constraints are not clearly formulated, or the optimization model is influenced by external uncontrollable disturbances of various kinds, etc. Such models are referred to as fuzzy OSP problems. In Ref. [31], a classification of fuzzy OSP problems is proposed in accordance with which elements of the problem have fuzziness. There are distinguished two main types of fuzzy OSP problems: the first one is with fuzzy elements of the problem that are not subject to optimization (problems of optimal set partitioning with fuzzy parameters); the second one is with fuzzy elements of the problem that are subject to optimization (fuzzy optimal partitioning problems).

Since the solution of fuzzy OSP problems leads to the construction of fuzzy Voronoi diagrams, then by analogy with the classification of fuzzy OSP problems, two main types of fuzzy Voronoi diagrams, mentioned above, can be distinguished: Voronoi diagrams with fuzzy parameters and Voronoi diagrams, in which the sets of points forming Voronoi cells are fuzzy sets (fuzzy cells). Besides, the solution of fuzzy OSP problems, as well as for deterministic OSP problems, leads to the construction of fuzzy Voronoi diagrams of two main types: Voronoi diagrams with fuzzy parameters and diagrams with fuzzy Voronoi cells.

Let us formulate a continuous problem of optimal partitioning of a set from E_n into subsets in the presence of fuzzy parameters in the objective functional and constraints, the solution of which leads to the construction of a Voronoi diagram with fuzzy parameters.

Problem 2 (*with fuzzy parameters in the objective functional and constraints*) [17]. Find

$$\min_{\{\Omega_1,\ldots,\Omega_N\}\in\Sigma_\Omega^N} \sum_{i=1}^{N} \int_{\Omega_i} c(x, \tau_i, \theta_i)\rho(x, \theta_0)dx$$

under the conditions

$$\int_{\Omega_i} \rho(x, \theta_0)dx \leq b_i, \quad i = 1, \ldots, N,$$

where $\Sigma_\Omega^N = \{ (\Omega_1, \ldots, \Omega_N) : \bigcup_{i=1}^{N} \Omega_i = \Omega,\ mes(\Omega_i \cap \Omega_j) = 0, i, j = 1, \ldots, N (i \neq j)\}$; $x = (x^{(1)}, \ldots, x^{(n)}) \in \Omega$, $\tau_i = \left(\tau_i^{(1)}, \ldots, \tau_i^{(n)}\right) \in \Omega_i, i = 1, \ldots, N$; $c(x, \tau_i, \theta_i)$ is a given real function bounded on $\Omega \times \Omega \times R_1$ and measurable by x on Ω for any fixed $\tau_i \in \Omega$ and $\theta_i \in R_1$ $(i = 1, \ldots, N)$; $\rho(x, \theta_0)$ is a given non-negative function bounded and measurable by x on Ω for any fixed $\theta_0 \in R_1$; $\theta_0, \theta_1, \ldots, \theta_N$ are given parameters, the values of which are described indistinctly in the form of fuzzy subsets of the universal set R^1, having the following form:

$$\theta_i = \{(\theta_i, \mu_i(\theta_i)),\ \theta_i \in R_1;\ \mu_i{:}R_1 \to [0, 1]\}, \quad i = 0, 1, \ldots, N;$$

$b_1, \ldots, b_N$ are given positive numbers.

The designation "min" has the meaning of making a rational choice of a partition $\{\Omega_1, \ldots, \Omega_N\} \in \Sigma_\Omega^N$, which, in a sense, "minimizes" the given objective functional.

The mathematical model of Problem 2 is an optimal solution to Voronoi diagrams with fuzzy parameters. Let us formulate one of the possible particular statements of Problem 2, in which the parameters in the objective functional are given indistinctly.

Problem 2.1 Find

$$\min_{\{\Omega_1,\ldots,\Omega_N\}\in\Sigma_\Omega^N} \sum_{i=1}^{N} \int_{\Omega_i} (c(x, \tau_i)/\tilde{\omega}_i + \tilde{a}_i)\rho(x)dx,$$

where $x = (x^{(1)}, \ldots, x^{(n)}) \in \Omega$, $\tau_i = \left(\tau_i^{(1)}, \ldots, \tau_i^{(n)}\right) \in \Omega_i, i = 1, \ldots, N$; $c(x, \tau_i)$ is a given real function bounded on $\Omega \times \Omega$ and measurable by x on Ω for any fixed $\tau_i \in \Omega$ $(i = 1, \ldots, N)$; $\rho(x)$ is a given non-negative function bounded and measurable on Ω; $\tilde{a}_1, \ldots, \tilde{a}_N$ and $\tilde{\omega}_1, \ldots, \tilde{\omega}_N$ are fuzzy parameters.

As noted above, the requirement to find an unambiguous (clear) partition of the set Ω from E_n may turn out to be rather rough and tough when solving problems with illy or poorly structured initial information, i.e. problems in which the uncertainty is of a fuzzy-possibilistic nature. Weakening of this requirement is carried out by introducing into consideration fuzzy subsets $\Omega_1, \ldots, \Omega_N$ of the set Ω and the corresponding membership functions, taking values from the interval [0, 1]. Then one of the variants of the fuzzy OSP problem becomes the problem of finding such a fuzzy partition of the set Ω into its fuzzy subsets $\Omega_1, \ldots, \Omega_N$, which, in a sense, "minimizes" the objective functional. This problem is reduced to finding degrees of membership of the set's Ω elements to the desired fuzzy subsets $\Omega_1, \ldots, \Omega_N$, which together define a fuzzy partition (fuzzy covering) of the set Ω.

Next, we present an algorithm for constructing one of the types of fuzzy Voronoi diagrams, namely, diagrams with fuzzy Voronoi cells both for a fixed set of generating points and with finding their optimal placement in a given set. We also demonstrate how the proposed algorithms work on model problems.

3 Results

3.1 Statement of Problems for Constructing Voronoi Diagrams with Fuzzy Cells

Before formulating the problem statements and algorithms for constructing fuzzy Voronoi diagrams, based on methods of the theory of optimal set partitioning, the theory of fuzzy sets and fuzzy logic, we give the definition of a fuzzy partition of a crisp or clear set $\Omega \subset E_n$ [21].

Definition 3.1 *A fuzzy partition of a crisp set* Ω from E_n, where Ω is a bounded, Lebesgue measurable, convex set, is a system of fuzzy subsets

$$\Re(\Omega) = \left\{\tilde{\Omega}_i : \quad \tilde{\Omega}_i \subseteq \tilde{\Omega},\ i = 1, \ldots, N\right\}$$

for which three conditions are satisfied:

(1) $\tilde{\Omega}_i \subseteq \tilde{\Omega}, \forall i = 1, \ldots, N$, where $\tilde{\Omega} = \left(\Omega, \mu_{\tilde{\Omega}}(x)\right), \quad \mu_{\tilde{\Omega}}(x) = 1,\ x \in \Omega$, i.e., this is a clear set Ω, considered as a special case of a fuzzy one;

(2) $\sum\limits_{i=1}^{N} \mu_{\tilde{\Omega}_i}(x) = \mu_{\tilde{\Omega}}(x) = 1, \quad \forall x \in \Omega;$

(3) $h_C < 1, \quad C = \tilde{\Omega}_i \cap \tilde{\Omega}_k,\ \forall i, k = 1, \ldots, N (i \neq k)$, where $h_C = \sup\limits_{x \in \Omega} \mu_C(x)$ is the height of the fuzzy set C; $\mu_C(x)$ is the membership function: $\mu_C : \Omega \to [0, 1]$.

Remark In Definition 3.1, the symbol "Ω" denotes a clear set, the symbol "$\tilde{\Omega}$" denotes a fuzzy set with the membership function $\mu_{\tilde{\Omega}}(x)$. Symbols "Ω_i" denote clear subsets, symbols "$\tilde{\Omega}_i$" denote fuzzy subsets with the membership functions $\mu_{\tilde{\Omega}_i}(x)$. Further, to simplify the notation, we will assume $\mu_{\tilde{\Omega}_i}(x) \equiv \mu_i(x)$ $(i = 1, \ldots, N)$.

It was shown in Ref. [17] that Definition 3.1 is equivalent to the following.

Definition 3.2 *A fuzzy partition of a crisp set* Ω from E_n, where Ω is a bounded, Lebesgue measurable, convex set, is a system of fuzzy subsets

$$\Re(\Omega) = \left\{\tilde{\Omega}_i : \quad \tilde{\Omega}_i \subseteq \tilde{\Omega},\ i = 1, \ldots, N\right\},$$

for which the condition $\sum\limits_{k=1}^{N} \mu_k(x) = \mu_{\tilde{\Omega}}(x) = 1, \ \forall x \in \Omega$ is satisfied, where $\tilde{\Omega} = \left(\Omega, \mu_{\tilde{\Omega}}(x)\right), \mu_{\tilde{\Omega}}(x) = 1, x \in \Omega$.

We denote the class of all possible fuzzy partitions of the set Ω into N fuzzy subsets by $\Re_{\Omega}^{N}$, i.e.

$$\Re_{\Omega}^{N} = \left\{ \left(\tilde{\Omega}_1, \ldots, \tilde{\Omega}_N \right) : \sum_{k=1}^{N} \mu_k(x) = \mu_{\tilde{\Omega}}(x) = 1, \forall x \in \Omega \right\}.$$

Note that the class Σ_{Ω}^{N} of all possible clear partitions of the set Ω is a subclass of the class $\Re_{\Omega}^{N}$. In turn, since an element of the class $\Re_{\Omega}^{N}$ is a collection of subsets $\tilde{\Omega}_1, \ldots, \tilde{\Omega}_N$, each of which belongs to Ω, then the fuzzy set $\Re_{\Omega}^{N}$ is a fuzzy subset of the set $\Omega^N = \underbrace{\Omega \times \ldots \times \Omega}_{N}$.

On the class $\Re_{\Omega}^{N}$ of all possible fuzzy partitions, we introduce the objective functional $F : \Re_{\Omega}^{N} \to R^1$ in the form

$$F\left(\left\{ \tilde{\Omega}_1, \ldots, \tilde{\Omega}_N \right\}, \tau \right) = \sum_{i=1}^{N} \int_{\tilde{\Omega}_i} (c(x, \tau_i)/\omega_i + a_i)\rho(x)dx, \tag{4}$$

where, as in clear Problem 1, $c(x, \tau_i)$ is given real function bounded on $\Omega \times \Omega$ and measurable by $x = \left(x^{(1)}, \ldots, x^{(n)}\right) \in \Omega$ for any fixed $\tau_i = \left(\tau_i^{(1)}, \ldots, \tau_i^{(n)}\right) \in \Omega$ $(i = 1, \ldots, N)$; $\rho(x)$ is a given non-negative function bounded and measurable on Ω; $a_1, \ldots, a_N$ are given non-negative numbers; $\omega_1, \ldots, \omega_N$ are given positive numbers.

Unlike clear Problem 1, here, first, $\tilde{\Omega}_1, \ldots, \tilde{\Omega}_N$ are fuzzy subsets of the set Ω with given coordinates of typical representatives (or centers) $\tau_1, \ldots, \tau_N$, respectively, and each center τ_i $(i = 1, \ldots, N)$ belongs to Ω; second, the measure of the set of boundary points of $\tilde{\Omega}_i, i = 1, \ldots, N$ is not necessarily zero.

Then the following problem is a fuzzy analogue of Problem 1.

Problem 3 [31]. Find a partition of the set Ω into N fuzzy subsets (some of which may be empty) that satisfies the conditions.

$$\bigcup_{i=1}^{N} \tilde{\Omega}_i = \Omega, \quad \underbrace{\sup}_{x \in \Omega} \mu_{\tilde{\Omega}_i \cap \tilde{\Omega}_j}(x) < 1, \ \forall i, j = 1, \ldots, N \ (i \neq j)$$

and, in a sense, "minimizes" the objective functional (4). Here, "minimization" can be understood as the choice of a fuzzy partition, which corresponds, in a sense, to the best fuzzy value of the objective functional.

If we draw an analogy with the classification of fuzzy mathematical programming problems, then the formulated Problem 3 can be attributed to the problems of optimization of a given ordinary functional $F : \Omega \times \ldots \times \Omega \to R^1$ on a given fuzzy set of feasible solutions (alternatives)

$$\Re_{\Omega}^{N} = \left\{ \left(\tilde{\Omega}_1, \ldots, \tilde{\Omega}_N \right) : \mu_{\Re_{\Omega}^{N}}\left(\tilde{\Omega}_1, \ldots, \tilde{\Omega}_N \right), \ \tilde{\Omega}_i \in \tilde{\Omega}, \ i = 1, \ldots, N \right\}$$

determined by the membership function $\mu_{\Re_\Omega^N} : \Omega \times \ldots \times \Omega \to [0, 1]$, which has the form (4) for Problem 3.

In turn, an element of the set $\Re_\Omega^N$ of feasible alternatives—a fuzzy partition $\left(\tilde{\Omega}_1, \ldots, \tilde{\Omega}_N\right)$—is a collection of fuzzy subsets $\tilde{\Omega}_i, i = 1, \ldots, N$ composed of points belonging to the set Ω. As you know, each fuzzy set $\tilde{\Omega}_i$ (as a collection of points x from Ω) is determined by its membership function $\mu_i(x)$, i.e. $\mu_i : \Omega \to [0, 1]$. Wherein, the value $\mu_i(x) = 1$ for $x \in \Omega$ means that the element x from Ω definitely belongs to the fuzzy set $\tilde{\Omega}_i$, which coincides with the value of the characteristic function $\lambda_{\Omega_i^+}(x) = 1$ for points of the kernel Ω_i^+ of the fuzzy set $\tilde{\Omega}_i$. In addition, the value $\mu_i(x) = 0$ means that the element x definitely does not belong to the fuzzy set $\tilde{\Omega}_i$, which coincides with the value of the characteristic function $\lambda_{\Omega_i^+}(x) = 0$ for the points of the kernel Ω_i^+ of the fuzzy set $\tilde{\Omega}_i$. If $0 < \mu_i(x) < 1$, then this means that the element x belongs to the fuzzy set $\tilde{\Omega}_i$ with the degree of membership $\mu_i(x)$; moreover, the points x, for which $0 < \mu_i(x) < 1$, form *the boundary of the fuzzy subset* $\tilde{\Omega}_i$ $(i = 1, \ldots, N)$.

Hence, in order to be able to identify a fuzzy partition $\left(\tilde{\Omega}_1, \ldots, \tilde{\Omega}_N\right)$ of the set Ω, it is necessary to know the degree of membership of each point from Ω to each of the subsets $\tilde{\Omega}_i, i = 1, \ldots, N$.

Thus, from all that has been said above, it follows that a fuzzy partition $\left(\tilde{\Omega}_1, \ldots, \tilde{\Omega}_N\right)$ is completely determined by its membership vector function of the form

$$\mu(x) = (\mu_1(x), \ldots, \mu_N(x)), \quad x \in \Omega. \tag{5}$$

It is also known that the characteristic function $\lambda_{\Omega_i}(\cdot)$ of an ordinary crisp set Ω_i is a special case of the membership function $\mu_i(\cdot)$. Then the concept of a fuzzy set $\tilde{\Omega}_i$ can be regarded as a generalization of the concept of an "ordinary" set Ω_i, and, in turn, the concept of a crisp set Ω_i – as a special case (narrowing) of the corresponding concept of a "fuzzy" set $\tilde{\Omega}_i$ $(i = 1, \ldots, N)$.

Let us rewrite Problem 3 in terms of membership functions as an extension of Problem 1.1 to the fuzzy case (fuzzy analogue of Problem 1.1) as follows.

Problem 3.1 (*of fuzzy optimal set partitioning without constraints with fixed centers of subsets*). Find a vector function $\mu^*(\cdot) = \left(\mu_1^*(\cdot), \ldots, \mu_N^*(\cdot)\right) \in \Gamma$ such that

$$I\left(\mu^*(\cdot)\right) = \min_{\mu(\cdot)\in\Gamma} I(\mu(\cdot)),$$

where

$$I(\mu(\cdot)) = \int\limits_{\Omega} \sum_{i=1}^{N} (\mu_i(x))^m (c(x, \tau_i) + a_i)\rho(x)dx,$$

$$\Gamma = \left\{ \mu(x) = (\mu_1(x), \ldots, \mu_N(x)) : \sum_{i=1}^{N} \mu_i(x) = 1, 0 \le \mu_i(x) \le 1, \quad i = 1, \ldots, N, \quad x \in \Omega \right\} \tag{6}$$

$x = \left(x^{(1)}, \ldots, x^{(n)}\right) \in \Omega$, $\tau_i = \left(\tau_i^{(1)}, \ldots, \tau_i^{(n)}\right) \in \Omega$, $i = 1, \ldots, N$; the coordinates $\tau_i^{(1)}, \ldots, \tau_i^{(n)}$ of the center $\tau_i \in \Omega$ $i = 1, \ldots, N$ are fixed; $a_1, \ldots, a_N$ are given non-negative numbers; m is a parameter called exponential weight.

Here, the designation "min" means that $\mu^*(\cdot)$, in a sense, "minimizes" the objective functional $I(\mu(\cdot))$. The requirement of equality in condition (6) is due to the fact that the desired fuzzy partition $\left(\tilde{\Omega}_1, \ldots, \tilde{\Omega}_N\right)$ must "cover" the ordinary crisp partition $(\Omega_1, \ldots, \Omega_N)$, which is at the same time an element of the fuzzy set $\Re_{\Omega}^{N}$ of feasible solutions and for which the values of the membership functions for each of the elements are equal to 1.

Note that if it is necessary to exclude the appearance of empty subsets in the desired fuzzy partition, then the condition

$$\int\limits_{\Omega} \mu_{\tilde{\Omega}_i}(x)dx > 0, \quad i = 1, \ldots, N$$

should be added to the constraints of Problem 3.1.

Similar to the above, the statements of Problems 1.2 and 1.3 can be extended. Let us formulate them.

Problem 3.2 (*of fuzzy optimal set partitioning without constraints with finding the coordinates of the centers of fuzzy subsets*). Find.

$$\min_{(\mu(\cdot),\tau)\in\Gamma\times\Omega^N} \int\limits_{\Omega} \sum_{i=1}^{N} (\mu_i(x))^m (c(x, \tau_i)/\omega_i + a_i)\rho(x)dx,$$

where

$$\Gamma = \left\{ \mu(x) = (\mu_1(x), \ldots, \mu_N(x)) : \sum_{i=1}^{N} \mu_i(x) = 1, 0 \le \mu_i(x) \le 1, \quad i = 1, \ldots, N, \quad x \in \Omega \right\},$$

$$\tau = \{\tau_1, \ldots, \tau_N\} \in \Omega^N.$$

Problem 3.3 (*of fuzzy optimal set partitioning under constraints in the form of equalities and inequalities with given coordinates of the centers of fuzzy subsets*). Find

$$\min_{\mu(\cdot)\in\Gamma_1} \int_{\Omega} \sum_{i=1}^{N} (\mu_{\tilde{\Omega}_i}(x))^m (c(x,\tau_i)/\omega_i + a_i)\rho(x)dx,$$

where

$$\Gamma_1 = \left\{\mu(x) = \left(\mu_{\tilde{\Omega}_1}(x), \ldots, \mu_{\tilde{\Omega}_N}(x)\right) \in \Gamma,\ x \in \Omega : \int_{\Omega} \rho(x)\mu_{\tilde{\Omega}_i}(x)dx \le b_i,\ i = 1, \ldots, N\right\},$$

$$\Gamma = \left\{\mu(x) = \left(\mu_{\tilde{\Omega}_1}(x), \ldots, \mu_{\tilde{\Omega}_N}(x)\right) : \sum_{i=1}^{N} \mu_{\tilde{\Omega}_i}(x) = 1,\ 0 \le \mu_{\tilde{\Omega}_i}(x) \le 1,\ i = 1, \ldots, N,\ x \in \Omega\right\};$$

$\{\tau_1, \ldots, \tau_N\} \in \Omega^N$ is a given set of the centers of the fuzzy subsets $\tilde{\Omega}_1, \ldots, \tilde{\Omega}_N$ of the set Ω, respectively; $a_1, \ldots, a_N$ and $b_1, \ldots, b_N$ are given positive numbers.

The following reasoning is carried out for an exponential weight $m = 2$.

3.2 *Algorithm for Constructing a Voronoi Diagram with Fuzzy Cells with Given Coordinates of Generating Points*

The Voronoi diagram with fuzzy cells with given coordinates of the generating points can be obtained as a solution to Problem 3.1 formulated above. The method for solving Problem 3.1 is theoretically substantiated in Ref. [31]; we briefly describe its essence.

To formulate the necessary and sufficient conditions for optimality for Problem 3.1, the Lagrange functional is introduced in the following form [31]:

$$h(\mu(x), \psi_0(x)) = I(\mu(x)) + \int_{\Omega} \psi_0(x)\left(\sum_{k=1}^{N} \mu_k(x) - 1\right)dx, \tag{7}$$

where $\psi_0(x)$ is a real function defined on Ω, $\psi_0(x) \in L_2(\Omega)$; $\mu(x)$ is a vector function from the set

$$M_1 = \{\mu(x) = (\mu_1(x), \ldots, \mu_N(x)) : \quad 0 \le \mu_i(x) \le 1, \forall i = 1, \ldots, N, \quad \forall x \in \Omega\}.$$

A pair $\left(\mu^*(x), \psi_0^*(x)\right)$ is called *a saddle point* of functional (7) in the domain $M_1 \times L_2(\Omega)$ if

$$h\left(\mu^*(x), \psi_0(x)\right) \le h\left(\mu^*(x), \psi_0^*(x)\right) \le h\left(\mu(x), \psi_0^*(x)\right)$$

for all $\mu(x) \in M_1,\ \psi_0(x) \in L_2(\Omega)$.

It is shown in Ref. [31] that Problem 3.1 can be reduced to the problem of finding the saddle point of functional (7).

Let us formulate an algorithm for constructing a diagram with fuzzy Voronoi cells with given coordinates of the generating points based on the method for solving Problem 3.1 in the case $\Omega \subset E_2$.

To solve the problem of finding the saddle point of the Lagrange functional, we cover the set Ω with a rectangular grid with nodes $x_j = \left(x_j^1, x_j^2\right)$, $j = 1, \ldots, q$ (q is a given number of grid nodes) and apply the iterative gradient method with respect to $\psi_0(x)$. To do this, we find the gradient of the Lagrange functional with respect to $\psi_0(x)$:

$$grad_{\psi_0(x)}(\mu(x)) = \sum_{k=1}^{N} \mu_k(x) - 1.$$

Algorithm. We choose a zero initial approximation $\psi_0^{(0)}(x)$, set $k = 0$, and calculate

$$\mu_i^{(0)}(x) = \left(-\frac{\psi_0^{(0)}(x)}{2\rho(x)(c(x, \tau_i) + a_i)}\right), \quad i = 1, \ldots, N;$$

$$\psi_0^{(1)}(x) = \psi_0^{(0)}(x) + \lambda_0 \cdot grad_{\psi_0(x)}\left(\mu^{(0)}(x)\right).$$

Let the values of $\psi_0^{(k)}(x)$, $\mu^{(k)}(x)$ be obtained because of calculations after $k(k = 1, 2 \ldots)$ steps of the algorithm.

Let us describe the $(k + 1)$-th step of the algorithm.

1. Calculate for all $i = 1, \ldots, N$

$$\mu_i^{(k)}(x) = \begin{cases} \left(-\frac{\psi_0^{(k)}(x)}{2\rho(x)(c(x,\tau_i)+a_i)}\right), & \text{if}\left(-\frac{\psi_0^{(k)}(x)}{2\rho(x)(c(x,\tau_i)+a_i)}\right) \in [0, 1]; \\ \frac{1}{2}\left[1 - sign\, grad_{\mu_i}\left(\mu_i^{(k-1)}(x)\right)\right], & \text{otherwise}; \end{cases} \tag{8}$$

2. Calculate

$$grad_{\psi_0(x)}\left(\mu^{(k)}(x)\right).$$

3. Find

$$\psi_0^{(k+1)}(x) = \psi_0^{(k)}(x) + \lambda_k \cdot grad_{\psi_0(x)}\left(\mu^{(k)}(x)\right).$$

4. If the condition for completing the iterative process

$$\max_{j=1,\ldots,q} \left| grad_{\psi_0}\left(\mu^{(k)}(x_j)\right)\right| \le \varepsilon \tag{9}$$

is not satisfied, where $\varepsilon > 0$ is the specified accuracy, then pass to the $(k+2)$-th step, otherwise pass to step 5.

5. Accept $\psi_0^*(x) = \psi_0^{(k+1)}(x)$ and find $\mu_i^*(x),\ i = 1, \ldots, N$. According to the Formula (8).

The algorithm is described.

Remark 1 (about the condition to complete the iterative process). The iterative process is performed with respect to $\psi_0(x)$, while at each step we find an optimal $\mu(x)$ corresponding to a given $\psi_0(x)$. In fact, $\psi_0(x)$ is "responsible" for the fulfillment of condition (6). From these positions, condition (9) can be interpreted as achieving the fulfillment of condition (6) with a given accuracy. Condition (9) is more stringent than, for example, the condition of equality to zero of the gradient norm with a given accuracy. This condition ensures the actual fulfillment of condition (6) at each grid point with an accuracy $\varepsilon > 0$.

Remark 2 (about the choice of the step factor λ_k). The step factor λ_k is chosen from the condition of monotonicity of the function:

$$h\left(\mu(x), \psi_0^{(k+1)}(x)\right) \ge h\left(\mu(x), \psi_0^{(k)}(x)\right),\ k = 0, 1, \ldots$$

3.3 *Examples of Constructing a Voronoi Diagram with Fuzzy Cells with Given Coordinates of Generating Points*

Let us describe the results of numerical experiments on fuzzy partitioning of a unit square from E_2 with the Euclidean metric using the above algorithm on the 250×250 grid.

In the figures below, the subset center, specified by the initial data, is denoted by symbol "•", and the gray color indicates the fuzzy border of the subsets.

To interpret the results obtained, namely, to refer each grid node either to a certain subset or to a fuzzy boundary, we introduce the following auxiliary concept of the distrust degree (DD).

The distrust degree DD $\in [0, 1]$ is the minimum value of the membership function of some fuzzy set, at which a given point can be confidently assigned to this fuzzy set, in other words, at which we consider this point to belong to this set. This indicator can also be interpreted as the minimum degree of reliability of the fact "a given point x belongs to a given subset", sufficient for its acceptance. The distrust degree is our requirement for the clarity of the partition. The larger DD is, the fuzzier the partition is, and the wider the areas of the fuzzy boundary are. Obviously, for DD $= 0$ we get a clear partition, and for DD = 1 we get the fuzziest partition, in which only the center and the kernel points (if such appear) will be assigned to a certain subset.

Example 1 Voronoi diagrams with fuzzy cells for six generating points (Tables 1 and 2).

Figures 1, 2 and 3 show fuzzy optimal partitioning of a unit square into six subsets.

In Fig. 1, partitioning into six subsets repeats the results for a crisp case with a low distrust degree. Since all $a_i, i = 1, \ldots, 6$ are equal to each other, straight lines serve as the boundaries between the cells. In Figs. 2, 3, with an increase in the distrust degree, the cells begin to "round", deforming according to the influence of their neighbors.

Table 1 Initial data for six generating points

Subset No.	Coordinates of generating points		a_i	Subset No.	Coordinates of generating points		a_i	Accuracy
	x	y			x	y		
1	0.165	0.750	0	4	0.165	0.250	0	0.0001
2	0.500	0.750	0	5	0.500	0.250	0	
3	0.835	0.750	0	6	0.835	0.250	0	

Table 2 Results of partitioning into six fuzzy subsets

Value of $\max_{j=1,\ldots,q} \left\| grad_{\psi_0}(\mu^{(k)}(x_j)) \right\|$	Functional value	Number of iterations
5,8092E-5	0,0501	66

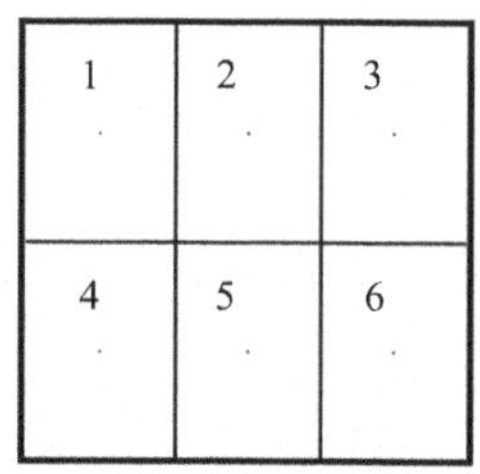

Fig. 1 Fuzzy partitioning into six subsets (DD = 0.01)

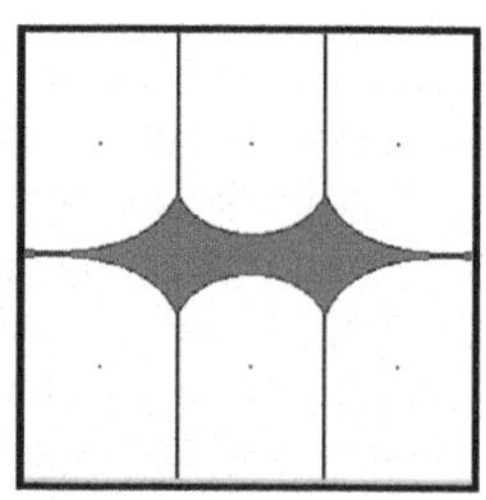

Fig. 2 Fuzzy partitioning into six subsets (DD = 0.28)

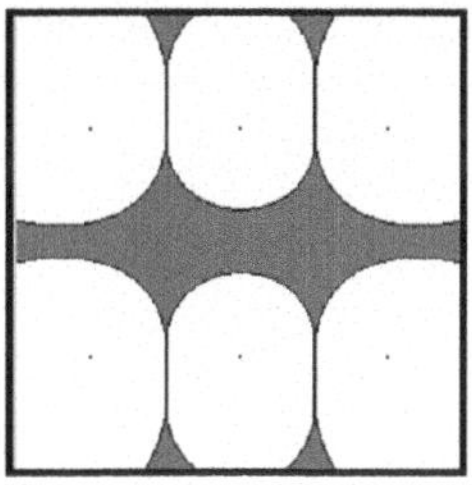

Fig. 3 Fuzzy partitioning into six subsets (DD = 0.3)

Example 2 Additively weighted Voronoi diagrams with fuzzy cells for six generating points, the coordinates of which are given in Table 3, and for

$$a_1 = 0.2;\ a_2 = 0.2;\ a_3 = 0.2;\ a_4 = 0.2;\ a_5 = 0.1;\ a_6 = 0.1$$

Figures 4, 5 and 6 show fuzzy optimal partitioning of a unit square into six subsets in the case of unequal $a_i,\ i = 1, \ldots, 6$.

In Fig. 4, partitioning into six subsets with different $a_i,\ i = 1, \ldots, 6$ at the low distrust level DD = 0.01 demonstrates hyperbolic boundaries between subsets associated with unequal a_i, and boundaries in the form of straight lines between

Table 3 Results of partitioning into six fuzzy subsets for unequal $a_i,\ i = 1, \ldots, 6$

Value of $\max\limits_{j=1,\ldots,q} \lvert grad_{\psi_0}(\mu^{(k)}(x_j)) \rvert$	Functional value	Number of iterations
7.2546E-5	0.0734	30

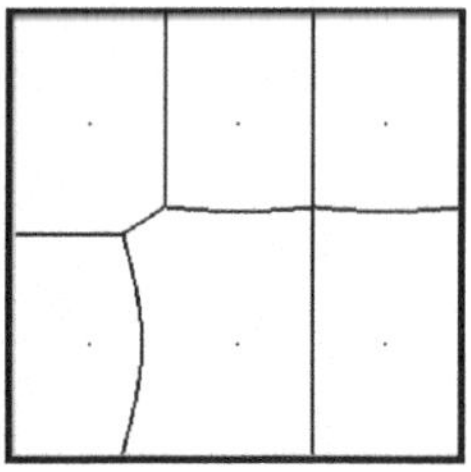

Fig. 4 Fuzzy partitioning into six subsets for unequal a_i (DD = 0.01)

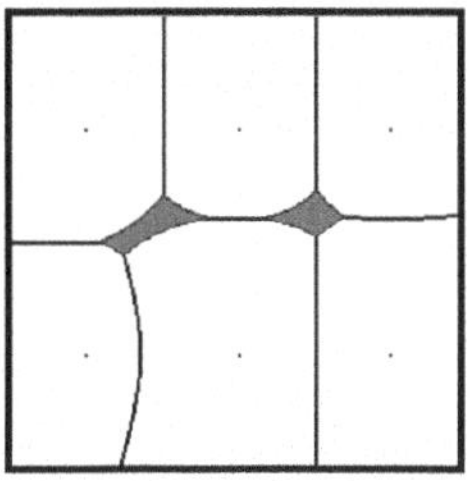

Fig. 5 Fuzzy partitioning into six subsets for unequal a_i (DD = 0.22)

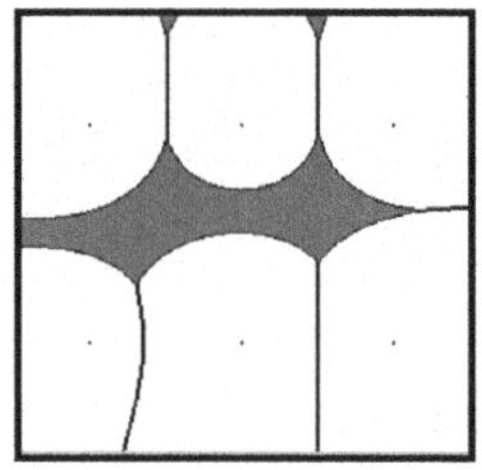

Fig. 6 Fuzzy partitioning into six subsets for unequal a_i (DD = 0.25)

subsets associated with equal a_i, in the absence of a fuzzy boundary. Note that in the case of the boundary between subset 4 and subset 5, and the boundary between subset 2 and subset 5, the hyperbola curvature depends on the value a_i $(i = 1, \ldots, 6)$: these boundaries have different curvature values.

In Figs. 5, 6, as the distrust degree increases, we see the emergence of a fuzzy boundary. For the first time, this fuzzy area appears between subsets 1 and 5. Their large a_i values reduce the measure of "confidence" in them, and therefore the points between these subsets are primarily questionable.

Example 3 Additively weighted Voronoi diagrams with fuzzy cells for eight generating points with unequal a_i, $i = 1, \ldots, 8$ (Tables 4 and 5).

Figures 7, 8 and 9 show fuzzy optimal partitioning of a unit square into eight subsets.

In Fig. 7, partitioning into eight subsets with different a_i, $i = 1, \ldots, 8$ at the low distrust level DD = 0.01, as in the case of six subsets, demonstrates hyperbolic boundaries between subsets with unequal a_i, and boundaries in the form of straight lines between subsets with equal a_i, in the absence of a fuzzy boundary.

In Figs. 8, 9, as the distrust degree increases, a fuzzy boundary emerges. For the first time, this fuzzy area appears between subsets associated with larger a_i values,

Table 4 Initial data for eight generating points

Subset No.	Coordinates of generating points		a_i	Subset No.	Coordinates of generating points		a_i	Accuracy
	x	y			x	y		
1	0.165	0.835	0.2	5	0.750	0.500	0.1	0.0001
2	0.500	0.835	0.2	6	0.835	0.165	0.1	
3	0.835	0.835	0.1	7	0.500	0.165	0.1	
4	0.250	0.500	0.1	8	0.165	0.165	0.1	

Table 5 Results of partitioning into eight fuzzy subsets

Value of $\max\limits_{j=1,\ldots,q} \left\| grad_{\psi_0}(\mu^{(k)}(x_j)) \right\|$	Functional value	Number of iterations
7.8321E-5	0.0927	101

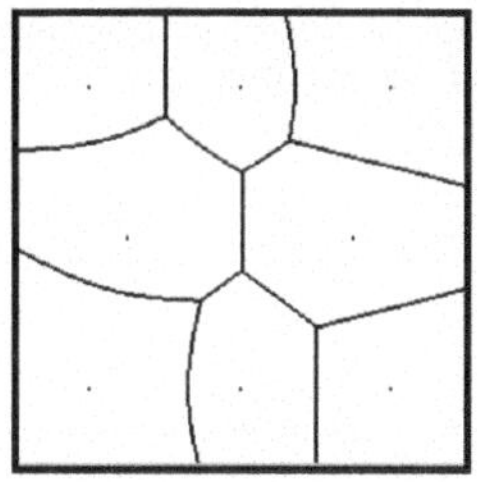

Fig. 7 Fuzzy partitioning into eight subsets for unequal a_i (DD = 0.01)

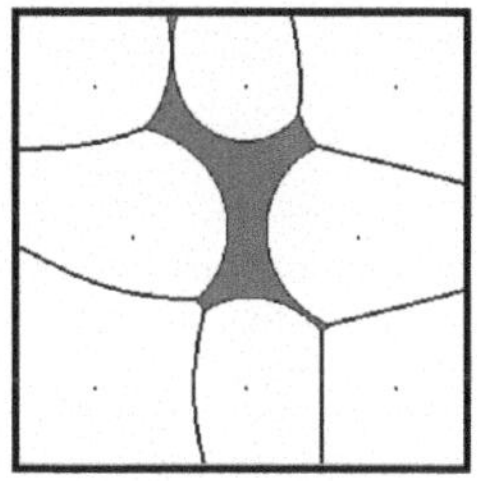

Fig. 8 Fuzzy partitioning into eight subsets for unequal a_i (DD = 0.20)

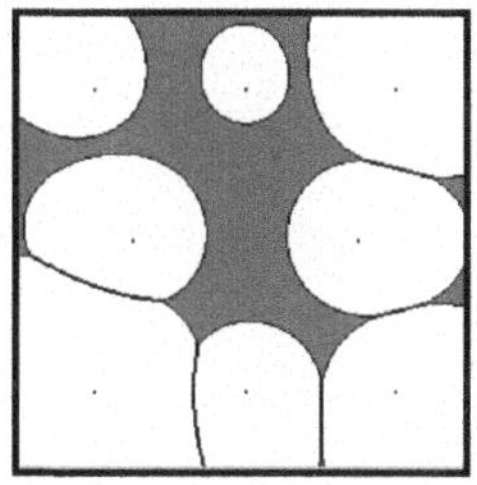

Fig. 9 Fuzzy partitioning into eight subsets for unequal a_i (DD = 0.25)

which reduces the measure of "confidence" in them, and therefore the points between such subsets are the first to be questioned.

Example 4 Voronoi diagrams with fuzzy cells for eighty-five generating points (Table 6).

Initial data: $a_i,\ i = 1, \ldots, 85$ are equal to each other and equal to 0; accuracy is 0.0001; the centers $\tau_i,\ i = 1, \ldots, 85$ are set by the rule:

for i: = 0 to 6 do
for j: = 0 to 6 do
x = 0.05 + i*0.15, y = 0.05 + j*0.15;
for i: = 0 to 5 do
for j: = 0 to 5 do

Table 6 Results of partitioning into eighty-five fuzzy subsets

Value of $\max\limits_{j=1,\ldots,q} \left\| grad_{\psi_0}(\mu^{(k)}(x_j)) \right\|$	Functional value	Number of iterations
4.9612E-5	0.0927	1565

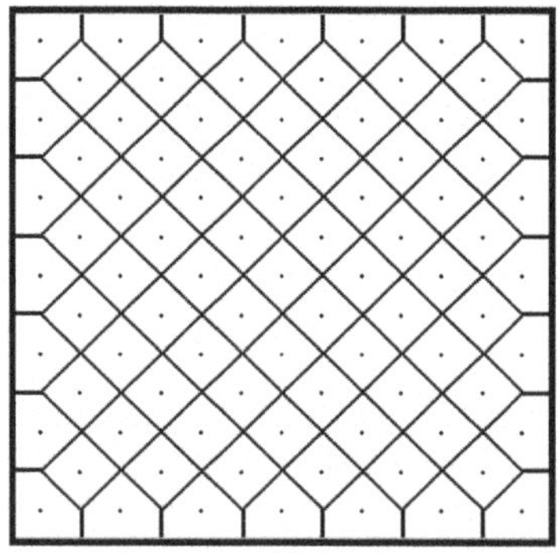

Fig. 10 Fuzzy partitioning into eighty-five subsets (DD = 0.01)

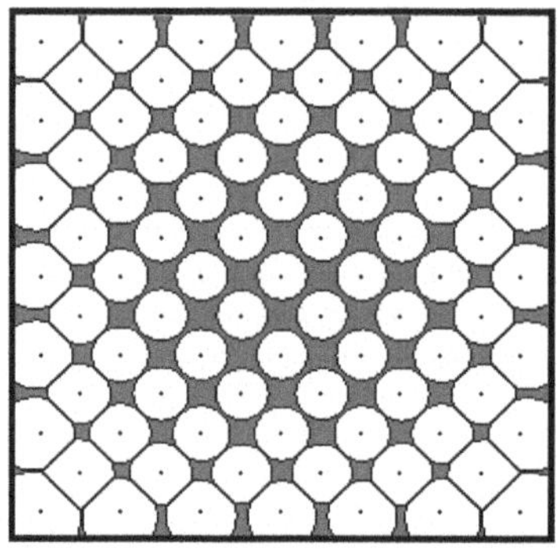

Fig. 11 Fuzzy partitioning into eighty-five subsets (DD = 0.08)

x = 0.125 + i*0.15, y = 0.125 + j*0.15.

Figures 10 and 11 show fuzzy optimal partitioning of a unit square into eighty-five subsets.

This example demonstrates partitioning into a large number of subsets, while the complexity of the implementation of the algorithm does not change.

4 Discussion and Conclusion

Voronoi diagrams shown in Figs. 1, 2, 3, 4, 5, 6, 7, 8, 9, 10 and 11 indicate that with the appropriate formulation of the continuous linear problem of optimal set partitioning, the solution of this problem leads to one or another version of the Voronoi diagram of a given number of points. Table 7 shows the correspondence between a specific version of the Voronoi diagram and the mathematical model of the continuous problem of optimal set partitioning, as a result of the solution of which this diagram can be obtained.

Table 7 can be continued by changing the parameters of the OSP problems and obtaining new generalizations of the Voronoi diagram that are not presented in the scientific literature.

Table 7 Models of continuous OSP problems and Voronoi diagrams, which are the optimal solutions to these problems

Continuous OSP problem model	Voronoi diagram
Problem 1.1: $\rho(x) = 1, \forall x \in \Omega$	
$c(x, \tau_i) = r(x, \tau_i)$, $a_i = 0$ $i = 1, ..., N$	Standard diagram
$c(x, \tau_i) = r(x, \tau_i)$, $a_i \neq 0$, $i = 1, ..., N$	Additively weighted diagram
$c(x, \tau_i) = r(x, \tau_i)/w_i, w_i > 0$, $a_i = 0, i = 1, ..., N$	Multiplicatively weighted diagram
$c(x, \tau_i) = \beta_i r(x, \tau_i)$, $\beta_i, a_i > 0$, $i = 1, ..., N$	Multiplicatively and additively weighted diagram
$c(x, \tau_i) = r^2(x, \tau_i)$, $a_i = A_i^2$, $i = 1, ..., N$	Power diagram (Laguerre diagram) with the weights A_i^2
$c(x, \tau_i) = -r(x, \tau_i)$, $a_i = 0$, $i = 1, ..., N$	Furthest point diagram
$c(x, \tau_i) = f_i(x, \tau_i)$, $a_i = 0$, $i = 1, ..., N$	Efficiency-based diagram
Problem 1.1: $\rho(x)$ is arbitrary in Ω; $c(x, \tau_i) = r(x, \tau_i)$, $a_i = const, i = 1, ..., N$	Other weighted diagrams
Problem 1.2: $\rho(x) = 1, \forall x \in \Omega$	
$c(x, \tau_i) = r(x, \tau_i)$, $a_i = 0$, $i = 1, ..., N$	Diagram of a finite number of generating points, optimally placed in Ω
$c(x, \tau_i) = r(x, \tau_i)$, $a_i \neq 0$, $i = 1, ..., N$	Additively weighted diagram of a finite number of generating points, optimally placed in Ω
Problem 1.3: $\rho(x) = 1, \forall x \in \Omega$	
$c(x, \tau_i) = r(x, \tau_i)$, $a_i = 0$, $b_i \neq 0$, $i = 1, ..., N$	Diagram with constraints on the powers of generating points
$c(x, \tau_i) = r(x, \tau_i)$, $a_i \neq 0$, $b_i \neq 0$, $i = 1, ..., N$	Additively weighted diagram with constraints on the powers of generating points
$c(x, \tau_i) = r(x, \tau_i)/w_i$, $w_i > 0$, $a_i \neq 0$, $b_i \neq 0$, $i = 1, ..., N$	Multiplicatively and additively weighted diagram with constraints on the powers of generating points
Problem 2.1: $\rho(x) = 1, \forall x \in \Omega$; $c(x, \tau_i) = r(x, \tau_i)/\tilde{w}_i$, $\tilde{w}_i > 0$, $\tilde{a}_i \neq 0$, $i = 1, ..., N$	Multiplicatively and additively weighted diagram with fuzzy parameters
Problem 3.1: $\rho(x) = 1, \forall x \in \Omega$; $c(x, \tau_i) = r(x, \tau_i)$, $a_i = 0$, $i = 1, ..., N$	Diagram with fuzzy cells at the given coordinates of a finite number of generating points
Problem 3.2 : $\rho(x) = 1, \forall x \in \Omega$; $c(x, \tau_i) = r(x, \tau_i)$, $a_i = 0$, $b_i \neq 0$, $i = 1, ..., N$	Diagram with fuzzy cells at the given coordinates of generating points and with constraints on their powers
Problem 3.3 : $\rho(x) = 1, \forall x \in \Omega$; $c(x, \tau_i) = r(x, \tau_i)$, $a_i = 0$, $i = 1, ..., N$	Diagram with fuzzy cells of a finite number of generating points, optimally placed in Ω

In view of the above, it can be argued that a unified approach (method) for constructing various variants of the Voronoi diagram is an approach based on the formulation of continuous problems of optimal partitioning of a set with a partition quality criterion that provides the corresponding type of Voronoi diagram. The approach also involves the subsequent application of the mathematical and algorithmic apparatus developed in Refs. [14, 15] to solve such problems.

The versatility of this approach to constructing Voronoi diagrams is also confirmed by the following facts:

- models and methods for solving continuous problems of optimal set partitioning can be generalized to the case of a fuzzy assignment of the initial parameters of the problem or the requirement of a fuzzy partition of a set, as a result of which the obtained Voronoi diagrams can also be fuzzy;
- among the continuous problems of optimal set partitioning, a class of problems is distinguished, in which, along with the partition, it is necessary to find also the optimal placement of the centers of subsets. Thus, along with the problem of constructing a Voronoi diagram, one can pose the problem of finding the optimal (in a sense) coordinates of the points generating this diagram;
- the complexity of implementation of algorithms for constructing Voronoi diagrams based on the described approach does not change significantly with an increase in the number of generating points;
- the result of this approach is the ability to construct not only the already known Voronoi diagrams, but also to develop new ones.

References

1. Kiseleva, E.M.: The emergence and formation of the theory of optimal set partitioning for sets of the n-dimensional Euclidean space. Theory and application. J. Autom. Inf. Sci. **50**(9), 1–24 (2018)
2. Voronoi, G.: Nouvelles applications des parametres continus Ala Theorie des formes quadratiques. Journal für die Reine und Angewandte Mathematik, 97–178 (1907).
3. Thiessen, A.H.: Precipitation averages for large areas. Mon. Weather Rev. **39**(7), 1082–1084 (1911)
4. Fejes, T.G.: Multiple packing and covering of spheres. Acta Math. Acad. Sci. Hungary **34**(1–2), 165–176 (1979)
5. Ziman, J.: Models of Disorder. Theoretical Physics of Uniformly Disordered Systems. Mir, Moscow (1982) (in Russian)
6. Klein, R.: Concrete and abstract Voronoi diagrams. In: Goos, G., Hartmanis, J. (eds.) LNCS, vol. 400, pp. 1–169. Springer-Verlag, Berlin (1989)
7. Klein, R., Mehlhorn, K., Meiser, S.: Randomized incremental construction of abstract Voronoi diagrams. Comput. Geom. **3**, 157–184 (1993)
8. Yoshiaki, O.: A geometrical solution for quadratic bicriteria location models. Eur. J. Oper. Res. **114**, 380–388 (1999)
9. Preparata, F., Shamos, M.: Computational Geometry: An Introduction, 1st edn. Springer-Verlag, New York (1985)
10. Bublik, B.N., Kirichenko, N.F.: Fundamentals of Control Theory. Vishcha shkola, Kyiv (1975) (in Russian)

11. Butkovskiy, A.G.: Methods of Control of Systems with Distributed Parameters. Nauka, Moscow (1975). (in Russian)
12. Yakovlev, S.V.: Formalizing spatial configuration optimization problems with the use of a special function class. Cybern. Syst. Anal. **55**(4), 581–589 (2019)
13. Kiseleva, E.M., Kadochnikova, Ya.E.: Solving a continuous single-product problem of optimal partitioning with additional conditions J. Autom. Inf. Sci. **41**(7), 48–63 (2009)
14. Kiseleva, E.M., Koriashkina, L.S.: Theory of continuous optimal set partitioning problems as a universal mathematical formalism for constructing Voronoi diagrams and their generalizations I. Theor. Found. Cybern. Syst. Anal. **51**(3), 325–335 (2015)
15. Kiseleva, E.M., Koriashkina, L.S.: Theory of continuous optimal set partitioning problems as a universal mathematical formalism for constructing Voronoi diagrams and their generalizations. II. Algorithms for constructing Voronoi diagrams based on the theory of optimal set partitioning. Cybern. Syst. Anal. **51**(4), 489–499 (2015)
16. Kiseleva, E., Prytomanova, O., Padalko, V.: An algorithm for constructing additive and multiplicative Voronoi diagrams under uncertainty. In: Babichev, S., Lytvynenko, V., Wójcik, W., Vyshemyrskaya, S. (eds.) ISDMCI 2020. Lecture Notes in Computational Intelligence and Decision Making, vol. 1246, pp. 714–727. Springer, Cham (2020)
17. Kiseleva, E.M., Shor, N.Z.: Continuous Problems of Optimal Partitioning of Sets: Theory, Algorithms, Applications. Naukova Dumka, Kyiv (2005). (in Russian)
18. Kiseleva, E.M., Prytomanova, O.M., Zhuravel, S.V.: Algorithm for solving a continuous problem of optimal partitioning with neurolinguistic identification of functions in target functional. J. Autom. Inf. Sci. **50**(3), 1–20 (2018)
19. Shor, N.Z.: Nondifferentiable Optimization and Polynomial Problems. Kluwer Academy Publication, Boston-Dordrecht-London (1998)
20. Stetsyuk, P.I.: Shor's r-algorithms: theory and practice. In: Butenko, S., Pardalos, P.M., Shylo, V. (eds) Optimization Methods and Applications: In Honor of the 80th Birthday of Ivan V. Sergienko, vol. 130, pp. 495–520. Springer, Cham (2017)
21. Zadeh, L.A.: Fuzzy sets. Inf. Control **8**, 338–353 (1965)
22. Lee, D.T., Drysdale, R.: Generalization of Voronoi diagrams in the plane. SIAM J. Comput. **10**, 73–87 (1981)
23. Aurenhammer, F., Klein, R., Lee, D.-T.: Voronoi Diagrams and Delaunay Triangulations. World Scientific, Singapore (2013)
24. Trubin, S.I.: Information Space Mapping with Adaptive Multiplicatively Weighted Voronoi Diagrams. Oregon State University, Corvallis (2007)
25. Balzer, M.: Capacity-constrained Voronoi diagrams in continuous spaces. In: 6th International Symposium on Voronoi Diagrams in Science and Engineering, pp. 79–88. IEEE, Copenhagen, Denmark (2009)
26. Kiseleva, E.M., Shor, N.Z.: Investigation of an algorithm for solving one class of continuous partition problems. Cybern. Syst. Anal. **1**, 84–96 (1994)
27. Bakolas, E., Tsiotras, P.: The Zermelo-Voronoi diagram: a dynamic partition problem. Automatica **46**(12), 2059–2067 (2010)
28. Jooyandeh, M., Mohades, A., Mirzakhah, M.: Uncertain Voronoi diagram. Inf. Process. Lett. **109**(13), 709–712 (2009)
29. Kiseleva, E., Hart, L., Prytomanova, O., Kuzenkov, O.: An algorithm to construct generalized Voronoi diagrams with fuzzy parameters based on the theory of optimal partitioning and neuro-fuzzy technologies. In: 1st International Workshop on Modern Machine Learning Technologies and Data Science, vol. 2386, pp. 148–162. CEUR Workshop Proceedings, Aachen, Germany (2019)
30. Prytomanova, O.: Solving the optimal partitioning set problem with fuzzy parameters in constraints. Visnyk of the Lviv University. Series Appl. Math. Comput. Sci. **27**, 97–107 (2019)
31. Kiseleva E.M., Hart L.L., Prytomanova O.M., Baleiko N.V.: Fuzzy Problems of Optimal Partitioning of Sets: Theoretical Foundations, Algorithms, Applications: Monograph. Lyra, Dnipro (2020) (In Ukrainian)

Chapter 16
Integrated Approach to Financial Data Analysis, Modeling and Forecasting

Petro Bidyuk and Nataliia Kuznietsova

Abstract This work is focused on developing appropriate approach to improving the quality of data analysis, modeling and forecasting. Actually financial data are increasing daily, updating in real time, and it is needed evaluation and scoring of dependable variables. So good high-speed methods and accurate models are needed for usage in financial data evaluation and forecasting. We develop the integrated models based on assumption that the combination of appropriate methods which understand the nature of data can give us more precise estimations. Our integrated models were tested on various actual financial data and were used for solving financial management tasks. Further was developed the dynamic approach which is aiming to evaluate risks in dynamic mode and forecasts main categories such as rating and degree of risk as the time function. This approach gives an opportunity to forecast the probability and loss at the concrete moment of time. The approach was tested for financial data analysis and technical risks forecasting and it gave really high accurate forecasts.

Keywords Integrated approach · Integrated models · Survival functions · Dynamic models · Financial data analysis · Risk forecasting

1 Introduction

The emergence of new problems with large amounts of input data (and it is not known in advance which of them are significant), which can't be solved using existing methods, necessitates the development of new integrated models and approaches. The mathematical apparatus of data mining, which is now rapidly developing, contains numerous methods, models and approaches that allow enough efficiently to process

P. Bidyuk · N. Kuznietsova (✉)
Institute for Applied System Analysis of the National Technical University of Ukraine "Igor Sikorsky Kyiv Polytechnic Institute", Kyiv, Ukraine
e-mail: natalia-kpi@ukr.net

P. Bidyuk
e-mail: pbidyuke_00@ukr.net

M. Zgurovsky and N. Pankratova (eds.), *System Analysis & Intelligent Computing*, Studies in Computational Intelligence 1022, https://doi.org/10.1007/978-3-030-94910-5_16

a wide variety of data [1–3]. The problem of choosing adequate models for analysis, depending on the field of application, creation technologies for using these models, data preparation, the results of quality assessment and preparing a holistic information and analytical platform for solving serious financial problems now is really urgent. The paper proposes an integrated approach based on Bayesian networks (BN) for financial data analysis and integrated models that avoid the input data limitations and more accurately and effectively establish causal relationships between individual characteristics-factors of the problem, their impact on the output variable and specific gravity.

This allows approach to increase the validity of decision-making and improve the accuracy of the forecast of the source data. The integrated models developed were compared with the models built on known methods, and showed high quality results and their competitiveness in solving financial problems such as analysis of enterprise financial data and forecasting the loan repayment probability. In the future the proposed integrated models will be used in new information technologies and decision support systems.

The possibility of using both static and dynamic models in the integrated approach is shown by the example of predicting financial risks at the time of their occurrence and their volume of loss.

2 Materials and Methods

2.1 The Idea of an Integrated Approach

The main idea of integrated approach is that combining appropriate methods of analysis organized in some consequence could improve the results of estimation, prediction and accuracy of financial forecasts [4, 5]. Because at the stage of collecting statistics for any task, the significant factors that cause the impact are not known in advance, financial institutions try to collect as much of information as possible. Having a large set of data on a particular event, it is necessary to build a model that will determine an adequate influence of factors on the result. Significant disadvantages of existing data analysis methods are certain restrictions on the amount and type of input data.

The integrated approach involves a combination of used and tested methods for avoiding the disadvantages mentioned in cases where other methods cannot be applied. When building a forecasting model the question arises how to formalize the collected financial data and identify which of them is significant. For some tasks it is proposed first to build a Bayesian network that will establish causal relationships between variables [5] that correspond to the factors, determine the strength of the relationships between these variables and identify variables that are not related to the resulting event ("hanging variables"). Based on the built network and established connections it could be significantly reduce the number of factors that should be included in the next step in a model building. It is known that for logistic regression, reducing the number of factors included in the model usually leads to a decreasing in the model quality. Therefore it is necessary to use the Bayesian network only as a tool to reduce the factors number that will be included in the model, and not as a tool that will identify the most important factors, and reject all others.

The sequence of an integrated approach implementation to data analysis

Step 1. Statistical data collection that sufficiently characterize the research process.
Step 2. Formalize the collected data and identify which ones are relevant. A Bayesian network is built and studied on the basis of statistical data, which reveals significant variables and causal relationships between them. This makes it possible to reduce the factors number to be considered in the next step of the model building.
Step 3. The identified set of significant factors and the corresponding variables are included in the model which is based on the chosen method (logistic regression, decision tree, cluster analysis etc.).
Step 4. Analysis of the results achieved, checking the model quality. In the case of an acceptable quality of the model it will be used for forecasting.
Step 5. Based on the constructed model it is calculated the forecast estimate and the recommendations are issued.

The models based on this approach are an integrated model based on the Bayesian network and the decision tree (IMBD) and an integrated model based on the Bayesian network and logistic regression (IMBL).

In Fig. 1 the integrated approach procedure is shown. The dotted line shows a module that is optional, that is, when it is possible to build several integrated models. If this is not possible, or it is known in advance which integrated model is best suited for the data, then only one module of this scheme is used [4].

The integrated approach can be applied in the reverse order: first significant factors are identified, and then the BN is built.

2.2 Integrated Models Construction

Based on the described above integrated approach different models were developed for financial data analysis: integrated model based on Bayesian network and decision tree, integrated model based on Bayesian network and logistic regression and integrated model based on decision tree and BN.

An integrated model based on the Bayesian network and the decision tree (IMBD) is a model where in the first stage, the Bayesian network is used to reduce the number of variables, and the decision tree is used to analyze financial data [4].

Construction of Integrated Model Based on the Bayesian Network and Decision Tree

Step 1. Determining the function, $y = F(x_i)$, where $i = 1...n$, n is the number of characteristics (variables) of the problem.

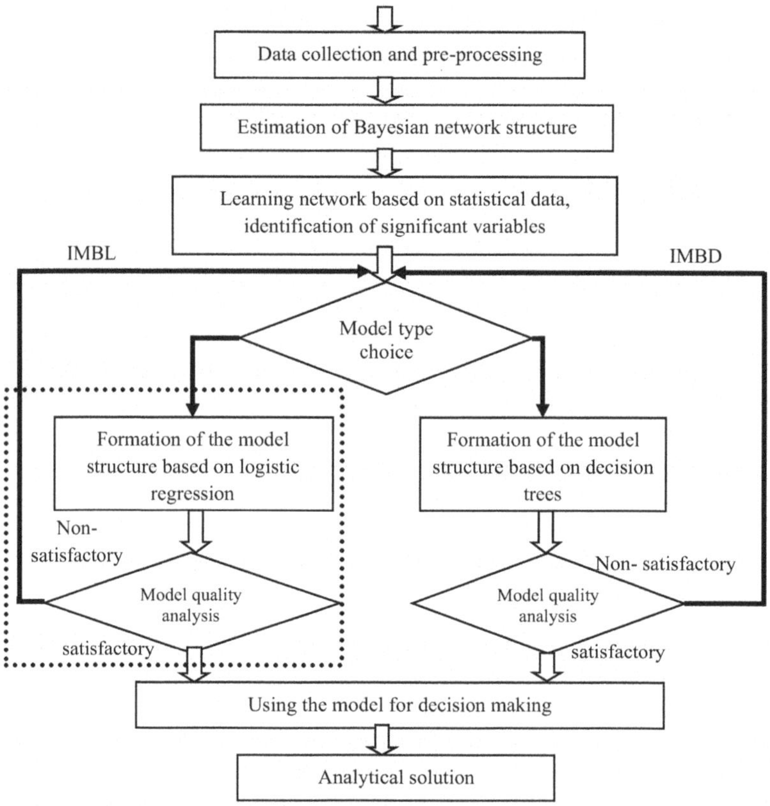

Fig. 1 The integrated approach procedure

Step 2. Construction of BN model and definition of essential variables:

$$y = F_1(x_k, G, J) = \underbrace{\sum\nolimits_s \cdots \sum\nolimits_r}_{k} p\left(x_1^s, \ldots, x_k^r\right) \cdot p(D|x_1^s, \ldots, x_k^r),$$

where x_k are significant variables; J is a probabilistic distribution for the variables x_k; G is a directed acyclic graph, the nodes of which correspond to random variables x_k of the simulated process; x_k^r are states of the variable x_k, x_1^s are states of the variable x_1; $k < j$.

Step 3. An integrated model based on the Bayesian network and decision tree forming: $y = F_2(x_k), k < n$.

An integrated model based on the Bayesian network and logistic regression (IMBL). Since the use of logistic regression is strictly limited by a small number of factors, it is proposed to apply an integrated approach, according to which first a

Bayesian network that defines significant variables is built and then only significant variables are included as the factors in the logistic regression model.

Construction of an Integrated Model Based on BN and Logistic Regression

Step 1. Determining the function, $y = F(x_i)$, where $i = 1..n$, n is the number of characteristics (variables) of the problem.
Step 2. Construction of BN model and definition of essential variables:

$$y = F_1(x_k, G, J) = \underbrace{\sum_s \ldots \sum_r}_{k} p\left(x_1^s, \ldots, x_k^r\right) \cdot p(D|x_1^s, \ldots, x_k^r),$$

where x_k are significant variables; J is a probabilistic distribution of variables x_k; G is a directed acyclic graph, the nodes of which correspond to random variables of the simulated process; x_k are the states of the variable, x_k^r are the states of the variable x_1; $k < j$.
Step 3. An integrated model based on the Bayesian network and binary logistic regression formation:

$$y = F_3(x_k) = \frac{\exp(\beta_0 + \beta_1 x_1 + \ldots + \beta_k x_k)}{1 + \exp(\beta_0 + \beta_1 x_1 + \ldots + \beta_k x_k)}, k < n.$$

The integrated approach described above can be generalized to the application of another method in the second step to identify significant factors such as logistic regression or decision tree, and in the third step to use Bayesian networks to build the model.

Integrated model based on BN and neural networks (IMBNN). Firstly, the Bayesian network is used to determine the causal relationship between the characteristic variables and the resulting variable. Then the significant variables determined by BN are used for modeling by neural network.

Integrated model based on decision tree and Bayesian network (IMDB). If the number of factors influencing the key variable is small then the proposed "reverse" integrated approach can be applied. That is, a decision tree is built for the problem, which determines the variables directly affect the result, and then this information is used for the Bayesian network construction. When building a network structure it could be specified in whole or partially by using expert data. After that the network structure construction continues and the final network structure obtained reflects the causal relationships between the variables. It should be noted that this approach cannot "block" relationships between variables, even if they are not detected by the decision tree, because the decision tree does not provide a deep understanding of the causal relationships between variables.

2.3 An Example of Applying an Integrated Approach to Credit Risk Analysis

The bank provided statistical data on borrower credit histories and set such task: to develop an effective mathematical model for predicting the default probability [6]. A number of models were built on the basis of an integrated approach. At the initial stage a Bayesian network was built which on the basis of bank statistics (credit histories) establishes causal links between variables—customer characteristics and result—loan repayment or non-repayment. When selecting significant factors, first exclude "hanging" variables that do not affect anything at all and are redundant. Then these variables that have no (neither direct nor indirect) impact on the resulting variable—loan repayment are excluded.

The Bayesian network has a set variables that do not in any way affect the resulting variable—loan repayment. Therefore, they can be excluded from the list of variables and then for building a decision tree the number of variables will be smaller. Factors that do not directly affect the resulting variable were further reduced. Among the significant factors (only 17 of them left) such significant variables as age, gender, marital status, income, type of employment, loan amount, type of residence (own or leased), etc. were identified. They were included in the decision tree in the next step. The results of constructing the IMBD model are shown in Table 1.

For the constructed models, the overall accuracy, errors of the 1st and 2nd kind for different cut-off thresholds were calculated. The cut-off threshold for the model in our case is the value of the default probability, above of which the client is considered unreliable and it is recommended not to provide him with credit. The threshold can be determined by the probability value of the loan repayment. Then, for clients whose probability value is higher than this value, it is recommended to issue a loan.

The Bayesian network reduced the number of factors from 26 to 17. It was possible to continue reduction the number of significant factors but it is obvious that this will lead to a deterioration in the model quality. Therefore, the backward stepwise method is used, which one by one excludes the variables that have the least significance. The

Table 1 Comparative table of characteristics for different models

Model	Index GINI	AUC	Accuracy	Model quality
Binary logistic regression	0.676	0.838	0.79	Very high
Decision tree	0.572	0.786	0.75	Acceptable
Neural Network	0.646	0.823	0.76	Very high
Bayesian Network	0.678	0.839	0.75	Very high
IMBD	0.664	0.832	0.795	Very high
IMBL	0.58	0.799	0.715	Acceptable
IMBNN	0.684	0.842	0.775	Very high
IMDB	0.568	0.784	0.75	Acceptable

results of applying the integrated approach to build an integrated model based on BN and logistic regression (IMBL) are shown in Table 1.

The IMBL model, based on an integrated approach, showed better detection of defaults and stronger distrust of unreliable borrowers, so the overall accuracy of the model and the GINI index were lower. This approach will be relevant for a bank that pursues a strict policy of cutting off unreliable customers, i.e. weeding out borrowers, for example, for whom the probability of loan repayment is set at 0.8 [7].

Next an integrated model based on the Bayesian network and the neural network was built. The training time of the neural network was reduced to 7 min 28 s, which is 3–4% of the total time. As a result of using the integrated approach, the accuracy of the model has increased, the number of errors has decreased and the GINI index has increased. The total results received on the experiments for the models quality based on decision trees, binary logistic regression, Bayesian networks and integrated models of IMBD, IMBL, IMBNN and IMDB are shown in the Table 1 [7].

The integrated model based on the decision tree and Bayesian network showed a fairly good identification of insolvent customers in the case of conservative policy of the bank. The number of the first kind errors, i.e. the defaults omission is 5% which is better result in comparison to the decision trees usage. It should be noted that the model is prone to reinsurance. The overall accuracy of the model is lower comparing to the decision tree. Such model will be useful for those banks that pursue a conservative policy and cut off customers with a loan repayment probability lower than 0.85–0.9.

The integrated approach allows the bank to pursue a more balanced credit policy—when setting fairly strict thresholds (at 0.1 and 0.15), the integrated model eliminates fewer reliable borrowers. Application of this approach will be especially relevant when the banks decide to resume mass retail lending. In such case the banks will should to conduct a very careful and cautious policy of checking the client and issuing a loan.

An integrated approach can be used to solve a variety of problems in different areas and domains. The main requirement for its application in practice is the possibility of combining methods for this type of task, i.e. the ability to process data by different methods and the technical feasibility of their implementation. Next, we will consider the possibility of using integrated models to analyze the financial data of a company.

2.4 *Experimental Modeling of the Company's Profitability Based on Integrated Models*

For the practical research the actual Intel Corporation data were selected such as: reports of its financial results for 1975–2019 and financial ratios, calculated on their basis. The following indicators were analyzed [8]:

Y—Net revenue—volume of goods and services multiplied by their price;

$X1$—Earning per Share (EPS)—the company's net profit divided by the weighted average number of shares issued during the accounting period;
$X2$—Gross Margin—the ratio of gross profit or margin—part of the total revenue of the company, remaining after deducting direct costs associated with the production and sale of goods or services;
$X3$—EBIT Margin used to determine profitability;
$X4$—Interest coverage ratio reflects the borrower's ability to cover interest on loans and bonds;
$X5$—Return on Assets (ROA)—an indicator of return on assets, characterizes the efficiency of available assets to generate revenue;
$X6$—Return on Equity (ROE)—an indicator of return which characterizes the efficiency of using not all the capital of the enterprise but the part that belongs to its owners;
$X7$—Assets Turnover Ratio—an indicator of asset turnover, determines the efficiency of production resources usage.

Yield performance forecasting was performed by different data mining methods [1–3, 8–11]. Partial autocorrelation function analysis of the pro-logarithmic and differentiated series showed that lags 0, 1, 2, and 3 are significant for the model, so the corresponding lags should be included in the model and the corresponding orders of the models should be analyzed. Since the visual analysis of the series showed the presence of a trend, a number of autoregression models with an integrated moving average were constructed [9–11]. The specificity of the selected input data indicates the presence of a certain seasonal component. Therefore, it was decided to take into account the seasonal component and build a set of seasonal autoregressive models. The results of the quality analysis of the constructed autoregressive models are given in Table 2.

Thus, the best model was a seasonal autoregressive model with an integrated moving average SARIMA (0,2,1) (2,0,1) (4) [4]. The equation of the seasonal autoregressive model after the transformation has the form: $\left(1 - \Phi_1 B^4 - \Phi_2 B^8\right)(1 - B)^2 Z_t = (1 - \theta_1 B)\left(1 - \vartheta_1 B^4\right)e_t$ with the following values of its parameters for seasonal and non–seasonal components, respectively: $sar1 = 0.9005$, $sar2 = 0.0506$, $ma1 = -0.9849$, $sma1 = -0.7649$ [8].

The next step of the study was to test the model for the constancy of its residues variance. The best model according to the set of criteria $AIC = -2.4106$, and,

Table 2 Comparative analysis of constructed autoregressive models

Model	AIC	BIC
ARIMA (1,1,1)	−373.157	−363.6626
ARIMA (2,1,1)	−371.1632	−358.5041
ARIMA (0,1,2)	−373.2225	−363.7282
ARIMA (0,2,1)	−373.0245	−366.7064
ARIMA (1,2,0)	−315.3137	−308.9956
SARIMA (0,2,1)(2,0,1) [4]	−406.4964	−390.7011

$BIC = -2.3566$ was the model ARCH (2) with the characteristics: $omega = 0.004598$, $alpha1 = 0.018598$, $alpha2 = 0.101433$ [8].

Next Net Revenue was forecasted for the next 4 quarters using regression models: the selected best model SARIMA (0,2,1) (2,0,1) [4], integrated model SARIMA (0,2,1) (2,0,1) [4] and ARCH (2) to forecast the absolute value of profitability and describe the balances of the SARIMA model to forecast the profitability of Intel Corporation for 2019. The forecasting results obtained by all the methods and their comparison with actual values are shown in Table 3.

Thus, the best results of forecasting the company's profitability without pre-processing and smoothing were obtained using the seasonal model SARIMA (0,2,1) (2,0,1) [4]. The best model after pre-treatment was the model based on the combination of Kalman filter [12] and integrated model of seasonal autoregression with integrated moving average and autoregressive neural network (SARIMA + NNAR), i.e. the use of neural networks to predict SARIMA model residues.

The simulation results confirmed the feasibility of pre-processing data and integrated models, the quality of forecasting financial indicators has increased by several orders of magnitude and reached the accuracy of the fifth decimal place. So one more

Table 3 Comparison different methods by the quality estimates forecasting the rate of return

Model	MSE	MAE	MAPE	U
Without pre-processing and data smoothing				
SARIMA(0,2,1)(2,0,1) [4]	0.00386	0.05109	0.521%	0.00632
SARIMA(0,2,1)(2,0,1) [4] + ARCH(2)	0.00413	0.05152	0.526%	0.00653
GDMH	0.00948	0.08847	0.9%	0.00991
Autoregressive neural network	0.00874	0.09063	0.926%	0.00955
SARIMA(0,2,1)(2,0,1) [4] + NNAR	0.0045	0.05144	0.00526	0.00687
Using methods of data processing and smoothing				
Holt–Winters method and SARIMA(1,2,2)(0,0,1) [4]	0.01882	0.13532	1.364%	0.01383
Holt–Winters method and SARIMA(1,2,2)(0,0,1) [4] + ARCH(4)	0.02647	0.16069	1.615%	0.01636
Holt-Winters method and GDMH	0.00402	0.05246	0.533%	0.00645
Holt-Winters method and autoregressive neural network	0.00367	0.05085	0.517%	0.00617
Kalman filter and SARIMA(2,2,3)(1,0,1) [4]	4.2e–05	0.00485	0.049%	0.00066
Kalman filter and SARIMA(2,2,3)(1,0,1) [4] + ARCH(4)	6.04e–05	0.00568	0.058%	0.00079
Kalman filter and GDMH	0.00095	0.02568	0.2632%	0.00316
Kalman filter and autoregressive neural network	0.00045	0.01633	0.167%	0.00217
Kalman filter and SARIMA(0,2,1)(2,0,1) [4] + NNAR	3.22e–05	0.00454	0. 0463%	0.00058

time the integrated models approved their efficiency and usability for solving the financial problems.

3 Integrated Dynamic Approach to Financial Risk Assessment

The statistical approach to risk assessment is currently used based on the calculation of the risk occurrence probability and the potential losses amount based on the IRB-approach (Internal Rating-Based Approach) [13]:

$$EL = \sum_{i=1}^{N} P(R_i) \cdot CE_i \cdot LGD_i,$$

where $P(R_i)$ is the probability (expected frequency) of the manifestation of the i-th type of risk (for example, the risk of reduced financial stability), which becomes important in the segment [0,1]; CE is threat due to the realization of risk—the loss amount (debts due to the risk realization); LGD is risk coverage by insurance (if any), collateral or effectiveness of precautionary measures which takes a value from 0 (risk fully covered by collateral) to 1 (risk not covered by collateral); N is a number of types of risks.

3.1 Methodology of Financial Risk Processing

The methodology of financial risk processing should summarize some very important points related to the need to process uncertainties and unstructured input data, develop and build models of data mining and use integrated models and combined methods to obtain higher estimates. The verification of the described models is carried out according to a set of statistical criteria in order to determine the best of the candidate models.

In Fig. 2 the methodology of financial risk reduction and management is presented.

Here at the first stage the financial risk is assessed, the financial risks inherent in the enterprise are identified and models for their quantitative assessment are developed. Assessment of financial risks makes it possible to determine the possible losses from market fluctuations and amount of capital that must be reserved to cover these losses [13].

Today many methods have been developed to assess financial risks, including various variations of VaR (Value-at-Risk), methods based on the IRB-approach, Shortfall, LDA (Linear Discriminant Analysis), methods using Bayesian programming and fuzzy logic. VaR methods estimate the risk as the expected maximum loss over a set period of time with a set level of probability; Shortfall is a more

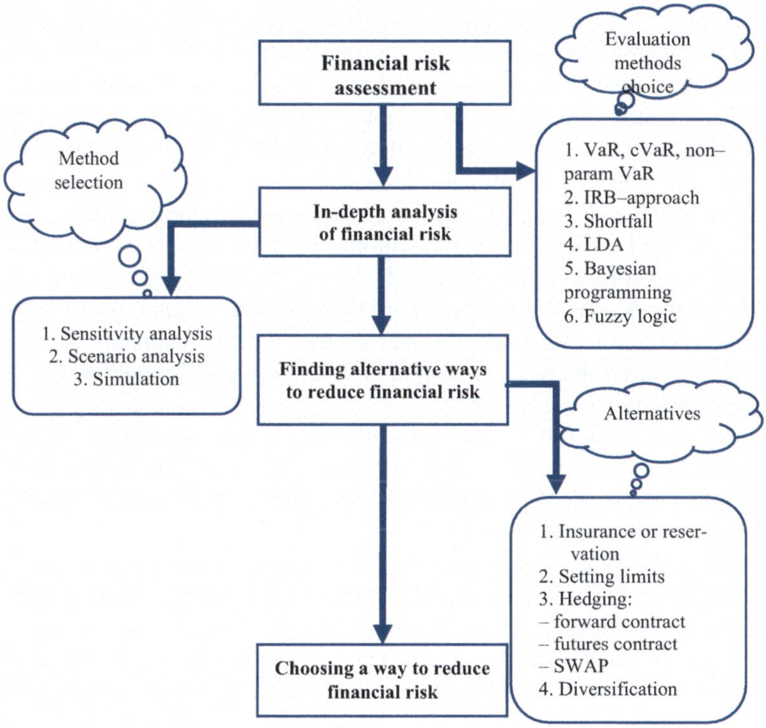

Fig. 2 Methodology of financial risks statistical processing

conservative method of risk assessment than VaR, i.e. the risk assessment that the actual return on investment will be less than the expected return; the LDA method is used to estimate the distribution of losses to calculate the amount of bank capital for operational risk.

Bayesian programming and fuzzy logic methods are more universal and can be applied to different types of financial risk, with the ability to establish rules and criteria and draw conclusions about the level of risk (fuzzy logic) [13]. Bayesian programming includes a set of methods such as Bayesian networks for predicting the risk probability, Bayesian regression for estimating the level of risk and possible losses, granular (particle) filtering and Bayesian classifier for processing input data and modeling multidimensional distributions.

The next step is an in-depth analysis of financial risks, including the identification of key parameters that are used for the business success evaluation (sensitivity analysis), analysis of various scenarios, considering alternative sets of input that may appear in the real situation; simulation etc.

At the third stage there is a search for alternative ways to reduce financial risk: through insurance or provisions, setting limits on transactions, hedging and diversification [13]. At the last, fourth stage, a management decision on financial risk is made. Based on all the analyzed alternatives, methods for reducing losses the best alternative (or even a combination of alternatives) is selected which allows obtain the lowest financial losses and thus increase profitability.

The most important for assessing the losses level is the moment in which the risk level passes and changes from acceptable to catastrophic. The risk probability is also estimated and changed over time and the moment at which the risk probability increases sharply can also be estimated [14].

So, it is desirable to fulfill the dynamic risk assessment as the probability risks and loss assessment and forecasting the time (moment) of risk transition to a higher (critical) level in terms of probability or loss $DE = <PR, Losses, t, S(t|x), \lambda(t|x)>$, where ***PR*** is the risk probability; ***Losses*** are the level of maximum possible losses; ***t*** is time; ***S*(*t*|*x*)** is the function of conditional survival, i.e. the financial system further functioning even after the risk manifestation; **λ(*t*|*x*)** is a conditional level of danger, i.e. the level of losses at the time ***t***.

Dynamic assessment differs from static by the ability to assess risks explicitly in the dynamics, i.e. forecasting the function of losses and the risk probability (transition to a higher degree: critical, catastrophic)—as a function of time. Formally, it is constructing of survival functions [14].

Establishing the level of danger and key points in time that characterize the acceptable, critical and catastrophic risk level is a task of system analysis which must be addressed within each type of risk, regardless of the type of risk and industry in which it is observed. The authors proposed an approach based on the definition of company losses as acceptable $\lambda(t_1|x) = c_1$, critical $\lambda(t_2|x) = c_2$ and catastrophic $\lambda(t_3|x) = c_3$, where (c_1, c_2, c_3)—certain constants that are determined by the company depending on its financial turnover, capacity, etc. (for example, the amount of equity). To determine the time points, (t_1, t_2, t_3), a method of dynamic risk assessment based on dynamic survival models and algorithms for determining time points based on allowable losses and probabilities were proposed [14].

We propose a new approach to the time moment of risk assessment. It is the moment in which the risk level passes and changes from acceptable to catastrophic, is the most important for assessing the losses level. The risk probability is also estimated and changes over time, and the moment at which the risk probability increases sharply can also be estimated. Figure 3 shows the characteristics of the main risk areas due to the risk probability and the amount of its possible losses.

According to the level of financial losses risks could be separated on:

- *tolerable financial risk*—financial loss on risk do not exceed the estimated amount of profit on the financial transaction;
- *critical financial risk*—financial loss on risk do not exceed the estimated amount of gross income for the financial transaction;
- *catastrophic financial risk*—financial loss on risk are determined by partial or complete loss of equity (this type of risk may be accompanied by the loss of borrowed capital).

The strategy that the company chooses in the future for its work will significantly depend on its financial capabilities and risk tolerance, i.e. what level of risk and,

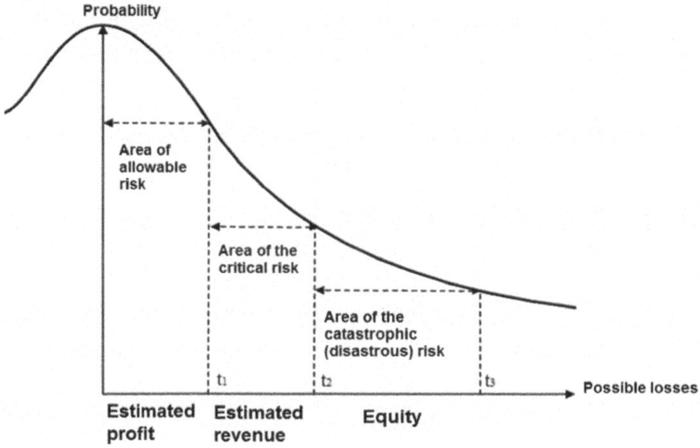

Fig. 3 Levels of financial risks

consequently, possible losses, can take. Formally, this is a set of curves—survival functions.

3.2 Algorithms for Predicting the Time of Risk Onset

According to the survival functions graphs it could be visually determined the time at which the risk probability reaches a certain value. However, this approach is not always convenient and accurate. For the tasks where data slices are received in a time interval daily, hourly, minutely, and the accuracy of time forecasting is critical, it is necessary to develop algorithms for setting time. Here one can hire several different approaches. If the risk function is determined in the modeling process through a parametric, nonparametric distribution, it is possible to calculate the time through the derivative of the risk function [14]. In Refs. [14, 15] the algorithms for calculating the time transition to a higher risk degree using the function derivation, using probability limits and based the losses occurrence were proposed. The developed algorithms make it possible to determine not only the degree and level of risk, as expected in the static assessment, but also to predict the time when the level or degree of risk changes dramatically.

3.3 Method of Dynamic Risk Assessment

To model risks in dynamics the authors proposed a method of dynamic risk assessment and forecasting, which involves integration of different types dynamic models for individual strata, determining the best of them, and using such a model to predict the time, level and degree of risk [15]. The method can be represented as a sequence of steps described below.

Step 1*. Determining the significance of variables-characteristics of financial risks—models parameters based on criteria:*

- correlation estimates R^2, χ^2, ACF, $PACF$,
- weight of evidence (variable category): $WoE_i = \ln\left(\frac{g_i}{b_i}\right)$,

where $\sum_{i=1}^{k} g_i = 1$ is the distribution of unit values of the target variable, and $\sum_{i=1}^{k} b_i = 1$ is the distribution of zero values of the target variable [6];

- information value (IV): $IV = \sum_{i=1}^{k} (g_i - b_i) \ln\left(\frac{g_i}{b_i}\right) = \sum_{i=1}^{k} (g_i - b_i) WoE_i$.

Step 2*. Development of different kind of survival models.*

If possible, it could be made assumptions about the risks distribution (proportional risks, with time covariants, non-parametric models, etc.). Then only one type of survival model will be built. If it is impossible to clearly determine the model type then the set of selected types of models is built and the best of them is determined.

2.1 Cox-proportional risk model

2.1.1 The assessment of the conditional level of danger function is defined as:

$$\hat{\lambda}(t|x) = \hat{\lambda}_0(t)\exp(x^T\hat{\beta}),$$

where $\hat{\lambda}_0(t)$—assessment of the basic function of the danger level $\lambda_0(t)$ [16–18], and $\hat{\beta}$—assessment of the vector of parameters β, and the probability of risk:

$$P\hat{R}^{PHM}(t|x) = \frac{\hat{F}_{\hat{\beta}}(t+b|x) - \hat{F}_{\hat{\beta}}(t|x)}{1 - \hat{F}_{\hat{\beta}}(t|x)} = 1 - \frac{\hat{S}_{\hat{\beta}}(t+b|x)}{\hat{S}_{\hat{\beta}}(t|x)},$$

where $1 - \hat{F}_{\hat{\beta}}(t|x) = \hat{S}_{\hat{\beta}}(t|x) = \exp(-\hat{\Lambda}(t|x))$.

2.1.2. The integral function of basic risk $\Lambda_0(t)$, is estimated as follows:

$$\hat{\Lambda}_0(t) = \sum_{i=1}^{n} \frac{1\{Y_i \le t, \delta_i = 1\}}{\sum_{j=1}^{n} 1\{Y_j \ge Y_i\}},$$

2.1.3 The parameter β is evaluated as $\hat{\beta}^{PHM} = \arg\max_{\beta} L(\beta)$, where the partial likelihood function is given by the expression [18]:

$$L(\beta) = \prod_{i=1}^{n} \frac{\exp(x_i^T \beta)}{\left(\sum_{j=1}^{n} 1_{\{Y_j > Y_i\}} \exp(x_j^T \beta)\right)}.$$

2.1.4 The estimation of the conditional integral risk function is determined by the formula [15]:

$$\hat{\Lambda}(t|x) = \int_0^t \hat{\lambda}(s|t)ds = \exp(x^T \hat{\beta}^{PHM})\hat{\Lambda}_0(t).$$

2.2 *Construction of a generalized linear model for risk assessment*

Assumptions about the proportionality of risks are not always acceptable and therefore it is possible to represent the risk function and survival functions through generalized linear regression. The estimation of the parameters is determined as a result of maximizing the logarithmic likelihood function:

$$\hat{\theta}^{GML} = \arg\max_{\theta} l(\theta).$$

2.3. Construction of a non-parametric model of financial risk assessment

$$\hat{S}_h(t|x) = \prod_{i=1}^{n} \left(1 - \frac{1_{\{Y_i \le t, \delta_i = 1\}} B_{ni}(x)}{1 - \sum_{j=1}^{n} 1_{\{Y_j < Y_i\}} B_{nj}(x)}\right),$$

where $B_{ni}(x) = \frac{K((x - X_i)/\text{h})}{\sum_{j=1}^{n} K((x - X_j)/\text{h})}$, $1 \le i \le n$,

$$P(\hat{R}^{NPM}(t|x)) = \frac{\hat{F}_h(t+b|x) - \hat{F}_h(t|x)}{1 - \hat{F}_h(t|x)} = 1 - \frac{\hat{S}_h(t+b|x)}{\hat{S}_h(t|x)}.$$

Step 3. *Comparison of models and selection of the best based on the set of statistical criteria.*

3.1. Selection of the best model based on the criteria of GINI, AUC, misclassification rate (MR), Bayes–Schwartz (BSC), Kolmogorov–Smirnov (KS): $AUC = \int_{\tilde{y} \in \tilde{Y}} Se(\tilde{y}) dFPR(\tilde{y})$, $GINI = 2 \cdot AUC - 1$, $MR = \frac{FP+FN}{N}$, $BSC = N \ln\left(\sum_{k=1}^{N} e^2(k)\right) + n \ln(N)$, $KS = \max_{x \in X} |F_B(x) - F_G(x)|$, etc.

3.2 Test the hypothesis of equal distribution of risk functions.

The following criteria are used: Kaplan-Mayer estimate, LogRank, Wilcoxon to compare the survival functions that were constructed for different executions [18–21].

Step 4. *Determining the time of risk.*

Depending on the problem statement, i.e. the need to determine the moment of risk transition to a higher degree or level, one of the described algorithms is used.

Step 5. *Determining the probability of financial risk and possible losses.*

As defined in the previous step t, you can calculate: $\lambda(t|x)$, $P(t|x)$ and the expected losses EL for: the allowable, critical and catastrophic level of risk at times (t_1, t_2, t_3) and the corresponding survival functions.

Performing this step in the model will be included only those variables that are statistically important. This reduces computational and operating costs when analyzing large amounts of financial data. These variables directly affect the resulting variable, which characterizes the emergence of financial risk.

4 Credit Risks Dynamic Modeling Example

The problem of credit risk modeling was solved in several stages. At the first stage the model was built that determined the customer's credit card behavior by various data mining methods. The best of them was selected and used for further evaluation. This is the task of the so-called classical scoring (evaluation) to determine: (a) further clients' actions (default/non-default); (b) determination of optimal limits and possible loss calculation due to the risk of client non-fulfillment financial obligations on a given credit card. From the task of personalized credit card assessment, the transition was also made to the credit risks assessment in the portfolio as a whole and the provision of recommendations for further credit policy. The second task that has been modeled is directly the task of dynamic risk assessment, i.e. to determine if there are problems with credit card services. It is also possible to solve the problem of detecting fraud as a deviation from the classic customer behavior and, accordingly, to prevent credit card fraud by blocking such transactions in real time.

Statistics describing the client and his credit cards were collected. The peculiarities of solving this problem were described in Ref. [22]. The credit card model contained the following data, which are illustrated in Table 4.

Table 4 Credit card model

Slots	Description
beginDate	Starting date
finshDate	Nominal expiration date
closeDate	Actual expiration date
gracePeriod	Grace period
setLimit()	Setting a limit
getLimit()	View the current limit
Balance	Balance
postDuePayments	Number of delayed payments
setLoan()	Granting a loan
getOverdue()	Obtaining the amount of delay
getOutstanding()	Obtaining the amount of urgent debt
getPayment()	Receiving a minimum monthly payment

Table 5 Models characteristics

Models	M1	M2	M3	Logit
AUC	0.919	0.888	0.829	0.804
AIC	8946.005	7819.972	7495.905	2334.007

Credit card monitoring and fraud tracking can be based on the average model of customer behavior which is characterized by a number of parameters, deviations from which the online monitoring system may perceive as fraud and refuse to conduct the transaction. To provide the customer with credit card services, the bank allows to adjust the credit card model by setting: credit limits, number of transactions per day, the ability to perform transactions on the Internet, payments abroad and more.

The procedure for predicting the credit card holder behavior is a statistical reflection of changes in credit limits, the transactions volume and money spent over time. Then the bank will consider fraud actions with a credit card that "fall out" of the customer behavior basic model and this will be a reason to block such suspicious transactions.

4.1 *Results and Discussion*

For the experiment 55,286 observations of 4037 credit cards with monthly details were selected. During the whole period, 714 cases of default were recorded. The models with lags 1, 2, 3 in which instead of current values outbody, ovd, montake, monpay values with a certain lag were used based on the list of parameters of credit and client were built (Table 5, Fig. 4) [22].

The target variable, i.e. the variable that described the fact of default/non-default, was formed by the criterion [22]:

- overdue more than UAH 100;
- more than three consecutive overdue payments.

For further comparison the methods of neural networks, linear regression and logistic regression were chosen. An algorithm for constructing a multilayer perceptron with three hidden layers was used for neural networks. According to various statistical criteria, the models quality on the training and test sample was compared (Tables 6 and 7). The best model according to the set of statistical criteria is the neural network but the results even in the training sample showed the need to improve the models.

The constructed models [22] allow to determine the probability transition to a state of default on the basis of some credit card information (parameters) and, accordingly, the payments history and loan servicing for previous periods [22]. This approach is a classic scoring application used to make credit (credit card) decisions. To solve the

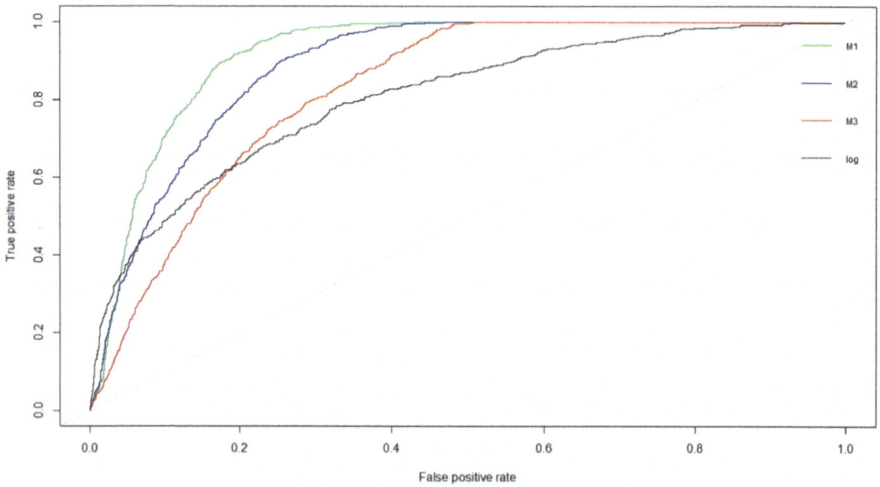

Fig. 4 Graphs of ROC-curves for models M1, M2, M3

Table 6 The results of modeling in the training sample

Model	Misclassification rate	IKA	SBC	GINI	KS
Neural network	0.26	5772.87	5997.73	0.634	0.481
Logistic regression	0.272	5922.28	5988.42	0.609	0.465
Linear regression	0.273	9357.37	9291.24	0.609	0.462

Table 7 Simulation results on the test sample

Model	Misclassification rate	GINI	KS
Neural network	0.271	0.615	0.461
Logistic regression	0.278	0.6	0.449
Linear regression	0.279	0.6	0.448

second problem of modeling, namely the analysis of customer behavior in order to dynamically assess them, a dynamic method of risk assessment was used.

Construction of Dynamic Models of Risk Assessment

To predict the period of credit problems, two types of survival models were built: parametric and semi-parametric.

From the initial sample in 55,286 observations, where 714 cases of default were recorded, in 3837 there were delays. 760 credit cards were censored, accounting for 16.53% of total credit cards.

Next, two types of dynamic models were built on the basis of survival theory: parametric (using the Lifereg procedure: exponential, logistic and Weibull distribution) and semi-parametric model (Cox proportional hazards) [18–23]. Among the

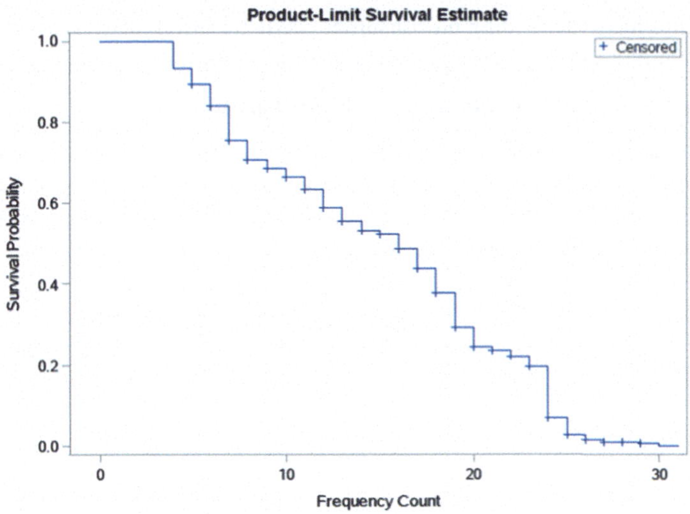

Fig. 5 Risk modeling by credit card survival functions

parametric models, the model using the Weibull distribution turned out to be the best.

The next step was to determine the period when there are problems with loan service. In Fig. 5 survival model for the whole set of customers is shown. There is a situation when the first 3 months the credit cards are well served [22] but with the growth of the "life" of the credit card more and more cases of loan delays happen.

Upon reaching the credit card service period of 12 months, the probability of "survival" of the credit card is more than 0.6, but then the probability of survival decreases significantly. There is also a rather sharp decline in the percentage of "ideal service" of the loan after 18 months.

The risk function presented in Fig. 6 was used for dynamic risk assessment. There is a similar situation where up to 18 months of loan service the risk level is acceptable (up to 0.1), because the functions of risk and survival are interrelated. A small spike in the values of the risk function is observed after 8 months, which then equalizes, but over time the risk increases significantly. Having determined the acceptable level of risk, one can set the optimal period of trouble-free loan service. To do this, use the degree of risk determined by the bank and described in this section, the algorithm for calculating the critical time of risk.

The constructed dynamic models (parametric and semi-parametric) allowed to determine the period in which servicing the credit cards debts problem occurs, as well as to determine the probability of such event at a certain stage of lending. This allows us to talk about the amount of banks possible losses as a result of negative dynamics on credit cards, as well as to apply real means of combating delays at earlier stages (before the loan expiration). Cox's proportional hazard models, which showed the best results for the studied problem, can be used by banks to develop new credit products (depending on consumer needs for different service periods and with different interest rates) and plan fiscal actions in the case of late loan servicing.

Nonparametric Regression and Kaplan–Meier Estimate

The Kaplan–Meier estimate was constructed on the basis of the selected sample (Fig. 7). However, such model does not allow evaluation each client separately, as it considers the entire population as a subject. However, it allows qualitatively assess the loan portfolio for different types of loans.

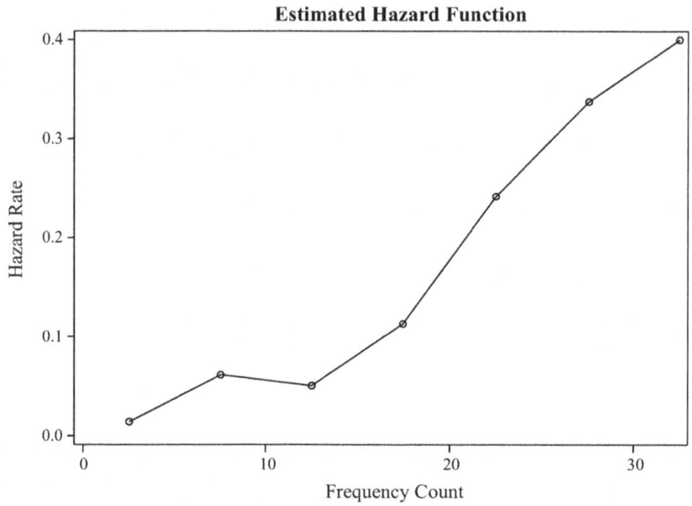

Fig. 6 Estimated risk function as a time function

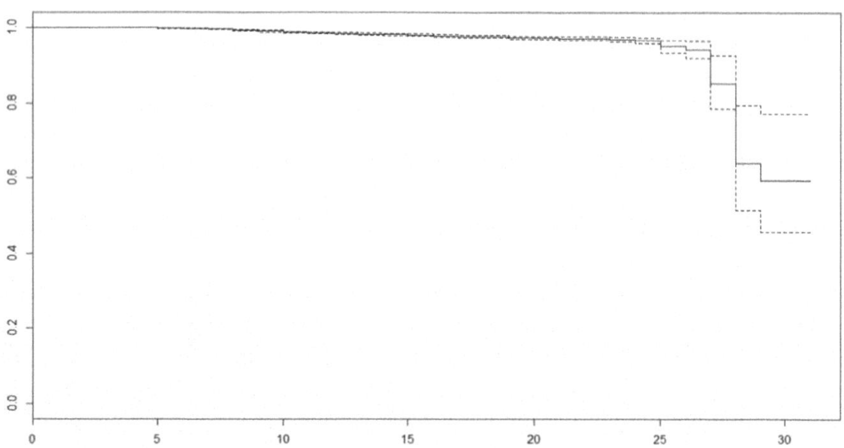

Fig. 7 Kaplan-Meyer evaluation graph

In the sample there were different years for loans, it is advisable to make a cut in accordance with the issue dates, comparing the models for different years based on estimates of KM as it was shown in [19].

Dynamic risk assessment is today a relevant and necessary approach for technical, economic, social spheres of life. It is advisable to use it for banks and financial companies operating in the market in transition and forced to respond quickly to external influences and actions of competitors. A method of dynamic risk assessment provides for the different types survival models construction and forecasting the transition time to a higher risk degree or level. It uses the advantages of ideologically different methods in integrated approach and allows to estimate possible losses in the form of point assessments and risk probabilities. The conducted experimental study showed the expediency of dynamic method of risk assessment applying for practical tasks of credit risks modeling and telecommunications company risks [24, 25].

5 Conclusions

Our work suggested the importance and perspective for the integrated modeling approach usage. Some proposed integrated models' combinations can be built in financial analytical systems as the base models and then be a part of bank analytics working places. We showed the integrated models in static and dynamics and emphasized on their necessity at the statistical data collection and choice. Our experiments with real data demonstrated the high efficiency of dynamic models in forecasting the time and losses from financial risks. Further on we'll present an information analytical system for financial tasks solving [26] based on integrated approach to financial data analysis, modelling and forecasting.

Developed integrated approach was tested on actual financial data for functioning financial companies and banks. It was shown that integrated models provide for more accuracy in forecasting the probability of default, financial indicators for the company, behavioral scoring etc. While it was needed to solve these tasks in dynamics it was proposed to adapt and use the idea of models integration and usage for dynamic method. That was developed as a basis for dynamic risk predictions and it was tested on real financial data in banks, telecommunication industry and even distributed on technical risks in prediction the fault-tolerance of the systems. It gave the possibility to predict the risk as a function of time and to receive different time-slices for the degree and level of risks based on settled cut-off. It is the effective mechanism in risk monitoring it is possible to control the changing of risks losses and probability in time and to forecast on pointed period how will the level or degree of risk change.

References

1. Tuffery, S.: Data Mining and Statistics for Decision Making. Wiley, New York (2011)

2. James, G., Witten, D., Hastie, T., Tibshinari, R.: An Introduction to Statistical Learning with Applications in R. Springer-Verlag, New York (2013). https://doi.org/10.1007/978-1-4614-7138-7
3. SAS: Analytics, artificial intelligence and data management, https://www.sas.com/engb/home. Last accessed 22 Sept 2021
4. Kuznietsova, N.V., Bidyuk, P.I.: Information technology for financial data analysis based on the integrated method (in Ukrainian). Syst. Res. Inf. Technol. **1**, 22–33 (2011)
5. Bidyuk, P., Polozhaenko, S., Kuznietsova, N., Levenchuk, L.: Probabilistic data analysis in non-stationary processes forecasting. In: 2020 IEEE 2nd International Conference on System Analysis and Intelligent Computing, SAIC 2020 (2020)
6. Siddiqi, N.: Credit Risk Scorecards: Developing and Implementing Intelligent Credit Scoring. Wiley, Hoboken (2005)
7. Kuznietsova, N.V., Bidyuk, P.I.: Neural and Bayesian networks in the credit risk analysis problem. Registration Storage Data Process. **17**(2), 61–71 (2015)
8. Kuznietsova, N.V., Chernysh, Z.S.: Regression models application for analysis and forecasting of the financial activity quality indicators of the company. Syst. Res. Inf. Technol. **2**, 67–81 (2020)
9. Papageorgiou, N.S., Radulescu, V.D., Repovs, D.D.: Noninear Analysis—Theory and Methods. Springer, Cham, Switzerland (2019)
10. Rassias, T.M.: Applications of Nonlinear Analysis. Springer, Cham, Switzerland (2018)
11. Beran, J.: Mathematical Foundations of Time Series Analysis. Springer, Cham Switzerland (2017)
12. Kalman, R.E.: A new approach to linear filtering and prediction problems. J. Basic Eng. **82**(1), 35–45 (1960)
13. Kuznietsova, N.V., Bidyuk, P.I.: Dynamic modeling of financial risks (in Ukrainian). Inductive Model. Complex Syst. **9**, 122–137 (2017)
14. Havrylovych, M., Kuznietsova, N.: Survival analysis methods for churn prevention in telecommunications industry. In: CEUR Workshop Proceeding, pp. 47–58 (2020). http://ceur-ws.org/Vol-2577/paper5.pdf
15. Kuznietsova, N.V.: Dynamic method of risk assessment in the system of financial management. Registration Storage Data Process. **21**(3), 85–98 (2019). https://doi.org/10.35681/1560-9189.2019.21.3.183724
16. Cox, D.R.: Regression models and life-tables. J. Roy. Stat. Soc. B **34**(2), 187–220 (1972)
17. Cox, D.R.: The regression analysis of binary sequences (with discussion). J. Roy. Stat. Soc. B **20**, 215–242 (1958)
18. Allison, P.D.: Survival Analysis Using SAS: A Practical Guide, 2nd edn. SAS Institute Inc., Cary, NC (2010)
19. Kaplan, E.L., Meier, P.: Non-parametric estimation for incomplete observations. J. Am. Stat. Assoc. **53**, 457–481 (1958)
20. Cao, R., Vilar, J.M., Devia, A.: Modelling consumer credit risk via survival analysis. SORT **33**(1), 3–30 (2009)
21. Beran, J., Dja, A.K.: Credit risk modeling based on survival analysis with immunes. Stat. Methodol. **4**, 251–276 (2007)
22. Kuznietsova, N.V., Bidyuk, P.I.: Modeling of credit risks on the basis of the theory of survival. J. Autom. Inf. Sci. **49**(11), 11–24 (2017)
23. Dabrowska, D.: Non-parametric regression with censored survival time data. Scand. J. Stat. **14**(3), 181–197 (1987)
24. Kuznietsova, N., Bidyuk, P., Kuznietsova, M.: Data mining methods, models and solutions for big data cases in telecommunication industry. In: Babichev, S., Lytvynenko, V. (eds.) Lecture Notes on Data Engineering and Communications Technologies. ISDMCI 2021. Lecture Notes on Data Engineering and Communications Technologies, vol. 77, pp. 107–127. Springer, Cham (2022)

25. Kuznietsova, N.V., Bidyuk, P.I.: Theory and Practice of Financial Risk Analysis: Systemic Approach. Lira-K, Kyiv (2020)
26. Kuznietsova, N., Bidyuk, P.: Intelligence information technologies for financial data processing in risk management. In: Babichev, S., Peleshko, D., Vynokurova, O. (eds.) Data Stream Mining & Processing. DSMP 2020. Communications in Computer and Information Science, vol. 1158, pp. 539–558. Springer, Cham (2020)

Chapter 17
Unranked Fuzzy Logic and Reasoning

Anriette Michel Fouad Bishara and Mikheil Rukhaia

Abstract One of the main tools in knowledge representation is ontology, which is a collection of logic-based formal language sentences. These sentences are used by automated reasoning programs to extract new knowledge and answer to the given questions. Although ontology languages are standardized by W3C, there are still many problems remaining. One of the most important problems is related to ontologies, where information is vague and incomplete. Such kind of ontologies have applications in many different fields, such as medicine, biology, e-commerce and the like. They require concepts from both, fuzzy and unranked theories. This is a follow-up paper of a one by the authors, where an unranked fuzzy logic was introduced. Here we develop a tableau method for reasoning over such logic. The unranked fuzzy logic is an extension of many-valued logics with sequence variables and flexible-arity function and predicate symbols. The unranked fuzzy tableau calculus corresponds to Hájek's witnessed fuzzy logics and is therefore complete.

Keywords Unranked logic · Fuzzy logic · Tableau method · Reasoning

1 Introduction

Artificial Intelligence is developing using statistical and logical methods (machine learning, automated deduction, natural language processing, etc.). Logical methods are core in automated reasoning, where it is vital to represent knowledge in a format understood by computers. Let us consider a simple problem from [1]:

> Players *A* and *B* are flipping a coin. Heads *A* wins. Tails *B* loses.
> Prove that *A* always wins.

A. M. F. Bishara
Faculty of Business and Technologies, International Black Sea University, Tbilisi, Georgia

M. Rukhaia (✉)
Laboratory of Applied Logic and Programming, Institute of Applied Mathematics, Tbilisi State University, Tbilisi, Georgia
e-mail: mrukhaia@logic.at

M. Zgurovsky and N. Pankratova (eds.), *System Analysis & Intelligent Computing*,
Studies in Computational Intelligence 1022, https://doi.org/10.1007/978-3-030-94910-5_17

It is very easy for a human to solve this problem, but the solution relies on the background knowledge—on the fact, which is not given in the problem description. In particular, we have to know, that after flipping the coin, there is either Heads or Tails and no other case is possible. Knowledge representation methods are used to develop such background knowledge bases for reasoners.

One of the main tools in knowledge representation is ontologies [2, 3], which are a collection of logic-based formal language sentences. These sentences are used by automated reasoning programs to extract new knowledge and answer the given questions.

The widest practical application of knowledge representation is in so-called Web 3.0. It is a collection of different kinds of technologies and the core among them is the semantic web. The name Semantic Web was introduced by Tim Berners-Lee, who was a founder of this scientific direction [4]. The main idea of the semantic web is to have the knowledge available for wide auditory (the purpose of the WWW itself) and to use this knowledge by implementing systems for searching, browsing, and evaluation. Thus, the main technologies in the semantic web are knowledge representation formats and different forms of knowledge.

Semantic Web is a collection of different technologies, where most of them are already standardized [5]. The main purpose of these technologies is to describe the semantic content of the web, i.e., their meaning and sense, in the format understood by computers. Such descriptions are called ontologies and the languages, on which these descriptions are written the ontology languages [6].

Nowadays, the main research is concentrated on the ontology, logic, and proof layers [7, 8]. The scientific modeling techniques are used to create ontologies, i.e., to describe objects according to the domain of interests.

The ontology itself is composed of class definitions, object definitions, and relation definitions. For example, an ontology for a modern online book library must contain the notion of a book (together with other notions), which will have attributes like author, title, publisher, and the like. The authors and publishers are separate terms, characterized by different properties, such as a name for example. There are numerous books about knowledge modeling (see, e.g. [9]). Nowadays the biggest ontologies are used in medicine since automatized medical tools are based exactly on such background knowledge bases [10]. An example of the biggest medical ontology is SNOMED CT.[1]

Although the ontology languages are standardized by W3C (e.g., OWL by [6]), there are still many problems remaining. One of the most important problems is related to, so-called, fuzzy ontologies. These are ontologies, where information is vague and imprecise [11, 12]. Fuzzy ontologies are obtained by integrating fuzzy logic with ontologies [13–15]. Such kind of ontologies have applications in many different areas, such as medicine, biology, e-commerce just to name a few (see e.g. [16–18]).

Fuzzy logic [19–21] is a multi-valued logic, where truth values instead of 0 (false) and 1 (true) can be any real number from the interval [0, 1]. The truth values can be

[1] SNOMED CT ontology homepage: http://www.snomed.org.

finite (finitely-valued logics) or infinite (infinitely-valued logics). The latter is often confused with probabilistic logic, but there is a fundamental difference between them. In fuzzy logic, any sentence is true to some degree, but the truth value itself is fixed. In probabilistic logic, there is a probability that a sentence is true, but the sentence itself can be either true or false and no other value is possible. In general, fuzzy logic is very important in modeling vague statements like "young scientist", "hot weather", "full glass", etc. In [22] the importance of fuzzy reasoning in web-related applications is explained.

Nowadays, on the logic layer of the semantic web, the formal languages are considered, which are based on unranked alphabets. This means that functional and/or predicate symbols do not have a fixed arity. Such languages can naturally model XML documents and operations over them [23] and many more. These kinds of languages are called unranked languages. One of the most interesting formalisms, based on an unranked alphabet, is Common Logic, which is used to exchange information between different systems and networks. It was given ISO standard in 2007 [24]. This language is more and more often used in knowledge representation and ontologies are written based on this language. (See e.g., COLORE[2] and OntoHub[3]).

The main novelty of this paper is to develop a tableau method for unranked fuzzy logic, which is an extension of many-valued logics with sequence variables and flexible-arity function and predicate symbols. Such an extension is of great importance for modelling knowledge, which is vague and incomplete. The following simple example illustrates this: assume a fuzzy ontology of clothes is given, with three instances of coat: X has property long with value 0.8, Y has property long with value 0.4 and Z has no property long at all. Note that not having the property means that the value of this property is unknown and cannot be assumed that its value is 0. To model such ontologies, both concepts, fuzziness and unranked is necessary. Without unranked formalism, we cannot directly omit the property for Z. Thus, such an extension will allow us to write unranked fuzzy ontologies, demand on which is increasing [22, 25, 26]. Having a reasoning method over unranked fuzzy language will broaden knowledge engineering capabilities in different fields like geography (e.g., GISS systems), medicine, biology, etc.

2 Unranked Fuzzy Logic

In [27] we presented an unranked fuzzy logic, called Ł$\Pi\forall$u, which is a combination of first-order unranked predicate logic given in [28, 29] and many-sorted first-order logic Ł$\Pi\forall$ given in [21, 30]. The main difference between Ł$\Pi\forall$ and Ł$\Pi\forall$u is that the

[2] Common Logic Ontology Repository, available at http://stl.mie.utoronto.ca/colore/.

[3] There are more than 2500 ontologies available in this repository, which are written on Common Logic, taking second place after CASL and leaving OWL on third place with about 2200 ontologies. For more details see https://ontohub.org/logics.

latter supports sequence variables and flexible-arity function and predicate symbols. We describe syntax and semantics of the unranked fuzzy logic below.

2.1 Syntax

In our language we consider an alphabet $\mathcal{A}$ consisting of the following pairwise disjoint sets of symbols:

- the set $\mathcal{V}_\mathrm{I}$ of individual variables, denoted by $x, y, z, \ldots,$
- the set $\mathcal{V}_S$ of sequence variables, denoted by $\overline{x}, \overline{y}, \overline{z}, \ldots,$
- the set $\mathcal{F}_r$ of fixed arity (ranked) function symbols, denoted by f_r,
- the set $\mathcal{F}_u$ of flexible arity (unranked) function symbols, denoted by f_u,
- the set $\mathcal{P}_r$ of ranked predicate symbols, denoted by p_r,
- the set $\mathcal{P}_u$ of unranked predicate symbols, denoted by p_u,
- the propositional constant 0 (falsum),
- logical connectives and quantifiers $\rightarrow_L, \rightarrow_\Pi, \&_\Pi, \forall_\mathrm{i}, \forall_\mathrm{s}$,
- auxiliary symbols: parentheses and the comma.

Each ranked symbol has a unique arity (rank) associated with it. Unranked symbols do not have a fixed arity. Each set of variables, function symbols and predicate symbols is countably infinite. The letter $\mathcal{F}$ denotes a set of function symbols and is a disjoint union of the sets $\mathcal{F}_u$ and $\mathcal{F}_r$. Respectively, the letter $\mathcal{P}$ denotes a set of predicate symbols and is a disjoint union of the sets $\mathcal{P}_u$ and $\mathcal{P}_r$; finally, $\mathcal{V}$ denotes a set of all kinds of variables and is a disjoint union of the sets $\mathcal{V}_\mathrm{I}$ and $\mathcal{V}_S$. We use the letter f to denote elements of $\mathcal{F}$, the letter p for the elements of $\mathcal{P}$, and the letter v for the elements of $\mathcal{V}$.

The terms and sequences over $\mathcal{A}$ are defined inductively as follows:

$$t ::= x \mid f_r(t_1, \ldots, t_n) \mid f_u(\overline{s}) \quad \text{terms}$$
$$\overline{s} ::= t \mid \overline{x} \mid (\overline{s}_1, \ldots, \overline{s}_n) \quad (n \geq 0) \quad \text{sequences}$$

An *atom* is a formula of the form $p_r(t_1, \ldots, t_n)$ or $p_u(\overline{s})$, where $p_r \in \mathcal{P}_r$ is an n-ary predicate symbol and $p_u \in \mathcal{P}_u$ is a flexible arity predicate symbol. *Formulas* are built as usual from atomic formulas, propositional constant 0, logical connectives $\rightarrow_L, \rightarrow_\Pi, \&_\Pi$, and quantifiers $\forall_\mathrm{i}, \forall_\mathrm{s}$:

$$A ::= 0 \mid p_r(t_1, \ldots, t_n) \mid p_u(\overline{s}) \mid A \rightarrow_L A \mid A \rightarrow_\Pi A \mid A \&_\Pi A \mid (\forall_\mathrm{i} x)A \mid (\forall_\mathrm{s} \overline{x})A$$

Quantifications are allowed on both, individual as well as on sequence variables. We will use $\forall$ to denote both $\forall_\mathrm{i}$ or $\forall_\mathrm{s}$, when the context is clear. For any variable v, $\neg_L(\forall v)\neg_L$ is denoted by $(\exists v)$, where $\neg_L$ is a negation in Łukasiewicz sense, defined as given in Fig. 1. Notions of free and bound variables are defined in a standard way. We use the letters $A, B, \ldots$ to denote formulas.

$$
\begin{array}{rcll}
 & & A \to_L B & \varphi \to_L \psi = min(1, 1-\varphi+\psi) \\
 & & A \to_\Pi B & \varphi \to_\Pi \psi = min(1, \frac{\psi}{\varphi}) \text{ if } \varphi > 0 \text{ otherwise } 0 \\
 & & A \&_\Pi B & \varphi \&_\Pi \psi = \varphi \cdot \psi \\
\neg_L A & \stackrel{def}{\equiv} & A \to_L 0 & \neg_L \varphi = 1 - \varphi \\
\neg_\Pi A & \stackrel{def}{\equiv} & A \to_\Pi 0 & \neg_\Pi \varphi = 1 \text{ if } \varphi = 0 \text{ otherwise } 0 \\
1 & \stackrel{def}{\equiv} & \neg_L 0 & \\
\Delta A & \stackrel{def}{\equiv} & \neg_\Pi \neg_L A & \Delta\varphi = 1 \text{ if } \varphi = 1 \text{ otherwise } 0 \\
A \&_L B & \stackrel{def}{\equiv} & \neg_L(A \to_L \neg_L B) & \varphi \&_L \psi = max(0, \varphi+\psi-1) \\
A \oplus B & \stackrel{def}{\equiv} & \neg_L A \to_L B & \varphi \oplus \psi = min(1, \varphi+\psi) \\
A \ominus B & \stackrel{def}{\equiv} & A \&_L \neg_L B & \varphi \ominus \psi = max(0, \varphi-\psi) \\
A \wedge B & \stackrel{def}{\equiv} & A \&_L (A \to_L B) & \varphi \wedge \psi = min(\varphi, \psi) \\
A \vee B & \stackrel{def}{\equiv} & (A \to_L B) \to_L B & \varphi \vee \psi = max(\varphi, \psi) \\
A \to_G B & \stackrel{def}{\equiv} & \Delta(A \to_L B) \vee B & \varphi \to_G \psi = 1 \text{ if } \varphi \leq \psi \text{ otherwise } \psi \\
A \leftrightarrow_L B & \stackrel{def}{\equiv} & (A \to_L B) \&_L (B \to_L A) & \varphi \leftrightarrow_L \psi = 1 - |\varphi - \psi| \\
A \leftrightarrow_\Pi B & \stackrel{def}{\equiv} & (A \to_\Pi B) \&_\Pi (B \to_\Pi A) & \varphi \leftrightarrow_\Pi \psi = min(\frac{\varphi}{\psi}, \frac{\psi}{\varphi}) \\
A \leftrightarrow_G B & \stackrel{def}{\equiv} & (A \to_G B) \wedge (B \to_G A) & \varphi \leftrightarrow_G \psi = 1 \text{ if } \varphi = \psi \text{ otherwise } min(\varphi, \psi)
\end{array}
$$

Fig. 1 Logical connectives and corresponding algebraic operations

Note, that the Ł$\Pi\forall$ logic, and thus the Ł$\Pi\forall$u logic as well, subsumes the product and Gödel logics. The full list of connectives of those logics, together with the corresponding algebraic operations, are given in Fig. 1.

A *substitution* is a finite set of distinct variable bindings, where the variable binding is either expression $x \mapsto t$ or $\overline{x} \mapsto \overline{s}$. Substitutions are denoted by σ (possibly with indices) and the empty substitution is denoted by ϵ. Application of a substitution σ on term t and formula A is defined in the intuitive way, similar to [28], and is denoted by $t\sigma$ and $A\sigma$, respectively.

The Ł$\Pi\forall$u logic is defined using the following axioms (we use * as a placeholder for L and Π):

(ŁΠ1) $(A \to_* B) \to_* ((B \to_* C) \to_* (A \to_* C))$

(ŁΠ2) $A \&_* (A \to_* B) \to_* B \&_* (B \to_* A)$

(ŁΠ3) $(A \&_* B \to_* C) \to_* (A \to_* (B \to_* C))$

(ŁΠ4) $(A \to_* (B \to_* C)) \to_* (A \&_* B \to_* C)$

(ŁΠ5) $((A \to_* B) \to_* C) \to_* (((B \to_* A) \to_* C) \to_* C)$

(ŁΠ6) $0 \to_* A$

(Ł) $\neg_L \neg_L A \to_L A$

$$
\begin{array}{ll}
(\Pi) & \neg_{\Pi} A \vee ((A \rightarrow_{\Pi} A \&_{\Pi} B) \rightarrow_{\Pi} B) \\
(\text{Ł}\Delta) & \Delta(A \rightarrow_{L} B) \rightarrow_{L} (A \rightarrow_{\Pi} B) \\
(\Pi\Delta) & \Delta(A \rightarrow_{\Pi} B) \rightarrow_{L} (A \rightarrow_{L} B) \\
(D) & A \&_{\Pi} (B \ominus C) \leftrightarrow_{L} (A \&_{\Pi} B) \ominus (A \&_{\Pi} C) \\
(\forall 1) & (\forall v) A(v) \rightarrow_{*} A(t) \\
(\forall 2) & (\forall v)(A \rightarrow_{*} B) \rightarrow_{*} (A \rightarrow_{*} (\forall v) B) \\
(\forall 3) & (\forall v)(A \vee B) \rightarrow_{*} (A \vee (\forall v) B)
\end{array}
$$

Note that, in the ($\forall 1$)–($\forall 3$) axioms, v represents either individual or sequence variable and t in ($\forall 1$) is either a term or a sequence, respectively, substitutable for v in A. In ($\forall 2$), ($\forall 3$) we require that v is not free in A.

The deduction rules in Ł$\Pi\forall$u are modus ponens (for all implications L, Π, and G), Δ-necessitation (from A infer ΔA) and generalization (from A infer $(\forall v) A$):

$$
\frac{A \quad A \rightarrow_{L} B}{B}\mathrm{MP_L} \qquad \frac{A \quad A \rightarrow_{\Pi} B}{B}\mathrm{MP_\Pi} \qquad \frac{A \quad A \rightarrow_{G} B}{B}\mathrm{MP_G}
$$

$$
\frac{A}{\Delta A}\mathrm{Nec} \qquad \frac{A(x)}{(\forall_{\mathrm{i}} x) A}\mathrm{Gen_1} \qquad \frac{A(\overline{x})}{(\forall_{\mathrm{s}} \overline{x}) A}\mathrm{Gen_2}
$$

The notions of proof, theorem, and derivability $\vdash$ are defined in the usual sense. In [31], it was proved that all implications are equivalent w.r.t provability in Ł$\Pi\forall$. We show that a similar result holds in our logic Ł$\Pi\forall$u, i.e., if the theory proves one implication, then it proves the other two as well.

Proposition 1 *Let $\mathcal{T}$ be a theory of* Ł$\Pi\forall$u, *then the following sentences hold:*

1. $\mathcal{T} \vdash A$ *iff* $\mathcal{T} \vdash \Delta A$.
2. $\mathcal{T} \vdash A \rightarrow_{L} B$ *iff* $\mathcal{T} \vdash A \rightarrow_{\Pi} B$.
3. $\mathcal{T} \vdash A \rightarrow_{L} B$ *iff* $\mathcal{T} \vdash A \rightarrow_{G} B$.

□

Proof

1. If $\mathcal{T} \vdash A$ then using the Nec rule we get $\mathcal{T} \vdash \Delta A$. Converse follow directly from the definition of Δ in Fig. 1.
2. If $\mathcal{T} \vdash A \rightarrow_{L} B$, then according to the statement (1), $\mathcal{T} \vdash \Delta(A \rightarrow_{L} B)$, hence using the (Ł$\Delta$) axiom and the $\mathrm{MP_L}$ rule we get $\mathcal{T} \vdash A \rightarrow_{\Pi} B$. Converse follows analogously using the ($\Pi\Delta$) axiom and the $\mathrm{MP_L}$ rule.
3. Directly follows from the definition of $\rightarrow_{G}$ in Fig. 1.

□

From now on, unless otherwise stated, we use $\rightarrow_{L}$ as the "default" implication.

2.2 Semantics

The unranked fuzzy logic Ł$\Pi\forall$u has precisely defined algebraic semantics. For this purpose, we recall notions of MV-algebra and Π-algebra.

An MV-algebra is a structure $(L, \oplus, \neg, 0)$, where $\oplus$ is a binary operation on L, $\neg$ is a unary operation on L and 0 a distinguished constant satisfying the following equations:

1. $\phi \oplus (\varphi \oplus \psi) = (\phi \oplus \varphi) \oplus \psi$ (associative)
2. $\varphi \oplus \psi = \psi \oplus \varphi$ (commutative)
3. $\varphi \oplus 0 = \varphi$
4. $\neg\neg\varphi = \varphi$
5. $\varphi \oplus \neg 0 = \neg 0$
6. $\neg(\neg\varphi \oplus \psi) \oplus \psi = \neg(\neg\psi \oplus \varphi) \oplus \varphi$

A Π-algebra (sometimes also called residuated lattice in the literature) is a structure $(L, \wedge, \vee, \rightarrow, \otimes, 0, 1)$, which has the following properties:

1. $(L, \wedge, \vee, 0, 1)$ is a lattice with the smallest element 0 and greatest element 1 (i.e., $\varphi \leq \psi$ iff $\varphi \wedge \psi = \varphi$, as usual).
2. The operation $\otimes$ is usually called product and it is associative, commutative and $\varphi \otimes 1 = \varphi$.
3. $\varphi \otimes \psi \leq \phi$ iff $\varphi \leq \psi \rightarrow \phi$ (residuation)

The notions of Ł$\Pi\forall$-algebra and M-evaluation are generalizations of the ones defined in [30].

An Ł$\Pi\forall$-algebra is a structure $\mathcal{L} = (L, \oplus, \neg_L, \rightarrow_\Pi, \&_\Pi, 0, 1)$, such that:

- $(L, \oplus, \neg_L, 0)$ is an MV-algebra
- $(L, \wedge, \vee, \rightarrow_\Pi, \&_\Pi, 0, 1)$ is a Π-algebra
  $\varphi \&_\Pi (\psi \ominus \phi) = (\varphi \&_\Pi \psi) \ominus (\varphi \&_\Pi \psi)$

We consider a standard Ł$\Pi\forall$-algebra over the domain [0, 1] with the operations as stated in Fig. 1.

Let $\mathcal{L}$ be an Ł$\Pi\forall$-algebra. An $\mathcal{L}$-structure M is defined as

$$(\mathrm{D}, (P_r^M)_{p_r \in \mathcal{P}_r}, (P_u^M)_{p_u \in \mathcal{P}_u}, (F_r^M)_{f_r \in \mathcal{F}_r}, (F_u^M)_{f_u \in \mathcal{F}_u}).$$

where:

- D is a non-empty set, called the domain of M. It is a disjoint union of two sets D_I, D_S and $D_\mathrm{I} \neq \emptyset$,
- P_r^M is an n-ary relation $D_\mathrm{I}^n \rightarrow L$ for each $p_r \in \mathcal{P}_r$,
- P_u^M is an unranked relation $D^\infty \rightarrow L$ for each $p_u \in \mathcal{P}_u$,
- F_r^M is an n-ary function $D_\mathrm{I}^n \rightarrow D_\mathrm{I}$ for each $f_r \in \mathcal{F}_r$,
- F_u^M is an unranked function $D^\infty \rightarrow D_\mathrm{I}$ for each $f_u \in \mathcal{F}_u$.

An M-evaluation is a mapping e, which assigns to each individual variable x an element $e(x) \in D_{\mathrm{I}}$ and to each sequence variable $\overline{x}$ an element $e(\overline{x}) \in D^{\infty}$. Further, $e[x \mapsto d]$, for $d \in D_{\mathrm{I}}$, is an M-evaluation such that $e[x \mapsto d](x) = d$ and $e[x \mapsto d](y) = e(y)$ for each individual variable y different from x. Analogously, $e[\overline{x} \mapsto (d_1, ..., d_n)](\overline{x}) = (d_1, ..., d_n)$ and $e[\overline{x} \mapsto (d_1, ..., d_n)](\overline{y}) = e(\overline{y})$ for $d_1, ..., d_n \in D$ and for each sequence variable $\overline{y}$ different from $\overline{x}$.

The value of a term and the truth value of a formula in an $\mathcal{L}$-structure M with respect to M-evaluation e are defined in Fig. 2.

$$
\begin{aligned}
\|x\|_{M,e}^{\mathcal{L}} &= e(x)\\
\|\overline{x}\|_{M,e}^{\mathcal{L}} &= e(\overline{x})\\
\|f_r(t_1,\ldots,t_n)\|_{M,e}^{\mathcal{L}} &= F_r^M(\|t_1\|_{M,e}^{\mathcal{L}},\ldots,\|t_n\|_{M,e}^{\mathcal{L}})\\
\|f_u(\overline{s})\|_{M,e}^{\mathcal{L}} &= F_u^M(\|\overline{s}\|_{M,e}^{\mathcal{L}})\\
\|(\overline{s}_1,\ldots,\overline{s}_n)\|_{M,e}^{\mathcal{L}} &= (\|\overline{s}_1\|_{M,e}^{\mathcal{L}},\ldots,\|\overline{s}_n\|_{M,e}^{\mathcal{L}})
\end{aligned}
$$

$$
\begin{aligned}
\|p_r(t_1,\ldots,t_n)\|_{M,e}^{\mathcal{L}} &= P_r^M(\|t_1\|_{M,e}^{\mathcal{L}},\ldots,\|t_n\|_{M,e}^{\mathcal{L}})\\
\|p_u(\overline{s})\|_{M,e}^{\mathcal{L}} &= P_u^M(\|\overline{s}\|_{M,e}^{\mathcal{L}})\\
\|0\|_{M,e}^{\mathcal{L}} &= 0\\
\|A \circ B\|_{M,e}^{\mathcal{L}} &= \|A\|_{M,e}^{\mathcal{L}} \circ \|B\|_{M,e}^{\mathcal{L}} \text{ for } \circ \in \{\rightarrow_L, \rightarrow_\Pi, \&_\Pi\}\\
\|(\forall_{\mathrm{i}} x)A\|_{M,e}^{\mathcal{L}} &= \inf\{\|A\|_{M,e[x\mapsto d]}^{\mathcal{L}} \mid d \in D_{\mathrm{I}}\}\\
\|(\forall_{\mathrm{s}} \overline{x})A\|_{M,e}^{\mathcal{L}} &= \inf\{\|A\|_{M,e[\overline{x}\mapsto(d_1,\ldots,d_n)]}^{\mathcal{L}} \mid d_1,\ldots,d_n \in D\}
\end{aligned}
$$

Fig. 2 Semantics of Ł$\Pi\forall$u logic

Note, that if the infimums do not exist, the truth values of the quantified formulas are undefined. An $\mathcal{L}$-structure M is safe iff $\|A\|_{M,e}^{\mathcal{L}}$ is defined for each formula A and M-evaluation e.

Finally, the truth value of a formula A in an $\mathcal{L}$-structure M is the infimum of its values with respect to all M-evaluations e, i.e.: $\|A\|_{M}^{\mathcal{L}} = \inf\{\|A\|_{M,e}^{\mathcal{L}}\}$. A is an $\mathcal{L}$-tautology iff $\|A\|_{M}^{\mathcal{L}} = 1$ for each safe $\mathcal{L}$-structure M. The latter is an $\mathcal{L}$-model of a theory $\mathcal{T}$ iff $\|A\|_{M}^{\mathcal{L}} = 1$ for all $A \in \mathcal{T}$.

The syntax and semantics of unranked fuzzy logic are bound together by the completeness theorem given below.

Theorem 1 (Completeness) *Let A be a formula and $\mathcal{T}$ a theory. $\mathcal{T} \vdash A$ iff $\|A\|_{M}^{\mathcal{L}} = 1$ for each Ł$\Pi\forall$u-algebra $\mathcal{L}$ and each safe $\mathcal{L}$-model M of $\mathcal{T}$.*

Proof The proof is similar to the analogous completeness theorem for the Ł$\Pi\forall$ logic given in [32].

The Ł$\Pi\forall$u logic is capable of modeling incomplete and vague information, which appears in medicine, e-commerce, and the like. □

Example 1 It is well known, that the friend-of-a-friend (foaf) graphs are very important in e-commerce and marketing. In [29], reasoning on incomplete foaf graphs was demonstrated using unranked formalism, but the foaf relation is vague in the real world. In our new settings of Ł$\Pi\forall$u logic, we can attach fuzzy values to friend-of (p_u^{fo}) and thus to friend-of-friend (p_u^{fof}) relations.

Consider an adapted version of the Clique of Friends example from [33]. Given a collection of address books, where each address book has an owner and a set of entries, some of which are marked as friend by the owner of the address book. The collection contains two address books, the first owned by Donald Duck and the second by Daisy Duck. Donald's address book has two entries, one for Scrooge, the other for Daisy, and only Daisy is marked as friend. Daisy's address book again has two entries, both marked as friend. Then we could express the following sentences:

– Donald Duck and Daisy Duck are best friends:

$$\|p_u^{fo}(\text{DonaldDuck, DaisyDuck})\|_{M,e}^{\mathcal{L}} = 0.9$$

– Daisy Duck and Gladstone Duck are good friends:

$$\|p_u^{fo}(\text{DaisyDuck, GladstoneDuck})\|_{M,e}^{\mathcal{L}} = 0.7$$

– Daisy Duck and Ratchet Gearloose are friends:

$$\|p_u^{fo}(\text{DaisyDuck, RatchetGearloose})\|_{M,e}^{\mathcal{L}} = 0.5$$

If we define friend-of-friend relation as

$$p_u^{fof}(x, y) ::= \neg_L p_u^{fo}(x, y) \&_\Pi \exists z \big(p_u^{fo}(x, z) \&_\Pi p_u^{fo}(z, y)\big)$$

and assume (GD, RG stands for GladstoneDuck, RatchetGearloose respectively)

$$\|p_u^{fo}(\text{DonaldDuck, GD})\|_{M,e}^{\mathcal{L}} = 0.1, \quad \|p_u^{fo}(\text{DonaldDuck, RG})\|_{M,e}^{\mathcal{L}} = 0$$

then

$$\|p_u^{fof}(\text{DonaldDuck, GD})\|_{M,e}^{\mathcal{L}} = 0.9 \cdot 0.9 \cdot 0.7 = 0.567,$$
$$\|p_u^{fof}(\text{DonaldDuck, RG})\|_{M,e}^{\mathcal{L}} = 1 \cdot 0.9 \cdot 0.5 = 0.45$$

3 Unranked Fuzzy Tableaux

There are so-called signed and unsigned tableaux calculi defined in the literature. In classical logic, signed and unsigned tableaux are dual to each other, thus choice in between is a matter of taste. When it comes to many-valued logics, as [34] explains, signs are a crucial part of the tableaux calculus. Note that in most of the many-valued logics there is no negation normal form exists. Although this is not the case for our Ł$\Pi\forall$u logic (i.e. NNF exists), nevertheless we developed the signed tableaux for it. The reason is that fuzzy tableaux are usually semantic driven, in contrast to tableaux for classical logic, which are driven by Gentzen's sequent calculus LK.

Definition 1 (Signed Formula) *A signed formula is an expression of the form* $\boldsymbol{T}A$ *or* $\boldsymbol{F}A$*, where A is a formula and* $\boldsymbol{T}$ *corresponds to the truth value 1 and* $\boldsymbol{F}$ *corresponds to the truth value 0. Additionally, the pair* $(\boldsymbol{T}A, \boldsymbol{F}A)$ *is called a complementary pair and* $\boldsymbol{T}$, $\boldsymbol{F}$ *complementary signs.*

Remark 1 Note, that for the propositional constants, the signed formulas $\boldsymbol{T}0$ and $\boldsymbol{F}1$ are evaluated to 0 and $\boldsymbol{T}1$ and $\boldsymbol{F}0$ are evaluated to 1.

We follow [35] to define labeled tableaux calculus for Ł$\Pi\forall$u. A label is an expression built over the structure $(V, 0, +, \dot{-})$, where V is a set of arithmetical variables. The labels are interpreted as rational numbers (intuitively expressing truth value of a formula), where $+$ is an ordinary addition and $\dot{-}$ is a modified subtraction: $m \dot{-} n = 0$, for $m \leq n$. We assume that for each atomic formula p there is a variable $\varphi_p \in V$. Analogously, the variable $\varphi_0 \in V$ for the propositional constant 0.

Definition 2 (Labeled formula) *A labeled formula is a signed formula with label* $\iota : \boldsymbol{T}A$ *or* $\iota : \boldsymbol{F}A$.

Remark 2 In [35], a labeled formula is either a signed formula with label or an inequality of labels ($\iota \leq \kappa, \iota < \kappa, \iota \geq \kappa, \iota > \kappa$). Here, we do not consider the latter, and instead of solving sets of inequalities, we will use unranked fuzzy unification to close tableaux.

To optimize the number of rules in our tableaux calculus, we define the tableaux expansion rules only for the basic connectives $\rightarrow_L$, $\&_\Pi$, $\forall_\mathrm{i}$ and $\forall_\mathrm{s}$. It is clear, that the formula containing defined connectives can be preprocessed and represented by the basic ones (including $\rightarrow_\Pi$ by the Proposition 1) before applying the tableaux rules.

The unranked fuzzy tableaux calculus consists of the following rules:

$$\frac{\iota : \mathbf{T}A \rightarrow_L B}{\kappa : \mathbf{F}A \mid \iota + \kappa : \mathbf{T}B}(\mathbf{T} \rightarrow_\mathrm{L}) \qquad \frac{\iota : \mathbf{F}A \rightarrow_L B}{\begin{array}{c}\varphi : \boldsymbol{T}A \\ \iota + \varphi : \boldsymbol{F}B\end{array}}(\mathbf{F} \rightarrow_\mathrm{L})$$

$$\frac{\iota : \mathbf{T}A \&_\Pi B}{\begin{array}{c}\iota : \boldsymbol{T}A \\ \iota : \boldsymbol{T}B\end{array}}(\mathbf{T}\&_\Pi) \qquad \frac{\iota : \mathbf{F}A \&_\Pi B}{\iota : \mathbf{F}A \mid \iota : \boldsymbol{F}B}(\mathbf{F}\&_\Pi)$$

$$\frac{\iota : \mathbf{T}(\forall_\mathrm{i} x)A}{\iota : \mathbf{T}A[x \mapsto t]}(\mathbf{T}\forall_\mathrm{i}) \qquad \frac{\iota : \mathbf{F}(\forall_\mathrm{i} x)A}{\iota : \mathbf{F}A[x \mapsto f]}(\mathbf{F}\forall_\mathrm{i})$$

$$\frac{\iota : \boldsymbol{T}(\forall_\mathrm{s}\overline{x})A}{\iota : \boldsymbol{T}A[\overline{x} \mapsto \overline{s}]}(\boldsymbol{T}\forall_\mathrm{s}) \qquad \frac{\iota : \boldsymbol{F}(\forall_\mathrm{s}\overline{x})A}{\iota : \boldsymbol{F}A[\overline{x} \mapsto (f_1, \ldots, f_n)]}(\boldsymbol{F}\forall_\mathrm{s})$$

where κ is an arbitrary label, $\varphi \in V$ is an eigenvariable not occurring in the tableaux, t is an arbitrary term, $\overline{s}$ is an arbitrary sequence, and f, $f_1, \ldots, f_n$ are new (ranked or unranked) skolem constants not occurring in the tableaux.

Definition 3 (Branch closure) *A tableau branch is closed if it contains a complementary pair* ($\iota : \mathbf{T}A, \iota : \mathbf{F}A$)*. Moreover, if all branches of a tableau are closed, then the tableau is closed as well.*

Example 2 Let us demonstrate how unranked fuzzy tableaux works by proving the (∀2) axiom defined in Sect. 2.1. To prove that a formula A is valid, the tableaux of 0:$\mathbf{F}A$ should be closed. Thus, we have:

$$\frac{0 : \mathbf{F}(\forall_s\overline{x})(A \rightarrow_L B(\overline{x})) \rightarrow_L (A \rightarrow_L (\forall_s\overline{x})B(\overline{x}))}{\varphi : \mathbf{T}(\forall_s\overline{x})(A \rightarrow_L B(\overline{x}))}(\mathbf{F} \rightarrow_L)$$

$$\frac{0: \mathbf{F}(\forall_s\overline{x})(A \rightarrow_L B(\overline{x})) \rightarrow_L (A \rightarrow_L (\forall_s\overline{x})B(\overline{x}))}{\begin{array}{c}\varphi: \mathbf{T}(\forall_s\overline{x})(A \rightarrow_L B(\overline{x})) \\ \dfrac{0+\varphi: \mathbf{F}A \rightarrow_L (\forall_s\overline{x})B(\overline{x})}{\psi: \mathbf{T}A}(\mathbf{F} \rightarrow_L)\end{array}}(\mathbf{F} \rightarrow_L)$$

$$\dfrac{\dfrac{\dfrac{0+\varphi+\psi: \mathbf{F}(\forall_s\overline{x})B(\overline{x})}{0+\varphi+\psi: \mathbf{F}B(\overline{f})}(\mathbf{F}\forall_s)}{\varphi: \mathbf{T}A \rightarrow_L B(\overline{x})}(\mathbf{T}\forall_s)}{\kappa: \mathbf{F}A \quad | \quad \varphi+\kappa: \mathbf{T}B(\overline{x})}(\mathbf{T} \rightarrow_L)$$

Now, κ can be an arbitrary label, so we can take ψ instead and $\overline{x}$ can be instantiated by an arbitrary term, so we can choose $\overline{f}$. Then we obtain the following complementary pairs: (ψ:$\mathbf{T}A$, ψ : $\mathbf{F}A$) and ($0 + \varphi + \psi$:$\mathbf{F}B(\overline{f})$, $\varphi + \psi$:$\mathbf{T}B(\overline{f})$). Hence, both branches are closed and therefore the tableaux as well.

Finally, we show that the unranked fuzzy tableau calculus is sound and complete. For this purpose, we define an interpretation I of a label over the structure $(V, 0, +, \dot{-})$, such that:

- A labeled formula ι:$\mathbf{T}A$ holds iff $M, e, I(\iota) \vdash A$.
- A labeled formula ι:$\mathbf{F}A$ holds iff $M, e, I(\iota) \nvdash A$.

Theorem 2 (Soundness) *The unranked fuzzy tableau calculus is sound, i.e., if a formula A is provable, then A is valid.*

Proof The proof is by contradiction: show that if a formula A is not valid, then it is not provable. That means the tableau of 0:$\mathbf{F}A$ is open. We proceed by induction on the tableau construction, and by case distinction on the rules, show that there is a satisfiable branch. We consider the case of the ($\mathbf{T}\forall_s$) rule, the other cases are similar.

Let ι : $\mathbf{T}(\forall_s\overline{x})A$ in a branch be expanded by ι:$\mathbf{T}A[\overline{x} \mapsto \overline{s}]$. By the induction hypothesis there exists an M-evaluation e and an interpretation I such that M, e, I (ι) $\vdash$ $(\forall_s\overline{x})A$. Then, according to the semantics of ŁΠ∀u defined in Fig. 2, $M, e, I(\iota) \vdash A[\overline{x} \mapsto \overline{s}]$ for an arbitrary term sequence $\overline{s}$. □

Theorem 3 (Completeness) *The unranked fuzzy tableau calculus is complete, i.e., if a formula A is valid, then A is provable.*

Proof By routine induction on the structure of a formula A. We show the case when the formula $A \equiv (\forall_s \overline{x})A'$, the other cases are similar.

Assume that $A \equiv (\forall_s \overline{x})A'$ is valid. Then we can construct a term sequence $(f_1, \ldots, f_n)$, such that none of $f_1, \ldots, f_n$ occurs in A' and $A'[\overline{x} \mapsto (f_1, \ldots, f_n)]$ is valid. By the induction hypothesis there exists a refutation R of 0: $\mathbf{F}A'[\overline{x} \mapsto (f_1, \ldots, f_n)]$. Let us assume that R consists of the rules $R_1, \ldots, R_m$, then the sequence of rules $(\mathbf{F}\forall_s)$, $R_1, \ldots, R_m$ is a refutation of 0: $\mathbf{F}A$. □

4 Unranked Fuzzy Unification

Solving equations between logic terms is a fundamental problem with many important applications in mathematics, computer science, and artificial intelligence. It is needed to perform an inference step in reasoning systems and logic programming, to match a pattern to an expression in rule-based and functional programming, to extract information from a document, to infer types in programming languages, to compute critical pairs while completing a rewrite system, to resolve ellipsis in natural language processing, etc. Unification and matching are well-known techniques used in these tasks.

As it was discussed above and illustrated in the Example 2, we need unranked fuzzy unification to close the branches of a tableau. In this section we define unranked fuzzy unification problem and discuss techniques to solve it. For this purpose, we need notion of label substitution, which is defined as mapping from arithmetical variables to labels. Label substitutions are denoted by θ (possibly with indices).

Unranked fuzzy unification problem $\mathcal{P}$ is a finite set of equational problems $\iota : A =^? \iota' : A'$, where ι,ι' are labels and A, A' are atoms (signs $\mathbf{T}$,$\mathbf{F}$ are ignored if any). A solution for $\mathcal{P}$ is a substitution pair (θ, σ) such that for all problems in $\mathcal{P}$ we have $\iota\theta : A\sigma = \iota'\theta : A'\sigma$.

To solve unranked fuzzy unification problem, it is divided into two parts: *unranked unification problem* and *arithmetic unification problem*. Unranked unification is well studied and the algorithm, together with some finitary fragments, was given in [36]. The arithmetic unification problem of labels can be solved using a standard linear programming approach. There are some more efficient modifications in the literature as well (see e.g. [37–39]).

Example 3 The refutation from the Example 2 leads us to the unranked fuzzy unification problem:

$$\left\{\psi\colon \boldsymbol{T}A =^? \kappa\colon \boldsymbol{F}A;\ 0 + \varphi + \psi\colon \boldsymbol{F}B\left(\overline{f}\right) =^? \varphi + \kappa\colon \boldsymbol{T}B(\overline{x})\right\}$$

Note that A, B represent general schema of formulae here, thus notation is deviating from formal definition. Next, we divide it into two separate problems, and after ignoring the signs $\boldsymbol{T}$, $\boldsymbol{F}$, get:

$$\textbf{Arithmetic unification problem: } \{\psi =^? \kappa;\ 0 + \varphi + \psi =^? \varphi + \kappa\}$$
$$\textbf{Unranked unification problem: } \{A =^? A;\ B(\overline{f}) =^? B(\overline{x})\}$$

Clearly, a solution is a pair (θ, σ), where $\sigma = \{\overline{x} \mapsto \overline{f}\}$ and θ is either $\{\psi \mapsto \kappa\}$ or $\{\kappa \mapsto \psi\}$.

Unranked unification problem, and thus unranked fuzzy unification problem, is non-terminating in general. Consider a simple problem $\{p_u(f_r, \overline{x}) =^? p_u(\overline{x}, f_r)\}$. It has infinitely many solutions $\{\overline{x} \mapsto ()\}$, $\{\overline{x} \mapsto (f_r)\}$, $\{\overline{x} \mapsto (f_r, f_r)\}$. In [36] several terminating fragments were identified:

KIF fragment: sequence variables are allowed only as the last argument in a predicate (corresponds to Knowledge Interchange Format, see e.g. [5, 9]).
Unitary fragment: a concrete variable must occur only once in a predicate.
Matching fragment: one side of each equation is always ground (i.e., does not contain variables). Usually, query answering falls into this fragment, where knowledge base contains ground terms (facts) and the query contains variables.

Clearly, having the similar syntactic restrictions on unranked fuzzy logic, makes the unranked fuzzy unification terminating. Even more, the unitary fragment makes unranked fuzzy tableau method terminating (i.e., decidable).

5 Conclusions and Future Work

We have developed unranked fuzzy logic and corresponding reasoning method. Namely, we presented unranked fuzzy tableau calculus and proved some of its properties. The unranked fuzzy tableau calculus corresponds to Hajek's witnessed fuzzy logics and is therefore complete. This result is important, since first order t norm based calculi with infimum/supremum setting and infinitely many values are usually incomplete (with exception of infinitely valued Gödel logics).

Our future plans include further development of unranked fuzzy logic by covering a wider class of logics. We have introduced sequence variables in the first-order logic $\text{Ł}\Pi\forall$, but it would be an interesting study whether this can be done in other fuzzy logics as well.

Next point will be the proof-theoretical study of unranked fuzzy theory. For this purpose, we could introduce sequent calculi and study cut-elimination properties of the language (whether Herbrand's theorem holds). Another interesting problem should be investigation of Skolemization, especially Andrew's Skolemization, since sometimes it can lead to a nonelementary speedup w.r.t. the length of proofs.

Acknowledgements This work was supported by Shota Rustaveli National Science Foundation of Georgia under the project no. YS-19-367.

We would like to thank Matthias Baaz for his useful remarks about the work.

References

1. Russell, S.J., Norvig, P.: Artificial Intelligence: A Modern Approach. Pearson Education Limited (2016)
2. Brewster, C., O'Hara, K.: Knowledge representation with ontologies: present challenges—future possibilities. Int. J. Hum Comput. Stud. **65**(7), 563–568 (2007)
3. Stephan, G., Pascal, H., Andreas, A.: Knowledge representation and ontologies. In: Semantic Web Services: Concepts, Technologies, and Applications, pp. 51–105 (2007)
4. Berners-Lee, T., Hendler, J., Lassila, O.: The semantic web. Sci. Am. **284**(5), 34–43 (2001)
5. Shadbolt, N., Berners-Lee, T., Hall, W.: The semantic web revisited. IEEE Intell. Syst. **21**(3), 96–101 (2006)
6. Antoniou, G., Van Harmelen, F.: Web ontology language: Owl. In: Handbook on Ontologies, pp. 67–92. Springer (2004)
7. Janjua, N.K., Hussain, F.K.: Development of a logic layer in the semantic web: research issues. In: 2010 Sixth International Conference on Semantics, Knowledge and Grids, pp. 367–370. IEEE (2010)
8. Maedche, A.: Ontology Learning for the Semantic Web, vol. 665. Springer Science & Business Media (2012)
9. Sowa, J.F.: Knowledge Representation: Logical, Philosophical and Computational Foundations. Brooks/Cole Publishing Co. (1999)
10. Benson, T.: Principles of Health Interoperability HL7 and SNOMED. Springer, London (2010)
11. Calegari, S., Ciucci, D.: Fuzzy ontology, fuzzy description logics and fuzzy-owl. In: International Workshop on Fuzzy Logic and Applications, pp. 118–126. Springer (2007)
12. Bobillo, F., Straccia, U.: Fuzzy ontology representation using owl 2. Int. J. Approximate Reasoning **52**(7), 1073–1094 (2011)
13. Gao, M., Liu, C.: Extending owl by fuzzy description logic. In: 17th IEEE International Conference on Tools with Artificial Intelligence (ICTAI'05), pp. 562–567. IEEE (2005)
14. Calegari, S., Ciucci, D.: Integrating fuzzy logic in ontologies. ICEIS **2**, 66–73 (2006)
15. Baziz, M., Boughanem, M., Loiseau, Y., Prade, H.: Fuzzy logic and ontology-based information retrieval. In: Fuzzy Logic, pp. 193–218. Springer (2007)
16. Parry, D.: A fuzzy ontology for medical document retrieval. In: Proceedings of the Second Workshop on Australasian Information Security, Data Mining and Web Intelligence, and Software Internationalisation, vol. 32, pp. 121–126. Australian Computer Society, Inc.
17. Lee, C.S., Jian, Z.W., Huang, L.K.: A fuzzy ontology and its application to news summarization. IEEE Trans. Syst. Man Cybern. Part B (Cybernetics) **35**(5), 859–880 (2005)
18. Lee, C.S., Wang, M.H., Hagras, H.: A type-2 fuzzy ontology and its application to personal diabetic-diet recommendation. IEEE Trans. Fuzzy Syst. **18**(2), 374–395 (2010)
19. Zadeh, L.A.: Fuzzy sets. Inf. Control **8**(3), 338–353 (1965)
20. Sanchez, E.: Fuzzy Logic and the Semantic Web. Elsevier (2006)
21. Hájek, P.: Metamathematics of Fuzzy Logic, vol. 4. Springer Science & Business Media (2013)
22. Straccia, U.: The quest for fuzzy logic in semantic web languages. In: Foundations of Fuzzy Logic and Semantic Web Languages (Open Access), pp. 1–8. Chapman and Hall/CRC (2016)
23. Coelho, J., Dundua, B., Florido, M., Kutsia, T.: A rule-based approach to xml processing and web reasoning. In: International Conference on Web Reasoning and Rule Systems, pp. 164–172. Springer (2010)
24. ISO/IEC 24707: Information Technology—Common Logic (CL): A Framework for a Family of Logic-Based Languages. International Organization for Standardization (ISO) Geneva, Switzerland (2007)
25. Mossakowski, T., Lange, C., Kutz, O.: Three semantics for the core of the distributed ontology language. In: Proceedings of the Twenty-Third International Joint Conference on Artificial Intelligence, pp. 3027–3031. AAAI Press (2013)
26. Imran, M., Young, B.: The application of common logic based formal ontologies to assembly knowledge sharing. J. Intell. Manuf. **26**(1), 139–158 (2015)

27. Bishara, A., Rukhaia, M.: Towards unranked fuzzy theory. In: IEEE 2nd International Conference on System Analysis & Intelligent Computing (SAIC), pp. 1–4. IEEE Xplore Digital Library (2020)
28. Kutsia, T., Buchberger, B.: Predicate logic with sequence variables and sequence function symbols. In: International Conference on Mathematical Knowledge Management, pp. 205–219. Springer (2004)
29. Dundua, B., Kurtanidze, L., Rukhaia, M.: Unranked tableaux calculus for web related applications. In: 2017 IEEE First Ukraine Conference on Electrical and Computer Engineering (UKRCON), pp. 1181–1184. IEEE (2017)
30. Cintula, P.: The ŁΠ and ŁΠ$\frac{1}{2}$ propositional and predicate logics. Fuzzy Sets Syst. **124**(3), 289–302 (2001)
31. Esteva, F., Godo, L.: Putting together Lukasiewicz and product logics. Mathware Soft Comput. **6**(2–3), 219–234 (1999)
32. Hájek, P., Cintula, P.: On theories and models in fuzzy predicate logics. J. Symbolic Logic **71**(3), 863–880 (2006)
33. Schaffert, F.S.: Xcerpt: A Rule-Based Query and Transformation Language for the Web. Ph.D. thesis, Ludwig-Maximilians-Universität München (2004)
34. Hähnle, R.: Automated Deduction in Multiple-valued Logics. Clarendon Press, Oxford (1993)
35. Olivetti, N.: Tableaux for lukasiewicz infinite-valued logic. Stud. Logica. **73**(1), 81–111 (2003)
36. Kutsia, T., Marin, M.: Solving, reasoning, and programming in common logic. In: 2012 14th International Symposium on Symbolic and Numeric Algorithms for Scientific Computing, pp. 119–126. IEEE (2012)
37. Badros, G.J., Borning, A., Stuckey, P.J.: The cassowary linear arithmetic constraint solving algorithm. ACM Trans. Comput.-Human Interact. (TOCHI) **8**(4), 267–306 (2001)
38. Dutertre, B., De Moura, L.: A fast linear-arithmetic solver for DPLL(T). In: International Conference on Computer Aided Verification, pp. 81–94. Springer (2006)
39. Hristakiev, I., Plump, D.: A unification algorithm for GP 2. Electron. Commun. EASST **71** (2015)

Chapter 18
Fractal Analysis and Its Applications in Urban Environment

Alexey Malishevsky

Abstract Fractal sets have been known for more than a century. However, only in 1975, Mandelbrot gave them the name "fractal" and mathematically defined them as sets whose Hausdorff dimension exceeds the topological dimension. While initially fractals were a pure mathematical phenomenon, afterwards fractal-like properties have been discovered in many natural and artificial objects and processes. The recently developed fractal analysis and theory of fractals have been applied in many different areas, including biology, health care, urban planning, environmental studies, geology, geography, chemistry, ecology, astronomy, computer science, social science, music, literature, art to name a few. This chapter gives a literature survey of the fractal theory and analysis applications to solve different problems in various areas. We describe main concepts and frequently used methods to compute a fractal dimension. Finally, we apply the fractal analysis to study the geography and infrastructure of Kyiv and facilitate decision making in urban planning.

Keywords Fractal · Fractal analysis · Fractal dimension · Hausdorff dimension · Information dimension · Correlation dimension · Ruler method · Box counting method · Power law · Multifractal · Geography · Street network · Parks · Road tunnels · Urban planning · Decision making

1 Introduction

The name "fractal" was given by Mandelbrot to structures that were composed of parts which were similar to the whole. Fractal is "a rough or fragmented geometrical object, which can be partitioned into parts, each of them is (at least approximately) reduced copy of the whole" [1]. Fractals are usually self similar and do not depend on scale. Mandelbrot mathematically defined fractals as sets whose Hausdorff dimension exceeds the topological dimension.

Several examples of fractals such as Menger sponge, Nova fractal, Mandelbrot set, Koch snowflake, Barnsley fern, and fractal tree are shown in Fig. 1. Nature itself

A. Malishevsky (✉)
Institute for Applied System Analysis, Igor Sikorsky Kyiv Polytechnic Institute, Kyiv, Ukraine

M. Zgurovsky and N. Pankratova (eds.), *System Analysis & Intelligent Computing*, Studies in Computational Intelligence 1022, https://doi.org/10.1007/978-3-030-94910-5_18

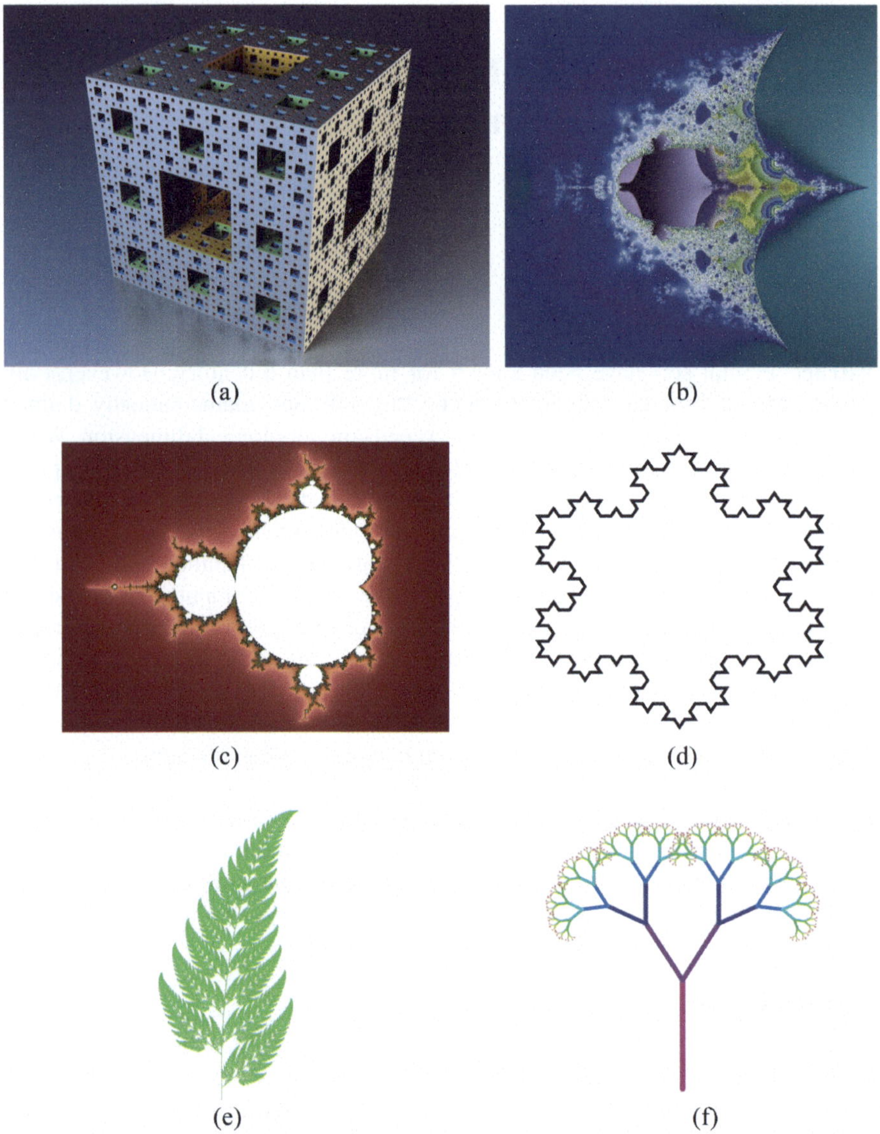

Fig. 1 Examples of fractals: **a** Menger sponge [2], **b** Nova fractal, **c** Mandelbrot set, **d** Koch snowflake (7 iterations), **e** Barnsley fern, **f** fractal tree

has many fractals-like objects including geological fractures, shore lines, mountains, various boundaries, trees, lungs, blood capillaries, particle paths. Along with geometrical objects, there are many processes that possess fractal properties such as spectra of signals from nonlinear complex systems (including the heart and respiratory rates), spectra of nuclear magnetic resonance signals, etc. A fractal analysis can be applied to different objects including actual physical objects and processes that possess fractal characteristics.

The fractal analysis computes the complexity of an object which can be anything including a dataset. The fractal analysis mostly uses the fractal dimension to evaluate the complexity. In general, the dataset can include almost anything such as processes, natural phenomena, signals, images, behavioral patterns, cultural works, geological data, living organisms, etc. For physical objects, the dataset may include their shape, movement, development, signals, or even the behavior. Given a fractal set, its dimension can be defined in many different ways. To calculate the generalized fractal dimension D_q, the data are placed into n-dimensional lattice with grid size r and the frequency p_i is calculated with which data points are placed in i-th cell (Eq. 1).

$$D_q = \frac{1}{q-1} \lim_{r \to 0} \frac{\log \sum_i p_i^q}{\log r} \tag{1}$$

Hausdorff fractal dimension (q = 0), information dimension ($\lim_{q \to 1} D_q$), and correlation dimension (q = 2) are frequently used. Information and correlation dimensions are often utilized by data mining [3, 4].

If a system under study has complex dynamics, a single exponent (the fractal dimension) may not be sufficient to describe it. Such systems include the brain activity, heartbeat dynamics, meteorology, turbulence, etc. In such cases, instead of a single fractal dimension, a generalization of a fractal system—a multifractal system with a continuous spectrum of exponents is employed [5].

This chapter is organized as the following: in Sect. 2, we will talk about applications of fractal analysis in various areas including biology, geology, health care, chemistry, computer science, urban studies; in Sect. 3, we will describe methods for computing the fractal dimension; in Sect. 4, we present three case studies of applying fractal analysis to geography and infrastructure of Kyiv city along with evaluation of alternatives for tunnels construction. In Sect. 5, we present the conclusions.

2 Applications of Fractal Analysis

In recent years, fractal analysis methods have been gaining popularity. They are used in various areas to solve different problems in health care, ecology, urban research, astronomy, computer science, chemistry, music, literature, arts, and so on. The fractal analysis usually measures the complexity of an object which is related to certain object's properties of interest or is connected with properties of another object or system that affects this object. For example, by analyzing the spectrum complexity of heartbeat intervals, the state of body's regulatory system may be inferred. Alternatively, by measuring the capillary network complexity, the destructive effects of diabetes can be inferred. With the fractal analysis, the changes in the complexity can be also tracked over time to infer how underlying processes affect it. For example, the complexity of geological faults changes due to erosion. Similarly, the complexity of urban areas changes over time as the city develops. In addition, the complexity can be used as a metric on an object for classification purposes.

The most prominent and important application of fractal analysis is in health care. The majority of biological systems possess fractal properties. These include not only geometrical shape of objects (for instance, lungs and blood vessels), but also their hidden properties (for instance, heart rate spectrum). Fractal nature of objects and their complexity are interconnected and frequently offer an insight on objects' useful properties. For example, if we consider the geometry of lungs, their high fractal dimension depicts the high surface area where the gas exchange occurs. In similar fashion, if we consider a biological process, the high fractal dimension of its spectrum reveals the high complexity of a process itself and consequently unveils the high complexity of a regulatory control. Diseases can affect not only the geometrical shape of organs, but also the complexity of biological processes, thus making the fractal analysis an invaluable tool for early detection of health problems.

Researchers have studied processes have fractal properties. For example, Suki et al. studied the process of the peripheral pathways opening in lungs [6]. This process conforms to the power law and can be described by the fractal lung model. Opening times of peripheral pathways are influenced by lung diseases. The authors constructed model that allowed to precisely describe such processes. In their work, the air resistance is measured which decreases stepwise by jumps. The probability distributions of both the sizes of such jumps and time between them conform to the power law.

Alzheimer's disease causes the degradation of the brain which was studied using fractal analysis by Warsi [7]. To accomplish this, the author analyzed the fractal dimension of signals, obtained using functional magnetic resonance (fMRI) [7]. It was shown that the disease could be detected at early stages using fMRI signals' complexity.

Stanley et al. employed the fractal analysis to ECG signals to identify heart problems. Heart problems affect the fractal dimension of the signal spectrum [8]. The dynamics of complex physiological oscillations were studied by focusing on the variability of heart beat intervals to determine non homeostatic physiological variability. Being under direct neuroautonomous control, the heart rate dynamics can reflect the state of the complex regulatory system of a patient. The fractal analysis can detect changes in signal's complexity which may point to failing neuroautonomous control that in turn predicts the progression of a disease.

In the area of clinical echocardiography, Captur et al. utilized the fractal analysis to compute the fractal dimension for transthoracic echocardiographic images [9]. In a biological organism, complex nonlinear regulatory systems result in fractal characteristics of a signal. The reduction in the fractal component reflects the failure of such regulatory systems due to the illness. Fractal properties can be considered not only for signals, but also for geometrical shapes of certain organs. When organs grow, failing regulatory systems can affect their geometry. Consequently, failing regulatory systems inhibit fractal characteristics of organs. Thus, the regulatory systems state is exposed by the fractal analysis.

In addition to exposing the underlying condition affecting an organ during its development or functioning, the fractal analysis can uncover the degradation of an organ over time. Uahabi and Atounti used images of eye's retina to calculate the fractal dimension of blood vessels for the diagnosis of diabetes [10]. The decrease in the fractal dimension of the blood vessels network is caused by the damage from diabetes.

The fractal analysis has been applied in ecology. In [11], the authors study the geography of the animal population distribution along with the time series of population numbers [11]. Both the geography and time series have fractal properties. By considering the fractal dimension of an area occupied by a population, the authors discovered that the corresponding fractal dimension was small for healthy populations and large for dying populations. Furthermore, the local extinction of the species can be predicted by the fractal model using time series for an animal population.

The fractal analysis can be successfully applied to geophysical networks. In particular, Lovejoy et al. considered meteorological networks while evaluating their empirical dimension and comparing it with the dimension of space these networks tried to cover [12]. The main idea is to compare an intrinsic dimension of a sample to the dimension of the space this sample was taken from. Their difference imposes certain constraints on detectable phenomena. Thus, fractal models can be used to determine for a given sample whether it is adequate in detecting events of a given dimension. For example, such a sample can be a monitoring stations network.

The fractal analysis has also found its place in astronomy. While studying scaled distribution of galaxies in the observable universe, Ribeiro and Miguelote devised the fractal model to describe such a distribution [13].

The fractal analysis of cities attracts a lot of research. To name a few, Shen considered urbanized areas and their growth by calculating three types of fractal dimensions of cities based on the size, shape and scale of a city [14]. In another work, Giovanni and Caglioni performed the fractal analysis of Milan including its perimeter and urbanized areas [15]. To study the city perimeter, it was extracted using the method of the dilatation and its fractal dimension was calculated using the correlation analysis. On the other hand, to study the whole urbanized areas, the correlation, dilatation, and grid analyses were used. In still another work, Beijing and Hangzhou were studied and the relationship between the special entropy and fractal dimension in urban environments was considered [16]. Three types of fractal dimensions were computed which combined with the entropy could characterize the spatial complexity of cities.

Another area of fractal analysis application is chemistry. In [17], Bao et al. developed fractal models in order to describe the dissolution of particles. In their paper, fractal models were developed for particle population. These models were based on two following principal properties of the particle surface: its chemical reactivity and its fractal nature. The fractal dimension of the surface and the fractal dimension of the chemical reactivity are compared with each other. A lot of works have been published regarding the adsorption. For example, Kaneko et al. used three methods to determine the fractal dimension of a surface (microporous carbon fibre). The first method uses adsorption data on organic molecules of different molecular sizes. The second method uses nitrogen adsorption isotherms combined with Avnir–Jaroniec theory. Finally the third method uses small-angle X-ray scattering. The authors discovered that the fractal dimension could be used to estimate the effective surface for the adsorption [18].

The area which received a lot of attention is the application of the fractal analysis in data mining. Sajjipanon and Ratanamahatana suggested the fractal representation of time series data using the compass dimension, modified compass dimension, and correlation dimension [19]. Such a representation of a time series using just three real values was sufficient. It not only simplified the time series comparison, but also it speeded it up. In addition, based on these metrics, the clustering approach was suggested. Other research uses the fractal dimension of a data set to reduce the number of attributes in it. For example, as a good approximation to the intrinsic dimension of a data set, Traina et al. employed data set's "fractal" parameter [20]. Thus, for a given set of n-dimensional vectors, this parameter was used for fast selection of the most important attributes. Here, the attributes that only weakly influenced the fractal dimension of the data were discarded. In similar fashion, Barbara and Chen proposed the fractal clustering method using the fractal dimension of a data set [4]. Likewise, Tasoulis and Vrahatis also suggested clustering using the fractal dimension [21].

Another promising area where the fractal analysis has been used is geology. For example, Wilson et al. evaluated numerical geologic variables using the fractal properties. Such variables included fracture patterns, reflection travel times, structural relief, drainage, topographic relief, and active fault patterns [22]. The authors inferred the fractal properties of structural relief using seismic data and structural cross sections. These fractal properties in turn allowed characterizing and comparing complex structural patterns. Aviles et al. with the help of the fractal analysis performed the quantitative analysis of the San Andreas Fault geometry [23]. In addition, Chang et al. studied geological faults created as a result of an earthquake [24]. The authors computed the fractal dimension for different parts of geological faults. The fractal dimension depicts the complexity of an object. Because the erosion processes smooth an object out, its complexity decreases with age which consequently decreases the fractal dimension. Therefore, the fractal dimension of an object correlates well with object's age.

Substantial research has been done in applying the fractal analysis to cultural works [25], including music, paintings, literature, fashion, cultural networking processes [26], and social processes. The fractal analysis has been applied to music, including Indian songs classification based on their fractal dimension by converting each song into a set of time series and calculating its fractal dimension [27], performers classification who played the same melodies with multifractal analysis while calculating Holder exponents α for music samples and Hausdorff dimensions for their distributions [28], complexity evaluation of music compositions by representing each music piece as a time series of pitch variations and then measuring the global and local dynamics by utilizing phase space portraits, spectral analysis, and information entropy [29].

3 Computing Fractal Dimension

To compute the approximation of the fractal dimension, Walker's Ruler and box counting methods are the most frequently used [30].

3.1 *Box Counting Method*

Box counting method is based on the concept of "coverage" and can be used with various definitions of a dimension. The advantages of the box counting method include generality, high power, and the ease of the implementation. However, its main disadvantage is that this method may not be suitable for complicated contours or boundaries. Let us describe the box counting method. Let $N(\eta)$ be the number of objects with the linear dimension greater then η needed to cover a given set (or an object) under investigation, then $N(\eta) \sim \eta^{-D}$. Or it can be expressed alternatively as $\log(N(\eta)) = a - D \log(\eta)$, where a is some constant. In order to compute D, an object is covered with the grid of squares (or by n-dimensional hypercubes in an n-dimensional case) for several different values of size η. We start from size η_1, then the number $N(\eta_1)$ of squares (hypercubes) containing a part of an object is determined. Then, size η_2 is used, obtaining $N(\eta_2)$ squares (hypercubes). This step is repeated S times, while using decreasing sizes. At the end, a regression line is fit between independent variable $\log(\eta_i)$ and dependent variable $\log(N(\eta_i))$, where $i = 1, \ldots, S$. An absolute value of the slope of a regression line determines the fractal dimension D [30].

3.2 *Walker's Ruler Method*

The Walker's ruler method was initially used by Richardson [31] and Mandelbrot [32]. The advantage of this method is that it works well for complicated boundaries. On the other hand, this method is less general than the box counting method. It is also difficult to implement. This method can be used only if a line or a boundary is connected. The method uses a ruler of varying length and works as follows. Ruler's size starts from some x, its one end is fixed to the beginning of the line while rotating the ruler n_1 times until reaching the end of the line being measured. Thus, the first value L is calculated as $n_1{*}x$. In the next step, ruler's length is set to $x/2$ and, in similar matter, it is rotated n_2 times. This produces value $L = n_2 * x/2$. The same process is repeated M times with progressively smaller rulers. This process generates M pairs of $L(\eta_i)$ and η_i. These values are related to the power law $L(\eta_i) \sim \eta_i^{-D}$ that demonstrates the growth of L with decreasing η. If we apply logarithm to each side of the equation, we obtain: $\log(L(\eta)) = a + (1 - D)\log(\eta)$, where a is some constant. Finally, the regression line is fit between a dependent variable $\log(L(\eta_i))$ and an independent variable $\log(\eta_i)$, where $i = 1, \ldots, M$ and M is the number of different values of η. The fractal dimension D is computed as one minus the regression line's slope [30].

4 Case Studies

In our studies, we apply fractal analysis to the geography and infrastructure of Kyiv to facilitate decision making in urban planning.

To approximate Hausdorff's dimension, we used the box counting method. In the box counting method, we started with the grid of the largest cell size s and counted the number of cells $N(s)$ that cover all objects. Then, we reduced the grid size and counted the number of cells covering objects. We repeated the process several times. By plotting $\log(N(s))$ versus $\log(s)$ and fitting the regression line along the linear part of the curve, we calculated the fractal dimension D as the absolute value of the slope of this line. The grid size ranged between 3 and 400–500 pixels.

4.1 Study 1: The Parks of Kyiv

In this study, we consider a map of Kyiv where we compute the fractal dimension of parks. Substantial research has been done in exploring the urban development of cities including New York, Omaha, Baltimore [14], Milan [15], Beijing and Hangzhou [16]. The fractal dimension of urban areas was computed. Some studies included its changes over time. For example, the fractal dimension for Beijing varied between 1.85 and 1.96 depending on the year [16], the fractal dimension for New York was 1.70 [14], and the fractal dimension for Omaha was 1.28 [14].

We instead considered park areas. As an example, we used Kyiv city maps (downloaded from http://www.kiev-maps.com/images/maps/Kiev_auto_map_ukraine.jpg). We extracted park areas based on color and applied the box counting method to determine the fractal dimensions of parks. We also determined borders of those parks and computed their fractal dimension.

The original image was 5176×5176 pixels. After preprocessing, we started from grid size of 400 pixels and iterated until each cell held no more then 3×3 pixels. We tried both, exponential and linear scales for grid sizes which gave essentially the same results. The fractal dimension $D = 1.62$ was computed by fitting the regression line and finding its slope. Figure 2 shows the original maps, extracted park areas, extracted park borders, and the curve of $\log(N(s))$ versus $\log(s)$, where s is the grid size and $N(s)$ is the number of boxes covering parks and park borders. For the borders of parks, we used the same image and the same process, except that we first detected edges of parks. The fractal dimension was determined to be $D = 1.38$.

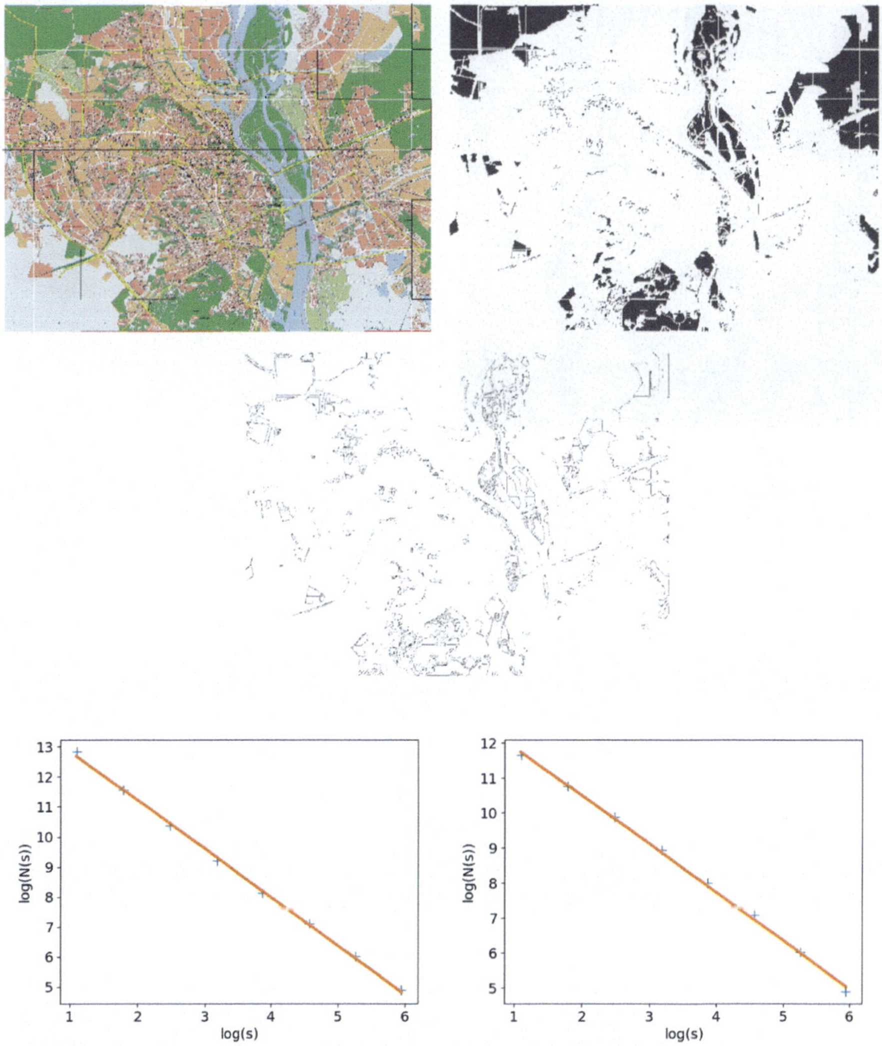

Fig. 2 Kyiv parks: original map (top left), parks (top right), park borders (middle), log(N(s)) versus log(s) plot for park areas with regression line from the box counting method (bottom left), log(N(s)) versus log(s) plot for park borders with regression line from the box counting method (bottom right)

4.2 Study 2: Kyiv Street Networks

In the second study, we considered the Kyiv street network (downloaded from https://misto.lun.ua/).

We manually cleaned the image to keep only the street network and extracted streets based on color. The box counting method was applied to determine the fractal dimension of the street network.

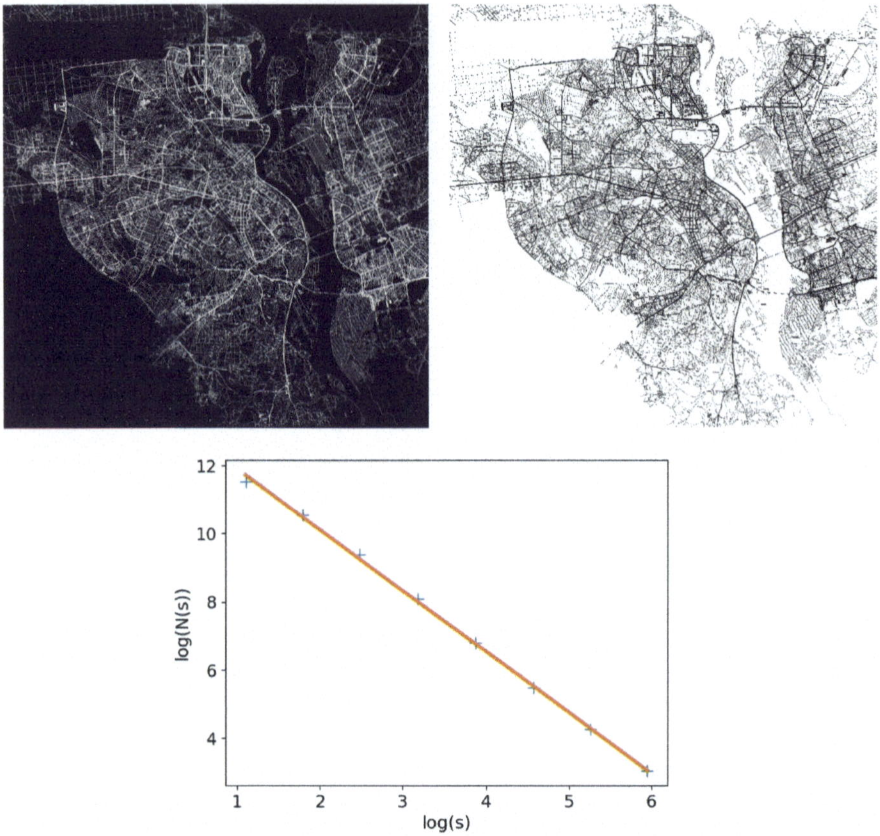

Fig. 3 Kyiv street network: original map (top left), processed map (top right), log(N(s)) versus log(s) plot with regression line from the box counting method (bottom)

The original image was 1600 × 1600 image of Kyiv. We started from grid size of 400 pixels and iterated until each cell held no more then 3 × 3 pixels using an exponential scale for grid sizes. The fractal dimension $D = 1.79$ was computed by fitting the regression line. Figure 3 shows original and processed maps along with the curve of $\log(N(s))$ versus $\log(s)$, where s is the grid size and $N(s)$ is the number of boxes covering streets.

4.3 Study 3: Urban Infrastructure Planning

In the third study, we demonstrated the usage of the fractal analysis in decision making. When developing urban infrastructure, many factors need to be considered in terms of uncertainty and risk in order to decide on the most appropriate development option that may be associated with the choice of the object type or location. In this case, it is necessary to calculate an objective assessment of each option, which should take into account the risks, cost, complexity of an option, required resources, etc. One of the important components of evaluating an alternative is the complexity of both the object itself and its location. The complexity can have both negative and positive consequences. The complexity of buildings and road network is detrimental for surface objects, but beneficial for underground infrastructure, such as tunnels, which can "unload" the surface. Thus, it is advisable to develop underground infrastructure in areas with high complexity above-ground. This is especially true for the transport infrastructure, where new construction or reconstruction due to the high complexity of the area can face many problems. Therefore, adapted fractal analysis methods can be used to support decision making.

We have studied eight location alternatives for building road tunnels in the city of Kyiv that included three road tunnels under the Dnieper river. For each corresponding location, we calculated its complexity in terms of buildings while using the fractal dimension as the complexity metric. The studied objects are presented in Table 1.

To evaluate the complexity of areas where tunnels were proposed, the following approach was used: high resolution Google Maps were used (level 17 which included buildings), corresponding locations were selected, and images were processed to select objects for which the complexity was calculated. Figures 4, 5, 6, 7, 8, 9, 10 and 11 show maps for areas above tunnels before and after image processing along with plots produced by the box counting method (with the regression line). Their complexity was computed using the fractal analysis. In the box counting method, grid sizes from 3 to 400 pixels were used. After building the regression line for each case, we obtained the following results (see Table 2).

Table 1 Studied objects in urban infrastructure planning

No.	Object
1	Tunnel №1—the road tunnel under the Botanical Garden from Naddnipryans'ke Highway to Saperno-Slobidska street with length of 1.5 km
2	Tunnel №2—the road tunnel from Frunze street to Chornovola street with length of 1.2 km
3	Tunnel №3—the north road tunnel across the Dnieper River with length of 4.3 km
4	Tunnel №4—the south road tunnel across the Dnieper River with length of 4.4 km
5	Tunnel №5—the tunnel across the Dnieper River from Peremohy square to Brovarskyi avenue with length of 7.0 km
6	Tunnel №6—the road tunnel from Naberezhno-Luhova street to Bohatyrska street with length of 1.8 km
7	Tunnel №7—the road tunnel from Frunze street to Kirovohradska street with length of 11.1 km
8	Tunnel №8—the road tunnel from the extension of Saksaganskoho and Zhylyanska streets to Naberezhne highway with length of 3.0 km

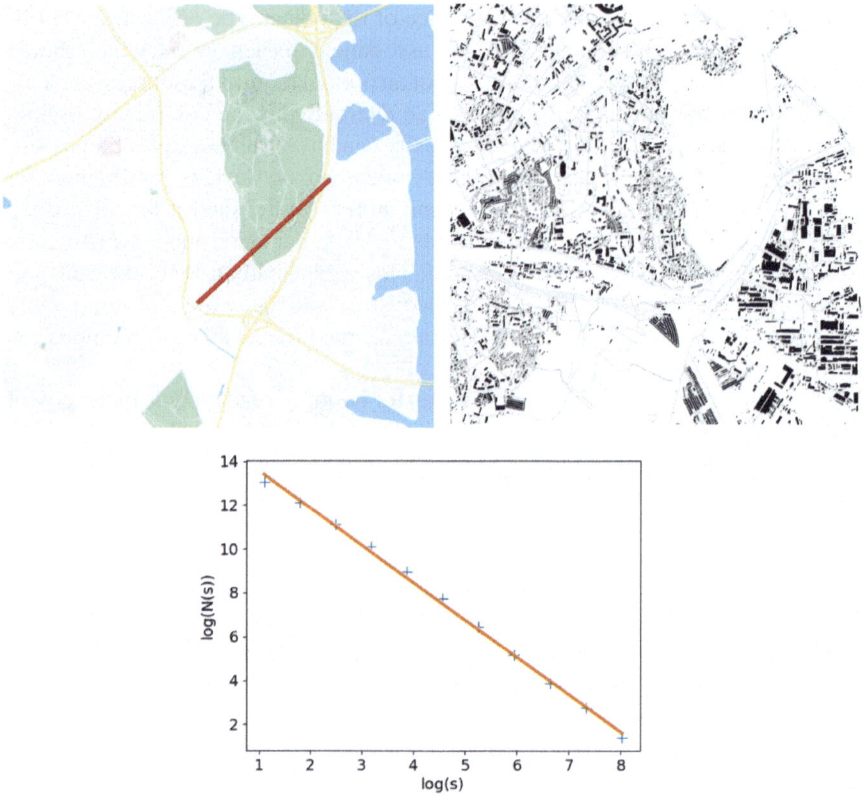

Fig. 4 Tunnel №1 (the map of the area where the tunnel is located (top left), the map after processing (top right), plot log(N(s)) versus log(s) from box counting method with the regression line (bottom))

The highest complexity of buildings and the road network is above tunnels №2 (D = 1.79), №7 (D = 1.81), and №8 (D = 1.82). On the other hand, the smallest complexity of buildings and the road network is above tunnel №3 (D = 1.48).

There are several conclusions that we can make. First, if the task is to decide on the choice of technology for the implementation of the object, the high complexity of above-ground infrastructure (buildings, roads) favors choosing an underground implementation. On the other hand, when the complexity of above-ground infrastructure is low, there is no longer a strong need to choose an expensive underground implementation: in this case, we can choose a cheaper alternative, such as building an overpass or highway. Second, if the task is to choose underground tunnels for the implementation of highways (in other words, to prioritize projects), then one of the factors influencing the choice of an alternative may also be the complexity of above-ground infrastructure. The greater this complexity is, the more important it is to choose this alternative, because first of all it is necessary to solve transportation problems in places where there is a great complexity of the existing infrastructure.

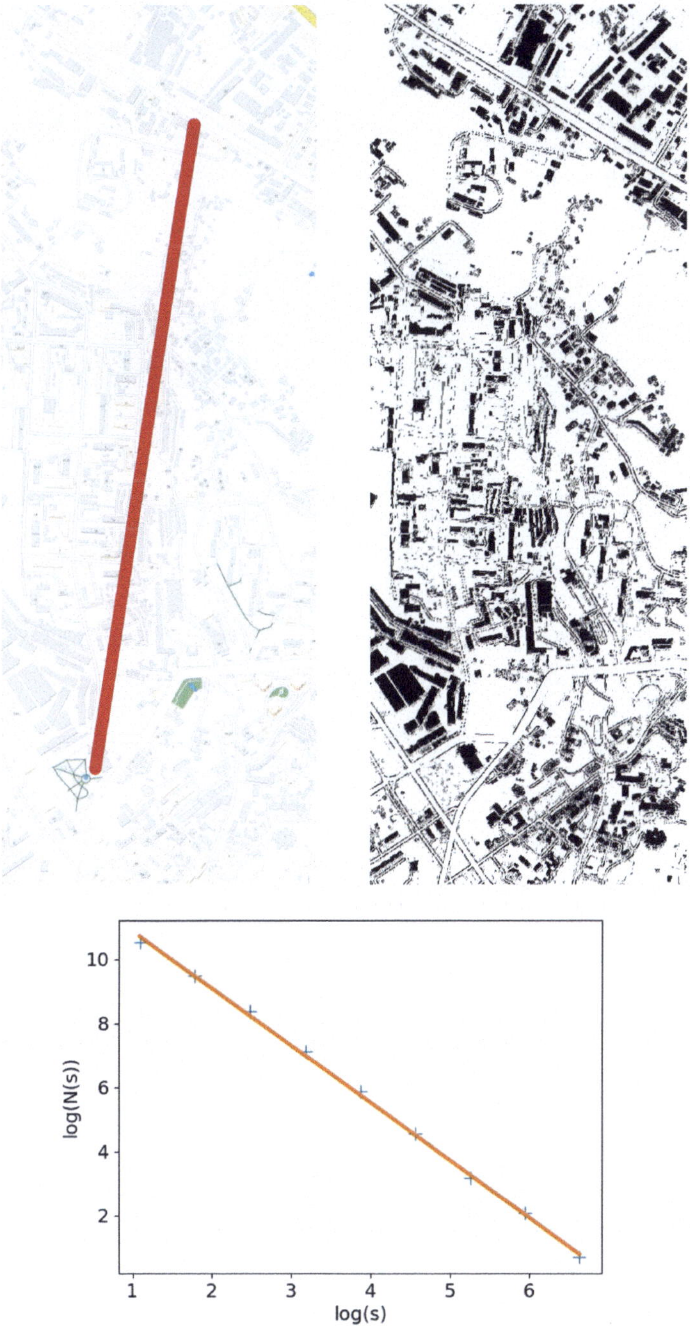

Fig. 5 Tunnel №2 (the map of the area where the tunnel is located (top left), the map after processing (top right), plot log(N(s)) versus log(s) from box counting method with the regression line (bottom))

Fig. 6 Tunnel №3 (the map of the area where the tunnel is located (top), the map after processing (bottom), plot log(N(s)) versus log(s) from box counting method with the regression line (bottom))

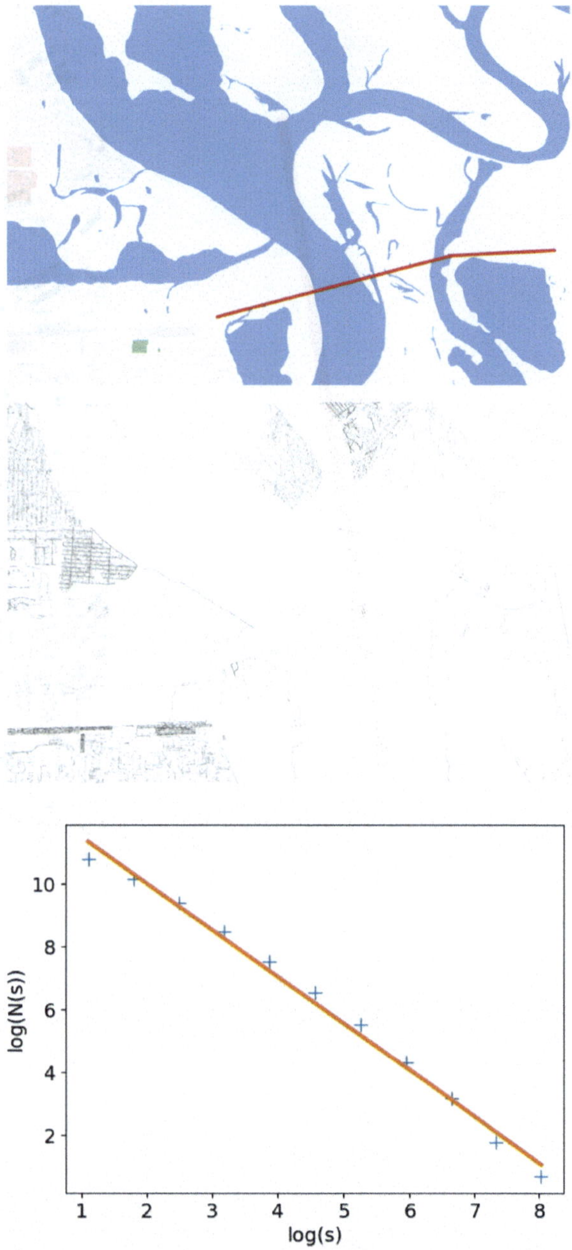

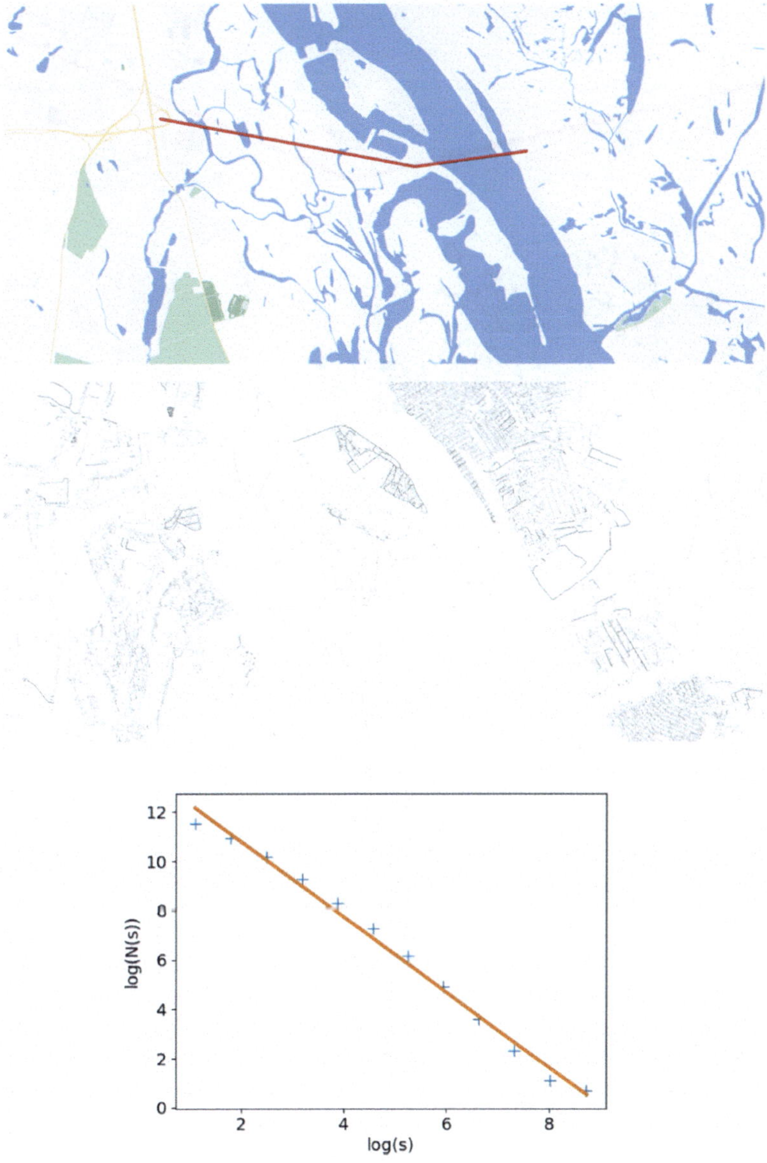

Fig. 7 Tunnel №4 (the map of the area where the tunnel is located (top), the map after processing (middle), plot log(N(s)) versus log(s) from box counting method with the regression line (bottom))

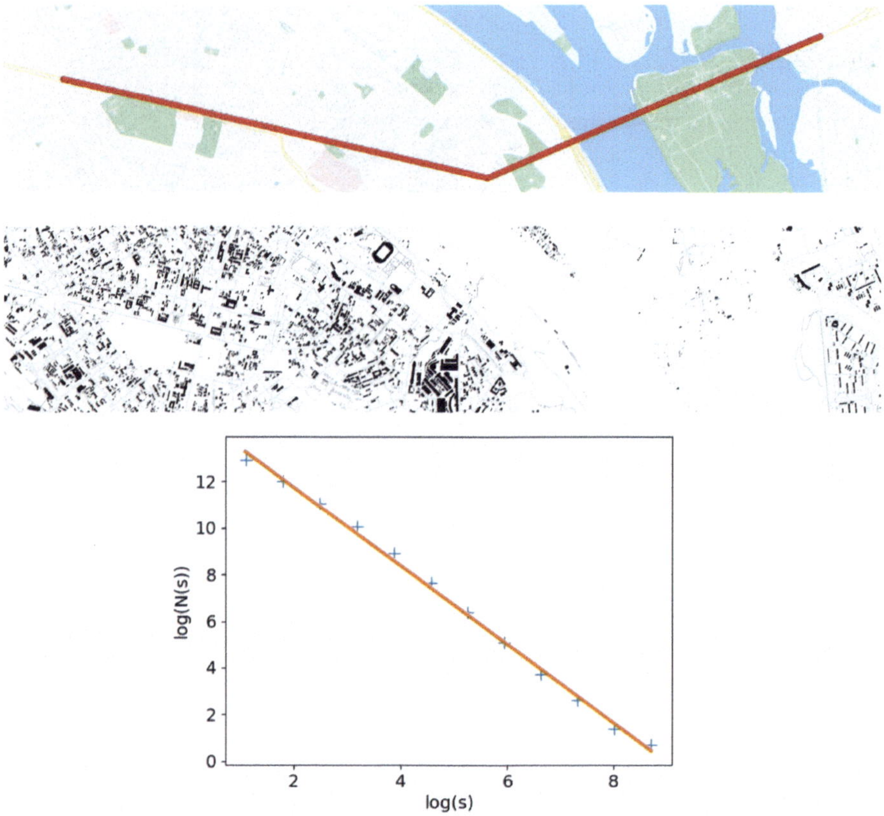

Fig. 8 Tunnel №5 (the map of the area where the tunnel is located (top), the map after processing (middle), plot log(N(s)) versus log(s) from box counting method with the regression line (bottom))

Thus, in the considered examples, we believe that from the point of view of the complexity that we have calculated, tunnels №2, №7 and №8 have a high priority for implementation compared to other tunnels. Regarding the choice of the technology, in terms of the calculated complexity, for locations above proposed tunnels №1, №2, №5, №6, №7, and №8, underground tunnels are the most beneficial, and for objects №3 and №4 above-ground alternatives can be considered.

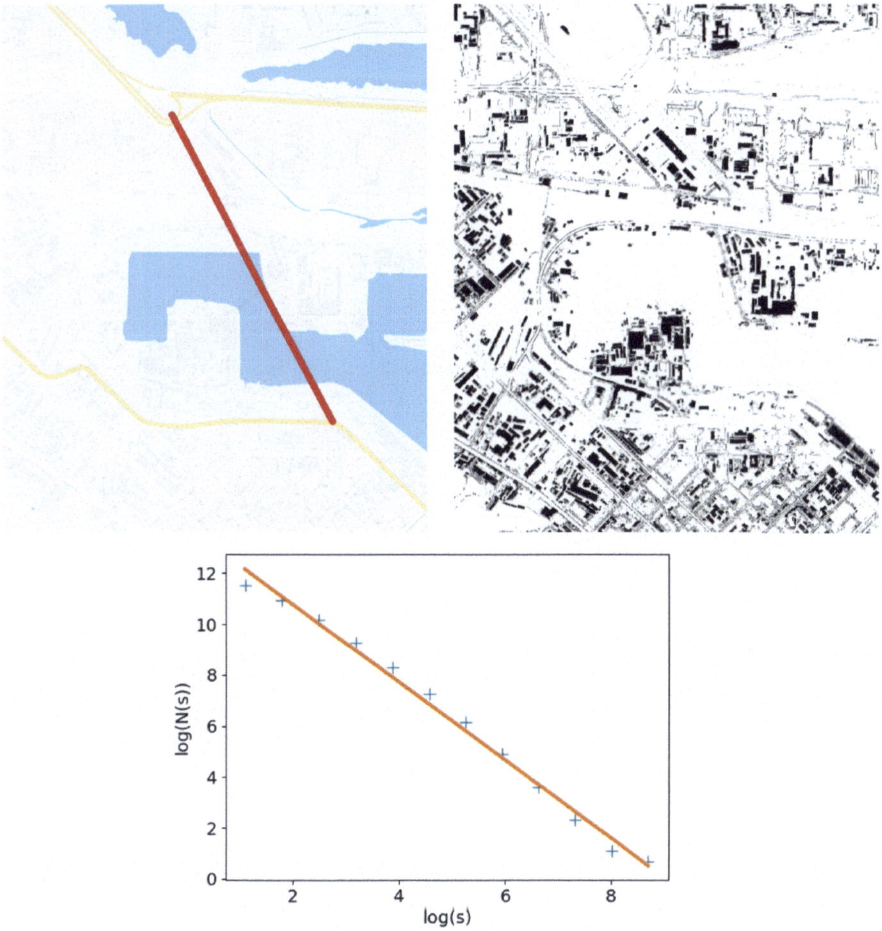

Fig. 9 Tunnel №6 (the map of the area where the tunnel is located (top left), the map after processing (top right), plot log(N(s)) versus log(s) from box counting method with the regression line (bottom))

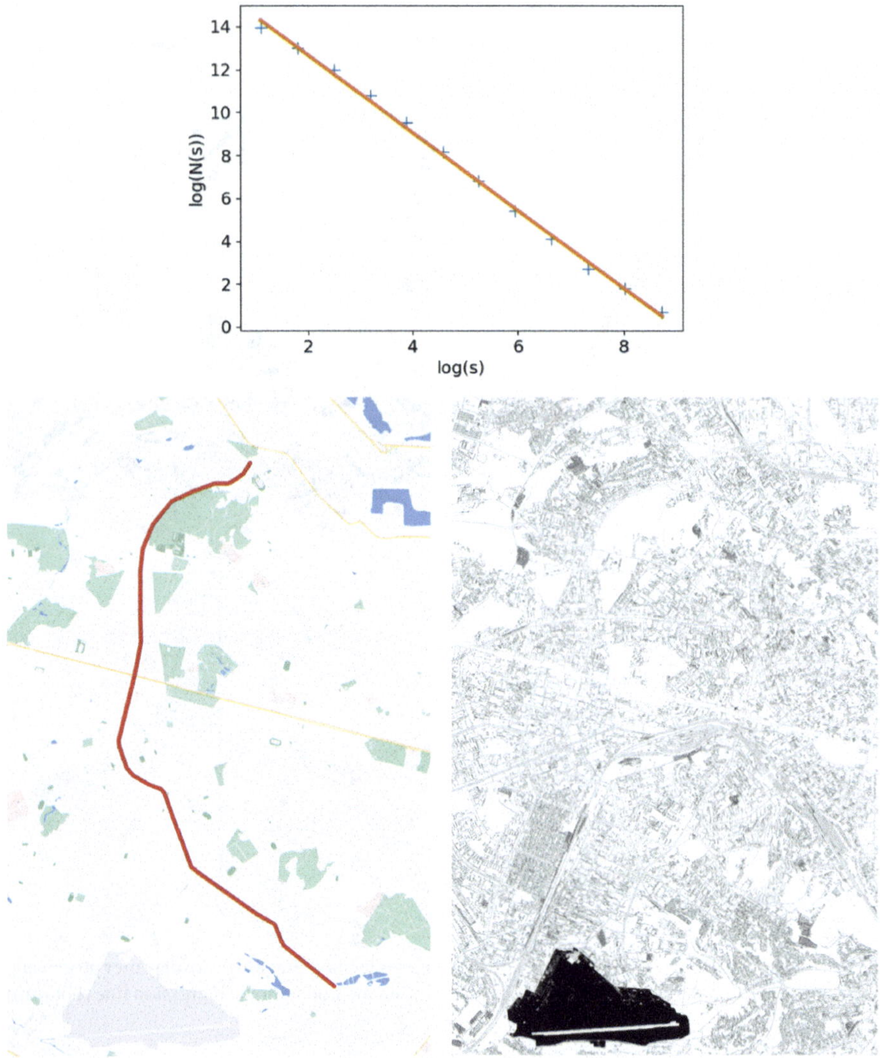

Fig. 10 Tunnel №7 (the map of the area where the tunnel is located (top left), the map after processing (top right), plot log(N(s)) versus log(s) from box counting method with the regression line (bottom))

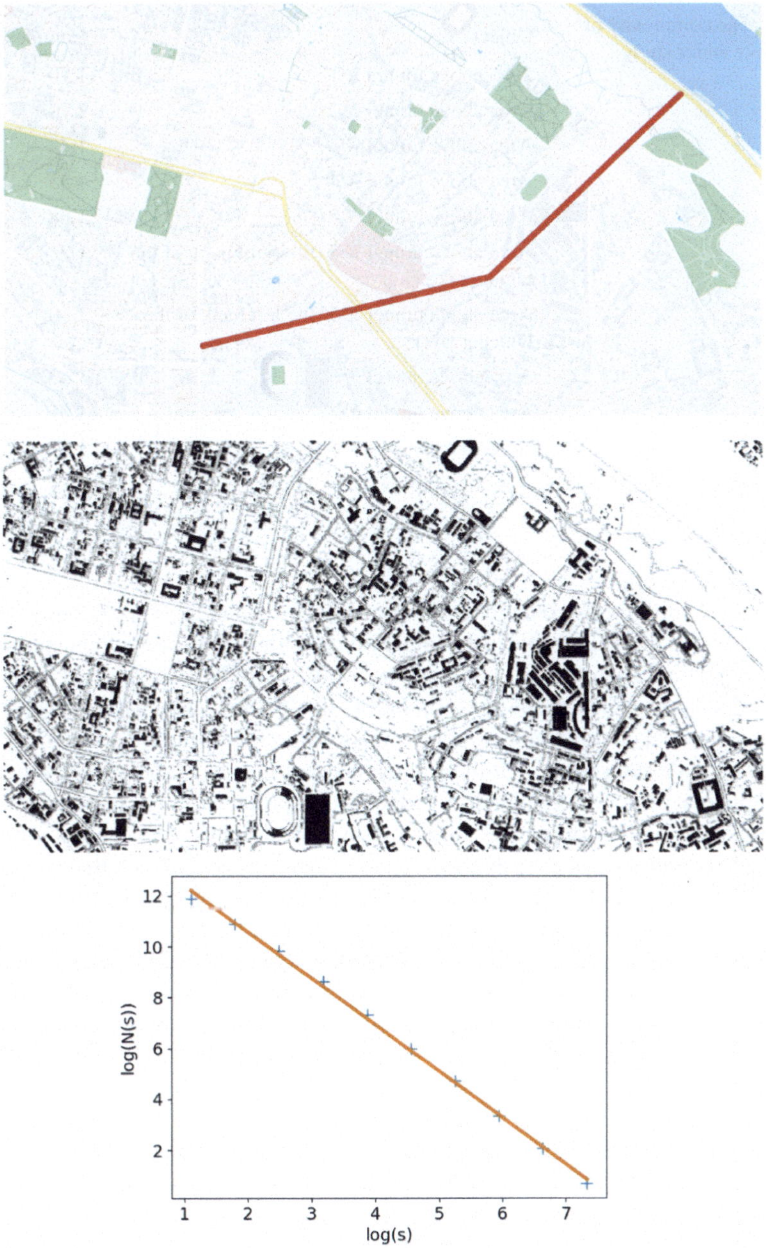

Fig. 11 Tunnel №8 (the map of the area where the tunnel is located (top), the map after processing (middle), plot log(N(s)) versus log(s) from box counting method with the regression line (bottom))

Table 2 The complexity of each object under study

Object	Complexity
Area above tunnel №1	1.70
Area above tunnel №2	1.79
Area above tunnel №3	1.48
Area above tunnel №4	1.53
Area above tunnel №5	1.68
Area above tunnel №5 (the right bank of the Dnieper river)	1.73
Area above tunnel №5 (the left bank of the Dnieper river)	1.60
Area above tunnel №6	1.69
Area above tunnel №7	1.81
Area above tunnel №8	1.82

5 Conclusions

Fractals were initially discovered as self similar mathematical objects. Only later, it became obvious that many objects in nature along with natural phenomena possessed fractal properties. Therefore, a need arose to apply the fractal analysis and often even multifractal analysis in various areas including health care, ecology, urban research, computer science, chemistry, astronomy, music, arts, literature to name a few.

It has been demonstrated that the fractal analysis is able to successfully measure the complexity of an object which is related to its properties. These properties can include certain hidden properties of a system that affect this object (for instance, biological regulatory systems that influence heartbeats or diabetes that causes the damage to blood vessels) or can be a good metric for objects classification.

We performed three case studies where the fractal analysis was applied to the geography and urbanized areas of Kyiv. We analyzed the parks of Kyiv, street network of Kyiv, and infrastructure of Kyiv for decision making purposes. In all cases, the fractal analysis successfully measured the complexity of corresponding objects. The fractal dimension calculated from Kyiv maps was compared with fractal dimensions computed in similar research for other cities. It corresponds well with other similar data. For example, it is greater then one computed for Omaha, but it is much lower then ones calculated for Beijing and New York. These results are expected due to the variability in complexities of cities. The third case study evaluated the above-ground infrastructure for various locations of the city to facilitate the decision making. Here, the complexity of the infrastructure (buildings and road networks) was assessed based on the fractal dimension to help choosing from a set of alternatives for the tunnel construction. However, complexity is only one of the factors used to make decisions. There are many other factors, including price, other priorities, the importance of a project or an area, the severity of transportation problems, and many other factors.

But the complexity of the infrastructure is an additional important factor in the decision-making process.

In addition to evaluating the complexity of above-ground objects, the underground infrastructure can also be evaluated. If the high complexity of above-ground infrastructure leads to decisions about the feasibility of tunnels, the high complexity of underground infrastructure can cause additional problems for building tunnels in the location, including the need to place tunnels at greater depths which increases construction costs. Thus, there is actually a compromise between the competing complexities of above-ground and underground infrastructure.

Acknowledgements The presented results were obtained in the National Research Fund of Ukraine project 2020.01/0247 «System methodology-based tool set for planning underground infrastructure of large cities providing minimization of ecological and technogenic risks of urban space».

References

1. Mandelbrot, B.B.: The Fractal Geometry of Nature. W.H. Freeman and Company, San Francisco, CA (1982)
2. Sponge, M.: https://upload.wikimedia.org/wikipedia/commons/5/52/Menger-Schwamm-farbig.png. Last accessed 01 June 2021
3. Barbara, D.: Chaotic mining: knowledge discovery using the fractal dimension. In: 1999 ACM SIGMOD Workshop on Research Issues in Data Mining and Knowledge Discovery, Philadelphia, USA (1999)
4. Barbará, D., Chen, P.: Using the fractal dimension to cluster datasets. In: Proceedings of the Sixth ACM SIGKDD International Conference on Knowledge Discovery and Data Mining, pp. 260–264. Association for Computing Machinery, Boston, Massachusetts, USA (2000). https://doi.org/10.1145/347090.347145
5. Harte, D.: Multifractals: Theory and Applications, 1st edn. Chapman & Hall/CRC (2001)
6. Suki, B., Barabasi, A.-L., Hantos, Z., Petak, F., Stanley, H.: Avalanches and power law behaviour in lung inflation. Nature **368**(6472), 615–618 (1994). https://doi.org/10.1038/368615a0
7. Warsi, M.A.: The fractal nature and functional connectivity of brain function as measured by BOLD MRI in Alzheimer's disease. Dissertation, McMaster University (2012)
8. Stanley, H.E., Amaral, L.A., Goldberger, A.L., Havlin, S., Ivanov, P. Ch., Peng, C.K.: Statistical physics and physiology: monofractal and multifractal approaches. Physica A **270**(1–2), 309–324 (1999)
9. Captur, G., Karperien, A.L., Hughes, A.D., Francis, D.P., Moon, J.C.: The fractal heart—Embracing mathematics in the cardiology clinic. Nat. Rev. Cardiol. **14**(1), 56–64 (2017). https://doi.org/10.1038/nrcardio.2016.161
10. Uahabi, K.L., Atounti, M.: Applications of fractals in medicine. Ann. Univ. Craiova Math. Comput. Sci. Ser. **42**(1), 167–174 (2015)
11. Sugihara, G., May, R.M.: Applications of fractals in ecology. Trends Ecol. Evol. **5**(3), 79–86 (1990)
12. Lovejoy, S., Schertzer, D., Ladoy, P.: Fractal characterization of inhomogeneous geophysical measuring networks. Nature **319**(6048), 43–44 (1986)
13. Ribeiro, M.B., Miguelote, A.Y.: Fractals and the Distribution of galaxies. Braz. J. Phys. **28**(2), 132–160 (1998). https://doi.org/10.1590/S0103-97331998000200007
14. Shen, G.: Fractal dimension and fractal growth of urbanized areas. Int. J. Geogr. Inf. Sci. **16**(5), 419–437 (2002)

15. Giovanni, R., Caglioni, M.: Contribution to fractal analysis of cities: A study of metropolitan area of Milan. Cybergeo: European Journal of Geography (2004).
16. Chen, Y., Wang, J., Feng, J.: Understanding the fractal dimensions of urban forms through spatial entropy. Entropy **19**(11), 600 (2017). https://doi.org/10.3390/e19110600
17. Bao, L., Ma, J., Long, W., He, P., Zhang, T., Nguyen, A.V.: Fractal analysis in particle dissolution: a review. Rev. Chem. Eng. **30**(3), 261–287 (2014)
18. Kaneko, K., Sato, M., Suzuki, T., Fujiwara, Y., Nishikawa, K., Jaroniec, M.: Surface fractal dimension of microporous carbon fibres by nitrogen adsorption. J. Chem. Soc., Faraday Trans. **87**(1), 179–184 (1991). https://doi.org/10.1039/FT9918700179
19. Sajjipanon, P., Ratanamahatana, C.A.: Efficient time series mining using fractal representation. In: Third International Conference on Convergence and Hybrid Information Technology, pp. 704–709. IEEE, Busan, the Republic of Korea (2008). https://doi.org/10.1109/ICCIT.2008.311
20. Traina, C., Jr., Traina, A., Wu, L., Faloutsos, C.: Fast feature selection using fractal dimension. J. Inf. Data Manage. **1**(1), 3–16 (2010)
21. Tasoulis, D.K., Vrahatis, M.: Unsupervised clustering using fractal dimension. Int. J. Bifurcation Chaos **16**(07), 2073–2079 (2006). https://doi.org/10.1142/S021812740601591X
22. Wilson, T., Dominic, J., Halverson, J.: Fractal interrelationships in field and seismic data. Technical Report 32158-5437, Department of Geology and Geography, West Virginia University, Morgantown, WV, United States (1997)
23. Aviles, C.A., Scholz, C.H., Boatwright, J.: Fractal analysis applied to characteristic segments of the San Andreas Fault. J. Geophys. Res. **92**(B1), 331–344 (1987). https://doi.org/10.1029/JB092iB01p00331
24. Chang, Y.-F., Chen, C.-C., Liang, C.-Y.: The fractal geometry of the surface ruptures of the 1999 Chi-Chi earthquake, Taiwan. Geophys. J. Int. **170**(1), 170–174 (2007). https://doi.org/10.1111/j.1365-246X.2007.03420.x
25. Vislenko, A.: Possibilities of fractal analysis application to cultural objects. Observatory Culture **2**, 13–19 (2015)
26. Song, C., Havlin, S., Makse, H.A.: Origins of fractality in the growth of complex networks. Nat. Phys. **2**(4), 275–281 (2006). https://doi.org/10.1038/nphys266
27. Das, A., Das, P.: Classification of different Indian songs based on fractal analysis. Complex Syst. **15**(3), 253–259 (2005)
28. Reljin, N., Pokrajac, D.: Music performers classification by using multifractal features: A case study. Archiv. Acoust. **42**(2), 223–233 (2017). https://doi.org/10.1515/aoa-2017-0025
29. Boon, J.-P., Decroly, O.: Dynamical systems theory for music dynamics. Chaos **5**(3), 501–508 (1995)
30. Gonzato, G., Mulargia, F., Marzocchi, W.: Practical application of fractal analysis: Problems and solutions. Geophys. J. Int. **132**(2), 275–282 (1998)
31. Richardson, L.F.: The problem of contiguity: an appendix of statistics of deadly quarrels. Gen. Syst. Yearbook **6**, 139–187 (1961)
32. Mandelbrot, B.B.: How long is the coast of Britain? Statistical self-similarity and fractional dimension. Science **156**(3775), 636–638 (1967)

Chapter 19
Fractal Analysis Usage Areas in Healthcare

Ebru Aydindag Bayrak and Pinar Kirci

Abstract This chapter presents a brief introduction to fractal analysis usage areas and their roles in the healthcare system. The definition of fractal, the properties of fractal, what is fractal dimension, most known methods used in fractal analysis and fractal analysis usage areas in healthcare have been examined under the different titles. The comparison of Euclidean geometry and fractal geometry has been explained. Accordingly, the relationship between fractal dimension (D) and Euclidean topological dimension (D_T) has been expressed. A range of scientific research about fractal analysis in the healthcare system have been reviewed. As a most known fractal analysis methods Box Counting method, Richardson's method, Dilation (pixel dilation) method and Mass (mass-radius) method have been explained briefly. Moreover, a glance of some usage areas for fractal analysis in healthcare system are COVID-19 disease, oncology, cardiology, brain imaging, neuroscience, dental, osteoporosis, ophthalmology and dermatology have been given. The reviewed studies about fractal analysis in many different areas showed that it can be used to obtain information about the severity and progression of the existing disease or early detection of a potential disease. The main goal of this study can be explained as briefly that giving an opinion to researchers about the usage areas of fractal analysis in healthcare.

Keywords Fractal · Fractal dimension · Fractal analysis · Healthcare

1 Introduction

Euclidean geometry has dominated for more than 2000 years. The shapes that appear in nature in this traditional geometry; it consists of lines and planes, circles and spheres, triangles, and conics. It is a well-known fact that the above-mentioned

E. A. Bayrak
Department of Engineering Sciences, Istanbul University-Cerrahpasa, Istanbul, Turkey

P. Kirci (✉)
Department of Computer Engineering, Bursa Uludag University, Bursa, Turkey
e-mail: pinarkirci@uludag.edu.tr

M. Zgurovsky and N. Pankratova (eds.), *System Analysis & Intelligent Computing*,
Studies in Computational Intelligence 1022, https://doi.org/10.1007/978-3-030-94910-5_19

abstract shapes are not sufficient to understand and model the complex structures of nature. Therefore, a new geometry is needed to better understand and model the nature. The name of the new geometry mentioned above is "fractal geometry". The universe, created by fractal geometry is not round or flat; it is a universe consisting of indented, broken, bent, intertwined, knotted, etc. shapes. This universe is not a boring or monotonous one as described by traditional Euclidean geometry; on the contrary, fractal geometry opens the doors of a different world to the researchers at every scale. When the researcher looks at a fractal object, witnesses how the concept of "infinity" in mathematics turns into concrete [1].

Fractals are irregular and complex geometric structures. They are used to characterize the shapes of most natural structures that are impossible to describe using traditional Euclidean geometry. Another important property of fractal geometry is a mathematical parameter which is called as the fractal dimension. No matter how much fractal object is enlarged, or how much the perspective of fractal object is changed, the fractals always remain the same. The fractal dimension is used to quantify the descriptive properties of an image or event, such as shape, texture, number, color, repetition, similarity, randomness, regularity, and heterogeneity [2].

However traditional approaches based on Euclidean geometry can be used to describe regular geometries, but it is insufficient to explain the complex and chaotic geometries. Fractal analyses comprise of many methods that are practical to measure a fractal dimension and other fractal characteristics to analyze datasets in all areas of science [3].

Traditional geometry (Euclidian geometry) was just related to the artificial realities of the first, second and third dimensions of objects. Fractal geometry is a new section of mathematics that is effective in explaining natural objects and structures whose dimensions are non-integer (fractional) values. The term of fractal geometry was coined by Benoit Mandelbrot and showed up in the 1970s. In fractal geometry, fractals are coming up at the end of an iterative or recursive construction with using an algorithm [4].

In the past 20 years, the concept of fractal geometry emerged as a new method to explain many natural phenomena in a basic and effective way. In addition, the fractal dimension has been applied to many different fields of biological science, among which are histology, molecular biology, normal and pathological anatomy, botany, and zoology [5].

Fractal analysis is especially used in the medical field to obtain information about the severity and progression of the pre-existing disease or early diagnosis of any potential disease.

Medical image analysis systems like human vascular system, nervous system, bones, organs, dental and breast tissue are so complicated and chaotic in pattern and most of the images are self-similar. For analysis of this complex medical images, most of the research specialists prefers the fractal analysis [6].

The aim of this study is to inform the researchers about fractal analysis usage areas in healthcare system. To point up this situation, a range of scientific research about fractal analysis in the healthcare system were reviewed. For the detection or early prediction or progression level of many diseases, fractal analysis used.

The presented chapter is organized as follows. In Sect. 2, what is fractal? the properties of fractal, and fractal dimension are explained. Also, the relationship between fractal dimension and Euclidian topological dimension has been expressed. In Sect. 3, information about fractal geometry is given and it has been compared with traditional Euclidean geometry. In Sect. 4, what is fractal analysis? and how is it applied? are explained briefly. Section 5 continues with the most known methods that are used in fractal analysis. In Sect. 6, fractal analysis usage areas in healthcare are explained with the several diseases. And the chapter ends with discussion and conclusion sections.

2 Fractal

The concept of fractal was firstly used by Mandelbrot in the literature. It was derived from the Latin word "fractus" that means fragmented and broken. Mandelbrot wrote on first pages of his book named "The Fractal Geometry of Nature", "Clouds are not spheres, mountains are not cones, coastlines are not circles, and bark is not smooth, nor does lightning travel in a straight line." [7].

The term 'fractal' is defined as uneven geometric form that can be sub divided into smaller fragments, where each part is similar the whole image. This situation indicates the most important property of fractal that is self-similarity. There are many best-known examples of fractals that are Koch curve, Sierpinski gasket, Julia set, and Cantor set etc. [8].

Fractals are complicated shapes, objects or patterns that are characterized by features like continuing to infinity and self-similarity. A part of a fractal object from each spatial direction, if it is statistically similar to whole of object, it is called as self-similarity. Also, the never-ending repetition of a fractal object is called recursiveness. Fractal dimension is pretty fruitful for explaining the irregularity and roughness of objects in nature [9].

The mathematics of fractal structures is based on a nonlinear equation, that has 'iterative' continuous repetition within itself. In fact, the geometry of nature is also "fractal geometry". Our veins that can travel around the world 6–7 times, the alveoli that cover an area as much as a few tennis courts; the DNA molecule, which is 2 m length when opened, and the nucleus of 100 trillion cells packaged into the micrometer-sized, the fact of fractal rules are lied behind [10].

Fractal geometry was devised to explain roughness differently. Scale invariance lie at the heart of fractal geometry, and it is also called as "symmetry under magnification." It means that fractals can be separated into smaller pieces and each of which closely resembles the whole of object. If the pieces of object scale isotropically, the object is called self-similar. If different scaling is used in different directions for fractal, the object is called self-affine [11].

Another question from Mandelbrot was, "What is the size of a ball of yarn?". Looking at the ball of yarn, the ball is a point, so its size is zero. When looking at the one-dimensional yarn of the ball with a magnifying glass, the yarn is in columns and these columns turn into one-dimensional fibers and the fibers into points. What is the actual size of the ball in this case? Mandelbrot said that objects that we cannot measure in any unit have a degree of roughness. According to him, when the scale changes, the degree of irregularity remains constant. Mandelbrot called this degree of roughness as the "fractal dimension". Objects with fractal dimensions are fractional and cannot be expressed with integers [12].

Natural fractal objects or structures are essentially characterized by four features [13]:

1. irregularity of shape,
2. self-similarity of structures,
3. non-integer (fractional) dimension,
4. scaling, which means that qualified features based on the scale at which they are qualified.

2.1 The Properties of Fractal

In fractals, details or patterns repeat in increasingly smaller scales (it is called recursive) and can persist indefinitely in purely abstract object. Such that, when each part of fractals is enlarged, it is still like the whole of the object. This is a very important feature that is also used in defining fractals [14].

This property of fractal can be named as "Self-similarity" and it is the most basic feature of fractals.

Self-similarity can take three different forms:

1. **Complete Self-Similarity**: It is the strongest type of self-similarity. No matter what scale we look at the fractal, it looks the same.
2. **Semi-Self-Similarity**: It is a weak form of self-similarity. At different scales, the fractal looks pretty much the same.
3. **Statistical Self-Similarity**: When a part of the fractal object is enlarged and examined, it is seen that the obtained parts are not the same as the original. The small part of whole is similar to the whole of object but is not its couple. This situation is called statistical self-similarity [12].

1. Perimeter lengths of fractal shapes cannot be measured in real terms. When you calculate, if the measuring tool gets smaller, you can calculate different results of same shape for every process. If measurement process is continued infinitely many times, results may end at the subatomic levels [10].

   Mandelbrot studied on the length of coastline with the ruler. When measured Great Britain coastline length with rulers of different sizes, he found that the length of the coastline increases indefinitely as it gets smaller [9].
2. The dimensions of fractal structures are not integer values, they are rational/fractional values. An example of the dimensions of geometric shapes can be given that a point is 0 (zero), a straight line has a dimension of 1, a square

has a dimension of 2, and a cube has 3 dimensions. But the sizes of fractal structures/shapes vary depending on how much area they are broken and bent. When fractals cover a plane, their size is between 1 and 2; if it is a protruding, curved structure, it takes between 2 and 3 value [15].

2.2 *Fractal Dimension*

A fractal is a system or structure which has self-similar patterns at different scales. It is identified in terms of a mathematical parameter that is called 'fractal dimension' [7].

The core of fractal analysis is in the notion of a fractal dimension. Fractal dimension is explained as a numerical value with explaining how the detail changes in pattern when the pattern is examined at different scales. This scaling is generally called as complexity. The higher of fractal dimension can be associated with the more complex the pattern of fractal object [16].

The fractal dimension can be explained as a ratio of change in structural detail to the change in scale and used to characterize the complexity of fractal objects or patterns. Also, it can be said that fractal dimension is a measure of the space-filling capacity of object [17].

The basic of the fractal dimension was laid by the famous mathematicians Hausdorff and Besicovitch. Mandelbrot defined fractals as a set with Hausdorff - Besicovitch dimension that exceeds the topological dimension [18].

Mandelbrot explained fractal dimension analytically, with the relationship among measuring scale (σ) and the length (L), that can be represented as:

$$L(\sigma) = K * \sigma(1 - D). \quad (1)$$

Here, K is a constant and D is called fractal dimension which is a fractional number. Several techniques are found to calculate fractal dimension and each method has some theoretical basics. But for all techniques, the process can be summarized as same concepts by the steps below:

- The dimension of the fractal object is calculated by various step sizes.
- According to the relationship between measured quantities and step sizes, log–log graph is plotted. Linear least-squares (LLS) best fit of regression line through the data points is calculated.
- Fractal dimension is calculated from the slope of the regression line [8].

The theory of fractal dimension is explained as follows: For a line, if we split the line in half then we have two parts of line to recreate the original line. If we split the line into four pieces, we have four of them to cover the line. We can write (1) generally, if we have a line the length of line s and then the number of parts that will cover the original line is given by:

$$N(S) = \left(\frac{1}{s}\right)$$

For a square, if we split into smaller squares each with the side length is to be 1/2 then we have 4 of these smaller parts to form the original square. If we split the square into more smaller squares each with the side length is to be 1/4 of, we have 16 of them to form the original square. As above we can write an expression (2) for the number of parts we need of size s to cover the original square, it is:

$$N(S) = \left(\frac{1}{s}\right)^2 \tag{2}$$

For a cube, this formula can be written as (3):

$$N(S) = \left(\frac{1}{s}\right)^3 \tag{3}$$

The exponents 1, 2, and 3 in the above examples are essential to explain the concept of the dimension of object. The equation can be generalized to:

$$N(S) = \left(\frac{1}{s}\right)^D \tag{4}$$

where D is the dimension, an integer as above but it need not be. If we take logarithms of both sides of equation $\log(N(s)) = D\log(1/s)$ had been. The dimension calculated by plotting $\log(N(s))$ versus $\log(1/s)$ from the slope of plot. If dimension is not an integer, then it's a fractal dimension [19].

The relationship between fractal dimension (D) and Euclidian topological dimension (D_T) is explained as following Eq. (4)

$$D > D_T \tag{5}$$

For points; $D_T = 0$.

For lines and curves; $D_T = 1$.

For surfaces; $D_T = 2$.

The kinds of fractal objects comprise of dusts, fractal curves, and fractal surfaces. A dust consists of a disconnected finite set of points. The fractal dimensions of dusts are generally between 0 and 1.

$$D_T = 0; D > 0$$

Fractal curves are irregular curves and fractal dimension of them greater than 1. A coastline can be an example of a fractal curve.

$$D_T = 1; D > 1$$

The fractal dimension value for a coastline is roughly calculated as $D = 3/2$. The fractal dimension of fractal surfaces is greater than 2 [20]:

$$D_T = 2;\ D > 2$$

3 Fractal Geometry

In Euclidean geometry, dimensions are expressed as integers, but in fractal geometry, integers are not sufficient to express the dimensions of objects. In this respect, the fractal dimension is very useful in explaining the complexity of a structure. The fractal dimension is not a topological concept, it is a metric concept. Fractal dimension is called as the fractional dimension or self-similarity dimension, and it is denoted by D. Also, the fractal dimension is identical with the Hausdorff dimension [12].

Euclidean geometry is only used to describe man-made objects. Fractal geometry is a method used to describe objects found in nature. It plays an important role in explaining dynamic systems such as meteorology, fluid dynamics, economy, biology, health, etc. The best answer to the question of what a fractal is good for: Fractals are a tool used to find answers to unsolved problems that are related to various branches of science [21].

Fractal geometry and Euclidean geometry can be compared as follows [22]:

- While the origin of Euclidean geometry dates back to 2000 years ago, studies in fractal geometry have accelerated with the development of computer technology.
- Euclidean geometry describes man-made objects, while fractal geometry describes objects in nature.
- The whole or part of a fractal cannot be expressed in terms of classical geometry. The properties of fractals cannot be described by simple geometric expressions. Fractals are not solution sets of simple equations.
- Despite their detailed structure, fractals are easy to identify. Traditional geometry elements are expressed with formulas. But fractals are created by recursion methods.
- The dimension of fractals cannot be defined by conventional measures. Fractals have a fractal dimension, which is larger than the Euclidean topological dimension.

In Euclidean geometry, the dimensions of objects are explained with integer values. In fractal geometry, fractal dimension is not an integer, is explained with fractional (non-integer) numbers (Fig. 1).

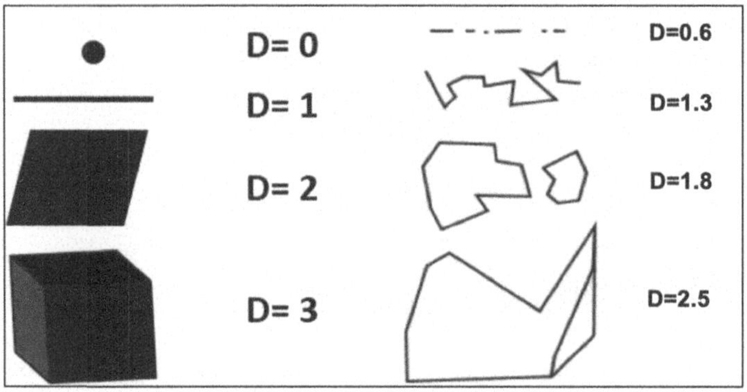

Fig. 1 The comparison of Euclidean dimension and Fractal dimension (Adapted from [1])

4 Fractal Analysis

Mandelbrot measured the degree of irregularity of objects whose measurements cannot be expressed in units. He stated that this degree of irregularity remains constant when the scale of these objects' changes. Fractal analysis is that developed from this point of view, a mathematical method. It enables to quantitatively describe complex structures and shapes that cannot be expressed with integral dimensions. It is expressed numerically with fractal dimension. Generally high on fractal dimension associates with the fractal structure is more complex. When fractal dimension is low, it means the structure has a simpler internal regularity [7].

Fractal analysis provides an approach to the analysis of many formless data sets. More importantly, it ensures meaningful explanation for the irregularity and complexity of our data. When Euclidean approach cannot give an answer, possibly fractal geometry may give an answer to the problem. According to the fractal geometry, our fractal data (image, sound, natural object, heart rates etc.) can be seem complex and rough, they have common rules of which we were unaware [23].

Fractal analysis is a mathematical method that is used for the analysis of complex shapes and structural formations. And the result of the analysis is obtained as a numeric value that is defined as the "fractal dimension". Early detection in bone changes, in the period where it is a method that helps, highlighted in many studies in the medical field [24].

5 Most Known Methods Used in Fractal Analysis

Many methods have been used to calculate fractal dimension in the literature. In the methods used, the applied steps are plotted on a logarithmic scale in general and a line is drawn in accordance with the obtained values. The slope of the drawn line gives the fractal dimension of the structure [25].

The methods used for fractal analysis can generally be grouped into two main groups; First method depends on the distance measurement between two points whose fractal dimension is calculated, and second method calculates fractal dimension based on volume measurement. In distance measurement, which is the most widely used method, fractal dimension calculates an edge of pixel as a unit of length, and it is used. In the second group that calculates the fractal dimension according to the volume measurement, the perimeter of the pixel is used as the volume unit [26].

The most known methods used in fractal analysis are Box Counting method, Richardson's method, Dilation method and Mass method. These methods and how they are applied are explained basically as the following:

5.1 Box Counting Method

The box-counting method is a way of calculating the fractal dimension of an object. Basically, it is measured by counting how many boxes of a specific length are required to cover the wanted to calculate object [27].

Box counting method (BCM) can be basically defined as splitting the image data into square boxes with same sizes and then counting the number of boxes which includes a part of the image data. This procedure is recurred with splitting the image data into smaller and smaller size of boxes. Then, the plot of the logarithm of the size of boxes versus the logarithm of the number of boxes counted is used to compute the fractal dimension of pattern or image. The best sitting line of the slope of plot gives the fractal dimension [28]:

$$FractalDimension = \lim_{r \to 0} \left[\frac{\log N(r)}{\log \frac{1}{r}} \right] \tag{6}$$

For objects or structures that are represented in the plane—for example, coast lines, mountain profiles, earthquake fault lines, rivers, fracture and cracking patterns, and growth of bacteria in stressed environments etc., the box counting method is used generally because of its easy computing. The box counting method was used for interpreted evaluating of fruitful insights for physical, chemical, and biological processes [11].

The general process of box counting method is shown in Fig. 2. For any fractal object, first three steps of the process are illustrated. In first step, it is shown that the analyzed fractal image is covered by a square box, the number of boxes (N) are counted that filled with image data. In second step of the box counting method, the image of fractal object is splitted with smaller box size and the number of boxes (N) are counted that filled with image data. In third step, in the same way, the image of fractal object is splitted with more smaller box size and again the number of boxes (N) are counted that filled with data.

The box counting method can be summarized with three steps briefly [30]:

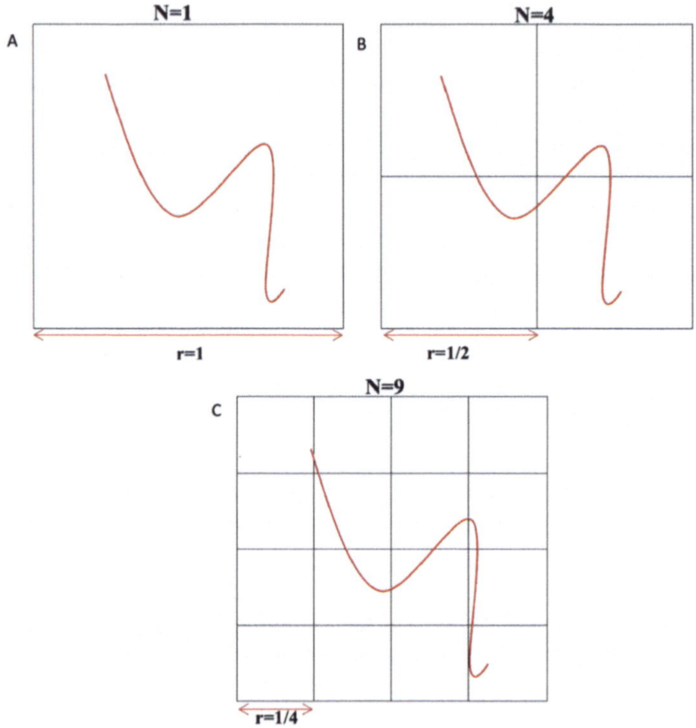

Fig. 2 The process of box counting method. **a** First step, the analyzed image is covered by a square box, the number of boxes (N) are counted that filled with data ($N = 1$). **b** In the second step of box counting, image is splitted with smaller box size ($r = 1/2$) and the number of boxes (N) are counted that filled with data ($N = 4$). **c** In the third step, image is splitted with smaller box size ($r = 1/4$) and again the number of boxes (N) are counted that filled with data ($N = 9$) (Adapted from [29])

Step 1. First, a set of box sizes δ (grid) is placed on the analyzing image. Each grid (box) has the size of $\delta \times \delta$.

Step 2. For each box size (δ), the number of boxes $N(\delta)$ needed to completely cover the object and it is counted.

Step 3. Finally, fractal dimension (D) is calculated from the slope of points of $\log(1/\delta) - \log(N(\delta))$.

The calculating fractal dimension of fractal object with box counting method is shown in Fig. 3. The relationship between log (box size)-log (the number of boxes) for calculating fractal dimension of object is illustrated.

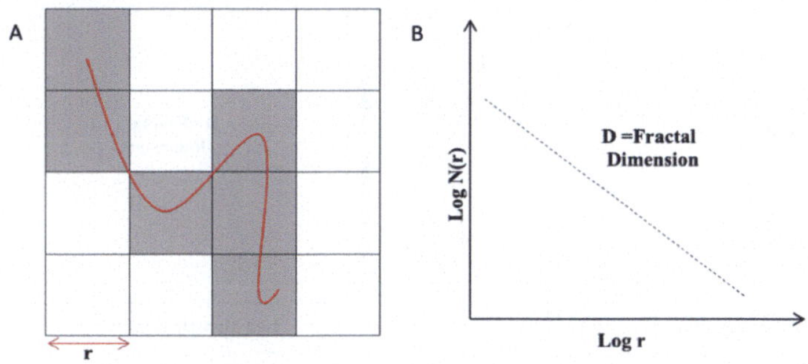

Fig. 3 The box counting method. **a** Calculating the fractal dimension of fractal object with box and the number of boxes $N(r)$ needed to completely cover the fractal object and it is counted. **b** Graph of resulting $\log(r) - \log(N(r))$ plot for calculating fractal dimension (Adapted from [31])

5.2 Richardson's (Ruler/Caliper/Divider) Method

Richardson's method that is a process for calculating dimensions with various measurements, first explained by British physicist Lewis Fry Richardson (1881–1953). In this technique, the perimeter of an object is measured with rulers of different lengths, and the dimension is calculated by plotting the slope. This technique is generally used when an object has a fractal structure but that is not exactly repeating, like a coastline. In this situation, the exact method of calculation cannot be used [32].

The Richardson's (divider) dimension method has a similar process to the box-counting method. The only difference of them, in place of boxes of varying sizes, steps of different perimeters of an object are used to measure. This method was used by Mandelbrot to calculate the fractal dimension of the coastline of Britain [33].

In this method, the total length of a contour or pattern or image is measured with different lengths of rulers. When using a large ruler, the small details would be unnoticed whereas using a small ruler, the least details would put into process [34].

Mandelbrot aimed to measure the length of the coastline of Great Britain with the ruler method or Richardson's method explained by Richardson (1961). When he reduced the size of the ruler for every step, he reached different measurement values for the length of the coastline. He associated the results with the power law and clarified with fractal dimension [9].

Mandelbrot measured the length of the coastline of Great Britain with the following equations:

$$P_i = N_i r_i \tag{7}$$

The length of coastline was measured with different ruler lengths in the Eq. (7) as to be, r_i: ruler length, N_i: number of steps and P_i: the length of coastline.

$$N_i = C / r_i^D \tag{8}$$

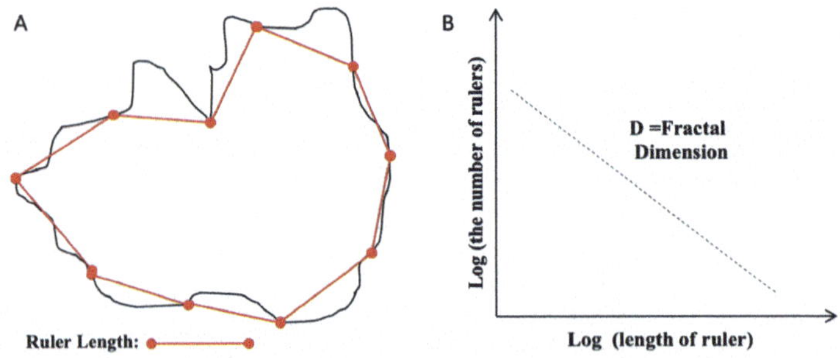

Fig. 4 The visual demonstration of Richardson's method. **a** Measuring the length of fractal object with ruler. **b** Graph of resulting log (length of ruler)—log (the number of rulers) plot for calculating fractal dimension (Adapted from [37])

C: ratio constant, D: fractal dimension when (8) equation replace in (7)

$$P_i = Cr_i^{1-D} \tag{9}$$

The Eq. (9) is obtained, and he calculated the fractal dimension of Great Britain coastline as $D \cong 1.25$ along the coastline [9].

One of the other methods for calculating fractal dimension of fractal object, Richardson's method is shown in Fig. 4. The illustration of measuring the length of fractal object with a ruler and the graph of relationship between log (length of ruler)-log (the number of rulers) plot for calculating fractal dimension are explained.

5.3 Dilation (Pixel Dilation) Method

The dilation (pixel dilation) method was first described by Flook (1978). In this algorithm, firstly, each border pixel of the analyzed image is replaced by a circle that its diameter ranged from 3 to 61 pixels (different sizes). The amount of dilation/expansion at the image boundary and the diameters of the circles are plotted on a logarithmic scale, and the slope of the line that fits the obtained points gives fractal dimension [35].

The dilation method can be seen as a kind of the Richardson's method, and it is easy use to apply on image analysis systems. To be watch out, when using computerized image analysis systems, the using of processing techniques like binary noise reduction. Because these techniques will cause to decreased by man-made effect in the calculated fractal dimension [33].

Dilation method was explained as briefly with the following steps for different scales of R according to [36]:

1. Extend (dilate) the structure of analyzed fractal object or pattern (profile, line, contour) by R using as a rule an isotropic structuring element

2. Qualify the area X of extended (dilated) structure
3. Find a length estimate (P) of the original structure as X/R where R is the gauge of the band
4. Finally it is plotted $\log(P)$ versus $\log(R)$ and fractal dimension (D) is calculated from the slope of the obtained straight line according to power law.

$$P \approx R^{1-D} \tag{10}$$

5.4 Mass (Mass-Radius) Method

The mass method is a version of box counting method. The fractal dimension is calculated according to the volume measurement and the perimeter of the pixel that is used as the volume unit. In this method, circles of different diameters are randomly placed in the image on the border and the pixels belonging to the image border inside the circles are counted. Then, the relationship between the log of the number of pixels within circle and the log of the measuring diameter is plotted. Fractal dimension is calculated from the slope of this log–log plot. The power relationship for calculating mass fractal dimension explains as the following [3]:

$$\mu(r) = Ar^D \tag{11}$$

$\mu(r)$ is the number of pixels (mass), r is the diameter of circle, A is pre-factor and D is mass fractal dimension.

The visual demonstration of Mass-radius method is shown in Fig. 5. Measuring the length of fractal object is demonstrated with the different sizes of radius. The graph of resulting log (the diameter of circle) – log (the number of pixels) plot for calculating fractal dimension are explained basically with the illustration.

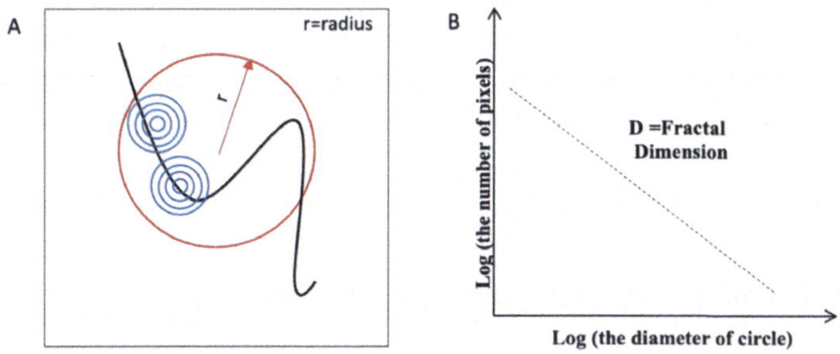

Fig. 5 The visual demonstration of Mass-radius method. **a** Measuring the length of fractal object with mass radius. **b** Graph of resulting log (the diameter of circle)—log (the number of pixels) plot for calculating fractal dimension (Adapted from [37])

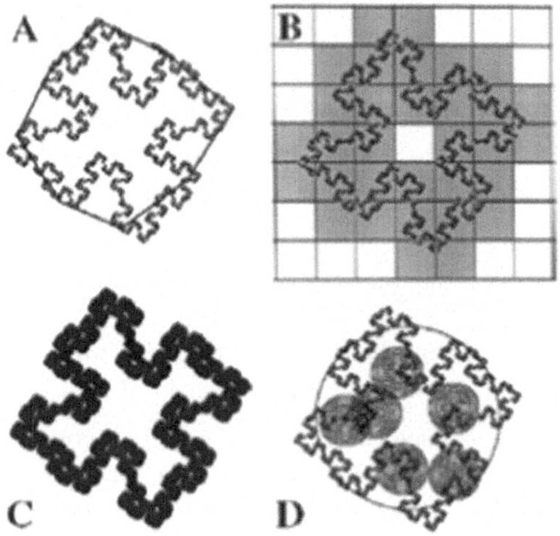

Fig. 6 Methods used in Fractal dimension calculation: **a** Richardson method, **b** box counting method, **c** dilation (pixel dilation) method, **d** mass method (quoted from [3])

A set of concentric circles with augmenting radius are scratched as centered on the point of the analyzing fractal object. M is to be the mass and r is to be as a given radius, $M(r)$ is a function of the size of the circle radius. $\log(M(r)) - \log(r)$ plot in other words $\log - \log M(r)$ versus r is plotted and from the slope of this relationship it is calculated the fractal dimension [19].

Mass dimension method is concerned with calculating how density scales according to size and it is based on to how the object is connected to its environment [11].

The methods used in Fractal Dimension Calculation are Richardson method, Box counting method, Dilation (Pixel dilation), and Mass method. They are shown in Fig. 6.

For calculating the dimension in fractal analysis, the box counting, Richardson and dilation methods are all measures length, while the mass method measures mass [35].

For physical objects, in three-dimensional space like physiological branching's (neural, circulatory, and respiratory), protein clusters, aggregate, dust balls, soot particles, and terrain maps. And mass dimension method is used generally as easier to compute [11].

6 Fractal Analysis Usage Areas in Healthcare

Although the objects in nature do not belong entirely to Euclidean geometry or wholly to fractal geometry, it is possible to examine this complex structure of forms using fractal geometry. There are rules of fractal geometry behind our blood vessels that can travel around the globe 6–7 times, lung alveolus covering an area of the size of a football field, and the packaging of the DNA molecule, which is over 2 m long, into a few micrometer nuclei in each of our 100 trillion cells [12].

Fractal analysis has been significantly applied to an extensive range of health-care area. It is especially used in the medical fields to obtain information about the severity and progression of the pre-existing diseases or early diagnosis of any potential diseases.

Fractal analysis methods are commonly used in several areas of signal and image process analysis. The applications of fractal analysis comprise the classification of histopathology slides in medicine, the measurement of fractal landscape or coastline, the calculation of object complexity, generation of several art form and new music, signal, and image compression and so on. In general fractal methods can be grouped as two. Fractal methods can be used for creation since art and music. Other methods are measurement (calculation) methods, and they can be used for comparison [38].

During the recent years, fractal geometry was applied commonly in medical image analysis for early detection of cancer cells in human body. Also, this method successfully applied in ECG signal, brain imaging for tumor detection and trabeculation analysis etc. [6].

In this section, we broadly reviewed many studies where fractal analysis plays a role. Especially, the terms of fractal and fractal analysis are active and an applicable area for medical image processing. Because medical images of human body and organs have a fractal character. We can see fractal structure on our lung, brain and alveolus, blood vessels and neurons.

6.1 Covid-19 Disease

Swapna et al. [39] applied the fractal analysis method as an image analysis on the scanning electron microscopic (SEM), transmission electron microscopic (TEM), and atomic force microscopic (AFM) images of 40 coronaviruses (CoV), by the normal and differential box counting algorithm. They showed the value of fractal dimension of COVID 19 viruses have a fractal structure with an average box counting and their fractal dimension is equal to 1.820 [39].

Namazi and Kulish [40] calculated the values of fractal dimension and Shannon's entropy for X-ray image for lungs using MATLAB. They examined on classifying the X-ray images of lungs for healthy subjects and patients with COVID-19, ARDS, cavitating pneumonia, pneumococcal pneumonia, chlamydia pneumonia, streptococcus pneumonia, pneumocystis pneumonia and aspiration pneumonia. According to the results, Shannon's entropy showed not a quantitative measure of the infection level, while fractal analysis gave better results on that purpose. Also, they explained that the fractal analysis works on well about distinguishing Covid-19 from other respiratory diseases [40].

Fernandes [41] investigated the growth factor of COVID-19 disease using non-linear techniques for better understanding the pattern of the spread of this disease.

They explained that the growth factor of COVID-19 shows a fractal pattern. The calculated fractal dimension was obtained as 1.1623 by ImageJ software [41].

Păcurar and Necula [27] analyzed COVID-19 spread based on fractal interpolation and fractal dimension. They evaluated the spread of the Covid-19 epidemic in countries like Germany, Spain, Italy, and Romania and analyzed the spread of the disease in those countries relying on their fractal dimension. At the end of study, they summarized that those fractals can be used as an assessment of epidemic outbreak [27].

6.2 Oncology

Fractal geometry can ensure a better measure of complex patterns of cancer than traditional Euclidean geometry. Because cancerous tumors are generally irregular and complex shape. Also, they show a certain randomness related with their growth situation [33].

Bayrak and Kirci [42] used fractal analysis to calculate fractal dimension and correlation coefficient values of thyroid ultrasound image database. Fractalyse Image Analysis toolbox was used to estimate fractal dimension and correlation coefficient values of thyroid ultrasound image database. The fractal dimension of images was calculated by the use of the box counting algorithm. The fractal dimension of thyroid images data calculated between the range of 1.59–1.86. The differences between calculated fractal dimensions explained as to be benign and malignant lesions located in different sizes on thyroid image data [42].

Iqbal et al. [43] studied on determination the efficacy of fractal dimension analysis in detecting malignancy potential of oral leukoplakia. ImageJ was used to analyze fractal dimensions of images and the results were compared with biopsy. Fractal dimension value calculated significantly higher in leukoplakia with dysplastic changes. They asserted that fractal analysis is a useful method in determination of complication in oral leukoplakia and can be used as an important tool at primary healthcare services for early response [43].

Qin et al. [44] aimed to use image-based fractal analysis for early detection of breast cancer cell images. The calculated fractal dimension values of cancer cells and healthy cells were analyzed and compared. They explained the results of preliminary test in which the mean of fractal dimension of healthy cells is smaller than (high tumor cellularity percentage of) cancer cells. Namely, they suggested that fractal analysis method can be useful for explaining abnormalities of breast cancer cells [44].

Bhandari et al. [17] studied on structural properties of colon cancer tissue microarray images. They analyzed five different cancer categories (normal, adjacent, begin, stage 1, and stage 2) and calculated their fractal dimension using ImageJ analysis software. They inferred that fractal analysis can be used as an efficient way to distinguish between normal and cancerous tissues at different stages. The results showed that cancerous colon tissues are more fractal contrasted with normal and benign tissues [17].

Czyz et al. [45] investigated on a secondary tool based on computer aid diagnosis in addition to traditional radiological techniques in detection of meningiomas with higher grade of malignancy. The average and maximum fractal dimension of the contrast-enhancing region of the tumor were calculated from preoperative brain MR images. The box-counting method and ImageJ 1.49 software were used in their study. The results showed that the fractal analysis of preoperative MR images could be useful in identifying meningiomas with potentially aggressive clinical behavior [45].

Marusina et al. [46] applied fractal analysis on the processing of liver magnetic resonance image (MRI) data. They used ImageJ software package for liver image processing and FracLac software was used for the calculation of fractal dimension using the box counting method. The calculated fractal dimension of the liver MRI data and the Hurst exponent were determined as diagnostic property for tomographic imaging. The experimental results showed significant differences of the fractal dimension and the Hurst exponent between the areas of healthy liver tissue and pathological cases (foci formations) [46].

Etehad Tavakol et al. [28] analyzed thermal images of breast using fractal analysis to determine the possible difference between malignant and benign tumors. The experimental results showed a significant difference in between malignant and benign cases according to fractal dimension value. They suggested that fractal analysis technique can be used in the detection of breast tumor [28].

Huang and Lee [47] used Fractal Analysis on Pathological Prostate Images for detection of prostatic carcinoma. The differential box-counting and entropy-based fractal dimension estimation techniques were applied as a feature extraction method to analyze differences of intensity and texture complexity in regions of prostate images. Then, each prostate image was classified with Support Vector Machine, Bayesian and k-NN machine learning classifiers [47].

Lv et al. [48] studied to determine computerized Characterization of Prostate Cancer from magnetic resonance images by Fractal Analysis extraction. The complexity features of regions of Interest selected from the peripheral zone were measured with Histogram fractal dimension and texture fractal dimension methods. For the used two fractal analysis methods, statistical significance was found between the normal and cancerous patient groups [48].

Ballerini and Franzen [49] studied on the detection of microscopic images of breast tissues by means of fractal geometry. The fractal dimension of binary microscopic images calculated using the Euclidian distance mapping. Also, Neural network was used as a classifier. With using fractal analysis, the sensitivity of classification of cancerous tissue from image data increased 90% to 92.5%. According to these results, they asserted that fractal analysis can be used to distinguish the normal from cancerous tissue images [49].

Esgiar et al. [50] utilized image processing techniques for the analysis and early diagnosis of colon cancer. Fractal analysis technique was used to distinguish malignant and normal colon images data from each other. The distribution of fractal dimension for all images in the cancer and normal groups were compared. The highly significant difference between groups (cancer and normal groups) was obtained [50].

6.3 Cardiology

Hamidi et al. [51] classified heart sound signals using two proposed methods that are fractal analysis and curve fitting. They used as a classifier K nearest neighbor machine learning algorithm with Euclidean distance. The performance of their proposed method is superior, particularly the efficiency of the fractal method is much better than the curve fitting method. For the used three data set, they achieved an overall accuracy of 92%, 81% and 98%, respectively [51].

Gavrovska et al. [52] analyzed the possibility of using the shape context and fractal theory in the S1 and S2 heart sound pattern characterization. The analysis of heart sound patterns was performed with support vector machine that show above 95% accuracy. The results showed that the proposed method improves the accuracy of analysis of the heart sound pattern is not using fitting and averaging approaches on the scales [52].

Michallek and Dewey [53] used fractal analysis of the ischemic myocardial transition region in perfusion imaging, to characterize pathomechanisms underlying myocardial ischemia in chronic ischemic heart disease. They showed that fractal analysis can increase right diagnosis, compared to traditional noninvasive myocardial perfusion analysis [53].

Captur et al. [54] explained that a different approach to measure left ventricular trabecular complexity using fractal analysis was applied to cardiovascular magnetic resonance. Fractal dimension value was obtained higher in left ventricular noncompaction patients when compared to healthy volunteers. It is emphasized that fractal analysis was right and repeatable technique when compared to current techniques [54].

Tapanainen et al. [55] examined the relationship between fractal analysis of heartbeat variability and patient prognosis that was investigated in a large sample size of patients with consecutive heart attacks patients. It was concluded that fractal analysis of heartbeat variability could be used to determine the mortality rate among patients with acute myocardial infarction [55].

Mäkikallio et al. [56] investigated the risk of sudden cardiac death in individuals aged 65 and over. They argued that the fractal structure of heartbeat variability in elderly individuals is to be the determinant of the cardiac mortality ratio. The results showed that altered short-term fractal scaling properties of heart rate in the random population of elderly subjects referred to an increased risk for cardiac mortality, especially cardiac death in suddenly [56].

6.4 Brain Imaging

Lenka et al. [57] examined on IoT-Cloud based fractal analysis for several brain tumor image analysis. They used different fractal analysis methods such as Box Counting, Radius Mass, Correlation, Dilation and Cluster. They explained that computed results of fractal dimension of brain tumor image can be helpful for the radiologists to carry out further processing and analysis. The results of box counting method showed both higher fractal dimension and lower error for each brain tumor image as compared with other fractal analysis method [57].

Namazi and Jafari [58] applied fractal analysis on the electroencephalography (EEG) signal recorded from the newborns along sleeping in different weeks of post conception. The fractal analysis results showed that the EEG signals for newborns in 45 weeks have the highest fractal dimension, whereas in 36 weeks for newborns, the lowest fractal dimension was obtained. They suggested that the complexity of brain development significantly changes with the newborn age [58].

Czyz et al. [59] focused on the usefulness of the fractal analysis based on brain MRI in the preoperative determination of a typical meningioma grade. Fractal analysis was performed with the use of ImageJ software. They inferred that fractal analysis of preoperative MR brain images appear to be a useful predictive tool in identifying meningiomas with potentially aggressive clinical behavior [59].

Akar et al. [60] aimed to apply fractal analysis using magnetic resonance imaging (MRI) data of brain, in which cerebellum extensions can be seen in the sagittal plane, and thus, the cerebellar gray matter (GM) morphological features can be seen. The results showed that the fractal dimension values of patients were higher than the values of healthy individuals. They highlighted the idea that fractal dimension values may be a useful sign in the diagnosis and evaluation of cerebellum disease, according to results [60].

King et al. [61] applied a modified fractal analysis technique to high-resolution T1 weighted magnetic resonance images to measure the variations in the shape of the cerebral cortex of Alzheimer's disease patients. The fractal analysis of the cortical ribbons was calculated using the box counting method. The mean of fractal dimension of the cortical ribbons from Alzheimer's disease patients were calculated lower than controls on six of seven profiles with same age range [61].

Zhang et al. [62] enhanced a three-dimensional box-counting method to calculate fractal dimension of human brain white matter interior structure, white matter surface and white matter general structure. For young individuals, the fractal dimension values of white matter skeleton and surface were obtained higher than old, and this result was related to the complexity of brain white matter structures in young people [62].

Iftekharuddin et al. [63] discussed existing fractal-based algorithms for identifying tumors in brain magnetic resonance images, also they offered three new improvements of fractal-based algorithms that are piecewise-threshold box-counting, piecewise-modified-box-counting (PMBC) and piecewise-triangular-prism-surface-area method (PTPSA). In PMBC and PTPSA methods, the determination of detection and location of the tumor in the brain MR images obtained more accurate [63].

6.5 Neuroscience

Appaji et al. [64] studied 277 participants (92 healthy volunteers, 98 schizophrenias, and 87 bipolar disorders) with different age ranges. The fractal dimension of retinal vascular images was calculated by the box-counting method. The average fractal dimension across the left and right eyes were calculated. Fractal dimension for both schizophrenia and bipolar disorder calculated higher compared to healthy volunteers [64].

Wolfson [65] used fractal analysis as a tool to measure the information complexity that was observed within electroencephalograph (EEG) signals and neurogenetic code. It is explained that fractal analysis can be a better method of diagnosis of autism spectrum disorder (ASD) [65].

Zaletel et al. [5] compared the box-counting method and modified Richardson's method in neuroscience. For analyzing, a total of 20 cortical pyramidal neurons from coronal sections were used and compared. The fractal dimension of apical dendrites of superficial and deep pyramidal neurons was calculated. They highlighted based on results that the fractal dimension value calculated by the modified Richardson's method has certain advantages over the box counting method [5].

Di Leva et al. [13] reviewed the basic concepts of fractal analysis and its main applications to the basic neurosciences. Many applications of fractal and fractal geometry in the clinical neurosciences were discussed. It was emphasized that the application of the principle of fractal geometry, on the contrary traditional Euclidean geometry, were allowed the measurement of the fractal dimension of almost all complex biological asset [13].

Ha et al. [66] aimed to explore three-dimensional fractal dimension of skeletonized cerebral cortical surface from magnetic resonance images from schizophrenia, obsessive–compulsive disorder (OCD) and normal controls. The results showed that the schizophrenic group had a significantly smaller mean fractal dimension than OCD group, and the OCD group than normal controls [66].

Fernandez and Jelinek [67] studied on the fractal theory in neuroscience. The used methods for calculating fractal dimension, advantages and potential problems of fractal dimension were examined. They reviewed lots of fractal theory studies about the complexity and irregularity of physiological and anatomical patterns. Also, they emphasized that fractal analysis methods were fruitful for image analysis in neuroscience [67].

6.6 Dental

Çoşgunarslan et al. [68] aimed to compare the mandibular bone structures of people with Chronic Kidney Disease (CKD) and healthy people with the fractal analysis method on panoramic radiographs. They found statistically significant difference between two groups as CKD and healthy people [68].

Soylu et al. [69] studied on that fractal analysis could be used as an effective indicator for determining osseointegration of dental implant. Fractal analysis was applied with the box-counting algorithm using White and Rudolph's method to Three regions of interest (ROIs) that are mesial, distal, and apical sites of the implants. It is highlighted that fractal analysis is a hopeful, noninvasive, and confidential method to diagnose osseointegration of dental implants based on two-dimensional dental radiographs, and it could abbreviate the total treatment time [69].

Çelebi [70] aimed to quantitatively reveal the changes in the jaw bones in patients with rheumatoid arthritis by obtaining fractal dimension from panoramic radiographs and to evaluate the correlation of fractal dimension with panoramic radio morphometric measurements. A total of 186 individuals (144 females, 42 males) aged 20–56 years were included in the study. Fractal dimensions were calculated in the mandible condyle, angulus, lower molar-premolar, mental foramen anterior region and canine-premolar region of the maxilla using the ImageJ program. She put forward that fractal dimension values from panoramic radiographs can demonstrate changes in the jawbone of rheumatoid arthritis patients and can potentially be used to refer these patients to appropriate medical investigation [70].

Franciotti et al. [71] used fractal analysis in dental images for detection of osteoporosis. They made a systematic review and used meta-analysis for assessing the reliability of fractal dimension measures differences between osteoporotic patients and healthy controls. Also, they intended to explain a standardized procedure of calculation of fractal dimension in dental radiographs [71].

Ersu et al. [72] analyzed the dental panoramic radiographs of the patients who declared to used systemic glucocorticoid. The mandibular bone structure of patients seen radiographs were evaluated with the fractal analysis method and the results were compared with the control group. Panoramic 4 different areas for measurements were chosen on radiographs ramus (IA1), angulus (IA2), anterior mental foramen (IA3), mandibular cortical area (IA4) manually selected. Fractal dimension was calculated with the ImageJ software. A significant decrease in fractal dimension was observed in IA4 in patients using glucocorticoids [72].

Kato et al. [73] reviewed the use of fractal analysis in dental images. They investigated lots of study and explained that fractal analysis in dentistry is used widely. Especially, Image J software and the box counting method were preferred to use in these reviewed studies [73].

Güleç et al. [15] aimed to investigate the relationship between fractal dimension (FD) values of mandibular trabecular bone with age, gender, and region of interest (ROI). The calculated of fractal dimension measurements in the regions selected from the condyle and angulus region of the lower jaw varied with age and sex, according to results [15].

Şener and Baksı [24] aimed to determine the mandibular bone tissue changes of patients under bisphosphonate treatment using fractal analysis method and compare them with the findings obtained from healthy patients. The rapid advancement of technology and the increase in the number of studies reporting parallel findings with the presented study, it was emphasized that dentists can have an active role in the early diagnosis of osteoporosis by applying the fractal analysis method on the radiographs for diagnostic purpose [24].

Updike and Nowzari [74] calculated fractal dimension on periapical radiographs of patients with severe periodontitis, mild periodontitis and healthy patients using the box-counting method. They found that the mean fractal dimension values of patients affected by periodontitis were significantly lower than the healthy group [74].

6.7 *Osteoporosis*

Zehani et al. [75] suggested a new technique for trabecular bone characterization using fractal analysis of X-Ray and MRI texture images for the diagnosis of osteoporosis disease. In the proposed fractal analysis model was used the differential box-counting method (DBCM) to estimate the fractal dimension after image preprocessing process. Their fractal model-based method showed the statistical results that had acceptable performance level of to discriminate healthy and unhealthy trabecular bone tissues [75].

Chen et al. [76] suggested that fractal dimension of bone matrix (less than 3) could be a complementary indicator to BMD to diagnose the osteoporosis. They arrested that with the collocation of BMD and fractal analysis, the screening of osteoporosis could be more accurate. Also, they explained the fractal dimension is more sensitive than BMD in specific cases according to literature reports [76].

Yeşiltepe et al. [77] aimed to evaluate the structure of the trabecular bone in the condyle head in rheumatoid arthritis (RA) patients with fractal analysis using Cone Beam Computed Tomography (CBCT) data. Fractal analysis were applied with Image J program and box counting method. The fractal dimension values showed statistically significant between RA patients and the healthy individuals for both right/left sides. The results of this study suggest that osteoporosis-related changes in the condyle heads of RA patients can be identified by fractal analysis on CBCT images [77].

Tafraouti et al. [78] used fractal Analysis and Support Vector Machine (SVM) for Diagnosis of Osteoporosis. Fractional Brownian Motion as a fractal analysis method was used the characterization of bone X-ray images. The performance of their method was resulted with using SVM classifier 95% accuracy classification rate [78].

Koh et al. [79] studied on prediction of age-related osteoporosis in postmenopausal women using fractal analysis on the trabecular pattern on panoramic radiographs. Fractal dimension values was calculated using the box counting method and the fractal dimension values of osteoporosis group compared with the normal group. It was resulted in that significant difference in the fractal dimension values between the osteoporotic and normal groups. The Statistical analysis of the fractal dimension values was performed using two-way ANOVA and the correlation coefficient was evaluated using the Pearson correlation coefficient [79].

6.8 Ophthalmology

Deepika et al. [80] investigated several fractal analysis methods to calculate fractal dimension of the retinal vessels. Box counting method, Hausdorff Fractal Dimension, Modified Hausdorff Fractal Dimension and Fourier Fractal Dimension methods were compared according to coefficient of variation. They suggested that Hausdorff Fractal Dimension can be one of the best methods to calculate fractal dimension for normal and retinal diseased images [80].

Yu and Lakshminarayanan [81] were reviewed scientific research on the relationship between retinal diseases and fractal analysis. Also, they used meta-analysis to specify whether different fractal dimension values are meaningful in retinal disease versus healthy individuals. They put forward that decreased fractal dimension values associated with presence of myopia, hypertension, and glaucoma but fractal dimension had incoherent results for diabetic retinopathy [81].

Lopez et al. [82] studied on fractal analysis on retinal vasculature. Fractal analysis was applied on macular optical coherence tomography angiography (OCTA) images to determine its role in variation microvascular differences between normal and diabetic eyes. Fractalyse image analysis program was used for calculating fractal dimensions using mass method on OCTA images. They suggested that fractal analysis method could be more forceful as a measure of microvascular dropout compared to 3×3 mm scans versus 6 mm $\times$ 6 mm scan [82].

Bhardwarj et al. [83] were used fractal analysis to investigate retinal vascular disease patterns in patients with diabetic retinopathy using optical coherence tomography angiography. The analyzed images were standardized using ImageJ program. Fractal analysis were applied using Fractalyse image analysis program. Fractal dimensions and correlation coefficient values of the deep and superficial capillary plexuses were check against between control eyes and those in different phases of diabetic retinopathy [83].

Fabrizii et al. [84] applied fractal Analysis that The box counting and multi fractal methods were used to calculate fractal dimension of Cervical Intraepithelial Neoplasia, to determine precursor lesions of cervical cancer. Statistically significant results were obtained between all grades of Cervical Intraepithelial Neoplasia. According to the confusion matrix was generated to evaluate differences between pathologist's grading and predicted group, fractal methods showed a value match of 87.1% [84].

Cheung et al. [85] examined the relationship between retinopathy in young individuals with type 1 diabetes and fractal analysis method. They aimed to use fractal analysis as a new method that measure the complexity of the retinal vascular branching model. Also, they desired to predict early detection of diabetic microvascular damage using fractal dimension. The huge fractal dimension values were associated with ascended complexity of retinal vasculature [85].

Jelinek and Fernandez [35] discussed how the calculated fractal dimension by several methods can be different and how fractal analysis can use for retinal ganglion cell characterization. They showed that same and different fractal analysis methods can measure varied values, but they are consistent [35].

Macgillivray et al. [86] used fractal analysis of the retinal vascular network in digital fundus images. They aimed to determine segment blood vessels and measure the fractal dimension of vascular skeletons. Mean fractal dimension values were obtained to be 1.40 from 18 images of the non-hypertensives subgroup and 1.398 from 20 images of the hypertensives sub-group [86].

6.9 Dermatology

Căliman et al. [87] aimed to propose a system based on the analysis of the color digital images for the psoriasis management. The proposed approach is based on the estimation of the local color fractal dimension of psoriasis lesion images. They suggested that fractal dimension was related the severity of psoriasis. The section of healthy skin, the skin appears smoother according to color texture and had small fractal dimension [87].

Przystalski et al. [38] used the box-counting dimension method of grayscale images of different skin lesion on was used to determine characteristic features in melanoma diagnosis. The results showed that some of the skin lesions had differential fractal dimension values. But also, they highlighted that more studies must be done the efficiency of fractal analysis on skin lesion determination [38].

Guarneri [88] have studied on a possible aid for dermatologist in detection skin tumors using fractals and computer vision. Open CV software was used the recognition of shape and the regularity of contours was calculated using fractal analysis. Using these methods, he aimed to the evaluation of variances in contour regularity between melanomas and nonmelanomas [88].

Mastrolonardo et al. [89] applied fractal analysis method to skin pigmented lesions using the genuine tool of the variogram technique. They explained the up-to-date method of the variogram and of fractal analysis broaden to all parts of interest of skin for identification of the malignant lesion from dermoscopic image, also they arrested that the results were satisfied [89].

Bianciardi et al. [90] studied on distinctive diagnosis between chronic dermatitis and mycosis fungoides using fractal analysis. They compared the diagnostic power of nuclear contour index of lymphoid cells and fractal geometry. Fractal analysis was applied using the box counting method. The correlation between nuclear contour index and fractal dimension were interpreted by linear regression analysis and one-way ANOVA test [90].

7 Discussion

In this chapter, it is aimed to present a specific information to the researchers by examining the studies on fractal analysis in the field of healthcare system. The application of fractal analysis in health were reviewed. The most striking of these studies are tried to be summarized. It is observed that the studies of fractal analysis have been conducted on many different health problems are COVİD-19 disease, oncology, cardiology, brain imaging, neuroscience, dental, osteoporosis, ophthalmology, and dermatology.

The study can be summarized as follows. The general information about what is fractal, the properties of fractal, and fractal dimension are explained. The relationship between fractal dimension and Euclidian topological dimension has been expressed. Fractal geometry and traditional Euclidean geometry has been compared. The properties of fractal geometry and its differences from Euclidean geometry are explained. What is fractal analysis means and how it can be applied on data have been explained. This chapter also focuses most known methods used in fractal analysis. These methods are Box Counting, Richardson's, Dilation and Mass methods and are seen that mostly preferred for calculating fractal dimension of medical data. The fractal analysis usage areas in healthcare system are explained with the several diseases.

Fractal analysis is especially used in the medical field to obtain information about the severity and progression of the pre-existing disease or early diagnosis of any potential disease.

8 Conclusion

This chapter presents a brief introduction to fractal analysis usage and their roles in the healthcare system. A range of scientific research about fractal analysis in the healthcare system have been reviewed.

This chapter provides an opinion of how fractal analysis can be used as aid diagnosis method and how medical data (image, signal etc.) can be used as a source for fractal analysis. The main aim behind this chapter is to provide an opinion and approach to researchers about fractal analysis which can be used for solving several health problems. It has been seen that fractal analysis is used widely detection of several diseases.

The reviewed studies about fractal analysis in many different areas showed that it can be used to obtain information about the severity and progression of the existing disease or early detection of a potential disease. The main goal of this study can be explained as briefly that giving an opinion to researchers about the usage areas of fractal analysis in healthcare.

The use of fractal analysis in the field of health is increasing day by day, as the fractal concept is easy to apply in complex structures such as the human body and organs. The positive results of the studies are increased the interest even more. Therefore, some limitations and disadvantages can occur in fractal application areas such as image analysis. In future research, any improvement of fractal analysis as applying can be better for good classification or diagnosis of several diseases.

References

1. Koçak, K.: Doğanın Geometrisi: Fraktal Geometri. https://web.itu.edu.tr/~kkocak/fraktal_yazi.htm. Last accessed 03 July 2021
2. Herbert, J.F., Cameron, J.L., Matthew, W.D.: Is there meaning in fractal analysis. Complex. Int **6** (1999)
3. Smith, T.G., Jr., Lange, G.D., Marks, W.B.: Fractal methods and results in cellular morphology—dimensions, lacunarity and multifractals. J. Neurosci. Methods **69**(2), 123–136 (1996)
4. Losa, G.A., Ristanović, D., Ristanović, D., Zaletel, I., Beltraminelli, S.: From fractal geometry to fractal analysis. Appl. Math. **7**(4), 346–354 (2016)
5. Zaletel, I., Ristanović, D., Stefanović, B.D., Puškaš, N.: Modified Richardson's method versus the box-counting method in neuroscience. J. Neurosci. Methods **242**, 93–96 (2015)
6. Nayak, S.R., Mishra, J.: Analysis of medical images using fractal geometry. In: Dey, N., Ashour, A., Kalia, H. Goswami, R., Das, H. (eds.), Histopathological Image Analysis in Medical Decision Making. IGI Global, Hershey (2019)
7. Mandelbrot, B.B.: The Fractal Geometry of Nature. 1st edn. New York (1982).
8. Kisan, S., Mishra, S., Rout, S.B.: Fractal dimension in medical imaging: a review. Int. Res. J. Eng. Technol. **4**(5), 1102–1106 (2017)
9. Mandelbrot, B.: How long is the coast of Britain? Statistical self-similarity and fractional dimension. Science **156**(3775), 636–638 (1967)
10. Narter, F., Köse, O.: Kanser geometrisi ve mesane kanserinde fraktallar. Galenos **12**(1), 11–17 (2013)
11. Mandelbrot, B.B., Frame M.: Fractals. In: Meyers R.A. (ed.) Encyclopedia of Physical Science and Technology, 3rd edn., pp. 185–207. Academic Press (2003)
12. Yılmaz, D.: Doğanın Fraktal Geometrisi, Yüksek Lisans Tezi, Afyon Kocatepe Üniversitesi. Fen Bilimleri Enstitüsü. (2013)
13. Di Ieva, A., Grizzi, F., Jelinek, H., Pellionisz, A.J., Losa, G.A.: Fractals in the neurosciences, part I: general principles and basic neurosciences. Neuroscientist **20**(4), 403–417 (2014)
14. Sağdıç, M.: Fraktal geometride boyut hesaplama teknikleri. Yüksek Lisans Tezi, İnönü Üniversitesi Fen Bilimleri Enstitüsü Matematik Anabilim Dalı. (2018)
15. Güleç, M., Taşsöker, M., Özcan, S.: Tıpta ve Diş Hekimliğinde Fraktal Analiz. Ege Üniversitesi Diş Hekimliği Fakültesi **40**(1), 17–31 (2019)
16. Karperien, A., Ahammer, H., Jelinek, H.: Quantitating the subtleties of microglial morphology with fractal analysis. Front. Cell. Neurosci. **7**, 3 (2013)
17. Bhandari, S., Choudannavar, S., Avery, E.R., Sahay, P., Pradhan, P.: Detection of colon cancer stages via fractal dimension analysis of optical transmission imaging of tissue microarrays (TMA). Biomed. Phys. Eng. Express **4**(6), 065020 (2018)
18. Mandelbrot, B.B.: Fractal: form, chance and dimension. W.H. Freeman, San Francisso (1977)
19. Bourke, P. (2003). Fractal Dimension Calculator. http://paulbourke.net/fractals/fracdim/. Last accessed 08 July 2021
20. Yackinous, W.S.: Understanding complex ecosystem dynamics: a systems and engineering perspective. In: Chapter 12-Fractals: The Theory of Roughness. Academic Press (2015)
21. Stevens, R.T.: Fractal Programming in C. IDG Books Worldwide, Inc. (1991)
22. Ürey, H.: Fraktal Geometri ve Uygulamaları. Yüksek Lisans Tezi. Afyon Kocatepe Üniversitesi. Fen Bilimleri Enstitüsü. (2006)

23. Brown, C., Liebovitch, L.: Fractal Analysis, vol. 165. Sage (2010)
24. Şener, E., Baksı, B.G.: Sağlıklı ve osteoporoz tanılı hastalarda fraktal boyut ve mandibular kortikal indeks değerlendirilmesi. Ege Üniversitesi Diş hekimliği Fakültesi Dergisi **37**(3), 159–167 (2016)
25. Lopes, R., Betrouni, N.: Fractal and multifractal analysis: a review. Med. Image Anal. **13**(4), 634–649 (2009)
26. Güleç, M., Taşsöker, M., Özcan, S.: Mandibular trabeküler kemiğin fraktal boyutu: Yaş, cinsiyet ve ilgi alanı seçiminin önemi nedir? Selcuk Dent. J. **6**(4), 15–19 (2019)
27. Păcurar, C.M., Necula, B.R.: An analysis of COVID-19 spread based on fractal interpolation and fractal dimension. Chaos, Solitons Fract. **139**, 110073 (2020)
28. Etehad Tavakol, M., Lucas, C., Sadri, S., Ng, E.Y.K.: Analysis of breast thermography using fractal dimension to establish possible difference between malignant and benign patterns. J. Healthcare Eng. **1**(1), 27–43 (2010)
29. Kırcı, P., Bayrak, E.A.: The application of fractal analysis on thyroid ultrasound images. Acta Infologica **3**(2), 83–90 (2019)
30. So, G.B., So, H.R., Jin, G.G.: Enhancement of the box-counting algorithm for fractal dimension estimation. Pattern Recogn. Lett. **98**, 53–58 (2017)
31. Miloevic, N.T., Rajkovic, N., Jelinek, H. F., Ristanovic, D.: Richardson's method of segment counting versus box-counting. In: 2013 19th International Conference on Control Systems and Computer Science, pp. 299–305. IEEE (2013)
32. Fractal Explorer Homepage, Chapter 4: Calculating Fractal Dimension. https://www.wahl.org/fe/HTML_version/link/FE4W/c4.htm. Last accessed 27 June 2021
33. Cross, S.S.: Fractals in pathology. J. Pathol.: J. Pathol. Soc. Great Britain Ireland **182**(1), 1–8 (1997)
34. Rangayyan, R.M., Nguyen, T.M.: Fractal analysis of contours of breast masses in mammograms. J. Digit. Imaging **20**(3), 223–237 (2007)
35. Jelinek, H.F., Fernandez, E.: Neurons and fractals: how reliable and useful are calculations of fractal dimensions? J. Neurosci. Methods **81**(1–2), 9–18 (1998)
36. Eins, S.: An improved dilation method for the measurement of fractal dimension. Acta Stereologica **14**(2), 169–178 (1995)
37. Jelinek, H., Cornforth, D.: Fractal analysis in clinical screening and investigation. In: Classification and Application of Fractals: New Research, pp. 277–301. Nova Publishers (2012)
38. Przystalski, K., Popik, M., Ogorzalek, M., Nowak, L.: Improved melanoma diagnosis support system based on fractal analysis of images. In: Proceedings of the 10th International Symposium on Operations Research and its Applications, pp. 203–211 (2011)
39. Swapna, M.S., Sreejyothi, S., Raj, V., Sankararaman, S.: Is SARS CoV-2 a multifractal? Unveiling the fractality and fractal structure. Braz. J. Phys. **51**(3), 731–737 (2021)
40. Namazi, H., Kulish, V.V.: Complexity-based classification of the coronavirus disease (COVID-19). Fractals **28**(05), 2050114 (2020)
41. E Fernandes, T.D. S.: Chaotic model for COVID-19 growth factor. Res. Biomed. Eng. 1–5 (2020)
42. Bayrak, E.A., Kırcı P.: Fractal analysis of thyroid ultrasound image data evaluation. In: 2020 IEEE 2nd International Conference on System Analysis & Intelligent Computing (SAIC), pp. 1–4. IEEE (2020)
43. Iqbal, J., Patil, R., Khanna, V., Tripathi, A., Singh, V., Munshi, M.A.I., Tiwari, R.: Role of fractal analysis in detection of dysplasia in potentially malignant disorders. J. Fam. Med. Primary Care **9**(5), 2448 (2020)
44. Qin, J., Puckett, L., Qian, X.: Image based fractal analysis for detection of cancer cells. In: 2020 IEEE International Conference on Bioinformatics and Biomedicine (BIBM), pp. 1482–1486. IEEE (2020)
45. Czyz, M., Radwan, H., Li, J., Filippi, C., Schulder, M.: Fractal analysis improves the preoperative identification of atypical meningiomas. Neuro-Oncology **20**(Suppl 1), i13 (2018)
46. Marusina, M.Y., Mochalina, A.P., Frolova, E.P., Satikov, V.I., Barchuk, A.A., Kuznetcov, V.I., Gaiukov, V.S., Tarakanov, S.A.: MRI image processing based on fractal analysis. Asian Pac. J. Cancer Prevent.: APJCP **18**(1), 51 (2017)

47. Huang, P.W., Lee, C.H.: Automatic classification for pathological prostate images based on fractal analysis. IEEE Trans. Med. Imaging **28**(7), 1037–1050 (2009)
48. Lv, D., Guo, X., Wang, X., Zhang, J., Fang, J.: Computerized characterization of prostate cancer by fractal analysis in MR images. J. Magn. Resonan. Imaging: Off. J. Int. Soc. Magn. Resonan. Med. **30**(1), 161–168 (2009)
49. Ballerini, L., Franzen, L.: Fractal analysis of microscopic images of breast tissue. WSEAS Trans. Circuit **11**(2), 7 (2001)
50. Esgiar, A.N., Naguib, R.N., Sharif, B.S., Bennett, M.K., Murray, A.: Fractal analysis in the detection of colonic cancer images. IEEE Trans. Inf Technol. Biomed. **6**(1), 54–58 (2002)
51. Hamidi, M., Ghassemian, H., Imani, M.: Classification of heart sound signal using curve fitting and fractal dimension. Biomed. Signal Process. Control **39**, 351–359 (2018)
52. Gavrovska, A., Zajić, G., Bogdanović, V., Reljin, I., Reljin, B.: Identification of S1 and S2 heart sound patterns based on fractal theory and shape context. Hindawi Complex, 1–9 (2017)
53. Michallek, F., Dewey, M.: Fractal analysis of the ischemic transition region in chronic ischemic heart disease using magnetic resonance imaging. Eur. Radiol. **27**(4), 1537–1546 (2017)
54. Captur, G., Muthurangu, V., Cook, C., Flett, A.S., Wilson, R., Barison, A., Sado, D.M., Anderson, S., McKenna, W.J., Mohun, T.J., Elliott, P.M., Moon, J.C.: Quantification of left ventricular trabeculae using fractal analysis. J. Cardiovasc. Magn. Reson. **15**(1), 1–10 (2013)
55. Tapanainen, J.M., Thomsen, P.E.B., Køber, L., Torp-Pedersen, C., Mäkikallio, T.H., Still, A.M., Lindgren, K.S., Huikuri, H.V.: Fractal analysis of heart rate variability and mortality after an acute myocardial infarction. Am. J. Cardiol. **90**(4), 347–352 (2002)
56. Mäkikallio, T.H., Huikuri, H.V., Mäkikallio, A., Sourander, L.B., Mitrani, R.D., Castellanos, A., Myerburg, R.J.: Prediction of sudden cardiac death by fractal analysis of heart rate variability in elderly subjects. J. Am. Coll. Cardiol. **37**(5), 1395–1402 (2001)
57. Lenka, S., Kumar, S., Mishra, S., Jena, K.K.: An IoT-cloud based fractal model for brain tumor image analysis. In: 2020 Fourth International Conference on I-SMAC (IOT in SOCIAL, MOIİLE, Analytics and Cloud) (I-SMAC), pp. 1–7. IEEE (2020)
58. Namazi, H., Jafari, S.: Estimating of brain development in newborns by fractal analysis of sleep Electroencephalographic (EEG) signal. Fractals **27**(03), 1950021 (2019)
59. Czyz, M., Radwan, H., Li, J.Y., Filippi, C.G., Tykocki, T., Schulder, M.: Fractal analysis may improve the preoperative identification of atypical meningiomas. Neurosurgery **80**(2), 300–308 (2017)
60. Akar, E., Kara, S., Akdemir, H., Kırış, A.: Beyincik Sarkması Tip-I Hastalarında Beyincik Gri Maddesinin Fraktal Boyut Analizi. Tip Teknolojileri Kongresi 2016. IEEE, Antalya (2016)
61. King, R.D., George, A.T., Jeon, T., Hynan, L.S., Youn, T.S., Kennedy, D.N., Dickerson, B.: Characterization of atrophic changes in the cerebral cortex using fractal dimensional analysis. Brain Imaging Behav. **3**(2), 154–166 (2009)
62. Zhang, L., Liu, J.Z., Dean, D., Sahgal, V., Yue, G.H.: A three-dimensional fractal analysis method for quantifying white matter structure in human brain. J. Neurosci. Methods **150**(2), 242–253 (2006)
63. Iftekharuddin, K.M., Jia, W., Marsh, R.: Fractal analysis of tumor in brain MR images. Mach. Vis. Appl. **13**(5), 352–362 (2003)
64. Appaji, A., Nagendra, B., Chako, D.M., Padmanabha, A., Hiremath, C.V., Jacob, A., Varambally, Kesavan, M.S., Venkatasubramanian, G., Rao, V.S., Webers, C.A.B., Berendschot T.J.M., Rao, N.P.: Retinal vascular fractal dimension in bipolar disorder and schizophrenia. J. Affect. Disord. **259**, 98–103 (2019)
65. Wolfson, S.: Diagnosing ASD with fractal analysis. Adv. Autism **3**(1), 47–56 (2017)
66. Ha, T.H., Yoon, U., Lee, K.J., Shin, Y.W., Lee, J.M., Kim, I.Y., Ha, K.S., Kim, S.I., Kwon, J.S.: Fractal dimension of cerebral cortical surface in schizophrenia and obsessive–compulsive disorder. Neurosci. Lett. **384**(1–2), 172–176 (2005)
67. Fernandez, E., Jelinek, H.F.: Use of fractal theory in neuroscience: methods, advantages, and potential problems. Methods **24**(4), 309–321 (2001)
68. Coşgunarslan, A., Aşantoğrol, F., Canger, E.M.: Kronik Böbrek Hastalarında Mandibular Kemik Kalitesinin Değerlendirilmesi. Türkiye Klinikleri. Dis hekimligi Bilimleri Dergisi **27**(1), 15–20 (2021)

69. Soylu, E., Coşgunarslan, A., Çelebi, S., Soydan, D., Demirbaş, A.E., Demir, O.: Fractal analysis as a useful predictor for determining osseointegration of dental implant? A retrospective study. Int. J. Implant Dent. **7**(1), 1–8 (2021)
70. Çelebi E.: Romatoid artritli hastalarda çene kemik yapısının fraktal analiz ile değerlendirilmesi. Diş Hekimliği Uzmanlık, Süleyman Demirel Üniversitesi, Diş Hekimliği Fakültesi, Ağız, Diş ve Çene Radyolojisi Ana Bilim Dalı. (2020)
71. Franciotti, R., Moharrami, M., Quaranta, A., Bizzoca, M.E., Piattelli, A., Aprile, G., Perrotti, V.: Use of fractal analysis in dental images for osteoporosis detection: a systematic review and meta-analysis. Osteoporosis Int. 1–12 (2021)
72. Ersu, N., Etöz, M., Akyol, R., Tanyeri, F.: Sistemik Glukokortikoid Kullanan Hastaların Mandibular Kemik Yapısının Fraktal Analiz ile Değerlendirilmesi. Osmangazi Tıp Dergisi Ağız Kanserleri Özel Sayısı 103–108 (2020)
73. Kato, C.N., Barra, S.G., Tavares, N.P., Amaral, T.M., Brasileiro, C.B., Mesquita, R.A., Abreu, L.G.: Use of fractal analysis in dental images: a systematic review. Dentomaxillofacial Radiology **49**(2), 20180457 (2020)
74. Updike, S.X., Nowzari, H.: Fractal analysis of dental radiographs to detect periodontitis-induced trabecular changes. J. Periodontal Res. **43**(6), 658–664 (2008)
75. Zehani, S., Ouahabi, A., Oussalah, M., Mimi, M., Taleb-Ahmed, A.: Bone microarchitecture characterization based on fractal analysis in spatial frequency domain imaging. Int. J. Imaging Syst. Technol. **31**(1), 141–159 (2021)
76. Chen, Q., Bao, N., Yao, Q., Li, Z.Y.: Fractal dimension: a complementary diagnostic indicator of osteoporosis to bone mineral density. Med. Hypotheses **116**, 136–138 (2018)
77. Yeşiltepe, S., Yılmaz, A.B., Kurtuldu, E., Sarıca, İ: Fractal analysis of temporomandibular joint trabecular bone structure in patients with rheumatoid arthritis on cone beam computed tomography images. Meandros Med. Dent. J. **19**(4), 345 (2018)
78. Tafraouti, A., El Hassouni, M., Toumi, H., Lespessailles, E., Jennane, R.: Osteoporosis diagnosis using fractal analysis and support vector machine. In: 2014 Tenth International Conference on Signal-Image Technology and Internet-Based Systems, pp. 73–77 (2014)
79. Koh, K.J., Park, H.N., Kim, K.A.: Prediction of age-related osteoporosis using fractal analysis on panoramic radiographs. Imaging Sci. Dent. **42**(4), 231–235 (2012)
80. Deepika, V., Jeya Lakshmi, V., Latha, P., Raman, R., Srinivasalu, S., Raman, S., Kandle, K.S.: Comparison of various fractal analysis methods for retinal images. Biomed. Signal Process. Control **63**, 102245 (2021)
81. Yu, S., Lakshminarayanan, V.: Fractal dimension and retinal pathology: a meta-analysis. Appl. Sci. **11**(5), 2376 (2021)
82. Lopez, J., Chiu, B., Chiu, H., Kumar, P., Hashmi, S., Gupta, A., Sarrafpour, S., Young, J.A.: Fractal dimension analysis of OCTA images in normal and diabetic eyes using the circular mass-radius method. Invest. Ophthalmol. Vis. Sci. **60**(9), 5345–5345 (2019)
83. Bhardwaj, S., Tsui, E., Zahid, S., Young, E., Mehta, N., Agemy, S., Garcia, P., Rosen, R.B., Young, J.A.: Value of fractal analysis of optical coherence tomography angiography in various stages of diabetic retinopathy. Retina **38**(9), 1816–1823 (2018)
84. Fabrizii, M., Moinfar, F., Jelinek, H.F., Karperien, A., Ahammer, H.: Fractal analysis of cervical intraepithelial neoplasia. PloS one 9(10), e108457 (2014).
85. Cheung, N., Donaghue, K.C., Liew, G., Rogers, S.L., Wang, J.J., Lim, S.W., Jenkins, A.J., Hsu, W., Lee, M.L., Wong, T.Y.: Quantitative assessment of early diabetic retinopathy using fractal analysis. Diabetes Care **32**(1), 106–110 (2009)
86. Macgillivray, T.J., Patton, N., Doubal, F.N., Graham, C., Wardlaw, J.M.: Fractal analysis of the retinal vascular network in fundus images. In: 2007 29th Annual International Conference of the IEEE Engineering in Medicine and Biology Society, pp. 6455–6458. IEEE (2007)
87. Căliman, A., Ivanovici, M., Richard, N.: Fractal feature-based color image segmentation for a healthcare application in dermatology. In: 2011 E-Health and Bioengineering Conference (EHB), pp. 1–4. IEEE (2011)
88. Guarneri, F.: Computer vision and fractals: a possible aid for the dermatologist in recognizing skin tumors? In: Communications to SIMAI Congress, vol. 3, pp. 268–1 (2009)

89. Mastrolonardo, M., Conte, E., Zbilut, J.P.: A fractal analysis of skin pigmented lesions using the novel tool of the variogram technique. Chaos Solitons Fractals **28**(5), 1119–1135 (2006)
90. Bianciardi, G., Miracco, C., De Santi, M.M., Luzi, P.: Differential diagnosis between mycosis fungoides and chronic dermatitis by fractal analysis. J. Dermatol. Sci. **33**(3), 184–186 (2003)

Chapter 20
Program Code Protecting Mechanism Based on Obfuscation Tools

Vadym Mukhin, Valerii Zavgorodnii, Yaroslav Kornaga, Ivan Krysak, Maxim Bazaliy, and Oleg Mukhin

Abstract This paper describes the methods for obfuscation of the server and client parts of program code. The server part is represented by two methods based on obfuscation algorithms and LLVM compilation algorithms. The first method is based on the optimal choice of software code obfuscation algorithms, and the second one on the LLVM compilation and toolkit technologies. Obfuscation of the client part of the program code is based on two models of machine learning. These models are based on periodically learned neural networks, which are in the form of tensors and they are integrated into binary applications.

Keywords Code generation · Obfuscation · LLVM · Learning · Neural networks

1 Introduction

Obfuscation is the action to create the source or machine code that is difficult for humans to understand it [1–3]. The obfuscation procedure is inteded to secure some information, and information protecion is important issue for the computer systems and networks [4–7].

It is known that JavaScript is compiled "just-in-time" programming language and designed to be used for client-side scripting. The main relevant to this article outcome of these features being selected is the fact that a major amount of source code is transferred over the network and is available to the end user. As the client-side compiler has to be able to access it in order to execute side code.

And so, the obfuscation came to be a pausable tool for protecting (at least somewhat) these enormous amounts of unprotected source code being sent over the network every minute [8–11].

V. Mukhin (✉) · Y. Kornaga · I. Krysak · M. Bazaliy · O. Mukhin
National Technical University of Ukraine "Igor Sikorsky Kiev Polytechnic Institute", Kiev, Ukraine
e-mail: v.mukhin@kpi.ua

V. Zavgorodnii
State University of Infrastructure and Technologies, Kiev, Ukraine

M. Zgurovsky and N. Pankratova (eds.), *System Analysis & Intelligent Computing*, Studies in Computational Intelligence 1022, https://doi.org/10.1007/978-3-030-94910-5_20

Obfuscation is a deliberate act to make an information difficult to understand whilst giving a full access to the person who doesn't necessary needs to have the access, as it could be harmful to either the person with the access (like the patient getting obfuscated note from the doctor) or the party owning the data (for example, a company trying to protect specific scripts developed for the website from potential rivals trying to plagiarize them) [12–15].

That being said, the theoretical importance of obfuscation is hard to dispute until the Internet stops to depend on "just-in-time" compiled scripts so hard or the enterprises stop trying to protect their company secrets from rivals, with latter being even more improbable than the former [10, 11].

Let's look at the methods of protecting the server part of the program code from analysis and unauthorized modification. Such methods do not require the binary code of programs and protect against automatic reverse engineering mechanisms, but at the cost of a small reduction in program speed. They are based on the principle of software "black box", the implementation of which involves the use of code transformation technologies during obfuscation [16–21].

Definition of the mandatory conditions that must comply with the algorithm for protecting the software code of computer systems (obfuscation):

1. The obfuscated code must perform the same functions as the source code.
2. The obfuscated code must run a polynomial number of times slower than the source code.
3. The obfuscated code must be a polynomial number of times larger than the source code.
4. There should be no obfuscation analysis algorithm more efficient than the assumption made based on analysis of inputs and outputs of program code.

As a result of the analysis of existing methods used to protect programs from reverse engineering and decompilation, it was concluded that it is necessary to create new and modify existing methods of program code protection based on obfuscation technologies. It is shown that the current methods for software code protection do not fully satisfy software developer needs. Primarily because, on the one hand, they did not require the program binary code and ensure a reduction in execution speed in less than twice, and on the other hand—they did not meet the conditions of resistance to deobfuscation algorithms that work for polynomial time.

2 Methods of Obfuscation of the Server Part

To effectively protect the software code, two obfuscation methods have been modified so that they take full advantage of the processor architecture and significantly complicate reverse engineering and decompilation. As a step forward, different types of algorithms are considered, and the effectiveness of their implementation and application is determined. The following algorithms are used to obfuscate the program code:

1. renaming the global variables and functions names (using common names from the libC library);
2. replacement and reordering of instructions (combination of mathematical operations when replacing the instructions);
3. flattening the program control flow graph (adding the basic blocks that are never executed);
4. obfuscation of program string literals (using a new string encryption algorithm for each program);
5. merging of functions (adding a switch as an input parameter to the resulting function);
6. adding junk code (combining junk code with mathematical operations);
7. creating aliases for global variables (use of several levels of aliases);
8. adding protection against dynamic analysis (use of a universal multiplatform approach).

Scheme of obfuscation of program code based on the above algorithms and using a universal system of analysis, transformation and optimization of computer programs (Fig. 1).

Steps of the which differs from existing in the order of transformation algorithms execution, that allows increasing the time of the analysis of the protected programs:

Step 1. Prepare and load the software code of the computer system.
Step 2. The program code is divided into logical blocks.
Step 3. Prepare and fill in the data that will be used in the process of obfuscation.
Step 4. Execution of the control flow flattening algorithm.
Step 5. Execution of the creating aliases for global variables algorithm.
Step 6. Execution of the replacement and reordering of instructions algorithm.
Step 7. Execution of the obfuscation of program string literals algorithm.
Step 8. Execution of the merging of functions algorithm.
Step 9. Execution of the adding junk code algorithm.
Step 10. Execution of renaming the global variables and functions names algorithm.
Step 11. Execution of adding a protection against dynamic analysis algorithm.
Step 12. Save the changes to the program code and pass the result code for further compilation.

To develop a method of processing program code using LLVM compiler and toolchain technologies, we define the scheme of the obfuscation process using the LLVM compiler annotations, which uses the following elements:

1. clang—the main compiler of the LLVM package.
2. opt—manager of intermediate representation (LLVM IR) code optimization modules.
3. llc—intermediate representation (LLVM IR) code compiler.

Let's define the main components of obfuscation:

X_1 program code size;
X_2 program execution time;

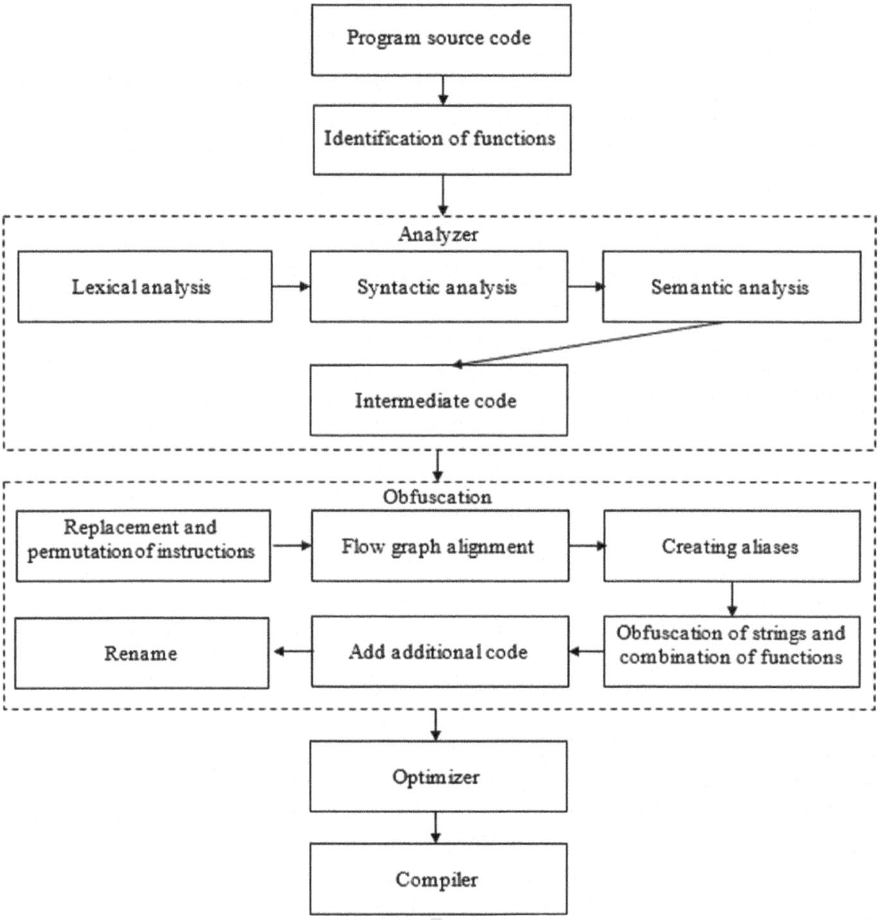

Fig. 1 General scheme of the program code obfuscation

X_3 time of the program obfuscation process;
X_4 program analysis time after obfuscation.

This solves the following optimization problem:

$$\begin{Bmatrix} F(X_1) \rightarrow \min \\ F(X_2) \rightarrow \min \\ F(X_3) \rightarrow \min \\ F(X_4) \geq T_{crit} \end{Bmatrix}$$

where T_{crit}—the allowable time of the program code analysis.

We will form the steps of the program code transformation method using LLVM compiler and toolchain technologies, the scientific novelty of which is that it is based on the mechanism of identification of the functions needed for obfuscation based on compiler annotations, which allows selective obfuscation of the main part of program code:

Step 1. Read the program code.
Step 2. Identification of functions for further obfuscation using compiler annotations.
Step 3. Lexical analysis (conversion of input code into a set of tokens).
Step 4. Syntax analysis (converting a set of tokens into an abstract syntax tree (AST)).
Step 5. Semantic analysis (presentation of the AST as input data and checking compliance with the rules and building a symbol table).
Step 6. Code generation (creation of intermediate representation code from AST).
Step 7. The analyzer reads the annotations from the source code and transmits this information to the obfuscator.
Step 8. Perform obfuscation according to the advanced method of obfuscation of the intermediate representation code.
Step 9. Running the optimization (The optimizer accepts an intermediate representation of the code, optimizes the speed, efficiency, and size of the code using optimization algorithms. At the output, the optimizer creates a new optimized intermediate representation of the code).
Step 10. Convert the optimized intermediate code representation into the assembly language of the corresponding processor.
Step 11. Running the linking step. (creating an executable module).

For the experimental studies, the obfuscation of program code was performed on an ×86 processor using a disassembler and analyzer IDA Pro. (Table 1).

For the experimental research, program code with a different number of functions (5, 10, 20, 50) was used and was determined the time of program analysis before obfuscation and using an advanced method of program code transformation using the LLVM compiler and toolchain technologies (Fig. 2).

While using the method of program code processing using the LLVM compiler and toolchain technologies, the analysis time of the computer system software code was measured. The result showed that with the increasing number of functions in the program code, the analysis speed increases significantly due to a significant

Table 1 The results of obfuscation

Program	Before obfuscation			After obfuscation		
	Size, Kb	Startup time, c	Analysis time, c	Size, Kb	Startup time, c	Analysis time, c
CPP_AES	159	0,082	1,044	223	0,158	1,373
Cryptopals	50	0,01	0,96	52	0,011	0,981
iOS jailbreak detector	230	0,16	3,443	1126	0,43	15,797

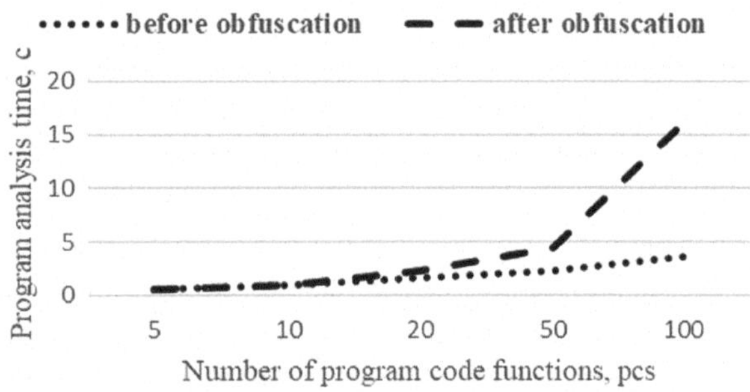

Fig. 2 Comparison of the average time of program code analysis

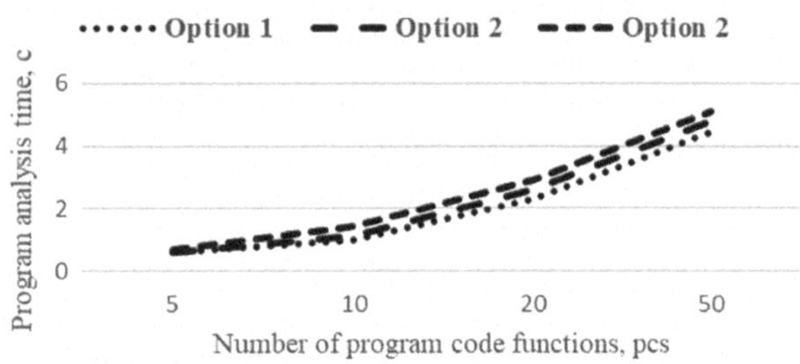

Fig. 3 Comparison of software code analysis time for different usage of obfuscation algorithms

increase in the number of confusing transformations. It is shown that for a program that contains more than 20 functions, the analysis time increases 1.4 times, with 50 functions—2 times, and with 100 functions—5 times. At the same time, the startup time of the program will increase slightly, which will be compensated by the speed of modern processors.

The optimal variant of running obfuscation algorithms for the improved method of obfuscation of program code is chosen. Comparison of options that give the highest compilation speed for 5, 10, 20, 50 functions in the program code (Fig. 3):

- option 1 includes the order of execution of algorithms: 3, 7, 2, 4, 5, 6, 1, 8;
- option 2 includes the order of execution of algorithms: 1, 4, 3, 6, 7, 2, 5, 8;
- option 3 includes the order of execution of algorithms: 7, 1, 5, 4, 3, 2, 6, 8.

It is shown that option 2 gives less analysis time, because after renaming the functions, other obfuscation algorithms may add new functions with common names or delete existing ones. It is shown that option 3 gives less analysis time, because we first make aliases for global variables. In the future, we perform other obfuscation modules that can add new global variables. In this case, not all global variables will have aliases, which will make obfuscation analysis easier.

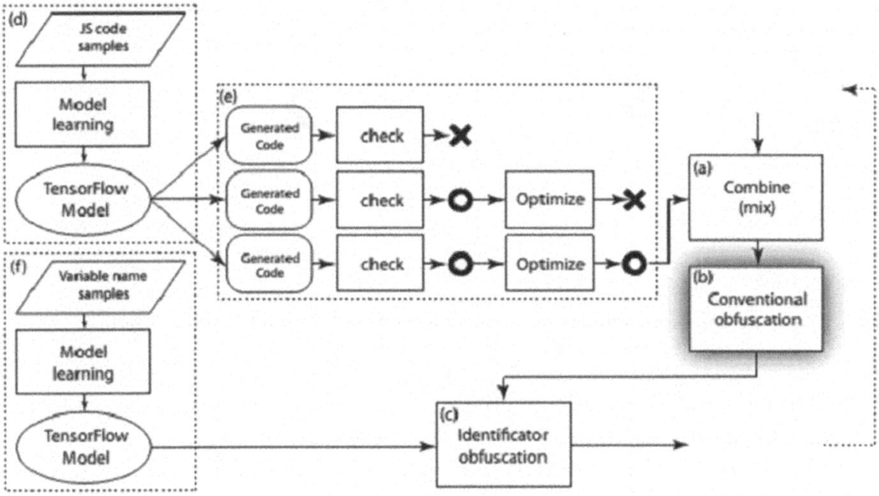

Fig. 4 Augmented with code generation and realistic identificator replacement obfuscation process

3 Methods of Obfuscation of the Client's Part

With wider than ever spread of cheap and powerful data processing devices, the perspective of using rather computationally intensive (and rather weakly optimized) "black-box" algorithm seems more appealing than ever. Even for tightly established fields like, for example, code obfuscation. So, a few questions arise: "what benefits can the usage of NNs yield?", "which NN type is the most suitable?" and "where to get an enormous amount of data to train them on?".

Answering the first question is rather straightforward, the NNs are capable of emulating data (and creating new instances based on what they were taught with). What data could we generate to improve code obfuscation? First of all, new code, to dilute the real code pool with fake functions hard to distinguish from real ones, making it even harder for humans to make sense of where to even begin. And secondly, generating identificators with little to no connection with code they are used could be even more devastating than randomized text strings [22–26].

This brings light to the second question which is now easier to answer thanks to the first one, recurrent neural network is taylor suited for working with sequences of data, like sounds or, more importantly, batches of text (or code, for that matter).

What's left is only a data search. Luckily, there are quite a few well formatted and preprocessed data sets ready to be used.

Figure 4 illustrates a mock-up structure of the obfuscation process featuring both techniques mentioned here.

Part (a) of the figure features the process of combining input with the code generated using a RNN (first suggested technique).

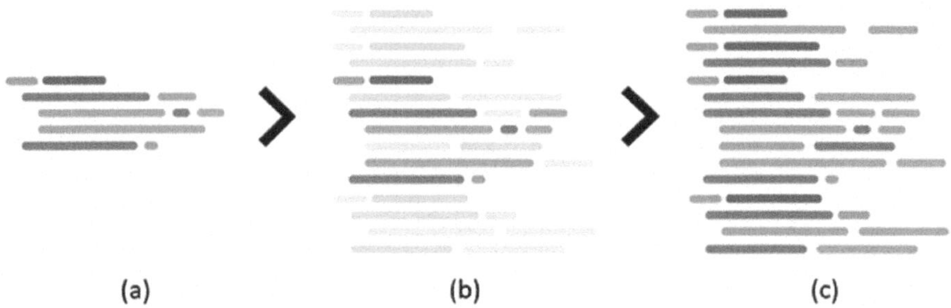

Fig. 5 Illustration of the combination of generated code padding and flow obfuscation

Part (b) represents conventional obfuscation techniques (flow and data obfuscation, etc.). Its inner structure is similar to that represented in Fig. 1.

Part (c) shows an alternative take on name obfuscation with using easily readable by humans by making no sense identificator names instead of commonly used randomly generated sequences of letters and numbers.

Block (d) represents the process of a recurrent neural network being taught using dataset [27] with the resulting ready to deployment TensorFlow model which is used in further steps.

Block (e) shows the fail-checks and tests to determine whether generated using the model code is suitable. First of all, it uses a typical compiler to make sure no errors are encountered and generated code is compilable, then optimizator (e.g. Google Closure) is used to determine whether code can be easily shrunken which would defeat the purpose of using generation. The process is repeated until a batch of compilable code which reacts well to the optimization is selected.

Block (f) is similar to block (d) in its structure and purpose even if it's quite a bit smaller. It's being taught using dataset [27] and the underlying model is the fraction of the size of the main model.

Essentially, using generated code for obfuscation purposes is rather similar to flow obfuscation except the added complications are found outside of execution blocks instead. For example, if we add three more "dummy" functions which are undistinguishable on the first look from the needed functions (especially if it's done on top of conventional obfuscation) understanding of "where to look" becomes even harder to achieve.

Figure 5 illustrates the effect of the combination of flow obfuscation and generated code insertion. The leftmost side of the figure presents inputed source code. In the middle inserted code can be seen, this image was provided to contrast the difference between "before" and "after" forms. And the rightmost one illustrates obfuscated code.

Part (a) of the figure shows the inputs, code being obfuscated.

Part (b) is added for creating a better visual clarity contrasting the differences between input and added "padding" code, which is presented with half the opacity.

Part (c) shows the final obfuscated code sample, where it's hard to determine which parts of the code were there initially and which of them were added artificially.

At first glance it seems excessive to expand the amount of data through the obfuscation so drastically but the main objective of the process is to make it harder for a human "reader" to make sense out of whatever they're seeing. So it seems like a logical step to increase the amount of the data so it's harder to even determine where to start from. With clever identificator fine-tuning a realisting seeming code can be achieved. It's important to repeat identificators as often as possible, for example fake functions should call real ones and blocks added using flow obfuscation should refer to fake ones. The process becomes even harder to follow for a human reader if all the identifiers look like gibberish or, getting ahead into the next technique, when identificators make no sense.

When looking at a piece of source code obfuscated using traditional techniques, it's really easy to see that it was tempered with. And the very first thing one probably notices is identificators, they are normally generated randomly or, sometimes, follow some hard for human to follow structure, for example "n240573257", "n20260572" and "nx25275027577" with a starting letter with a big randomized (or not) number following it. It surely is hard for a human to follow but it sells the idea of code being obfuscated at face value. So, let's look at an example. Let's compare listing 6, showing the final version of the code obfuscated using conventional techniques and listing, featuring the very same code but with identificators randomly selected from pieces of real code but making no sense in this context, they look realistically, but confuse a person even more, than long numbers and such:

```
function keyDown(unicodeFont, packer) {
  var vertexSize = (9 * 8) + (7 * 6) - (5 * 4) + ((3 * 2) / 1),
    bigInt = Math.pow(vertexSize, 1/2),
    spawnDepth = Math.pow(vertexSize + 8, 1/2),
    createFixture = spawnDepth + bigInt,
    attributes = spawnDepth - bigInt,
    shapeRenderer = Math.pow(createFixture, 1/3),
    loadGlyphs = Math.pow(attributes, 1/3);
  if (shapeRenderer < loadGlyphs).
    return unicodeFont * loadGlyphs +
    Packer * shapeRenderer;
  else.
    return unicodeFont * packer *
   (shapeRenderer - loadGlyphs);}
```

Listing shows further improvements of this technique hiding the weakest link, standart function calls (Math.pow() in the example) from the main function body into an additional "dummy" function, which makes function a little bit harder to understand (as you'd additionally need to find the used function in the masses of randomly generated code above and such, which is even more pronounced if there are a lot of such functions).

```
function compareTo(numComponents, iterable) {
  return Math.pow(iterable, numComponents)}.
```

```
/**... **/
function keyDown(unicodeFont, packer) {
  var vertexSize = 100,
    bigInt = compareTo(1/2, vertexSize),
    spawnDepth = compareTo(1/2, vertexSize + 8),
    createFixture = spawnDepth + bigInt,
    attributes = spawnDepth - bigInt,
    shapeRenderer = compareTo(1/3, createFixture),
    loadGlyphs = compareTo(1/3, attributes);
  if(shapeRenderer < loadGlyphs).
    return unicodeFont * loadGlyphs +
      Packer * shapeRenderer;
  else.
return unicodeFont * packer *
(shapeRenderer - loadGlyphs);}
```

At the first glance code looks almost normal (except for an abundance of variables (names of which don't make any sense)).

Here's the explanation of terms and blocks used in Fig. 6:

- DLLs—Deep Learning Layers—represents a conventional deep learning network. Number of layers can be varied.
- &—concatenation operation, where one of the vectors is seamlessly appended to the end of the other one.
- c—copying operation, where an additional copy of data vector is created.
- *—element wise multiplication operation.
- +—element wise addition operation.
- σ—sigmoid function.
- t—hyperbolic tangent function.

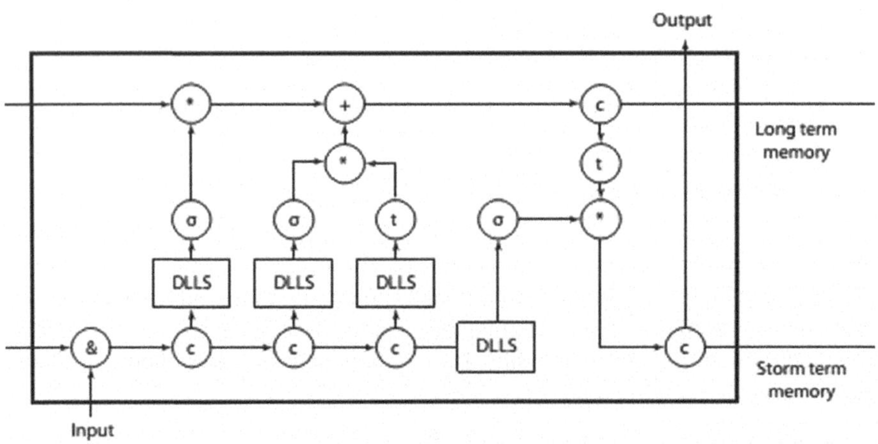

Fig. 6 Inner LSTM RNN structure

Inputs are represented as vectors going in the network-block from the bottom. Outputs are represented as vectors going out of the network-block at the top. Upper connecting blocks line represents long term memory, while lower one—short term.

4 Conclusion

It has been shown that when using advanced methods to measure the time of analysis of program code of a computer system, the experiment showed that with increasing number of functions in program code, the speed of code analysis decreases significantly due to a significant increase in confusing code. It is shown that for a program that contains more than 20 functions, the analysis time increases 1.4 times, for 50 functions—2 times, and for 100 functions—5 times. At the same time, the analysis time of the program increases slightly, which will be compensated by the speed of modern processors.

The presented technique utilizes two machine learning models (recurrently trained neural networks) represented as tensors and integrated into the application binaries.

The first one is used to generate "dummy code" and is more complex in its structure. It's trained using freely obtained source code of big projects such as jQuery or jxCore (dataset [23]). But as generated code is prone to be unreliable (and it always needs to be ignored by the compiler (cause no error, to be more specific), otherwise it cannot be used for obfuscation purposes), a well-designed testing system is necessary. It can be as simple as attached js-engine (V8 or SpiderMonkey, for example) used to compile-check the code, if any errors are caused code is discarded and a new batch is generated (see upper part of the figure). Next step is to optimize such code to look a bit more realistic (using tools like YUI Compressor or Closure Compiler). In case it fails, the code is discarded as well (see the figure). If generation was successful, the code is combined with the inputs and is subjected to the conventional obfuscation.

The second one only generates identificators, so it isn't limited to JavaScript as a language. That's why it can be taught using a dataset for any language. Created this way identificators are used to finalize the obfuscation (and give it a bit more human-like) appearance generating special conversion tables where each of the input identificators has a corresponding fake one.

The process can be repeated increasing complexity of the obfuscation while enlarging code base at the same time.

References

1. Sebastian, S.A., Malgaonkar, S., Shah, P., Kapoor, M., Parekhji, T.: A Study & Review on Code Obfuscation. In: Conference: 2016 World Conference on Futuristic Trends in Research and Innovation for Social Welfare (WCFTR'16). https://www.researchgate.net/publication/322520412_A_Study_Review_on_Code_Obfuscation

2. Barak, B.: Hopes, fears, and software obfuscation. Commun. ACM (2016)
3. Boyle, E., Chung, K.-M., Pass, R.: On extractability obfuscation. In: Theory of Cryptography Conference. Springer (2014)
4. Hu, Z., Mukhin, V., Kornaga, Y., Lavrenko, Y., Herasymenko, O.: Distributed computer system resources control mechanism based on network-centric approach. Int. J. Intell. Syst. Appl. **9**(7), 41–51 (2017)
5. Hu, Z., Mukhin, V., Kornaga, Ya., Herasymenko, O., Mostoviy, Ye.: The Analytical model for distributed computer system parameters control based on multi-factoring estimations. J. Netw. Syst. Manage. **27**(2), 351–365 (2019)
6. Mukhin, V., Volokyta, A., Heriatovych, Y., Rehida, P.: Method for efficiency increasing of distributed classification of the images based on the proactive parallel computing approach. Adv. Electr. Comput. Eng. **18**(2), 117–122 (2018)
7. Zhengbing, H., Mukhin, V.Y., Kornaga, Y.I., Herasymenko, O.Y.: Resource management in a distributed computer system with allowance for the level of trust to computational components. Cybern. Syst. Anal. **53**(2), 312–322 (2017)
8. Boyle, E., Pass, R.: Limits of extractability assumptions with distributional auxiliary input. In: International Conference on the Theory and Application of Cryptology and Information Security, Springer (2014)
9. Garg, S., Gentry, C., Halevi, S., Wichs, D.: On the implausibility of differing-inputs obfuscation and extractable witness encryption with auxiliary input. In: International Cryptology Conference. Springer (2014)
10. Schrittwieser, S., Katzenbeisser, S., Kinder, J., Merzdovnik, G., Weippl, E.: Protecting software throughobfuscation: can it keep pace with progress in code analysis? ACM Comput. Surveys (2016)
11. Horváth, M., Buttyán, L.: The birth of cryptographic obfuscation—A survey (2016)
12. Ananth, P., Boneh, D., Garg, S., Sahai, A., Zhandry, M.: Differing inputs obfuscation and applications. IACR Cryptology ePrint Archive (2013).
13. Ishai, Y., Pandey, O., Sahai, A.: Public-coin differing-inputs obfuscation and its applications. In: Theory of Cryptography Conference, Springer (2015)
14. Bitansky, N., Canetti, R., Kalai, Y.T., Paneth, O.: On virtual grey box obfuscation for general circuits. In: International Cryptology Conference, Springer (2014)
15. Abebe, S.L., Arnaoudova, V., Tonella, P., Antoniol, G., Gueheneuc, Y.: Can lexicon bad smells improve fault prediction? In: WCRE (2012)
16. Allamanis, M., Barr, E.T., Bird, C., Sutton. C.: Learning natural coding conventions. In: FSE (2014)
17. Allamanis, M., Sutton, C.: Mining source code repositories at massive scale using language modeling. In: MSR, IEEE Press (2013)
18. Allamanis, M., Sutton, C.: Mining Idioms from Source Code. In: FSE (2014)
19. Arnaoudova, V., Di Penta, M., Antoniol, G., Gueheneuc, Y.-G.: A new family of software anti-patterns: linguistic anti-patterns. In: CSMR (2013)
20. Arnaoudova, V., Eshkevari, L.M., Penta, M.D., Oliveto, R., Antoniol, G., Guéhéneuc, Y.: REPENT: analyzing the nature of identifier renamings. In: IEEE TSE (2014)
21. Arnaoudova, V., Penta, M.D., Antoniol, G.: Linguistic antipatterns: What they are and how developers perceive them. In: EMSE (2015)
22. Pascanu, R., Gulcehre, C., Cho, K., Bengio, Y.: How to construct deep recurrent neural networks (2013)
23. Mukhin, V., Kornaga, Ya., Zavgorodnii, V., Zavgorodnya, A., Herasymenko, O., Mukhin, O.: Social risk assessment mechanism based on the neural networks. In: IEEE International Conference on Advanced Trends in Information Theory, Kiev, pp. 179–182 (2019)
24. Musienko, A.P., Serdyuk, A.S.: Lebesgue-type inequalities for the de la Vallée poussin sums on sets of entire functions. Ukr. Math. J. **65**(5), 709–722 (2013)
25. Pashynska, N., Snytyuk V., Putrenko, V., Musienko, A.: A decision tree in a classification of fire hazard factors. Eastern-Eur. J. Enterprise Technol. **5/10**(83), 32–37 (2016)

26. Mukhin, V., Zavgorodnii, V., Barabash, O., Mykolaichuk, R., Kornaga, Y., Zavgorodnya, A., Statkevych, V.: Method of restoring parameters of information objects in a unified information space based on computer networks. Int. J. Comput. Netw. Inf. Secur. **12**(2), 11–21 (2020). https://doi.org/10.5815/ijcnis.2020.02.02
27. Raychev, V., Bielik, P., Vechev, M. Krause, A.: Learning programs from noisy data. In: Proceedings of the 43nd Annual ACM SIGPLAN-SIGACT Symposium on Principles of Programming Languages POPL'16, ACM (2016)

The manufacturer's authorised representative in the EU is Springer
Nature Customer Service Centre GmbH, Europaplatz 3, 69115 Heidelberg,
Germany. If you have any concerns regarding our products, please
contact ProductSafety@springernature.com

Printed and bound by CPI Group (UK) Ltd, Croydon, CR0 4YY
30/09/2026
02225498-0017